TREATISE ON SOLID STATE CHEMISTRY

Volume 1
The Chemical Structure of Solids

TREATISE ON SOLID STATE CHEMISTRY

Volume 1 • The Chemical Structure of Solids
Volume 2 • Defects in Solids
Volume 3 • Crystalline and Noncrystalline Solids
Volume 4 • Reactivity of Solids
Volume 5 • Changes of State
Volume 6 • Surfaces

TREATISE ON SOLID STATE CHEMISTRY

Volume 1
The Chemical Structure of Solids

Edited by
N. B. Hannay
Vice President
Research and Patents
Bell Laboratories
Murray Hill, New Jersey

PLENUM PRESS • NEW YORK-LONDON

Library of Congress Cataloging in Publication Data

Hannay, Norman Bruce, 1921–
The chemical structure of solids.

(His Treatise on solid state chemistry, v. 1)
Includes bibliographical references.
1. Solid state chemistry. 2. Chemical bonds. 3. Crystals – Defects. I. Title. II. Series.
QD478.H35 vol. 1 [QD461] 541'.042'1s [541'.042'1]
ISBN 0-306-35051-3 73-13798

Six-volume set: ISBN 0-306-35050-5

Published by Plenum Press, New York
A Division of Plenum Publishing Corporation
227 West 17th Street, New York, N.Y. 10011

United Kingdom edition published by Plenum Press, London
A Division of Plenum Publishing Company, Ltd.
Davis House (4th Floor), 8 Scrubs Lane, Harlesden, London, NW10 6SE, England

Printed in the United States of America

Foreword

The last quarter-century has been marked by the extremely rapid growth of the solid-state sciences. They include what is now the largest subfield of physics, and the materials engineering sciences have likewise flourished. And, playing an active role throughout this vast area of science and engineering have been very large numbers of chemists. Yet, even though the role of chemistry in the solid-state sciences has been a vital one and the solid-state sciences have, in turn, made enormous contributions to chemical thought, solid-state chemistry has not been recognized by the general body of chemists as a major subfield of chemistry. Solid-state chemistry is not even well defined as to content. Some, for example, would have it include only the quantum chemistry of solids and would reject thermodynamics and phase equilibria; this is nonsense. Solid-state chemistry has many facets, and one of the purposes of this *Treatise* is to help define the field.

Perhaps the most general characteristic of solid-state chemistry, and one which helps differentiate it from solid-state physics, is its focus on the chemical composition and atomic configuration of real solids and on the relationship of composition and structure to the chemical and physical properties of the solid. Real solids are usually extremely complex and exhibit almost infinite variety in their compositional and structural features.

Chemistry has never hesitated about the role of applied science, and solid-state chemistry is no exception. Hence, we have chosen to include in the field not only basic science but also the more fundamental aspects of the materials engineering sciences.

The central theme of the *Treatise* is the exposition of unifying principles in the chemistry, physical chemistry, and chemical physics of solids. Examples are provided only to illustrate these principles. It has, throughout, a chemical viewpoint; there is, perforce, substantial overlap with some areas of solid-state physics and metallurgy but a uniquely chemical perspective underlies the whole. Each chapter seeks to be as definitive as possible in its particular segment of the field.

The *Treatise* is intended for advanced workers in the field. The scope of the work is such that all solid-state chemists, as well as solid-state scientists and engineers in allied disciplines, should find in it much that is new to them in areas outside their own specializations; they should also find that the treatment of their own particulár areas of interest offers enlightening perspectives.

Certain standard subjects, such as crystal structures, have been omitted because they are so well covered in many readily available standard references and are a part of the background of all solid-state scientists. Certain limited redundancies are intended, partly because they occur in different volumes of the series, but mainly because some subjects need to be examined from different viewpoints and in different contexts. The first three volumes deal with the structure of solids and its relation to properties. Volumes 4 and 5 cover broad areas of chemical dynamics in bulk solids. Volume 6 treats both structure and chemical dynamics of surfaces.

N.B.H.

Preface to Volume 1

No aspect of chemistry is more fundamental to the science than is the study of the nature of the chemical bond. Solids exhibit the complete range of bonding behavior and offer opportunity, therefore, for gaining special insight into the nature of interatomic binding forces. The regularity of many solids facilitates experimental and theoretical examination of chemical bonds and allows the interpretation of the properties of solids in fundamental atomic terms. This volume is concerned with these aspects of solid-state chemistry. Thus it furnishes a fundamental basis for later volumes.

The ideal solid would be perfectly ordered and geometrically regular, chemically pure and stoichiometric. No real solid is found in this state, and some of the most interesting and important properties of solids depend upon departure from the ideal. The study and control of these deviations from the perfect solid are a principal concern of solid-state chemists. Later volumes of the *Treatise* will be concerned with defects in solids in relation to physical and chemical properties. Here we lay the foundation by examining the nature of defects and the equilibria controlling their concentrations. In addition, the characterization of solids is included; it would seem almost self-evident that the measurement of departures of a solid from the ideal chemical and geometrical state is a *sine qua non* in the study of the imperfect solid, yet it has all too often been overlooked or treated casually.

N.B.H.

Contents of Volume 1

Chapter 1
Chemical Bonds in Solids 1
J. C. Phillips

1. Why Solids Are Different from Molecules 1
    1.1. Quantum Theory of Chemical Bonds 5
    1.2. The Five Solid Types 6
    1.3. Bonds and/or Bands? 9
2. Crystal Structures and Cohesive Energies of the Elements 11
    2.1. Valence Groupings 11
    2.2. Shell Effects 12
    2.3. Transition Series 14
3. Binary Compounds and Alloys 17
    3.1. Minerals 17
    3.2. Semiconductors 22
    3.3. Intermetallic Solutions 28
4. Chemical Bonding and Physical Properties 30
    4.1. Classical Polarizabilities 31
    4.2. Dispersion 32
    4.3. Covalent and Ionic Energies 34
    4.4. Chemical Trends in Physical Properties 37
5. Summary 39
    References 40

Chapter 2
Energy Bands **43**
D. Weaire

1. Introduction 43
1.1. Historical Remarks 43
1.2. The Independent-Electron Approximation 45
2. Energy Bands in General 49
3. The Classical Descriptions of Energy Bands in Periodic Systems 52
3.1. Introduction 52
3.2. Two Classical Limits—Tight Binding and Nearly Free Electron 52
3.3. Tight Binding Theory 53
3.4. Wannier Functions 56
3.5. Nearly-Free-Electron Theory 57
3.6. Pseudopotentials 60
3.7. The Cellular Method 63
3.8. Orthogonalized Plane Wave, Augmented Plane Wave, and Related Methods 63
4. Approximations, Interpolations, Perturbations 66
4.1. Introduction 66
4.2. Moment Methods 67
4.3. Nearly-Free-Electron Perturbation Theory 69
4.4. The $\mathbf{k} \cdot \mathbf{p}$ Method 70
4.5. Small-$\mathbf{k}$ Expansions for KKR Theory 72
5. Some Relevant Experiments 73
5.1. Introduction 73
5.2. Soft X-Ray Emission and Absorption 75
5.3. Optical Spectroscopy 77
5.4. Fermi Surface Analysis 81
6. Typical Band Structures 83
6.1. Introduction 83
6.2. Simple Metals 84
6.3. Alkali Halides 85
6.4. Group IV Semiconductors 86
6.5. The III–V and II–VI Semiconductors 88
6.6. Silicon Dioxide 90
6.7. Transition Metals 92
6.8. Transition Metal Compounds 92
7. Disordered Solids 94
7.1. Introduction 94
7.2. Definition of Problems 95
7.3. The Density of States in an Alloy 96
7.4. The Anderson Problem 100

7.5. Topological Disorder 101
7.6. Applications 103
8. Conclusion 106
Acknowledgments 106
References 107

Chapter 3
Factors Controlling the Formation and Structure of Phases **115**
W. B. Pearson

1. Introduction 115
2. Practical Prediction of Phase Stability 119
2.1. Metals: Use of Thermodynamic Data 119
2.2. Valence Compounds: Use of Crystal Chemical Knowledge 123
3. General Structural Consequences of Bonding Types 124
3.1. Ionic Crystals 124
3.2. Compounds with Saturated Covalent Bonds 127
3.3. Metallic Phases 128
3.4. *A Priori* Separation of Structure Types 129
4. Atomic Size and Structural Constraint 133
5. Factors Influencing the Stability of Crystal Structures 136
5.1. Electrochemical Factor 136
5.2. Geometric Effects 137
5.3. Energy Band Effects 146
5.4. Environmental Factors 161
6. Distortions of Crystal Structures 164
6.1. Distortions Arising from Cation–Cation Bonds 165
6.2. Jahn–Teller Distortions 168
6.3. Spin–Orbit Coupling Distortions 169
6.4. Magnetic Exchange Energies 170
6.5. Mechanical Instability 170
7. Epilogue 171
Appendix—Structure Diagrams 171
Acknowledgments 172
References 172

Chapter 4
Structure and Composition in Relation to Properties **175**
J. H. Wernick

1. Magnetic Behavior 176
1.1. Introduction 176
1.2. The 3*d* Transition Elements 180

1.3. Rare Earth Metals 182
1.4. Role of Local Atomic Environment Regarding Development of Atomic Moments and Long-Range Order 187
1.5. Directional Ordering and Magnetic Anisotropy 194
1.6. Magnetic Oxides 199
1.7. Magnetic Semiconductors 205
1.8. Linear and Two-Dimensional Magnetic Behavior 214
1.9. Amorphous Magnetic Materials 217
1.10. Summary 218
2. Superconducting Behavior 219
2.1. Introduction 219
2.2. The Cr_3Si (β-W) and Transition Metal Nitride and Carbide Phases. Electron Concentration and Lattice Instability 225
2.3. Role of Stoichiometry and Atomic Order 234
2.4. Metastable Superconducting Phases 236
2.5. Paramagnetic Impurities in Superconductors 237
2.6. Ternary Superconducting Chalcogenides 238
2.7. Superconductivity of Degenerate Semiconductors 239
2.8. Summary 240
3. Dielectric Materials 240
3.1. Ferroelectrics 240
3.2. Piezoelectrics 247
3.3. Nonlinear Optical Materials 249
3.4. Electrooptic and Pyroelectric Materials 257
3.5. Summary 258
4. Mechanical Behavior 259
4.1. Introduction 259
4.2. Elastic Behavior 260
4.3. Plastic Behavior 267
4.4. Summary 271
Acknowledgments 272
References 272

Chapter 5
Introduction to Chemical and Structural Defects in Crystalline Solids **283**
Morris E. Fine

1. Introduction 283
2. Point Defects 287
3. Dislocations 291
4. Planar Defects 310
5. Volumetric Defects 322
Acknowledgments 329
References 330

Chapter 6
Defect Equilibria in Solids **335**
George G. Libowitz

1. Introduction 335
   1.1. Native Defects 335
   1.2. Law of Mass Action and Point Defects 337
   1.3. Electronic Defects 338
   1.4. Energetics of Defect Formation 339
2. Native Defects 339
   2.1. Defect Equilibria in Elemental Crystals 339
   2.2. Defect Equilibria in Binary Compounds 341
   2.3. Nonstoichiometry—Equilibria with External Phases 345
   2.4. Ionization of Defects 351
   2.5. Relationship between Mass Action Law and Statistical Thermodynamics 361
   2.6. Defect Interactions 367
3. Multicomponent Systems 371
   3.1. Equilibria Involving Foreign Atoms 371
   3.2. Multicomponent Compounds 378
4. Extended Defects 380
   Acknowledgment 383
   References 383

Chapter 7
Characterization of Solids—Chemical Composition **387**
W. Wayne Meinke

1. Introduction 387
2. Current Capability for Determination of Chemical Composition 388
   2.1. Introduction 388
   2.2. General Over iew 389
   2.3. Analytical Techniques: Present Status 391
   2.4. Precision and Sensitivity of Analytical Techniques 408
3. Application of Current Techniques to Characterization of Materials 408
   3.1. Characterization of Major Phase 408
   3.2. Characterization of Minor Phases and Impurities 413
   3.3. Characterization of Surfaces 415
4. Utilization of Existing Techniques 416
   4.1. Literature Examples 416
   4.2. Factors Determining Use 416
   Acknowledgments 426
   References 426

Chapter 8
Structural Characterization of Solids **437**
R. E. Newnham and Rustum Roy

1. Introduction ... 437
2. Structural Characterization by Optical Techniques ... 439
   2.1. Morphology ... 439
   2.2. Bulk Optical Properties ... 441
   2.3. Scattering Studies ... 444
   2.4. Surface Characterization ... 444
   2.5. Particle Size and Shape ... 448
3. Structural Characterization by X-Ray Diffraction ... 451
   3.1. X-Ray Powder Methods ... 451
   3.2. Single-Crystal X-Ray Methods ... 458
   3.3. Temperature and Pressure Experiments ... 467
   3.4. X-Ray Topography and Interferometry ... 474
4. Electron Methods for Materials Characterization ... 476
   4.1. Electron Microscopy ... 476
   4.2. Electron Diffraction ... 480
   4.3. Scanning Electron Microscopy ... 484
5. Neutron Scattering from Solids ... 487
   5.1. Neutron Sources ... 487
   5.2. Interactions with Matter ... 489
   5.3. Structure Analysis with Neutrons ... 490
   5.4. Magnetic Structure Analysis ... 491
   5.5. Lattice and Spin Dynamics ... 493
6. Spectroscopy and Local Symmetry ... 496
   6.1. Absorption Spectra in the Visible Range ... 497
   6.2. Infrared Absorption Spectroscopy ... 499
   6.3. Raman Spectra ... 501
   6.4. Soft X-Ray Spectra ... 501
   6.5. Electron Spin Resonance ... 503
   6.6. Nuclear Magnetic Resonance ... 506
   6.7. Mössbauer Effect ... 510
   6.8. Electron Spectroscopy ... 513
   6.9. Acoustic Spectroscopy ... 514
7. Physical Properties as Characterization Tools ... 515
   7.1. Introduction ... 515
   7.2. Some Crystal Physical Generalizations ... 516
   7.3. Dielectric Measurements ... 518
   7.4. Electrical Characterization of Solids ... 519
   7.5. Magnetic Measurements ... 521
   7.6. Calorimetric Measurements ... 527

Acknowledgment 529
References.. 529

Index **535**

1

Chemical Bonds in Solids

J. C. Phillips
Bell Laboratories
Murray Hill, New Jersey

1. Why Solids Are Different from Molecules

Most of the ideas about chemical bonds that are contained in the collective wisdom of science have been refined inductively from the great body of chemical knowledge concerning the structure, properties, and reactivities of molecules in gases. Therefore to discuss chemical bonding of atoms in solids, one begins by stressing how it differs from chemical bonding of molecules in gases. Above all, most solids are much denser than corresponding molecules. For example, in the diatomic molecule Na^+Cl^- each ion has one nearest neighbor, but in the Na^+Cl^- crystal each ion has six nearest neighbors. The diatomic molecule has a dipole moment, which is important in many properties. The diatomic crystal has no permanent dipole moment. Generally speaking, the symmetry of the crystal is much higher than that of the molecule. Most solids are "three dimensional"; those solids that are composed of layers and chains of atoms exhibit many molecular characteristics, while most of what we have to say applies to chemical bonds in "three-dimensional" solids. The cases of layers, chains, and surfaces are special and are treated as such elsewhere in these volumes.

The formation of chemical bonds between atoms brings about a redistribution of electronic charge relative to a simple superposition of atomic charge densities. One may characterize many of the properties of an atom in two ways: by its ground-state charge distribution (which determines the atomic size), and by the accessibility of its first few excited states. When a second atom interacts with the first one the interaction can be thought of primarily as an electric dipole one, which polarizes the first atom and leads to partial occupation of excited states. Much of the polarizability of an isolated atom is associated with excitations into the ionization continuum. In a molecule some of the oscillator strength associated with ionization of the atoms is transferred to bonding → antibonding excitations. Nonbonded or lone-pair electrons on the "outside" of the molecule retain a large part of their oscillator strength in ionization processes. (This accounts in part for some of their special properties.) In the solid, on the other hand, except for atoms on or very near the surface, almost all the oscillator strength is associated with excitation of the atoms *considered as part of the medium*. Thus while the identity and properties of atoms are partially preserved in low-polarizability molecules, great care should be exercised in extending atomistic concepts into the discussion of chemical bonding in much more highly polarizable solids.

As a simple illustration of this theme, consider the specific problem of the crystal structures of the alkali halides. These are the simplest salts known, and chemical theories of their properties are discussed at greater length in Section 3.1. However, the structures of these very simple ionic crystals illustrate in the most direct way how much different solids are from molecules.

The compounds of formula M^+X^- are expected to be well represented by electrostatic models because the electronic configurations of the cations M^+ and the anions X^- are isoelectronic to those of inert gases such as Ne and Ar. It now seems reasonable to discuss alternative crystal structures in terms of mechanical models based on electrostatic attractions of rigid spheres. So long as the cation and anion sphere radii are nearly equal, the crystal structure favored by this model will be one with the largest possible number of oppositely charged ions coordinated around each central ion. This is the CsI structure (eightfold coordination, with each ion surrounded by oppositely charged ions located at the corners of a cube centered on the first ion). However, a simple geometric calculation shows that when the radius ratio $\rho = r_+/r_-$ of the sphere radii is less than 0.732 the eight anions will come into mutual contact, thereby determining a larger cation–anion separation than would have been fixed by cation–anion repulsion alone. As a result, the NaCl structure (sixfold coordinated, nearest neighbors centered on cube faces) will be favored. Finally, when $\rho = 0.414$ the six anions come into contact, and again

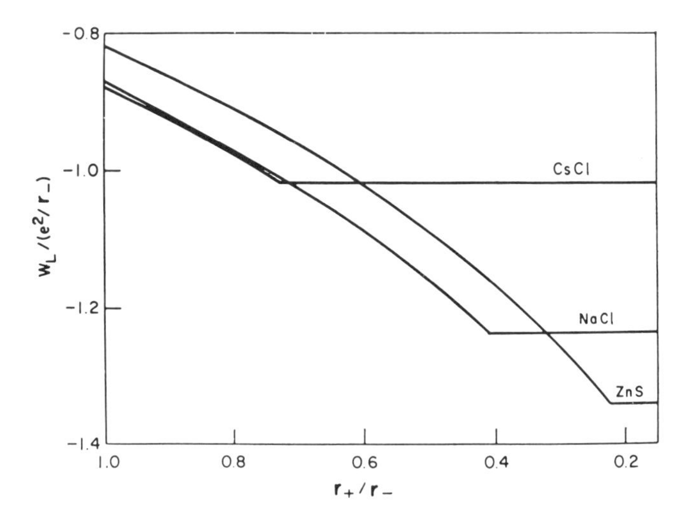

Fig. 1. The total energy of a cubic lattice of rigid anions and cations as a function of r_+ with r_- fixed, for different coordination configurations. When the anions come into mutual contact as a result of decreasing r_+ their repulsion determines the lattice constant and the cohesive energy becomes constant when expressed in terms of r_-. Thus near the values of r_+/r_- at which anion–anion contact takes place the mechanical model predicts phase transitions to structures of successively lower coordination numbers.

for smaller values of ρ still lower coordination numbers (e.g., four, as in the tetrahedrally coordinated sphalerite compounds) will be favored. The behavior of the total energy as a function of r_+ (for fixed r_-) is shown in Figure 1.

The argument that has been given here is familiar from discussions of chemical bonds in molecules. The mechanism of Figure 1, which differentiates among fourfold-, sixfold-, and eightfold-coordinated structures, is the steric factor which is important in determining the structures of many molecules, and in the case of closed-shell ions (i.e., those with electronic configurations isoelectronic to the inert gases) we expect it to be the only important factor, as covalent effects should be small. It is therefore surprising to see that for any reasonable choice of sphere radii the model gives a poor account of the observed crystal structures. These are shown in Figure 2, and the correlation of the observed structures with the predictions of the rigid-ion model is very poor indeed. The model predicts that several Li salts will have the sphalerite structure, whereas none actually do. It also predicts that if CsI is eightfold coordinated, then the other Cs

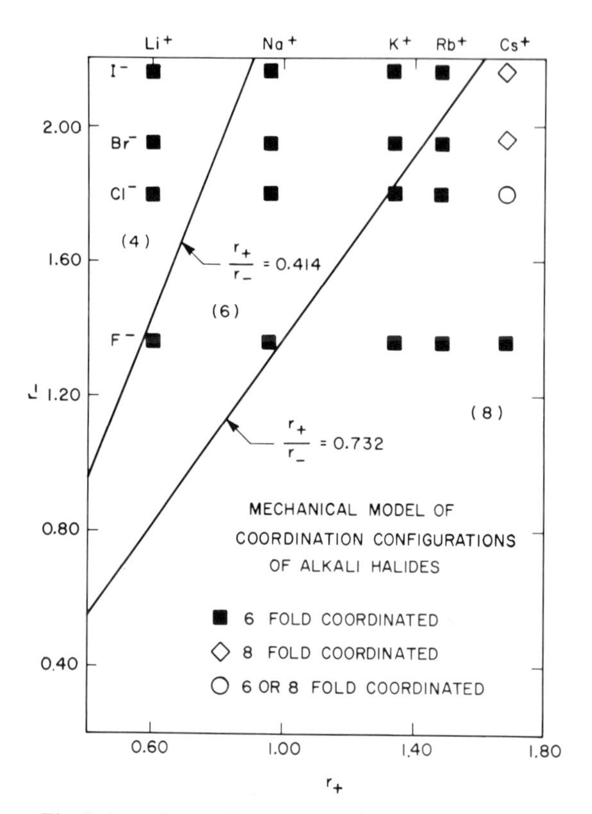

Fig. 2. Actual crystal structures of the alkali halides, as contrasted with the predictions of the mechanical or rigid-sphere model. The r_+ versus r_- plane is divided into three regions by the lines $r_+/r_- = 0.732$ and $r_+/r_- = 0.414$, corresponding to eightfold-, sixfold-, and fourfold-coordinated structures. However, these three regions show little correlation with observed crystal structures.

salts will also be eightfold coordinated, but this is not the case, either. In fact, the atomistic model does not give even a good qualitative account of the crystal structures of the alkali halides.

The picture just painted is much too black. Models to be discussed in Section 3.1 do give very accurate accounts of the cohesive energies of most of the alkali halides, a good result obtained by allowing the ions to be less than rigid. The main feature of Figure 2—that most of the alkali halides have the sixfold-coordinated NaCl structure—comes out of those deformable-ion calculations as well. (Actually, the calculations predict that all alkali halides have the NaCl structure, including the Cs salts, which do not.) But it is clear from the discussion that chemical bonding in crystals is a delicate matter, and that care should be exercised in extending atomistic concepts (such as ionic radii) into the area of quantitative theories of this bonding.

There is one characteristic energy of the solid, regarded as a medium, which is seldom discussed in a molecular context (and rightly so!): This is the plasma energy $\hbar\omega_p$, where ω_p is the plasma frequency. In simple models ω_p is given by

$$\omega_p{}^2 = 4\pi N e^2/m \tag{1}$$

where e and m are the electron mass and charge, and N is the number of effective valence electrons per unit volume. (In diamond, for example, N is $4/\Omega$, where Ω is the volume per carbon atom.) The meaning of ω_p is that it is the natural frequency of oscillation of the valence electrons when their center of gravity is displaced slightly from that of the positive ion lattice.

Many three-dimensional solids are highly polarizable, with electronic dielectric constants ε_0 large compared to unity. In all but the most ionic solids typical electronic excitation energies (e.g., associated with bonding $\rightarrow$ antibonding transitions) tend to be several times smaller than $\hbar\omega_p$. This means that many charge-transfer processes are screened by almost the full electronic dielectric constant ε_0, which continues to screen excitations at energies (or frequencies) up to $\hbar\omega_p$. Because of plasma screening, many charging effects that are important in discussing molecular properties are of secondary significance in solids, and this basic fact simplifies theories of chemical bonding in solids. An example of a charging effect is the difference between the local electric field at some point in the crystal and its average, or macroscopic, value. In all but the most ionic crystals local field corrections are usually small.

1.1. Quantum Theory of Chemical Bonds

Although in principle all the properties of a certain assembly of atoms, including the structure itself, can be derived from quantum mechanics, in practice prescriptions for doing so usually fail to yield results of chemical significance. One such prescription, for example, would recommend that we approximate the crystal potential of germanium by a superposition of atomic potentials and then solve the one-electron Schrödinger equation to find the energy levels of the crystal. Depending on technical details, this approach might predict that germanium is indeed a semiconductor, but it would be almost as likely to predict that germanium is a metal. It may be added that such prescriptions become even less satisfactory for binary or multicomponent compounds and their alloys. The problem here is not with the machinery of quantum mechanics and energy band theory (Chapter 2), but rather with the initial assumption of the integrity of atomic character and the atomic potential, which is supposed to carry over unchanged to the crystal.

The nature of the solid as a dense electronic medium can be incorporated into theory and experiment by considering groups or families of related solids (see also Chapter 3 of this volume). Examples of such families are the alkali halides, tetrahedrally coordinated semiconductors, and interstitial transition metal–metalloid compounds such as TiC. Just as Hückel theory was developed primarily to provide an accurate account of chemical trends in the electronic properties of the very large family of hydrocarbon molecules, so one may develop theories to account for chemical trends within each family of solids. The strength of such theories lies in the variety and differences of the constructs utilized to describe chemical trends in one family compared with those utilized for another family. Eventually one may hope to trace back the different constructs to the underlying quantum mechanics, but from the chemical point of view that exercise is of secondary importance. What is essential is that we can develop theories of sufficient accuracy to reproduce chemical trends within certain large families of materials, and these theories can be used to predict the properties of new compounds and alloys.

1.2. The Five Solid Types

Solids can be classified into five broad groups according to general physical and structural properties and according to the positions of the elements composing them in the periodic table. Simple metals are composed of electropositive nontransition elements such as Na and Al. Transition metals involve elements with $3d$, $4d$, or $5d$ valence electrons. They are often magnetic ($3d$ elements) or superconducting at low temperatures ($4d$ and $5d$ elements). The rare earth metals ($4f$ and $5f$ valence electrons) can be regarded as a special case of transition metals. Electrical and thermal conductivities tend to be higher in simple metals than in transition metals. Structurally metals tend to be close-packed.

Ionic crystals are composed of electropositive cations and electronegative anions satisfying formal valence rules. For example, the sum of the valences of the cations per formula unit equals the similar sum of the valences of the anions. Most of the electronic charge is localized on the anions. In strongly ionic crystals such as NaCl, the formal representation, Na^+Cl^-, is accurate and realistic. Each ion is isoelectronic to a rare gas atom (Na^+ to Ne, Cl^- to Ar). In less ionic crystals such as MgO, the actual charge distribution is probably closer to Mg^+O^- than it is to the formal charge distribution $Mg^{2+}O^{2-}$. Pure ionic crystals are transparent well into the ultraviolet, and are good insulators with low electrical and thermal conductivities at room temperature. At high temperature the ionic conductivities can be appreciable.

Semiconductors are situated, as the name implies, on a place intermediate between metals and insulators. The occupied electronic states (valence-band states) are separated from the unoccupied (conduction-band) states by a small energy gap of order 1 eV. As a result, semiconductors are transparent only toward the red, and absorb strongly beginning either in the visible or ultraviolet. Semiconductors have predominantly covalent structures which can be explained in terms of formal valence rules and hybridized directed orbitals. Most semiconductors with "three-dimensional" structures are tetrahedrally coordinated and average eight electrons per atom pair (the octet compounds, chemical formulas A^nB^{8-n} or $A^{n-1}B^{n+1}C_2^{8-n}$, etc.). Much of the technological value of semiconductors is derived from the fact that crystalline samples can be prepared commercially in states of high purity. These states can be further modified by addition in suitable geometries of controlled amounts of impurities with specific electrical and/or optical properties. These favorable material properties are made possible by the highly polarizable but structurally still quite stable covalent bonds of materials like germanium and especially silicon.

The last kind of solid type is composed of molecular crystals in which the molecular units are bound together by weak, rather long-range forces of the van der Waals type. Each unit has the electronic characteristic of having filled shells separated from unoccupied levels by a substantial energy gap, e.g., in solid Ar, or in solid anthracene. Because of their lack of structural stability, molecular crystals have found fewer applications than the other types of materials.

Useful as the foregoing classification is, there are many compounds with characteristics intermediate between those of the types listed. For instance, As, Sb, and Bi are semimetals with many of the characteristics of semiconductors, but with a small overlap of a few tenths of an eV (a small "negative" energy gap) between valence and conduction bands, giving rise to low electrical and thermal conductivities which persist to low temperatures. The column IB metals Cu, Ag, and Au behave in some respects (e.g., electrical conductivity) like simple metals with one conduction electron per atom. However, in other respects they behave like transition metals, because their last filled d shells are highly polarizable and contribute to chemical bonding.

Elements from the first (Li) period form crystals with significantly different properties from those of elements from the other nontransition periods. Some of these differences arise because of the small size of atoms from the first period. However, even after allowance has been made for size differences, first-period atoms form qualitatively different chemical bonds, especially with atoms from other periods. The reason for this appears to be connected with the presence of only $1s$ states in the cores of these atoms, and the correponding absence of p states. The $2s$ valence states of first-period atoms are prevented from penetrating the $1s$ core region by the exclusion principle, while the $2p$ core states are repelled from the same core region only by their angular momentum. In practice the latter mechanism is less effective than the former, and so (on a relative scale) the s–p energy difference is smaller for atoms from the first period than it is for atoms from other periods. Indeed, although the $2s$ energy of the atom always lies lower than the $2p$ energy, in the solid in certain situations energy levels associated with first-period atoms can be inverted, so that states having $2p$-like symmetry may lie lower than those having $2s$-like symmetry.

The extent and character of directed valence bonds depends on the completeness of s–p hybridization, and this in turn depends on the apparent s–p energy difference in the bonded state (*not* in the free atom). Some ways in which this apparent energy difference can be defined are discussed in the next section. In any case, it is interesting to utilize here the completeness of s–p hybridization as a means of illustrating the differences in chemical bonding in molecules and crystals. For molecules it is clear, from the richness and variety of organic compounds, that directed bonds are most easily formed with atoms from the first period. In crystals, on the other hand, perfect hybridization occurs most often with atoms from the second period (especially Si, P, and S, which are most likely to form directed bonds in solids). This difference arises primarily because of the greater density of the solid compared to the molecule, and secondarily because the solid is more "three dimensional." A complete justification for these qualitative statements would easily fill this volume, but partial support will appear in later sections.

As we go from the second period to the third and then the fourth nontransition period, the *s*-like valence energies drop compared to *p*-like ones. This process (which one may call dehybridization, or loss of directed bonding character, or metallization) is a gradual one until we reach the fifth period, which contains Cs, Pb, and Bi. Then, quite sharply, the 6*s* electrons are much more strongly bound than the 6*p* ones. (This accounts, for example, for the fact that Pb tends to be divalent rather than quadrivalent.) The quantum mechanical explanation for this involves relativistic energy shifts of the *s* electrons, which are of order 1 eV even for the 6*s* electrons.

1.3. Bonds and/or Bands?

The literature of crystal chemistry and metallurgy abounds with contradictory statements concerning the number of effective valence electrons per element, apparent atomic sizes, the importance of electronegativity differences and effective charges, the relative importance of different hybridization states in a given bond type, etc. These differences arise because the chemical bond is less well defined in crystals than in molecules. As we have noted above, there is no reason to expect that concepts developed to explain chemical trends in one family of compounds should be entirely consonant with concepts appropriate to other, and structurally quite different, materials. Nevertheless, at the risk of becoming somewhat technical, we may spend some time considering the reasons why the chemical bond is more easily defined in molecules, and whether, at least in principle, it is possible to produce equally useful definitions for the crystal.

Most atoms in molecules are surrounded over a large solid angle by vacuum. For this reason quantum mechanical calculations can be carried out by representing the molecular wave function as a superposition of orbitals centered on individual atoms. Such wave functions attenuate or decay rapidly in the vacuum directions and so satisfy easily the most important physical condition, that the wave function describe electrons confined to the molecule. This method of representing the molecular wave function is called the method of linear combinations of atomic orbitals (LCAO), but the atomic orbitals in question need not be the same as the orbitals of the free atoms. (They will actually be compressed by the effects of interatomic interactions.)

The atomic orbitals in the molecule will form a convenient basis for discussing chemical bonds in molecules, as demonstrated by the Hückel theory and its many refinements.[1] In crystals, on the other hand, atomic orbitals have proved to be a poor basis for discussing electronic structure as a property of the medium. Their chief virtue, of attenuating properly in vacuum, is relevant only at the crystal surface. Within the crystal the density is so high that atomic orbitals overlap far beyond nearest neighbors. For example, calculations of the energy bands of silicon and germanium have shown that to achieve the proper ordering of the lower unoccupied levels in the conduction band requires the inclusion of overlap to at least tenth-nearest neighbors. This means that it will be difficult to assign any special chemical significance to the overlap of atomic orbitals on nearest neighbors, even though it is just this overlap which is used to give quantum mechanical meaning to the notion of chemical bonding in molecules.

Let us suppose, for the moment, that all the energy levels of the crystal—its so-called energy bands $E_n(\mathbf{k})$, where $\hbar\mathbf{k}$ is the crystal momentum and n is the band index—are known to us in some fashion. (Methods for calculating these bands and adjusting them to fit experiment are discussed in Chapter 2). There is a wide range of data that shows that nearest-neighbor interactions determine most of the properties of a material. Isn't there some way that the chemical bond can be recovered from the apparent complexity of the energy bands of the crystal?

In 1960 most solid-state scientists would probably have answered this question negatively. However, since 1960 there have been substantial advances in crystal chemistry which show that we can define bonding in a quantitative way in crystals, at least in those families of materials which contain many members and for which there is some accumulation of measurements of material properties on single crystals. Thus, while we do not yet have (and probably never will have) a neat prescription for predicting the properties of an arbitrary compound, we do have a much improved picture of the overall mechanisms responsible for phase transitions and trends in material properties. And, what is equally important, we can describe these pictures in simple terms that do not require elaborate quantum mechanical calculations. In fact, the absence of a satisfactory universal basis set of wave functions (e.g., LCAO's in molecules) means that the most useful statements that can be made do not involve wave functions at all. These statements are, as a mathematician would put it, representation-independent, i.e., independent of the detailed wave functions which may be used in actual quantum mechanical calculations.

The general prescription for defining chemical bonds in crystals is therefore the following. We imagine that the energy bands are known, and then we perform some kind of average over these bands to define characteristic energies of chemical bonds. The details of the band structure, or of the wave functions, need not concern us so long as we can identify the proper average to be taken. Clearly this is a nontrivial task, but it has been solved in some cases, and there is no reason not to believe it should work in many others as well. An average bonding property could be cohesive energy, heat of formation, polarizability, interatomic electronegativity difference, ratio of directional to central nearest-neighbor forces, extent of metallic character, or something similar. Such average properties have been used by Pearson to describe structural transitions (Chapter 3) and he has been quite successful with this approach. A wide range of material properties of semiconductors has been analyzed in this way by J. A. Van Vechten and myself; some results are given in a later section.

2. Crystal Structures and Cohesive Energies of the Elements

The structures of the elements illustrate some of the five solid types discussed in Section 1. The cohesive energies of the elements illustrate chemical bonding in solids in its simplest form.

2.1. Valence Groupings

According to Hume-Rothery,[2] the crystal structures of the elements can be grouped as shown in Figure 3. Also listed with each element in this table is the heat of atomization at 300°K or at the melting point, whichever is lower.

It can be seen that most of the elements are metals in their elemental states (class I). The structures are largely close-packed (face-centered cubic or hexagonal close-packed) or body-centered cubic. In class II there are metals in which some kind of valence effect disturbs chemical trends and structural coordination; none of these metallic structures (including white Sn) can be analyzed simply, although their proximity to the valence structures (class III) explains the presence of partially covalent effects.

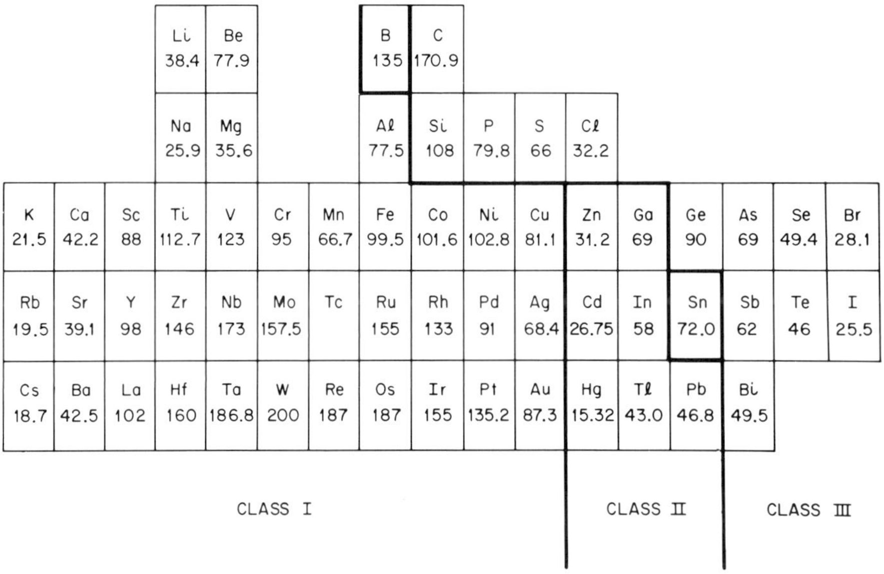

			Li 38.4	Be 77.9			B 135	C 170.9								
			Na 25.9	Mg 35.6			Al 77.5	Si 108	P 79.8	S 66	Cl 32.2					
K 21.5	Ca 42.2	Sc 88	Ti 112.7	V 123	Cr 95	Mn 66.7	Fe 99.5	Co 101.6	Ni 102.8	Cu 81.1	Zn 31.2	Ga 69	Ge 90	As 69	Se 49.4	Br 28.1
Rb 19.5	Sr 39.1	Y 98	Zr 146	Nb 173	Mo 157.5	Tc	Ru 155	Rh 133	Pd 91	Ag 68.4	Cd 26.75	In 58	Sn 72.0	Sb 62	Te 46	I 25.5
Cs 18.7	Ba 42.5	La 102	Hf 160	Ta 186.8	W 200	Re 187	Os 187	Ir 155	Pt 135.2	Au 87.3	Hg 15.32	Tl 43.0	Pb 46.8	Bi 49.5		

CLASS I CLASS II CLASS III

Fig. 3. Classification of crystal structures according to Hume-Rothery. Also shown are heats of atomization in kcal/g-atom at 300°K or at the melting point, whichever is lower. The three classes into which the elements are divided are discussed in the text. The cohesive energies are taken from compilations of thermodynamic properties in NBS reports (D. D. Wagman *et al.*).

Conventional chemical valence considerations explain the coordination configurations of the solids comprising class III. In most cases an atom of valence N has $8 - N$ nearest neighbors. This statement applies to all the elements in class III except As, Sb, and Bi, which are best regarded as slightly deformed simple cubic structures, similar to PbTe, but distorted because of the absence of the effect of partially ionic forces.

2.2. Shell Effects

With the addition of each new shell to the atomic core, or with the partial filling of a new valence shell, there are characteristic periodic effects, many of which are already evident in molecular structure (e.g., transition metals and transition metal ion complexes). Shell effects which are specific to solids are discussed here with respect to the cohesive energies of the elements.

Two of the simplest cases to discuss are the alkali metals and the group IV elements, which are shown in Figure 4. It is known from

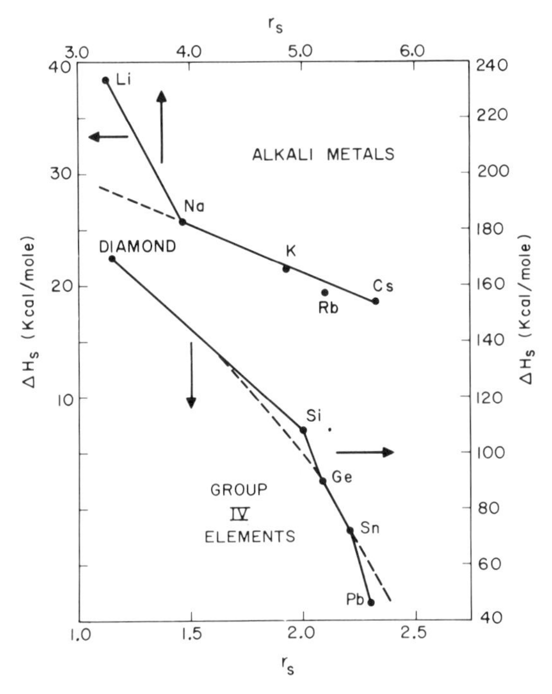

Fig. 4. Trends in cohesive energies (heats of atomization) for alkali metals (column IA) and tetravalent nontransition metalloids (column IVB) of the periodic table. The abscissa r_s is the mean interelectronic spacing of the valence electrons expressed in units of the Bohr radius a_0.

energy band studies (both theoretical and experimental) that the valence electrons of Na, K, Rb, and Cs are nearly free, and their cohesive energies are almost a linear function of electronic spacing or lattice constant. The hybridization configuration is roughly $s^{3/4}p^{1/4}$, compared to the free-atom ground state s^1. However, in Li the p states have relatively lower energy because there are no p states in the atomic core. Thus the cohesive energy of Li is about 10 kcal/mole greater than one would expect for an alkali metal with its lattice constant, and the hybridization configuration is roughly $s^{1/2}p^{1/2}$. It is the departure from the nearly-free-electron behavior of the other alkali metals that makes the cohesive energy of Li unusually large.

The diamond series, on the other hand, shows that Si has an anomalously large cohesive energy. The energy bands of Si are known to have the antibonding s and p states nearly degenerate in energy, a condition which optimizes sp^3 bonding. This is why ΔH_s is so great for Si. The figure also shows that ΔH_s(Pb) is about 10 kcal/mole lower than one would have expected from the cohesive energies of Ge and Sn. This is explained by the fact that the $6s^2$ subshell of Pb is chemically inactive and contributes relatively little to the cohesion of the solid. A similar analysis applied to the sequence Zn, Cd, Hg explains why the cohesive energy of the latter is also lower than expected by about 10 kcal/mole.

There are a number of interesting trends in the $3d$, $4d$, and $5d$ transition periods. First, note that the large cohesive energies of Cu, Ag, and Au suggest that their d-electron subshells, although complete in the free atom, still contribute substantially to ΔH_s, just as for the transition metals with partially filled d subshells. The cohesive energies of Cu, Ag, and Au alternate (higher–lowest–highest, respectively), and a similar alternation (of lesser magnitude) is observed for Ca, Sr, and Ba at the beginning of the transition metal series. The colors of Cu, Ag, and Au are correlated with the alternation of their cohesive energies. Spectroscopic studies have shown that the d bands of Cu, Ag, and Au start about 2, 4, and 2 eV, respectively below the Fermi energies, and give the metals their characteristic colors. The d-electron contribution to the cohesion of these metals is least in Ag because polarization of its d shell is less because of the larger energies required to produce a virtual excitation.

2.3. Transition Series

Within each transition series one would expect an additional contribution to the cohesive energy from partially filled shells. This is the case for the $5d$ series, where the cohesive energy peaks at W (outer valence configuration $6s^1\,5d^n$, with $n = 5$). In the $5d$ series the cohesive energy is approximately an even function of $5 - n$ near $n = 5$.

Almost all the transition metals favor body-centered cubic structures near $n = 5$ and face-centered cubic or haxagonal close-packed structures near $n = 2$ or $n = 9$. This effect can be explained microscopically through energy band calculations. The distribution of d-like energy states in bcc structures contains two peaks of equal area, corresponding to five electrons each per atom. These two peaks are similar to bonding and antibonding bands in covalent crystals. In the fcc structure, however, the density of d states is more nearly constant. As a result, the bcc cubic structure is stabilized when it is

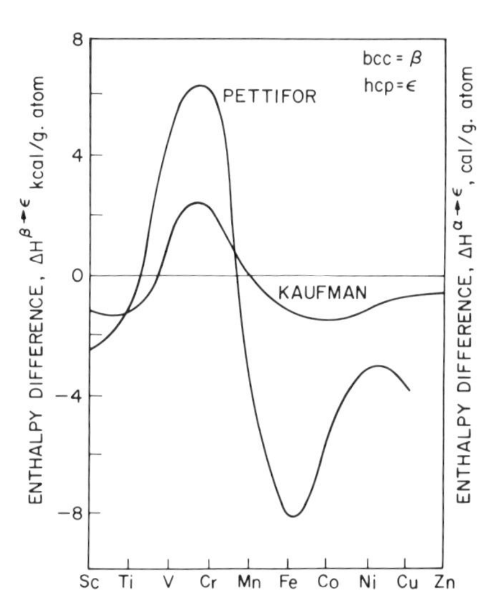

Fig. 5. Differences in cohesive energies of transition metals in fcc, hcp, and bcc structures as a function of column in the periodic table. The curve based on band structure calculations is valid near $T = 0°\text{K}$ (Pettifor, see Ref. 3), while the empirical curve[22] is based on phase transitions which take place at high temperatures ($T \sim 10^{3}°\text{K}$) near the melting point.

nonmagnetic and nearly half filled ($n = 5$), while the close-packed structures are stabilized nearer $n = 2$ or $n = 9$.

The scale of energy differences between the structures is determined by phase diagrams of alloys near the melting points. In Figure 5 energy differences between hcp and bcc transition metals are compared according to band theory and according to what is required to explain phase diagrams. The two curves are similar, but differ in scale, presumably because the band theory curve refers to $T = 0$ while the phase-diagram curve describes the situation at high temperatures T near the melting temperature T_m. At these temperatures T the effects of atomic vibrations already reduce the energy difference between the two structures considerably, e.g., because the anisotropy of the vibrations is different in the two structures. Nevertheless, the qualitative similarity of the two curves is striking.

In addition to the band effects illustrated in Figure 5, in the $3d$ series exchange interactions among the $3d$ electrons are large and

Hund's rule plays an important role. This means that electrons of like spin tend to line up when they are centered on the same atom. The Hund-rule or intraatomic exchange combined with the effect of the exclusion principle on interatomic interactions reduces the d-electron contribution to cohesion. The Hund-rule energy is large enough to produce local magnetic moments and reduce substantially the cohesive energies of Cr, Mn, Fe, Co, and Ni. In the last three elements the atomic moments have an attractive interatomic exchange interaction and the metals are ferromagnetic at room temperature. The interatomic interaction in Cr is weakly antiferromagnetic. However, in Mn there are local moments but no long-range magnetic order. These local moments make it possible for Mn to exist in four allotropic modifications; at high temperatures, near the melting point, thermal fluctuations reduce the effects associated with the atomic moments and the structure is the normal one (bcc). Because of the disordered local moments, which effectively reduce the valence of Mn, its cohesive energy is very low. The low cohesive energy of Mn also reflects the stability of atomic Mn in the $4s^2\ 3d^5$ configuration.

The local moment effects which strongly modify bonding in the $3d$ series, and are of little importance in the $5d$ series, are present to a lesser extent in the $4d$ series. Thus the element with the largest cohesive energy in this series is Nb ($5s^1\ 4d^4$), while Mo ($5s^1\ 4d^5$) has a cohesive energy which is lower than expected (from comparison with the $5d$ series) by about 30 kcal/mole. In the $3d$ series the loss in cohesion in similar elements near $n = 5$ is 50 kcal/mole or more.

In the d transition series the connection between local moments and valence is ambiguous because in most cases the d electrons have an itinerant or bandlike character, upon which there is superposed through intraatomic or Hund-rule interactions a greater ($3d$ series), moderate ($4d$ series), or minor ($5d$ series) degree of local or atomic character. In the pioneering days of the late 1920's and early 1930's a number of efforts were made to construct simple atomistic models to explain trends in atomic radii and ferromagnetic moments, especially in the $3d$ series. Some of these models employed itinerant electrons in bands, while others postulated complicated valence hydridization states to explain trends in atomic radii. Today it is known (at least near the Fermi energy) that the band model is correct, and careful studies of the geometry in **k** space (or crystal momentum space, as marked out with the help of Brillouin zones) of these bands has demonstrated the correctness of the band model in some detail. From band models one can state with considerable accuracy the effective ground-state hybridization configuration, and demonstrate, for example, that Cu has ten $3d$ electrons and one $4s$–$4p$ conduction electron. However, this band knowledge by itself does not erase the contribution of the $3d$ electrons to increasing the cohesion of Cu and to contracting its effective atomic volume. Efforts have been made to connect band theory with cohesion and atomic volume in metals, but so far these definitely have a preliminary character.

The $4f$ and $5f$ rare earth series elements exhibit integral local magnetic moments in magnetic compounds because the $4f$ and $5f$ electrons in partially filled shells in ionic states are tightly bound and are practically part of the ion core. Elements in the rare earth series show some dependence of cohesion on the number of $4f$ electrons; thus Zachariasen has estimated that a $4f$ electron is 5% as effective as a $5d$ electron in contributing to the cohesion of the element.

3. Binary Compounds and Alloys

Our knowledge of matter and its properties is continually evolving, from the primary task of gathering data, to its organization in reviews and bibliographies, and eventually to its explanation and systematization in theoretical terms. Often the last processes lag one or more decades behind the prior ones. The quantity of experimental data on binary alloys up to 1964 covers 2380 systems,[4] with new binary systems being studied at the (naturally diminishing!) rate of about 100 per year. This growing quantity of data poses difficult problems for bibliographers, and even more difficult problems for theorists. Broadly speaking, the theoretical ideas are simpler and more general when there are less data available! Thus most of the ideas presented in this section are not new, and are no longer accepted without qualifications of several kinds. However, some knowledge of these ideas is essential for navigation through an ever-increasing sea of data and interpretations. Our approach will be to mention an idea, together with a history of its origin, and then mention some recent developments which have modified the originally simple situation. Each idea deserves a review article of its own, but the purpose here is simply to expose the reader to the more frequently occurring themes and their refinements.

3.1. Minerals

In Section 1.2 we classified insulators such as NaCl, MgO, and SiO_2 as ionic compounds. For historical reasons we describe them here as minerals, to emphasize the role that circumstance has played in the development of qualitative theories of chemical bonding in ionic crystals. The pioneering X-ray work on structural determinations of binary compounds was carried out in the 1920's and early 1930's. At that time leaders in this work generally obtained their samples from mineralogists. (V. M. Goldschmidt, for example, was a geochemist by profession, and many of his samples were supplied by his Norwegian geological colleagues.) In those days few crystals were grown synthetically, and one's outlook on the structure of solids was mineralogical, as is apparent from the titles of several of Bragg's books.[5] Today, of course, almost all applications of solid-state chemistry involve samples prepared, at least in part, synthetically. This is true of semiconductors used in electronic devices, intermetallic alloys with specified magnetic properties, and complex ionic crystals used for optical materials.

What are the characteristics of a compound which make it readily available in crystalline form as a naturally occurring mineral? Above all, it must be stable chemically at rather high temperatures relative to other compounds involving the same elements. Thus it is likely to be a very good insulator, with an electronic configuration of a closed-shell type. Mechanically, it is likely to be relatively rigid, and its structure is likely to be well described in most cases in terms of mechanical packing of atoms or ions with definite spherical radii.

With these considerations in mind we can understand why tables of ionic radii constructed by Goldschmidt and Pauling proved so successful in accounting for the structures of many naturally occurring minerals. A simple and straightforward discussion of ionic radii has been given by Pauling,[6] which we shall not reproduce here in its entirety. As proposed by Madelung and Born around 1910, the forces between cations and anions in ionic crystals can be separated into pairwise electrostatic attractive forces, the largest of which by far is that between point charges, and a pairwise "hard-core" repulsive force varying like r^{-n}, where n is determined empirically for each period from compressibility data. (For an A^{+}–B^{-} interaction where A and B have different closed-shell configurations, the average of the n values of A and B is used.) In this way the equilibrium interatomic distances in the alkali halides are used to define univalent ionic radii R_1 for each alkali ion and each halogen ion. Ionic radii R_z for polyvalent ions of valence z are related to a Madelung constant A (which is a geometric constant of the crystal) and a repulsive constant B

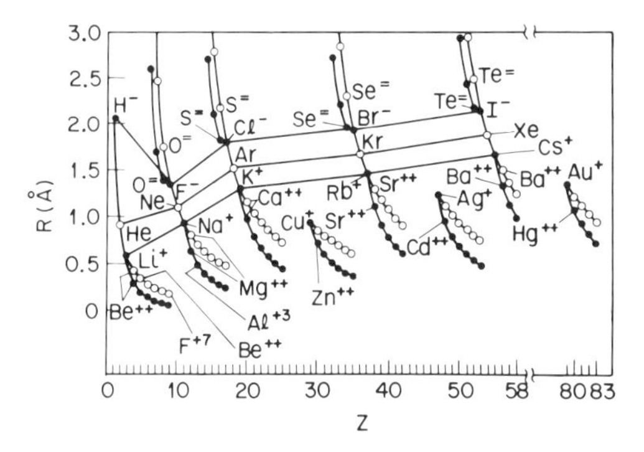

Fig. 6. Comparison of polyvalent radii R_Z and univalent radii R_1 of cations and anions according to Pauling's simple model. The univalent radii are the ones used in crystals containing polyvalent ions to make radius ratio arguments.

(which depends only on the rare gas configuration of the ion) through

$$R_z = R_1(A, B, n)z^{-2/(n-1)} \tag{2}$$

The importance of the discussion of polyvalent radii and univalent radii lies in the fact that the former are used to calculate interatomic separations, while only the latter exhibit simple chemical trends. Figure 6 compares univalent and polyvalent radii. In the early days of structural work Pauling's radii often proved to be more accurate (to within 0.01 Å, at least) than crude values obtained from X-ray studies of powder samples.

For strongly ionic compounds such as the alkali halides the closed-shell or rare gas approximation to the electronic structure of the ions is extremely accurate. Goldschmidt and Pauling have shown that for such compounds there exist strong correlations between many physical properties (crystal structure, bond lengths, compressibilities, heats of fusion and sublimation, melting and boiling points, solubility) and the ratio $\rho = R_1^{\ +}/R_1^{\ -}$ of cation univalent radius to anion univalent radius. The normal value for ρ is determined as the average value $\rho_0 = 0.75$ in the alkali halides (NaF, KCl, RbBr, and CsI) in which the ion cores of the cation and anion are isoelectronic (to Ne, Ar, Kr, and Xe, respectively). In these normal compounds the dominant interaction is the cation–anion one. The secondary interactions (cation–cation and anion–anion) are small and are of roughly equal importance. Deviations of ρ from ρ_0 imply that anion–anion interactions grow in importance ($\rho < \rho_0$) or cation–cation interactions become significant ($\rho > \rho_0$). The former case is by far the more common, and in extreme cases (e.g., LiI) low values of ρ mean that anion–anion contacts, together with cation–anion ones, cause a double repulsion which determines the lattice constant.

The simplest example of the importance of anion–anion interactions is the phase transition from the CsI to the NaCl structure (eightfold coordination to sixfold coordination). With eight anions cubically situated around a cation, anion contact takes place for $\rho = \sqrt{3} - 1 = 0.73$. (Similarly, six anions octahedrally disposed around a cation are in contact for $\rho = \sqrt{2} - 1$, and four tetrahedrally coordinated anions are in contact for $\rho = (\sqrt{3}/\sqrt{2}) - 1$.) Thus as ρ diminishes from the isoelectronic value of about 0.75, the closed-shell model predicts a series of phase transitions (assuming cubic coordination symmetry) from eightfold- to sixfold- to fourfold-coordinated structures.

The actual crystal structures of the alkali halides are not so simple as this argument would suggest. For LiI ($\rho = 0.28$) the model predicts fourfold coordination, but the only stable or metastable structure of LiI is the sixfold-coordinated NaCl structure. To get around this difficulty, one may replace the hard-sphere mechanical-contact criteria mentioned above by more realistic calculations of the repulsive interaction (e.g., r^{-n} with $n = 9$, the hard-sphere model corresponding to $n = \infty$). This rescues LiI, but it then predicts that the NaCl structure will always be more stable than the CsI one unless $n \gtrsim 30$, which is unrealistic. We conclude that the Goldschmidt–Pauling model can give a good account of chemical trends in properties of at least the alkali halides, but that it is not capable of predicting structural transitions.

More recent refinements of the Goldschmidt–Pauling approach have modified the functional form of the interatomic interactions. Dipole and quadrupole terms have been added to the electrostatic attractive interations, and the repulsive forces have been given the form required from quantum mechanical considerations ($B'e^{-\lambda r}$ rather than Br^{-n}). These refinements improve agreement between theory and experiment for cohesive energies and compressibilities of the alkali halides. However, to explain the observed crystal structures of the alkali halides,[7] and in particular the eightfold-coordinated structures of CsCl, CsBr, and CsI, it is necessary to assume (in effect) that B' and λ are different for Cs ions depending on whether they are sixfold- or eightfold-coordinated, a hint that the closed-shell model is not valid even for alkali halides when both ions have significant polarizabilities.

The closed-shell model is even less satisfactory for compounds containing polyvalent ions (e.g., MgO). The qualitative explanation for these cases is that the electronic distribution of each ion is not sufficiently close to that of the isoelectronic rare gas atom, e.g., the charge distribution of O^{2-} is greatly different from that of Ne. This point has been confirmed by empirical studies[8] of ionic polarizabilities based on the Clausius–Mossotti model. These show that the polarizability of the halide ions is nearly constant for each ion (independent of the compound in which it finds itself), but that the polarizability of O^{2-} varies by as much as a factor of three from one compound to another. It is possible that introduction of polarization data beyond the simple multipole model, in a manner similar to that used in successful treatments of covalently bonded compounds (see Section 4), would improve matters in these cases.

Current theoretical research on ionic crystals includes treatment of noncentral or three-body forces and their effects on shear forces (e.g., deviations from the Cauchy relation for elastic constants). Such calculations are of secondary interest from a chemical and structural point of view.

Recent structural research on complex ionic compounds still utilizes radius-ratio arguments to obtain qualitative guides to the structures of ternary and quaternary compounds containing polyvalent ions.[(9)] Whereas in simple ionic compounds there are many cases of symmetric cations and anions (e.g., NaCl and CsI), in the more complex minerals asymmetric valence is the rule. The simplest examples of asymmetric valence ionic compounds are CaF_2 and TiO_2. In the more complex structures one ordinarily finds that the cations M have higher valence than the anions X. This is because M^{2+} is more stable than X^{2-}, and M^{3+} and M^{4+} are at least formally stable, but X^{3-} and X^{4-} are rarely stable in ionic structures.

To deal with complex mineralogical structures such as the silicates, Pauling introduced in 1928 several rules which are still widely used. The first of these, and the most important, stresses the formation about each cation of a coordinated anion polyhedron. Allowable anion configurations are determined from radius-ratio considerations, such as those discussed above for simple ionic solids. Although these are not really precise enough to account in exact detail for all the structures of simple ionic crystals, they still provide an excellent guide for understanding the structure of minerals such

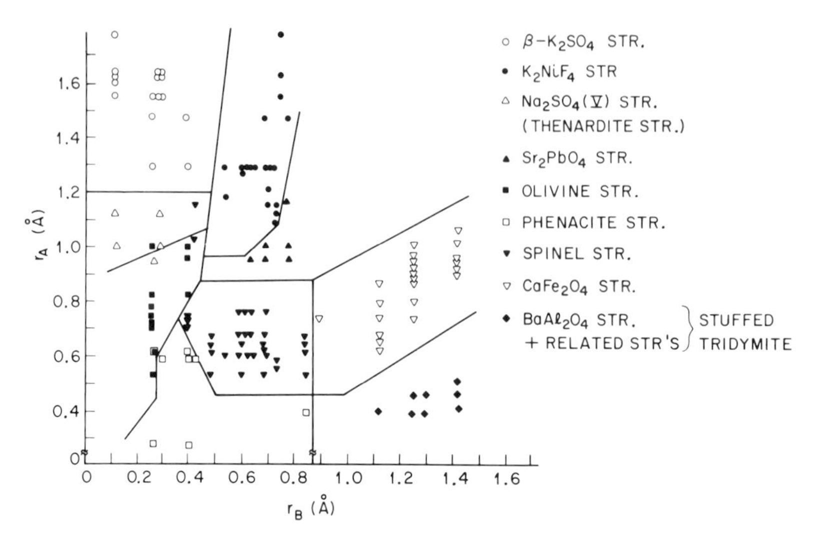

Fig. 7. Classification of the structures of A_2BO_4 ionic minerals. For further discussion of materials of this kind, see Ref. 23.

as mica, $KAl_3Si_3O_{10}(OH)_2$. Recent modifications include emphasis on polarizability effects and the way in which these affect Pauling's rules. For example, a grouping of greater cation charge on one side of a highly polarizable anion is favored by polarization energies.

Ternary ionic compounds commonly contain one anion and two cations. Minerals based on the same anion (usually O^{2-} or F^-) can be conveniently classified according to the radii r_A and r_B of the two cations in the compound. A map of this kind is shown in Figure 7 for A_2BO_4 compounds. From this map it can be seen that the choice of structure is not determined by mechanical contact considerations. However, the structures are well separated, which suggests that the rigid-ion model gives a better description of second-neighbor interactions than it does of nearest-neighbor ones.

3.2. Semiconductors

From a chemical point of view tetrahedrally coordinated semiconductors such as Si^4, Ge^4, Ga^3As^5, Zn^2S^6, and many other compounds with an average of four valence electrons per atom are members of a prototypical family of sp^3-bonded crystals. However, the crystal chemistry of these covalent materials receives little attention in most texts, which base their discussions of chemical bonding in crystals primarily upon mineralogical models constructed in early structural studies.

The first monograph giving a broad survey of structural trends and chemical bonding in tetrahedrally coordinated compounds is that of Parthe.[10] The condition that there be an average of four valence electrons per atom accounts for all possible binary compounds with every atom tetrahedrally coordinated. In addition to the *4*, *35*, *26* compounds mentioned above (this being an abbreviated notation for the classes of compounds containing Si^4, Ga^3As^5, and Zn^2S^6, respectively) there are also $17, 3_26, 3_37, 25_2, 2_37_2, 15_3$, and 1_26_3 tetrahedrally coordinated binary compounds. For example, ZnP_2 is a representative of the 25_2 class. The fact that so many compounds exist in tetrahedrally coordinated structures with such a wide variety of nominal valence states (from *1* to *7*) is another illustration of how dramatically chemical bonding in crystals differs from chemical bonding in molecules. Moreover, a large number of pseudobinary alloys with wide miscibility ranges can be prepared based on these materials.

Among binary compounds the condition of having an average of four valence electrons per atom is necessary but not sufficient for a compound to have a tetrahedrally coordinated structure. Thus none of the alkali halides are tetrahedrally coordinated, although according to the mechanical model (Figures 1 and 2), some of them (e.g., LiI) should be. The general situation among simple binary *17*, *26*, *35*, and *4* compounds is the following. All *4* and *35* compounds are tetrahedrally coordinated. Some *26* compounds have the ionic NaCl structure, while some have tetrahedral structures. Most *17* compounds have ionic structures, but the Cu halides and AgI are tetrahedrally coordinated. The situation is summarized in Figure 8, which gives the coordination configurations of tetrahedrally and/or octahedrally coordinated A^nB^{8-n} compounds.

The problem now is to construct a quantitative theory of covalent and ionic bonds which is sufficiently accurate to predict the coordination configurations shown in Figure 8. Plainly this cannot be done with the classical ionic model, which is only partially successful for the alkali halides, and which fails badly in treating the IB halides and *26* compounds such as ZnS. Tetrahedrally coordinated structures do not result from small values of $\rho = r_+/r_-$, as suggested by mechanical models, but rather from the formation of covalent bonds. No perturbation on rare gas electronic configurations can be expected to yield covalent structures, because the covalent bond describes sharing of valence electrons. The shared electronic charge (the bond charge)

	IA				IB	
	Li	Na	K	Rb	Cu	Ag
F	6	6	6	6	4	6
Cl	6	6	6	6	4	6
Br	6	6	6	6	4	6
I	6	6	6	6	4	4

	IIA			IIB				
	Ca	Sr	Ba	Be	Mg	Zn	Cd	Hg
O	6	6	6	4	6	4	6	6
S	6	6	6	4	6-4	4	4	6-4
Se	6	6	6	4	6-4	4	4	4
Te	6	6	6	4	4	4	4	4

Fig. 8. Tableau of the coordination configurations of simple binary compounds A^nB^{8-n} for $n = 1$ and $n = 2$. The entries four and six refer to tetrahedrally and octahedrally coordinated compounds, respectively. Chemical trends from covalent to ionic bonding are associated with the $4 \rightarrow 6$ transition. Several compounds can be prepared in either form at STP, corresponding to differences in free energy of order 2 kcal/mole.

cannot be separated into two parts, one belonging to the cation and one to the anion. A different approach altogether is required.

The first solution to this problem was produced phenomenologically by Mooser and Pearson. The solution for A^nB^{8-n} compounds is reproduced in Figure 9. Similar solutions apply not only to A^nB^{8-n} semiconductors and insulators, but also to many intermetallic compounds including transition metals. This work provides the first step toward explaining structural and phase transitions in chemically homologous families of binary crystals. It has made the question of the proper treatment of chemical bonding in crystals susceptible to theoretical analysis, whereas formerly work based on mechanical models (ionic compounds) or quantum mechanical perturbation theory (nearly-free-electron metals) made the same problem appear insoluble.

How has this revolution in attitude come about? The older theories laid great emphasis on the deductive approach, and especially on absolute calculations of cohesive energies and related properties. Pearson's approach recognizes that absolute calculations, while useful in filling up space in the literature, are actually of little chemical interest. What matters to most materials scientists is not whether a model (be it quantum mechanical or classical) yields correct order-of-magnitude results (without adjustable parameters) or detailed numerical agreement in a few cases (with suitably adjusted models), but whether, *without adjustable parameters*, it correctly predicts chemical trends in important observables. In a sense, chemical bonding is a problem in pattern recognition of trends in constitutive

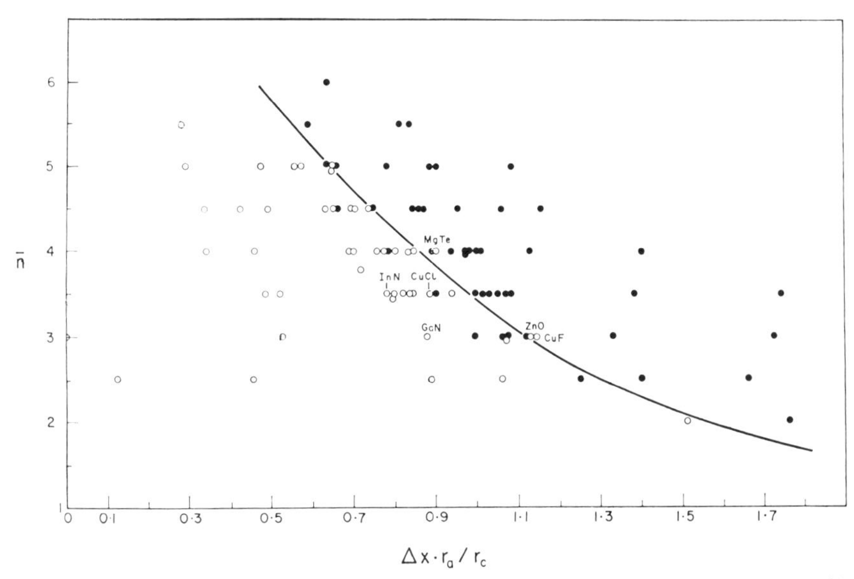

Fig. 9. Coordination configurations of A^nB^{8-n} compounds as plotted by Pearson.[24] Solid circles are compounds with sixfold coordination (NaCl structure), while open circles are fourfold-coordinated compounds. The smooth curve separates the two structures successfully in about 90% of the cases. Compare with Figure 2, where the mechanical model of the transformation from sixfold to eightfold coordination fails. The microscopic dielectric theory of covalent bonding is 100% successful in treating this phase transition (see Figure 13).

properties. What Pearson has shown is that in one of the most delicate properties (phase transitions), the proper choice of variables makes the problem of recognizing the actual pattern much easier.

The importance of this attitude toward chemical bonding in crystals cannot be overstated. However, we must recognize at the same time that Pearson's approach is but the first step toward a complete chemical theory of material properties. Each new property in each new family of chemically homologous materials requires new insight. As solutions accumulate one may expect them to overlap and to suggest further solutions to other cases. At present the only family of binary compounds to which this approach has been applied systematically is the A^nB^{8-n} family of semiconductors and insulators, which has been studied extensively because of its technological importance. From trends in chemical bonding and material properties in this family, however, we can expect to learn a great many things which apply to other solids as well. We also learn, and this may be equally important, that many approaches (such as mechanical models) should not be applied to discuss bonding questions.

The first step in recognizing the pattern of chemical bonding in a family of binary compounds is to identify appropriate chemical variables. For simplicity Pearson uses the Pauling electronegativity difference

$$\Delta X = X_A - X_B$$

and the average principal quantum number

$$\bar{n} = (n_A + n_B)/2$$

as variables, or in some cases $(\Delta X)(r_B/r_A)$. The ionicity of the chemical bond depends on ΔX, while metallic character increases with $\bar{n}$. Mechanical effects are, of course, included in the radius ratio $\rho = r_A/r_B$. These phenomenological variables may not be the best ones, but Pearson's results indicate that they are close to the best ones. In particular, the inclusion of metallic tendencies through the variable $\bar{n}$ is strikingly successful. The combination of the electrochemical variable ΔX or $(\Delta X)(r_B/r_A)$ with the metallurgical one, $\bar{n}$, shows that material properties may be best understood by combining the outlooks of several disciplines.

In a later section we show how for semiconductors one can use polarizabilities to separate covalent and ionic effects. When this is done one can define an elemental electronegativity of nontransition elements for use in tetrahedrally coordinated crystals. This electronegativity can be compared with Pauling's values, which are based largely on heats of formation of molecules. By adjusting a scaling factor and an additive constant, the two sets of values can be made to agree for elements from the first period. The two different scales of electronegativities for nontransition elements are compared in Figure 10, which also includes Gordy's metallic scale.

Is it really necessary to have three different scales of electronegativities? The answer is definitely yes. For example, according to Pauling's scale, $X(\text{Si}) = X(\text{Ge}) = X(\text{Sn}) = X(\text{Pb})$. On the dielectric scale $X(\text{Si}) > X(\text{Ge}) > X(\text{Sn}) > X(\text{Pb})$. Recalling Pauling's definition[6] of electronegativity as "the power of an atom in a bonded state to attract electrons to itself," we can begin to understand these differences. Offhand, one would expect larger values of X_A for smaller atoms A because potential energies are proportional to $1/r_A$. However, for the heavier elements part of the heats of formation (upon which Pauling bases his table) arise from polarization of the atomic cores. For the heavier elements the increase in core polarization energies is just sufficient to compensate for the decrease in charge-transfer

Li 1.00 (1.0) [(0.95)]	Be 1.50 (1.5) [(1.5)]	B 2.00 (2.0) [(2.0)]	C 2.50 (2.5) [(2.5)]	N 3.00 (3.0) [(3.0)]	O 3.50 (3.5) [(3.5)]	F 4.00 (4.0) [(3.95)]
Na 0.72 (0.9) [(0.9)]	Mg 0.95 (1.2) [(1.2)]	Al 1.18 (1.5) [(1.5)]	Si 1.41 (1.8) [(1.8)]	P 1.64 (2.1) [(2.1)]	S 1.87 (2.5) [(2.5)]	Cl 2.10 (3.0) [(3.0)]
Cu 0.79 (1.9) [(1.8)]	Zn 0.91 (1.6) [(1.5)]	Ga 1.13 (1.6) [(1.5)]	Ge 1.35 (1.8) [(1.8)]	As 1.57 (2.0) [(2.0)]	Se 1.79 (2.4) [(2.4)]	Br 2.01 (2.8) [(2.8)]
Ag 0.67 (1.9) [(1.8)]	Cd 0.83 (1.7) [(1.5)]	In 0.99 (1.7) [(1.5)]	Sn 1.15 (1.8) [(1.7)]	Sb 1.31 (1.9) [(1.8)]	Te 1.47 (2.1) [(2.1)]	I 1.63 (2.5) [(2.55)]
Au 0.64 (2.4) [(2.3)]	Hg 0.79 (1.9) [(1.8)]	Tl 0.94 (1.8) [(1.5)]	Pb 1.09 (1.8) [(1.9)]	Bi 1.24 (1.9) [(1.8)]		

Fig. 10. Comparison of scales of electronegativity for nontransition elements. For each element values are taken from the dielectric scale (sp^3 hybridized bonds) in line 1, from Pauling's thermochemical scale in line 2 (in parentheses), and from Gordy's analysis (based on work functions at surfaces and molecular data) in line 3 (in brackets and parentheses).

energies of valence electrons; this causes the constancy in Pauling's values for column IV elements noted above. In the dielectric method, on the other hand, core polarization effects are treated separately. For A^nB^{8-n} compounds composed of nontransition elements core polarization apparently has little influence on structural properties. In particular, the dielectric scale gives a more accurate description of the fourfold- to sixfold-coordinated phase transition, as shown in a later section. For intermetallic compounds containing both transition and nontransition elements, however, either Pauling's or Gordy's scale yields a much better description of phase transitions when used for both kinds of elements (Pearson, Chapter 3 of this volume). This result suggests that in the presence of *d* electrons core polarization of the nontransition elements plays an important role in determining structure and affecting the chemical bond. The magnitude of this effect is best measured by lumping it together with the valence effect on heats of formation. If we consider that electrons in partially filled *d* shells of transitions elements are in some respects intermediate between free *s–p* valence electrons and polarizable electrons in filled outer *d* shells on nontransition atoms, then this conclusion appears reasonable.

3.3. Intermetallic Solutions

At present the most widely known theory of alloys is that developed in the 1930's by Hume-Rothery and his associates.[2] Studies of alloys based mainly on Cu and Ag led Hume-Rothery to the conclusion that the formation of substitutional solid solutions between elements A and B is governed by three factors: differences in the atomic radii r_A and r_B; differences in Pauling electronegativities $X_A - X_B$; and differences in valence. A difference in radius of more than 15% reduced the solubility to 1% or less. This is illustrated in Figure 11 for solubility in Mg of various nontransition elements.[11] The same figure shows how differences in valence affect solubility limits.

A two-dimensional plot, as we saw in connection with Pearson's work, can be more informative than a one-dimensional one. In Figure 12 the Pauling electronegativity and atomic radius (in a 12-fold-coordinated structure, which is appropriate here) of nontransition

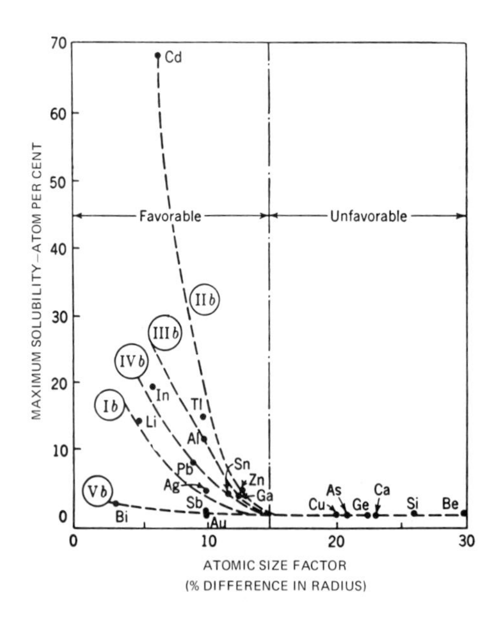

Fig. 11. Solubility of metals in Mg according to the differences in atomic size and the nontransition group to which the solute belongs.

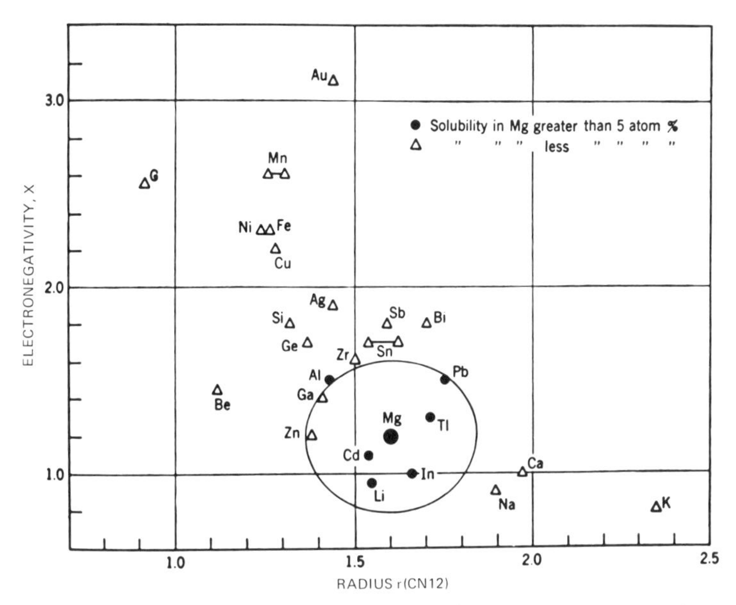

Fig. 12. A two-dimensional plot of metallic solubilities in Mg, showing the effect of size and electronegativity difference on whether a given metal is soluble or not, solubility being defined as more than 5 % soluble. The typical allowed size deviation is 15 %, the typical allowed electronegativity difference 0.4.

solutes dissolved in Mg is shown. This is known as a Darken–Gurry plot, and it is qualitatively successful in predicting mutual solubilities. Quantitative improvements are obtained when Pauling's molecular electronegativities are replaced by a modified scale for metals derived by Gordy from screening considerations, as discussed in a later section. With this modification it is found that an electronegativity difference of 0.4 unit has an effect in limiting solubility which is equivalent to a size difference of 15 %. If one arbitrarily chooses 5 at. % as the boundary separating extensive solubility from limited solubility, then a modified Darken–Gurry plot is successful about 75 % of the time in separating the two classes. A one-dimensional plot based on atomic size alone is successful about 65 % of the time. Thus, while the size effect is dominant in limiting solubilities, electrochemical effects still play a significant role.[12]

If we compare the 75 % successful Darken–Gurry theory of alloy solubilities with the 90 % successful Mooser–Pearson theory of compound structures, we see that the absolute successes of the two theories are comparable because of the fact that not all liquid alloys will form substitutional solid solutions. If allowance is made for tendencies toward compound formation (short-range order in the alloy), then this additional factor is probably sufficient to account for the reduced success of the Darken–Gurry analysis. Another point that limits the success of the alloy analysis is the neglect of factors other than the two considered (e.g., neglect of core polarization effects). In the case of compound formation $\Delta X_{AB} = X_A - X_B$ is generally of order 1.0, whereas ΔX_{AB} for extensive metal solubility is of order 0.2. When ΔX is large it is more important than the other factors, but when it is small the other factors become significant. Thus it is easier to discuss structural trends in compounds or to predict very limited solubilities (ΔX large) than it is to predict extensive solubility (ΔX small).

Some alloy phases which possess extensive solubility ranges because of similar atomic sizes and electronegativities also exhibit characteristic structural behavior; these phases are called electron phases. Examples of electron phases are the Hume-Rothery alloys based on the solvents Cu or Ag, with solutes from the Cu or Ag periods, respectively. Phase transitions in these alloys tend to occur at certain values of the electron/atom ratio, regardless of the solute; thus 20% Ga in Cu is similar to 40% Zn in Cu, and so on. Historically the existence of these alloys has made an impression in the literature somewhat out of proportion to the number of systems which can be legitimately described as primarily electron phases. One reason for this may be the ease with which the electron/atom (e/a) ratio can be calculated, and the graphical facility with which one can make (not necessarily successful) plots of material properties as a function of e/a.

In certain transition metal alloy systems the e/a ratio does appear to play an important part in determining physical properties and structural trends because d electrons tend to be rather well localized on the parent atom and are resistive to charge-transfer effects that arise from electrochemical differences between nontransition elements. In any case analyses of chemical trends in metallic alloys based entirely or primarily on e/a values should be treated with caution.*

4. Chemical Bonding and Physical Properties

The broad classification of materials into minerals, semiconductors, and metals discussed in the preceding section gives a good

* Analysis of the properties of alloys solely in terms of e/a ratios is particularly popular with metallurgists of the Viennese school, e.g., Schubert.[13]

account of differences in physical properties, at least with respect to order of magnitude. Today, however, solid-state chemistry is concerned with rather smaller differences in material properties. A good example is afforded by superconductivity. It is not enough to know that many metals become superconducting, or even that transition metals with $4d$ or $5d$ valence electrons combine with the heavier nontransition metals (such as Sn and Pb) to form good superconducting compounds. We would like to understand which particular combinations of elements will form structures with high superconducting transition temperatures T_c and which aspects of chemical bonding make these structures possible.

At present no microscopic theory exists which can answer questions like this, and there are few phenomenological descriptions either. What is badly needed is a common framework which will at least facilitate the development of incisive and informative phenomenological descriptions. I believe that polarizabilities can provide such a framework, and that they can be used to explain trends in chemical bonding and physical properties even when an absolute and rigorous connection is not demonstrated. For this to become so, however, it is necessary to understand the quantum mechanical meaning of polarizabilities in a more profound way than has generally been the case in the past. The story of how little this rather simple subject has been explored theoretically illustrates why it is that our understanding of the properties of materials is still at so primitive a level.

4.1. Classical Polarizabilities

The classical model of crystalline polarizabilities was introduced at least as early as 1850 by Mosotti, and discussed in an 1879 book by Clausius. The polarizability α_m of the medium is described in terms of a set of polarizability parameters α_j of the ions j in the crystal distributed in a unit cell of volume V_m. The relation between α_m and the parameters α_j is (per unit cell)

$$\alpha_m = \sum_j \alpha_j \tag{3}$$

This relation assumes that the individual atomic polarizabilities α_j are independent of the nearest neighbors of ion j. It can therefore be expected to be valid only in the extreme ionic limit. The model is, in fact, a variant of the mechanical model of ionic crystals. The latter fails, as we have seen in Sections 1 and 3.1, in describing differences in the crystal structures of the alkali halides.

Because of the cooperative interaction between the external electric field $\mathbf{E}$ and the internal polarization $\mathbf{P}$, the internal field $\mathbf{E}_{\text{eff}}$ acting to polarize a given ion is given by

$$\mathbf{E}_{\text{eff}} = \mathbf{E} + L\mathbf{P} \tag{4}$$

where normally $L = 4\pi/3$. With the relations

$$\mathbf{P} = \sum_j \mathbf{p}_j \tag{5}$$

$$\mathbf{p}_j = \alpha_j \mathbf{E}_{\text{eff}} \tag{6}$$

around 1900 Lorentz derived the result for $\varepsilon_0 = n^2$ (n is the index of refraction)

$$\alpha_m = \phi_m(\varepsilon_0 - 1)/[4\pi + L(\varepsilon_0 - 1)] \tag{7}$$

which has formed the basis for all subsequent analysis of the polarizabilities of ionic (and some not-so-ionic) crystals.

For the alkali halides AX one can show that a fixed set of parameters α_j and the Lorentz value $L = 4\pi/3$ reproduce experimental values of ε_0 to within about 3%. However, in all of these compounds the A–X bond is more than 90% ionic, according to the definition of ionicity given in the next subsection. For less ionic bonds the basic assumption of additivity, Eq. (3), is simply not valid. Moreover, for the less ionic crystals L rapidly tends to zero, and may be taken as zero for bonds which are less than 80% ionic. Thus, except for a small number of salts, the classical model can tell us little of chemical value about polarizabilities. Nevertheless, until 1968 the classical model was the only one which had been used to attempt to describe chemical trends in crystalline polarizabilities.

4.2. Dispersion

One of the weaknesses of the classical Clausius–Mosotti model is that it refers to the electronic polarizability $\alpha(\omega)$ only at frequency $\omega = 0$. However, associated with a single chemical bond jk between atoms j and k one has excitation energies ω_m^{jk} (in the case of a diatomic molecule), or a band of frequencies (for interacting chemical bonds in solids). As a simple first step, one can assign an average frequency $\bar{\omega}^{jk}$ to the jk bond. Then the real part ε_1 of the dielectric polarizability associated with the excitation of N valence electrons per unit volume is given by

$$\varepsilon_1(\omega) = 1 + (\hbar\omega_p)^2/[E_0{}^2 - (\hbar\omega)^2] \tag{8}$$

with $E_0 = \bar{\omega}^{jk}$ in crystals containing only one important cation–anion bond. The plasma energy $\hbar\omega_p$ is defined in terms of the plasma frequency

$$\omega_p{}^2 = 4\pi Ne^2/m \tag{9}$$

and it is a measure of the total oscillator strength of the N electrons per unit volume.

Analysis of dispersion data for a number of insulators containing only one anion species has been carried out[14] using (8) and introducing a "dispersion energy" defined by*

$$E_d = (\hbar\omega_p)^2/E_0 \tag{10}$$

When the analysis is carried out in this way it does not yield satisfactory results from a bonding viewpoint, because the values of E_d and E_0 that are obtained do not yield chemically meaningful values of $\hbar\omega_p$, i.e., the empirical values of $\omega_p{}^2$ do not correlate with N, as required by (9). (In this respect the arbitrariness of the parameters resembles that of the Clausius–Mossotti theory.) However, this approach does represent an improvement over the classical static model because there is established a proportionality between E_d and N_e, the number of valence electrons per cation–anion pair, Z_a, the anion valence, and N_c, the cation coordination number, which plays a crucial part in mechanical theories of structures of minerals; formally, we have

$$E_d \propto N_e Z_a N_c \tag{11}$$

In fact, this approach is sufficiently close in spirit to Pauling's resonating bond theory (where Z_a and N_c play important roles) that we may regard it as an electronic realization of some of his thermochemical ideas.

This approach, like Pauling's, is quite general and it works well for many insulators which have minimum gaps between valence and conduction bands of more than 1 eV. However, its generality is also

* Although of limited chemical value in crystals, the oscillator energy E_d is useful in describing the optical properties of amorphous semiconductors and glasses, as shown by Wemple.[15]

its chief weakness from a chemical point of view. The results obtained do not distinguish between different states of hybridization, nor do they permit one to obtain values of E_0 which exhibit identifiable chemical trends. Inasmuch as E_0 is the one parameter which measures bond strength, we may find this disappointing. However, the difficulties have arisen because attention has been focused not on the basic chemical parameters but on the most convenient optical ones.

A systematic classification of the optical parameters of solids is obtained by separating the dielectric function $\varepsilon(\omega)$ into its real and imaginary parts:

$$\varepsilon(\omega) = \varepsilon_1(\omega) + i\varepsilon_2(\omega) = [n(\omega) + ik(\omega)]^2 \tag{12}$$

where n and k are the real and complex parts, respectively, of the index of refraction. The electronic absorption associated with bonding → antibonding transitions is best described in terms of $\varepsilon_2(\omega)$, which describes in solids a band spectrum, in contrast to line spectra in atoms or vibration-broadened line spectra in small molecules.

The method of moments is convenient in many contexts, including spin resonance and infrared absorption. It can also be applied to band spectra. Define the moments M_r as

$$M_r = (2/\pi) \int_0^\infty (\hbar\omega)^r \varepsilon_2(\omega)\, d(\hbar\omega) \tag{13}$$

Then the energies E_d and E_0 in Eqs. (8) and (10) are given by

$$E_0{}^2 = M_{-1}/M_{-3} \tag{14}$$

$$E_d{}^2 = M_{-1}^3/M_{-3} \tag{15}$$

On the other hand, the plasma energy is given by

$$(\hbar\omega_p)^2 = M_1 \tag{16}$$

The connection between $\varepsilon_2(\omega)$ and the static polarizability $\varepsilon_1(0)$ is given by

$$\varepsilon_1(0) = 1 + M_{-1} \tag{17}$$

4.3. Covalent and Ionic Energies

The advantage of introducing optical moments is that by considering (14) and (15) one sees that the dispersion energies are defined in an asymmetric way which involves M_{-3}, and which therefore places too much emphasis on $\varepsilon_2(\omega)$ for small values of ω. This obscures chemical trends connected with the average bond energy $\overline{\hbar\omega^{jk}}$. A better choice of parameters has been made by Phillips[16] and Van Vechten[17] They introduce an average energy gap E_g to deal with bonding in A^nB^{8-n} compounds. It is defined by

$$\varepsilon_1(0) = 1 + (\hbar\omega_p)^2/E_g^{\,2} \tag{18}$$

or, according to (16) and (17),

$$E_g^{\,2} = M_1/M_{-1} \tag{19}$$

a result which is much more symmetric than (14) or (15).

The octet binary compounds A^nB^{8-n} include both NaCl and diamond, the prototypes of ionic an covalent crystals, respectively. They include about 80 compounds, and taken together they form the largest family of homologous solids, and by far the most thoroughly studied one. The subgroup of alkali halides, which has been the starting point for most previous chemical discussions of material properties of insulators, contains only 20 members, and these are the most ionic ones. By studying the entire family as a whole, and not merely its most ionic members, one can analyze quantitatively the transition from ionic to covalent bonding.

The chemical significance of the average energy gap E_g is brought out by separating it into its covalent and ionic parts E_n and C, respectively:

$$E_g^{\,2} = E_h^{\,2} + C^2 \tag{20}$$

where E_h depends on the bond length d:

$$E_h(\mathrm{AB}) = E_h(\mathrm{Si})(d_{\mathrm{Si}}/d_{\mathrm{AB}})^{2.5} \tag{21}$$

The ionic energy gap $C(\mathrm{AB})$ depends on the ion-core charges Z_A and Z_B of atoms A and B, on their covalent radii r_A and r_B, and on a Thomas–Fermi screening wave number k_s. The explicit expression for $C(\mathrm{AB})$ is

$$C = 1.5[(Z_\mathrm{A}e^2/r_\mathrm{A}) - (Z_\mathrm{B}e^2/r_\mathrm{B})]\exp(-k_s\bar{r}) \tag{22}$$

$$C = (r_\mathrm{A} + r_\mathrm{B})/2 \tag{23}$$

In Eq. (22) the constant $b = 1.5$ has been determined by using Eq. (18) and (20) to make a least-squares fit to experimental value of $\varepsilon_1(0)$. The value of b thus determined is accurate to about 10%. There is a simple interpretation of C in terms of elemental electronegativities, which is discussed below. The dependence of E_h on d is probably best regarded as a characteristic of tetrahedrally coordinated crystals that arises from the effects of constructive (destructive) interference of bonding (antibonding) wave amplitudes centered on nearest-neighbor atoms.

Because both E_h and C are obtained, in effect, from d and $\varepsilon_1(0)$ for each A^nB^{8-n} compound, we can define the fraction of ionic or covalent character as follows:

$$f_i = C^2/(E_h{}^2 + C^2) = C^2/E_g{}^2 \tag{24}$$

Note that f_i is dimensionless and that it lies in the interval $0 \le f_i \le 1$. Similarly, the fraction of homopolar (covalent) bond character f_h is

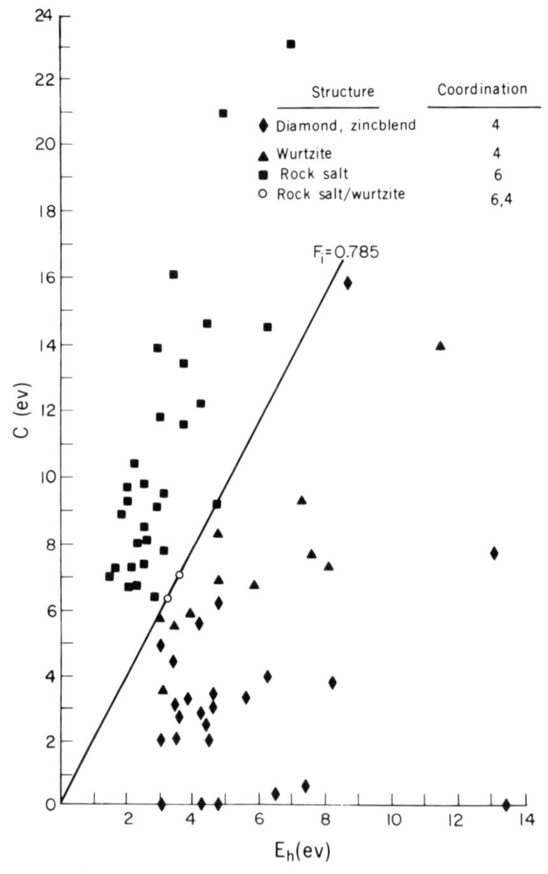

Fig. 13. A two-dimensional population distribution of A^nB^{8-n} compounds indexed by their covalent energy gaps E_h and ionic energy gaps C. The bond ionicity is defined as $f_i = C^2/(E_h{}^2 + C^2)$. The line $f_i = 0.785$ separates fourfold-coordinated compounds from sixfold-coordinated ones.

defined by

$$f_h = E_h^2/E_g^2 = 1 - f_i \tag{25}$$

One can plot the population distribution of all A^nB^{8-n} binary compounds in the (E_h, C) plane. According to the definition (24), curves corresponding to f_i = const are actually straight lines passing through the origin. Such a plot can be used to separate these compounds into those with fourfold-coordinated structures and those with sixfold-coordinated structures, the former lying below the line $f_i = 0.785$, the latter lying above it (see Figure 13). This demonstrates, in objective structural terms, that ionicities defined by the spectroscopic moments M_1 and M_{-1} are accurate to about 1% (80 structures correctly predicted).

4.4. Chemical Trends in Physical Properties

Spectroscopic ionicities defined by Eq. (24) can be made the basis for analyzing chemical trends in a wide range of physical properties of A^nB^{8-n} compounds. Space does not permit detailed discussion of these trends, but a list of some of the properties which can be understood in this way is of interest.

Both cohesive energies and heats of formation[18] of A^nB^{8-n} depend primarily on bond length and spectroscopic ionicity f_i. The cohesive energies (Gibbs free energy of sublimation of the solid at STP into atoms, ΔG_s) are found to be linear functions of f_i in isoelectronic sequences for which d is nearly constant (examples: diamond, cubic BN, BeO, or Ge, GaAs, ZnSe). This in itself is a striking result. Even more striking, however, is the result obtained by studying heats of formation of AB compounds. The latter are found to be proportional to f_i and to a dehybridization or metallization factor associated with the covalent–metallic transition which takes place in heavier elements or compounds, e.g., gray Sn $\leftrightarrow$ white Sn. By concentrating only on the ionicity f_i, one can establish the connection between heats of formation of A^nB^{8-n} tetrahedrally coordinated semiconductors and the Pauling electronegativity differences $X_A - X_B$. Especially for lighter atoms (where metallization effects do not complicate the analysis), this connection turns out to be the one implied by Eq. (22). It is

$$X_A \to (Z_A e^2/r_A) \exp(-k_s r_A) \tag{26}$$

whereas the correspondence used by Gordy is

$$X_A \rightarrow (Z_A e^2/r_A)\sigma_A \tag{27}$$

In (27) the number σ_A is a screening constant, originally derived by Slater from studying atomic spectra. It is dependent only on Z_A and not on the electron density or r_A. In situations where the integrity of an atom is well preserved (e.g., in molecules, at surfaces, or in transition metal compounds) Gordy's table of electronegativities is the most useful one. But in contexts such as A^nB^{8-n} semiconductors, where the valence electrons are nearly free, the dielectric table of electronegativities, based on Eq. (26), is to be preferred.

The spectroscopic ionicity f_i is useful for describing nonlinear dielectric properties[19] as well as linear dielectric properties. In A^nB^{8-n} semiconductors the nonlinear dielectric coefficient $\varepsilon^{(2)}$ is found to be proportional to $[f_i(1 - f_i)^5]^{1/2}$. The analysis can be applied with success to a wide range of semiconductors and insulators with complex crystal structures. In effect the study of chemical trends in the linear polarizability $\varepsilon^{(1)}$ is used as a bootstrap to predict values of $\varepsilon^{(2)}$ on many different materials.

An interesting application of ionicity lies in the analysis of interatomic force constants.[20] There are three elastic constants (c_{11}, c_{12}, c_{44}) for a given cubic crystal. Happily, these can be used to determine a bond-stretching force constant α and a bond-bending force constant β. The former is associated with central forces, the latter with directional or noncentral forces. Central forces are present regardless of whether the bond is ionic or covalent, but the strength of noncentral forces is a direct measure of covalency. This is illustrated in Figure 14, which shows β/α for a number of A^nB^{8-n} tetrahedrally coordinated compounds. The straight line is proportional to the fraction of covalent character $f_h = 1 - f_i$.

The discussion in this section has emphasized primarily chemical trends associated with the covalent → ionic phase transition. Many other trends can be explained by studying covalent → metallic phase transitions. For example, semiconductors melt to form metals, and one can predict melting temperatures and phase diagrams of many semiconductors[21] by introducing spectroscopic parameters to characterize metallic and covalent effects. Similarly, metallic and covalent bonding effects are important at metal–semiconductor interfaces (Schottky barriers).

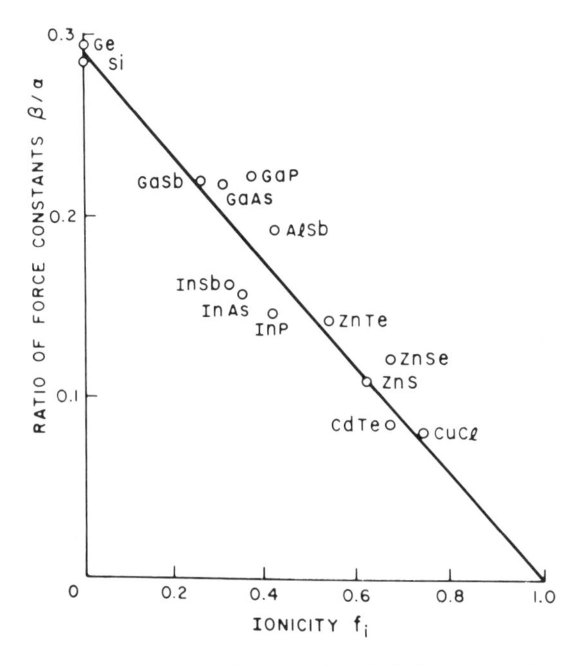

Fig. 14. Interatomic forces in A^nB^{8-n} compounds can be separated into bond-stretching forces (α) and bond-bending forces (β). The bond-stretching forces are central forces and change very little with ionicity for fixed bond length (e.g., in an isoelectronic sequence such as Ge–GaAs–ZnSe). The bond-bending forces are directional (noncentral) and are a direct measure of the amount of covalency $f_c = 1 - f_i$ of the AB bond. The ratio β/α is dimensionless, and is proportional to f_c, as shown here for many compounds.

5. Summary

At present there is a paucity of theories of chemical bonding in solids. The best-known models (mechanical models of the billiard-ball type) are quite limited in their range of application. Moreover, although these models are intuitively appealing, their actual ability to predict structural transitions is quite limited.

Within recent years, however, new approaches to structural properties of solids have been developed. Many of these appear to be capable of accounting, at least qualitatively, for phase transitions between different structures and between different types of bonding, e.g., ionic, covalent, or metallic. Similar theories are also useful in predicting chemical trends in solubility and physical properties of alloys. Some of these theories are discussed elsewhere in this volume (Chapter 3).

The connection between microscopic electronic structure (energy band theory, Chapter 2) and chemical bonding is also becoming clearer, at least in favorable cases such as semiconductors. We do not yet have a cut-and-dried prescription for calculating energy bands with sufficient accuracy to reproduce chemical trends. We do know, however, what kinds of trends must be present in the energy bands in order for them to be consistent with observed chemical trends in physical properties. This is a great step forward, as it represents a much more coherent understanding of the nature of matter. In the future we may expect development of intrinsically chemical theories of material properties (including phase transitions) for new groups of materials.

References

1. J. A. Pople and D. L. Beveridge, *Approximate Molecular Orbital Theory*, McGraw-Hill, New York (1970).
2. W. Hume-Rothery and G. V. Raynor, *The Structure of Metals and Alloys*, 4th ed., Institute of Metals, London (1962).
3. V. Heine and D. Weaire, *Solid State Phys.* **24**, 299 (1970).
4. M. Hansen, *Constitution of Binary Alloys*, McGraw-Hill, New York (1958); R. P. Elliott, First Supplement (1965); F. A. Shunk, Second Supplement (1969).
5. W. L. Bragg, *Atomic Structure of Minerals*, Cornell Univ. Press, Ithaca, N.Y. (1937).
6. L. Pauling, *Nature of the Chemical Bond*, 3rd ed., pp. 511–517, Cornell Univ. Press, Ithaca, N.Y. (1960).
7. M. P. Tosi, *Cohesion of Ionic Solids in the Born Model* (Solid State Physics, Vol. 16), Academic, New York (1964).
8. J. R. Tessman, A. H. Kahn, and W. Shockley, *Phys. Rev.* **92**, 890 (1953).
9. E. W. Gorter, *J. Solid State Chem.* **1**, 279 (1970).
10. E. Parthe, *Crystal Chemistry of Tetrahedral Structures*, Gordon and Breach, New York (1972).
11. L. S. Darken and R. W. Gurry, *Physical Chemistry of Metals*, McGraw-Hill, New York (1953).
12. J. T. Waber *et al.*, *Trans. Metall. Soc. AIME* **227**, 717 (1963).
13. K. Schubert, *Kristallstrukturen Zweikomponentiger Phasen*, Springer, Berlin (1964).
14. S. H. Wemple and M. DiDomenico, Jr., *Phys. Rev.* **3B**, 1338 (1971).
15. S. H. Wemple, *Phys. Rev.* **B7**, 3767 (1973).
16. J. C. Phillips, *Rev. Mod. Phys.* **42**, 317 (1970).
17. J. A. Van Vechten, *Phys. Rev.* **182**, 891; **187**, 1007 (1969).
18. J. C. Phillips and J. A. Van Vechten, *Phys. Rev.* **2B**, 2147 (1970).

19. C. L. Tang and C. Flytzanis, *Phys. Rev.* **4B**, 2520 (1971).
20. R. M. Martin, *Phys. Rev.* **B1**, 4005 (1970).
21. J. A. Van Vechten, *Phys. Rev. Letters* **29**, 769 (1972).
22. L. Kaufman, *Phase Stability of Metals and Alloys* (P. S. Rudman, ed.), McGraw-Hill, New York (1967).
23. R. Roy and O. Mueller, *Crystal Chemistry*, Springer-Verlag, Heidelberg (1973).
24. W. B. Pearson, *J. Phys. Chem. Solids* **23**, 103 (1962).

2

Energy Bands

D. Weaire
Department of Engineering and Applied Science
Becton Center, Yale University
New Haven, Connecticut

1. Introduction

1.1. Historical Remarks

The author of this chapter was not even born when the pioneering studies of the band structure of solids were being conducted by Slater, Wigner, Seitz, and others in the 1930's, and he will not presume to paraphrase those early achievements. Suffice it to say that most of the ground rules for the description of energy bands in crystals were worked out rather thoroughly at that time. Despite having such early and respectable antecedents, band structure theory has enjoyed a period of hectic activity since about 1960—not only hectic, but also successful, since, over a wide area of solid-state theory, the balance has tipped from puzzlement to ennui in a single decade. Why? The answer is not to be sought on the back of any theorist's envelope, but rather in two developments which indirectly facilitated and stimulated our theoretical understanding. These are the advent of high-speed computers and the refinement of experimental techniques, two aspects of the electronic revolution.

With the benefit of the hindsight acquired in this period of rapid progress we can see that some of the hypotheses which were hesitantly advanced by early workers do in fact have a wide range of validity. The neglect of many-body effects, which was no more than a pious hope, has turned out to be remarkably successful. Second, the nearly-free-electron description of solids has been justified, systematized, and

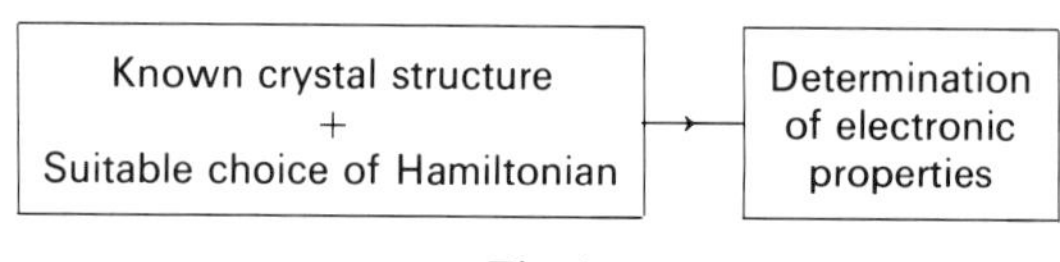

Fig. 1.

quantified by the development of pseudopotential theory. Pseudopotential theory has played a preeminent part in the rapid advances to which we refer. This is because, even with high-speed computers, it is highly desirable to have a theory in which the steps between the definition of the Hamiltonian and the prediction of properties are few and transparent. Pseudopotential theory has this simple virtue. For very accurate work from first principles, other methods may in practice prove to be better, but for qualitative and semiquantitative purposes it is likely to remain unchallenged for the systems to which it applies.

A somewhat rosy picture has been painted so far, and indeed there is a widespread feeling that all of the main problems in the study of band structure have been solved. This is simply not so unless one is interested only in formalism. The progress of which we have spoken consists of a rather satisfactory solution to the problem shown schematically in Figure 1.

We must not lose sight of the fact that nature, as presently understood, often poses the same question with an *unknown*, or only partially specified structure. For example, an impurity in a semiconductor may involve a large local distortion rather than an ideal substitution of the foreign atom. In the absence of an exact specification of this distortion, it must be determined as part of the solution to the problem. A similar situation is confronted in the study of amorphous semiconductors, and indeed the whole field of alloys, so we are not speaking of some abstruse or unusual systems but rather a wide range of solids.

Of course, there is no problem at all in principle. Structure is itself an "electronic property," in that it must be determined by the Coulomb forces of the electrons (and nuclei) of the system and should be derivable from the Hamiltonian. The troublesome "needle-in-haystack" aspect of structure determination is often a nuisance here, but even if we are lucky enough to be faced with a choice between a few reasonable structures (as in the impurity problem), the fact is that the description of the cohesive forces of solids is a difficult and delicate business. Perhaps the stumbling-block of structure determination has been overemphasized here, but there does not seem to be a wide realization of the problem and its importance to future progress.

The above remarks have been more of a setting of the scene than a historical survey. In the subsequent sections, topicality and experimental evidence are further emphasized, at the risk of an inadequate coverage of the classic calculations and eternal verities of the subject. It is partly out of respect for the available basic literature that such a policy was adopted. We should mention Slater[1] (for an authoritative and exhaustive survey), Callaway[2] (for a general account of the theory), and Ziman[3] (for a recent summary of various mathematical techniques and their interrelationships) as three such sources to which the reader may turn for guidance, as did the author on occasion. In particular, considerations of space and the spirit of this treatise have precluded the inclusion of much mathematical detail, and rather than setting forth an incomplete description of this, inviting misunderstanding, it has been avoided almost entirely.

1.2. The Independent-Electron Approximation

One of the reasons why so much progress has been possible in the analysis of band structures is the fact that many-body effects are, for many purposes, negligible. Most of the effects which demand the consideration of many-body effects are confined to very low temperatures (superconductivity, Kondo effect, etc.) or very near thresholds (soft X-ray edge singularities, etc.). A band structure calculated in the Hartree approximation, with corrections for exchange and correlation within this self-consistent independent-electron formalism, suffices for the interpretation of many properties.

The Hartree approximation is summarized in Figure 2 in schematic form. The heavy arrow denotes the basic band structure problem as stated in the previous section. The $\rho \rightarrow H$ arrow in the self-consistent loop denotes, in the unrefined Hartree method, Poisson's equation, by which the charge density ρ is converted into an electrostatic potential which is to be incorporated in the Hamiltonian. With

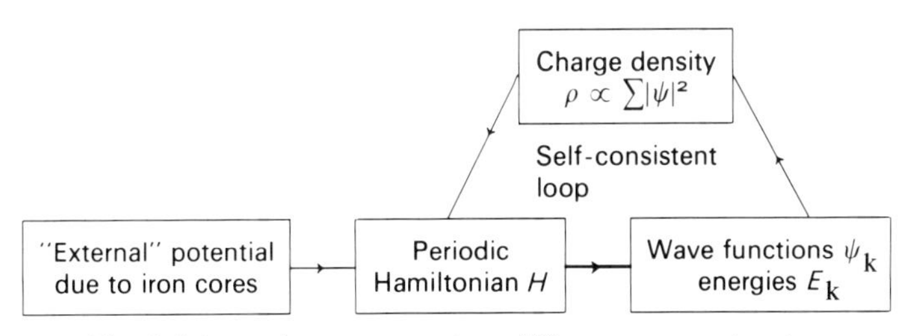

Fig. 2. Schematic representation of Hartree approximation.

certain refinements, to be outlined below, the attainment of the indicated self-consistency is the object of most band structure calculations. In pseudopotential theory it can be guaranteed (at least approximately) in advance, because perturbation theory can be used to screen the pseudopotential self-consistently at the outset (Section 3.6). In general, one must either cycle the calculation ($H \to \psi \to \rho \to H \to$ etc.) to convergence or rely on guesswork.

One might think that the next step in sophistication would be the Hartree–Fock method but in metallic solids this has unsatisfactory consequences, such as a zero density of states at the Fermi level.[4] To incorporate exchange without correlation is therefore a step backward; they tend to cancel each other in their effects, and it is necessary to consider them together. Recent studies have centered on the problem of adding corrections within the Hartree formalism which allow for exchange and correlation. A widely practiced approximation is the addition of a potential to the Hamiltonian which is dependent only on the electronic density $\rho(\mathbf{r})$. This idea is supported by a theorem due to Hohenberg and Kohn[5] which says that the Hamiltonian (and hence the ground-state energy) must be a unique functional of the electron density. Kohn and Sham[6] discussed the practical implications of this statement. The proof is so simple and elegant that it is worth reiterating here.* Suppose two Hamiltonians H and H' containing external potentials v and v' generate (many-body) wave functions Ψ and Ψ' and the *same* charge density $\rho(\mathbf{r})$. Then the minimum principle for the ground-state energy gives

$$E' = \langle\Psi'|H'|\Psi'\rangle < \langle\Psi|H'|\Psi\rangle = E + \int (v' - v)\rho \, d\mathbf{r} \tag{1}$$

Reversing the roles of the primed and unprimed systems and adding the two resultant inequalities leads to

$$E + E' < E + E' \tag{2}$$

Hence v and v', H and H', E and E' must be identical (by *reductio ad absurdum*).

This result can be used to develop a set of self-consistent one-electron equations which are *exactly* equivalent to the full many-body problem. There are two snags in this procedure. First, the result is purely formal—it does not tell us how to actually construct the

* It really depends on two assumptions: (a) that the ground state is nondegenerate, (b) that different Hamiltonians have different ground states.

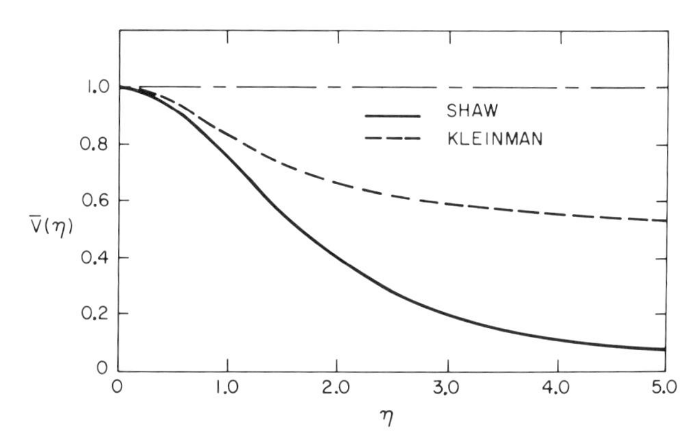

Fig. 3. The normalized effective interaction potential $\bar{V}(\eta)$, according to the models of Shaw[13] and Kleinman[14]; $\eta = q/k_F$.

explicit equations. One is forced to make approximations at this stage, resulting in a *local density correction*, by which we mean that the correction for exchange and correlation is a local potential related to the local electron density. This can be regarded as the first term in a perturbation expansion,[6] but in practice the higher terms have not been much investigated. A second objection is that "equivalent" here refers only to such properties as the ground-state energy and charge density. There is no apparent guarantee that the details of band structure so generated have any particular validity. This difficulty is discussed in a later paper by Sham and Kohn.[7]

The idea of a local density correction predates the Sham–Kohn work. In particular, Slater[8] proposed a local exchange potential of the form*

$$-(3/2\pi)[3\pi^2\rho(\mathbf{r})]^{1/3} \tag{3}$$

and this correction is widely used. Kohn and Sham's work suggested a different numerical factor in this expression as well as further corrections for correlation. For a history of this subject a review by Hedin and Lundqvist may be consulted.[9]

A different approach, usually associated with the work of Hubbard,[10] and used especially in the discussion of interatomic forces, results in a formalism which may most easily be conceived as

* Atomic units ($\hbar = m = e = 1$) will be used throughout, in the text. Note also that 1 Ry = 1/2 a.u. $\simeq$ 13.6 eV.

defining a new effective potential between electrons, replacing the Coulomb interaction. Explicitly,

$$(4\pi/\Omega q^2)[1 - f(q)] \tag{4}$$

is the Fourier transform of the potential (normalized to the atomic volume Ω), where $f(q)$ is the correction for exchange and correlation. Again, opinions differ as to the form of $f(q)$.[11] Recently Singwi *et al.*[12] have pointed out inconsistencies in earlier work (such as an electron–electron correlation function which had large negative values at small r) and proposed a better (but still somewhat arbitrary) scheme for the estimation of $f(q)$. This has been simplified by Shaw.[13] Figure 3 shows Shaw's calculation of $\bar{V} = [1 - f(q)]$ as a function of $\eta = q/k_F$,* compared with another recent theory, that of Kleinman.[14] This type of approximation has been applied in most pseudopotential studies.[11] It turns out that the calculated effects of exchange and correlation are quite significant in properties related to the cohesive energy (e.g., phonon spectra) but in band structures it is fairly small. It is therefore not easy to discriminate between different approximations on the basis of some limited comparison with experiment. Hence the question of which is the best choice among the plethora of proposed approximations is still an open one.

Readers more familiar with atomic and molecular physics may find these approximations unfamiliar. However, according to Kohn and Sham,[6] the approximations they propose are valid only for high density (e.g., in an ion core) or slowly varying density (e.g., the region just outside the ion core) but *not* in the outer "surface" region of atoms and molecules. It is the neglect of surface effects in solids that makes such an approach possible.

The situation presented by tightly bound d or f functions does not obviously conform to the above picture. If they are sufficiently tightly bound, one may expect electrons to be localized on individual atoms. This is the case in the f shells of the rare earths; however, transition metals and their compounds are difficult to categorize in terms of the "band" or "itinerant" picture rather than localization. Certainly the band picture has enjoyed considerable success for transition metals and alloys, but in the case of compounds it is still a matter of conjecture.

* $k_F = (3\pi^2 z/\Omega)^{1/2}$, where z denotes valence, Ω atomic volume.

2. Energy Bands in General

The following constitutes a skeletal review of the basic ideas which are common to the entire field of band structure, i.e., independent of the details of the techniques used or the substances studied (with the exception of aperiodic systems). The material will be quite familiar to many readers and is therefore covered in a cursory fashion. The standard textbooks of elementary solid-state theory[15,16] contain fuller descriptions of many of the matters which are reviewed.

Whereas the periodicity of a one-dimensional system is completely specified by a single repeat distance a, in three dimensions it is necessary to specify the three smallest (independent) vectors $\mathbf{a}_1, \mathbf{a}_2, \mathbf{a}_3$ which are such that the system is invariant under the displacements $\mathbf{a}_i$. When a given point is displaced by every possible combination of these a *Bravais lattice* is generated. Every periodic solid corresponds to a certain Bravais lattice, but only for simple elemental solids does the atomic arrangement itself correspond directly to such a lattice, i.e., it can be generated by placing a single atom at each point in the Bravais lattice. In more complicated solids the unit which is repeated consists of several atoms which may be regarded as residing in a unit cell, the precise shape and position of which is somewhat arbitrary. A convenient definition is that of the *Wigner–Seitz cell* obtained by drawing planes which bisect the lines joining a given point to the rest of the Bravais lattice—the polyhedron thus drawn around the point is defined to be the Wigner–Seitz cell. It is also sometimes convenient to make the same construction using the position of the atoms and if there is more than one per cell, the result is a subdivision of space into what we will call *atomic* cells. (In the context of disordered solids, the term Voronoi polyhedron is popular.)

Any function, such as a potential energy function, which has the periodicity defined by $(\mathbf{a}_1, \mathbf{a}_2, \mathbf{a}_3)$ can only have Fourier components associated with the following values of $\mathbf{q}$, the variable in the "reciprocal" space of the Fourier transform:

$$\mathbf{q} = l\mathbf{g}_1 + m\mathbf{g}_2 + n\mathbf{g}_3 \tag{5}$$

where l, m, n are integers, and

$$\mathbf{g}_i \cdot \mathbf{a}_j = 2\pi\,\delta_{ij} \tag{6}$$

This defines a reciprocal lattice with primitive vectors $\mathbf{g}_1, \mathbf{g}_2, \mathbf{g}_3$. The Wigner–Seitz construction, in reciprocal space, defines the *Brillouin zone*.

Bloch's theorem* states that wave functions $\psi(\mathbf{r})$ can always be chosen to have the periodicity of the Bravais lattice apart from a single multiplicative factor of $\exp(i\mathbf{k}\cdot\mathbf{r})$, i.e.,

$$\psi_{\mathbf{k}}(\mathbf{r}) = [\exp(i\mathbf{k}\cdot\mathbf{r})]u_{\mathbf{k}}(\mathbf{r}) \tag{7}$$

where $u_{\mathbf{k}}(r)$ is a periodic function.† When Fourier-transformed this means that the wave function contains only the Fourier components $\mathbf{k} + l\mathbf{g}_1 + m\mathbf{g}_2 + n\mathbf{g}_3$. If we wish to use $\mathbf{k}$ as a label for wave functions, the choice of $\mathbf{k}$ is arbitrary to within addition of multiples of $\mathbf{g}_i$. It is often convenient, therefore, to confine $\mathbf{k}$ to be within the Brillouin zone ("*reduced zone scheme*").

The principal properties of the energy as a function of $\mathbf{k}$ are as follows. Within the Brillouin zone $E(\mathbf{k})$ is a continuous function. It is, of course, a multiple-valued function in the reduced zone scheme. At a Brillouin zone plane the gradient of $E(\mathbf{k})$ must be in the plane, except in certain exceptional cases. Finally, the band structure must be symmetric under inversion, $\mathbf{k} \to -\mathbf{k}$. This is usually referred to as time-reversal symmetry, but for simple Hamiltonians without spin–orbit coupling it follows simply from complex conjugation of Schrödinger's equation.

At symmetry points defined by the meeting of three or more Brillouin zone places the above conditions necessitate a stationary point $[\nabla E(\mathbf{k}) = 0]$ in the band structure. There is a consequent singularity in the density of states at such points. Such singularities (again excluding pathological cases) are of square-root form and are known as Van Hove singularities.[17] Figure 4 illustrates the various types which are associated with different kinds of stationary points. (They need not necessarily be symmetry points.) Optical spectra, while not associated simply with $n(E)$ but rather with a joint density of states of valence and conduction bands (modified by matrix element effects), show such singularities. Photoemission spectra show (at least in principle) rather more complicated singularities.[18]

A common and convenient way of expressing some aspect of a given band structure, such as the band structure around some symmetry point in a given band, is by means of an effective mass. The band structure is compared with that of free electrons of mass m (= 1

* We have omitted the question of boundary conditions. The use of periodic boundary conditions on a finite solid composed of N atoms, which restricts $\mathbf{k}$ to a fine grid of points in reciprocal space, and the taking of the infinite limit $N \to \infty$, is discussed in Refs. 15 and 16. Whenever N is used in the text this limit is implied.

† Here and elsewhere u is normalized to unity in the Wigner–Seitz cell.

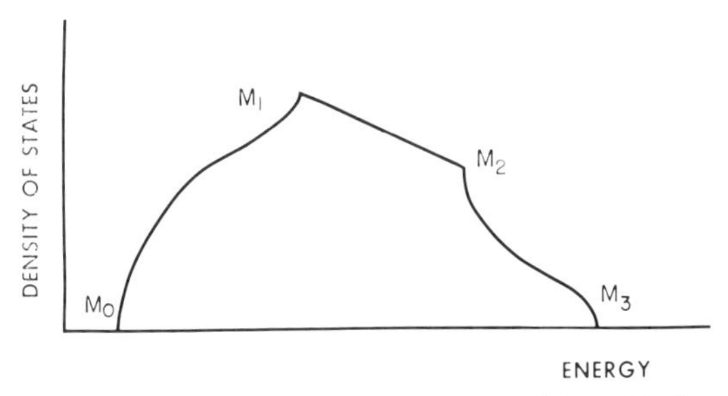

Fig. 4. Van Hove singularities are of four kinds, depending on whether the band structure involves a minimum (M_0), maximum (M_3), or saddle point (M_1, M_2) in **k** space at the energy of interest.

in the units used here):

$$E = (1/2m)k^2 + E_0 \tag{8}$$

and m, or $1/m$, is adjusted to fit the given situation. In the case referred to, the result is an effective mass tensor $(1/m)_{ij}$ since the band structure will not in general have the simple isotropic form (8), but rather must be written (for the case in which it is the band structure around $\mathbf{k} = 0$ that is of interest)

$$E = \tfrac{1}{2}(1/m)_{ij}k_i k_j + E_0 \tag{9}$$

Effective masses can also be defined (somewhat confusingly) by a variety of other comparisons. A fitting of the free-electron-like part of the optical absorption of a metal (the Drude term) defines an *optical* mass. The electronic specific heat of a metal gives a *thermal* mass. The latter is an example of an effective mass which is *not* related simply to some average or derivative of the band structure $E(\mathbf{k})$ as most such masses are. The thermal mass includes important contributions from many-body enhancements. In comparing different effective masses one must therefore consider not just their different definitions, but the different degree to which they are enhanced by such effects (Ref. 9, pp. 108–22).

Finally, we introduce the notion of an equal-energy surface, i.e., the surface in **k** space defined by $E = \bar{E}$, a constant. The most important case is that in which $\bar{E}$ is such that the states defined by $E < \bar{E}$ hold just the number of electrons dictated by the valences of the atoms in the system, so that $\bar{E}$ is, in fact, the Fermi energy and we have defined the Fermi surface. The determination of this surface is one of the principal experimental goals of the study of metals (Section 5.4). In insulators all bands are full or empty and hence there is no Fermi surface. In a semiconductor the gap is small and in a semimetal the bands touch or overlap at the Fermi level, giving properties intermediate between the two cases.

3. The Classical Descriptions of Energy Bands in Periodic Systems

3.1. Introduction

In the following sections various different frameworks for the calculation and interpretation of band structures are outlined. From the outset the discussion will be in the spirit of Section 1.2, i.e., it is assumed that exchange and correlation are incorporated as some effective potential. Little or nothing will be said about the treatment of spin–orbit coupling and relativistic effects, which become important for the heavier elements. Again, the core electrons are regarded as outside our area of interest. This separation of the electronic structure problem into core electrons (whose electronic structure is assumed not to change significantly on going from isolated atom to solid) and outer electrons, which are responsible for solid-state properties, must, at some level, break down. While it has been widely successful in straightforward band structure calculation, one wonders if it might be the root cause of the failure of most existing methods to properly describe properties related to the cohesive energy. In any case, we will accept this separation of the problem here. The problem then is: Given some arrangement of ions (e.g., Al^{3+}), what kind of bands are formed by the rest of the electrons?

3.2. Two Classical Limits—Tight Binding and Nearly Free Electron

There are two classical limits of band structure description—tight binding theory and nearly-free-electron theory. They may be regarded as opposite limits. When the potential in the neighborhood of some or all of the ions is very strong compared with that in the interstitial regions between them, one has a situation in which the isolated uncoupled atoms provide good first approximation to the solid. The eigenfunctions of the isolated atom [tight binding (TB) functions] are suitable basis states for the description of the band structure. At the opposite limit is the case in which the individual atomic potentials overlap so much that it is a better approximation to regard the potential as flat. The natural basis functions are then plane waves, which are weakly mixed by the potential, giving a nearly-free-electron (NFE) theory.

In reality, typical solids of interest occupy a spectrum between these two extremes, from the alkali halides, which are well described by TB theories, to Al, which is well described by NFE theories. (It is not, in fact, straightforward NFE theory, but rather pseudopotential theory, which is successful for Al, but for the moment this distinction will be ignored.) Intermediate between the two cases are, for instance, group IV semiconductors, in which both points of view have proved useful, and transition metals, in which different parts of the band structure conform to NFE and TB descriptions.

In describing NFE and TB theory, the emphasis will be on model calculations which convey the essence of these points of view, rather than practical detail.

3.3. Tight Binding Theory

In this, as in many other methods, after choosing basis functions (labeled by m) the problem of finding the energy levels E for given $\mathbf{k}$ amounts to the solution of a determinantal equation

$$\det|H_{nn'} - ES_{nn'}| = 0 \tag{10}$$

where $H_{nn'}$ and $S_{nn'}$ are, respectively, the matrix elements of the Hamiltonian and the unit operator between the basis functions. If the basis functions obey Bloch's theorem (for the chosen $\mathbf{k}$), the dimension of the matrices in (10) equals the number of basis functions *per unit cell*. In the case of tight binding theory (taking for simplicity the case of a structure which is a Bravais lattice) we choose basis functions of the type

$$\phi_{\mathbf{k}n} = (1/\sqrt{N}) \sum_{m} [\exp(i\mathbf{k} \cdot \mathbf{R}_m)] U_{mn} \tag{11}$$

where U_{mn} is the nth tight binding function localized on the mth atomic site at $\mathbf{R}_m$, and is normalized to unity. To take a specific example, the valence band of Si is describable in terms of one s and three p atomic functions so n in that case would run from one to four. Higher up in the band structure one would need to augment this "minimal basis set" with d functions.

The matrix elements $H_{nn'}$ and $S_{nn'}$ can be written, using (11), in terms of matrix elements of H and S between atomic functions U

localized on different sites. The calculation of these overlap integrals has a long history, mainly in the chemical literature.[19] Clearly they should decrease rapidly with increasing separation of the sites. Indeed, for qualitative and semiquantitative purposes such overlaps are often restricted to nearest neighbors only. For a discussion of such approximations as currently practiced in chemistry, see Refs. 20, 21.

Suppose, then, we have only interactions between nearest neighbors and only one basis function (an s state) per atom. These functions are assumed to be orthogonal. We then have a very simple Hamiltonian indeed, which may be written in Dirac notation as

$$H = V \sum_{\substack{m,m' \\ \text{neighbors}}} |U_m\rangle\langle U_{m'}| \tag{12}$$

Of course, with only one basis function per atom, the solution for $E(\mathbf{k})$ is now trivial for any structure with only one atom per unit cell. Using (11), we have immediately

$$E_{\mathbf{k}} = \langle \phi_{\mathbf{k}} | H | \phi_{\mathbf{k}} \rangle \tag{13}$$

$$= V \sum_m \cos \mathbf{k} \cdot (\mathbf{R}_m - \mathbf{R}_0) \tag{14}$$

where the summation runs over the $\mathcal{N}$ nearest neighbors of a given atom at $\mathbf{R}_0$. This generates a band which is clearly bounded by $\pm \mathcal{N} V$. Although the dispersion relation of $E(\mathbf{k})$ is comparatively simple, the problem of finding the density of states $n(E)$ for the corresponding bands is by no means trivial, and a good deal of careful numerical work has been expended on this problem, principally because such elementary Hamiltonians are of immediate interest in problems involving spin interactions. The results are shown in Figures 5–7 for simple cubic, fcc,[22–24] and diamond cubic[25] structures. Of these three cases, only the spectrum for simple cubic structures consists entirely of continuous pieces joined by the simple Van Hove singularities of the normal type (Section 2). In the case of an fcc structure there is a logarithmic singularity at the bottom of the band arising from a line degeneracy in the band structure.[22] The diamond cubic structure consists of two interpenetrating fcc lattices, and, as a consequence of this, there exists an analytic transformation[25,26] between the densities of states of diamond cubic and fcc structures for the operator (12). This is only one of a number of interesting properties of this operator, of which we will mention a few. Any structure which has two interpenetrating sublattices generates an *even* spectrum. Again, *all* close-

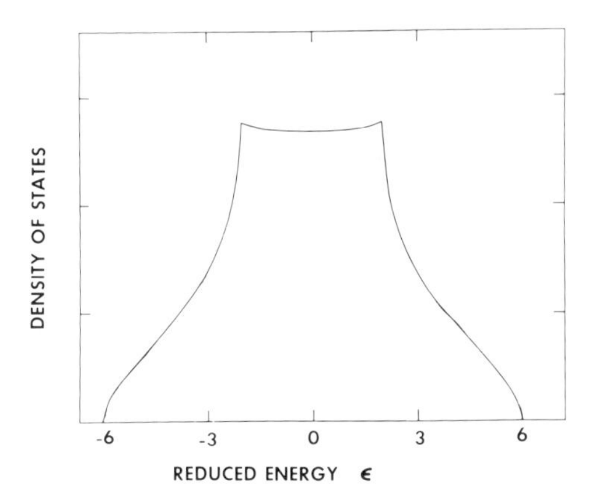

Fig. 5. Density of states for the simple tight binding Hamiltonian (12) as a function of the reduced energy $\varepsilon = E/V$ for the simple cubic structure.

packed structures generate the same density of states as the fcc[27] (and consequently all "polytype" structures such as wurtzite generate the same density of states as the diamond cubic[28]). These results give some indication of the extent to which the spectrum of such a simple

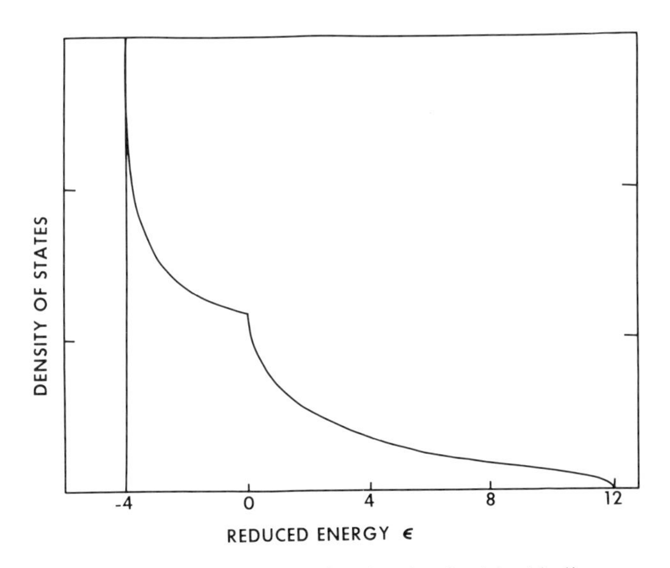

Fig. 6. Density of states for the simple tight binding Hamiltonian (12) as a function of the reduced energy $\varepsilon = E/V$ for the fcc structure.

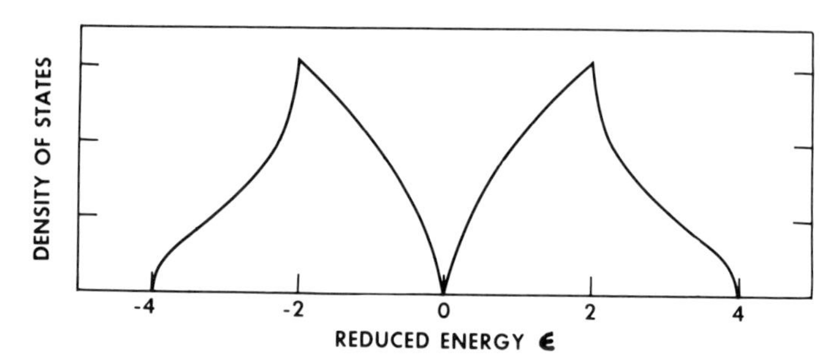

Fig. 7. Density of states for the simple tight binding Hamiltonian (12) as a function of the reduced energy $\varepsilon = E/V$ for the diamond cubic structure.

Hamiltonian can be taken apart and understood. At the frontiers of our understanding, in the study of amorphous solids and elsewhere, this transparent simplicity is worth the sacrifice, at least initially, of any attempt at a realistic description.

3.4. Wannier Functions

In the previous section extended wave functions were built out of localized basis functions with the aid of Bloch's theorem. In the simple case where only one localized function is used per unit cell, a single band is thus generated. Since the process by which the extended Bloch functions are formed from the localized ones is just a Fourier transform (12), it may be inverted to recover a localized function from the extended wave functions, according to

$$U_{mn} = (1/\sqrt{N}) \sum_{\mathbf{k}} [\exp(-i\mathbf{k} \cdot \mathbf{R}_m)] \phi_{\mathbf{k}n} \tag{15}$$

This process can be generalized to *any* given band structure. By performing (15) on a single band, localized functions may be obtained. They are called Wannier functions.[29] One might expect (from consideration of the tight binding approximation) that, if the band is well isolated in energy from others, the resulting functions would be well localized, and indeed this has been shown.[30]

The mathematics of Wannier functions involves a number of subtleties which will not be covered here. The choice of phase factors in (15),[31] the case of overlapping or degenerate bands,[32] and the possibility of more general transformations than (15)[33] are some of these. Some comments on past and potential utility of the use of Wannier functions follow.

First, it should be noted that the transformation of a given set of wave functions into Wannier functions is not an approximation—it generates an equivalent set of basis functions, if done exactly. Why might this be useful? It seems to be a step in the wrong direction (eigenfunctions → non-eigenfunctions). However, provided that the Wannier functions are well localized, it raises various interesting possibilities. First, the matrix elements between states in different bands, which are essential to the description of many phenomena, such as optical absorption (Section 5.3), must reduce to combinations of matrix elements between Wannier functions on a single site and its near neighbors, with appropriate phase factors. Thus a simpler "chemical" description with few parameters would replace the more direct and complicated description of such matrix elements. Another advantage of Wannier functions is the obvious fact that, given such functions, the band structure is recoverable by the tight binding formalism, and this formalism is the most appropriate for many applications. For example, the study of amorphous systems (Section 6) is generally pursued within a tight binding framework. Wannier functions might seem to offer an excuse for this in materials in which the tight binding *approximation* is not valid. However, this raises delicate questions about the transferability of Wannier functions from one structure to another.

In conclusion it might be said that Wannier functions, although they date back to (at least) 1937, have not yet been at all widely used. Since the topics mentioned above, for which they hold most promise, have been studied intensively only in the last few years, Wannier functions may yet come to the fore as indispensable practical tools in the application of band theory.[34]

3.5. Nearly-Free-Electron Theory

The nearly-free-electron theory, once thought of by many as largely irrelevant to real solids, has achieved a degree of respectability with the development of pseudopotential theory. For the moment the necessity of the use of a pseudopotential will be ignored and it will be assumed that the crystal potential $V(\mathbf{r})$ is sufficiently weak to justify this point of view. Plane waves

$$\psi_{\mathbf{k}} = \Omega^{-1/2} \exp(i\mathbf{k} \cdot \mathbf{r}) \tag{16}$$

are then defined as basis functions and the matrix element of the

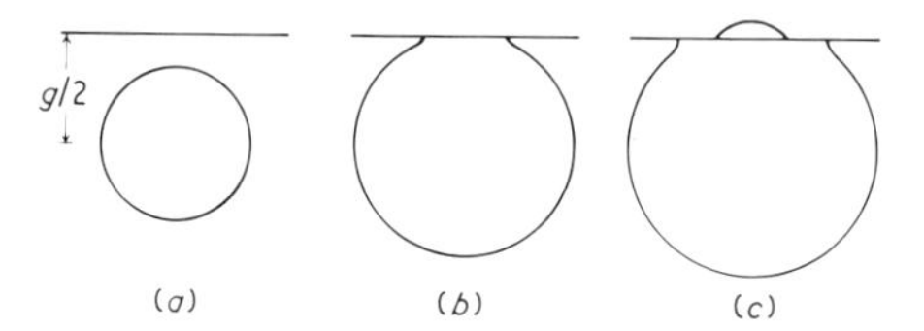

Fig. 8. Equipotentials for the case of a single Brillouin zone plane.

Hamiltonian between these is

$$\tfrac{1}{2}k^2\,\delta_{\mathbf{kk}'} + V(\mathbf{k} - \mathbf{k}') \tag{17}$$

where the potential energy term* has been assumed to be local, and

$$V(\mathbf{q}) = \Omega^{-1}\int_\Omega V(\mathbf{r})\exp(i\mathbf{q}\cdot\mathbf{r})\,d\mathbf{r} \tag{18}$$

What do we mean by "sufficiently weak"? First of all, if we are going to use perturbation theory, we would like

$$V(\mathbf{q}) \ll \tfrac{1}{2}k_F{}^2 \tag{19}$$

since the right-hand side would be a typical energy denominator. Whether we use perturbation theory or a finite secular determinant, we would like to have $V(q)$ converging strongly to zero at large q so that the perturbation series converges well or the secular determinant can be reasonably small. The ensuing discussion is predicated on these two assumptions which, as we have already mentioned, really demand the use of pseudopotentials.

To simplify things as much as possible, suppose that $V(\mathbf{q})$ has only a single nonzero component, corresponding to reciprocal lattice vector $\mathbf{g}$ (physically most unreasonable, but mathematically convenient). There is then only a single Brillouin zone plane as in Figure 8. Figure 9(a) shows the resultant band structure[36] for a particular choice of $V(\mathbf{g})$, as given by the secular equation

$$\begin{vmatrix} \tfrac{1}{2}k^2 - E_\mathbf{k} & V(\mathbf{g}) \\ V(\mathbf{g}) & \tfrac{1}{2}(\mathbf{k}-\mathbf{g})^2 - E_\mathbf{k} \end{vmatrix} = 0 \tag{20}$$

Also shown are the results of perturbation theory and the free-electron approximation. Perturbation theory gives a divergent result when the

* The total potential or pseudopotential is denoted by V; it is made up to contributions from the potentials of individual atoms, denoted by v, according to $V(\mathbf{g}) = S(\mathbf{g})v(\mathbf{g})$. See Ref. 11, pp. 269–70.

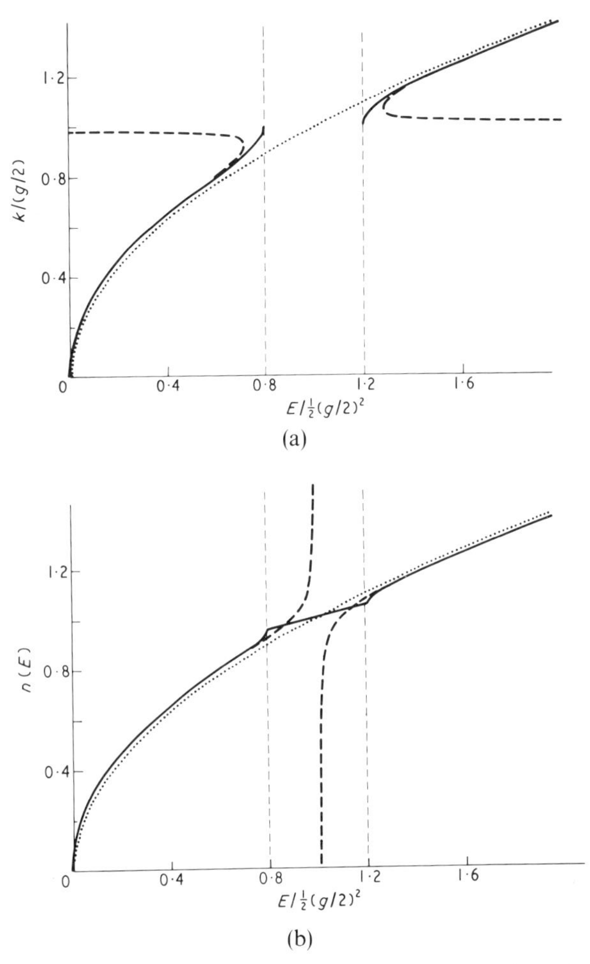

Fig. 9. (a) Band structure along the line normal to the Brillouin zone plane in a model in which only one such plane is considered. (b) The corresponding density of states. In each case, the full line is exact, the dashed line is given by perturbation theory, and the dotted line is the free-electron result.

states $\mathbf{k}, \mathbf{k} - \mathbf{g}$ become degenerate and would appear to be useful only for states which are not close to the Brillouin zone plane which is defined by this degeneracy. While this is true, it turns out that for quantities which involve integrals over all occupied states, such as the charge density ρ or the total energy U, the (principal part) integration of the perturbation formulas gives quite good results. This is essentially due to the cancellation of the positive and negative divergences in Figure 9(a).

Figure 9(b) shows the density of states for the same problem. The variation from the free-electron curve is misleadingly small because of the choice of only a single Brillouin zone plane (whereas in real cases there are, say, eight such equivalent planes). Again, perturbation theory diverges.

3.6. Pseudopotentials

In practice, the true crystal potential does not satisfy the criterion for the applicability of the nearly-free-electron approximation, but there are much weaker equivalent potentials, or pseudopotentials, which do. By equivalent we mean that they produce the same band structure for the valence and conduction bands (but not necessarily the same wave functions). The difference is that the potential must of necessity be strong enough to bind states at lower energies (core states, more or less) but the pseudopotential need not. The elimination of such bound states produces a potential which is much weaker in the region close to the ion cores. This "cancellation" of the strong inner part of the potential can be seen from many points of view[37,38] but will here be accepted as a fact of life for *s–p*-bonded systems.

It should be emphasized that while the utility of this approach is largely based on the NFE method, no such approximation is made at the outset.

Pseudopotentials have been derived from different points of view —transformation of the Schrödinger equation,[39] transformation of KKR theory,[40] the fitting of free-ion term values,[41] etc. The last of these possibilities, which introduces a judicious amount of empiricism, in that it sidesteps various difficulties concerning exchange and correlation on the core, is called the model potential method. Whichever method is used, the results are much the same if the pseudopotential is chosen to be as weak as possible within the chosen framework. A typical pseudopotential is shown in Figure 10. Note the remarkable consistency of the theory and experiment (see also Stafleu and de Vroomen[43,44]).

In practice a "bare" pseudopotential, equivalent to the potential produced by an ion core, is first calculated and then "screened," i.e., the potential due to the outer electron is added. In this case the self-consistent incorporation of a screening correction is particularly simple because of the use of NFE theory. In lowest-order perturbation theory, performed self-consistently, the relation is just

$$v_{\text{screened}}(q) = v_{\text{bare}}(q)/\varepsilon(q) \tag{21}$$

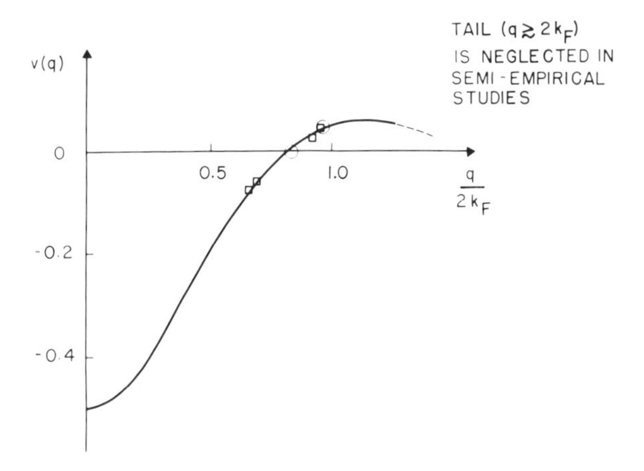

Fig. 10. A typical pseudopotential (for Sn), as calculated by Heine and Animalu,[41] shown together with points which have been fitted in semiempirical studies of (○) (semiconducting) gray Sn[42] and (□) (metallic) white Sn.[43]

where ε is the Lindhard dielectric function (Ref. 39, pp. 46–52):

$$\varepsilon(q) = 1 - (8\pi/\Omega q^2)\chi(q) \tag{22}$$

The function χ derives from the use of perturbation theory, being simply

$$\chi(q) = -\frac{3}{2} z k_F^{-2} \left(\frac{1}{2} + \frac{4k_F^2 - q^2}{8qk_F} \log \left| \frac{q + 2k_F}{q - 2k_F} \right| \right) \tag{23}$$

The dielectric function ε is often modified for exchange and correlation by the incorporation of an effective electron–electron interaction (Section 1.2) rather than the Coulomb interaction, which appears, in Fourier transform, as $4\pi/q^2$ in Eq. (23).

In fact, pseudopotentials are somewhat more complicated than this because they are energy dependent and nonlocal, i.e., we should write not $v(q)$ but $\langle \mathbf{k}|v(E)|\mathbf{k} + \mathbf{q}\rangle$. These complications are often avoided arbitrarily by (a) fixing the energy at E_F at the outset and (b) replacing $\langle \mathbf{k}|v(E)|\mathbf{k} + \mathbf{q}\rangle$ by $v(q)$, which is evaluated at the Fermi level, with $\mathbf{k}$ and $\mathbf{q}$ chosen according to the on-Fermi-sphere approximation of Figure 11.

The success of some of the earliest pseudopotential calculations is quite remarkable and is illustrated in Figure 12 for a large number of s–p-bonded elements.

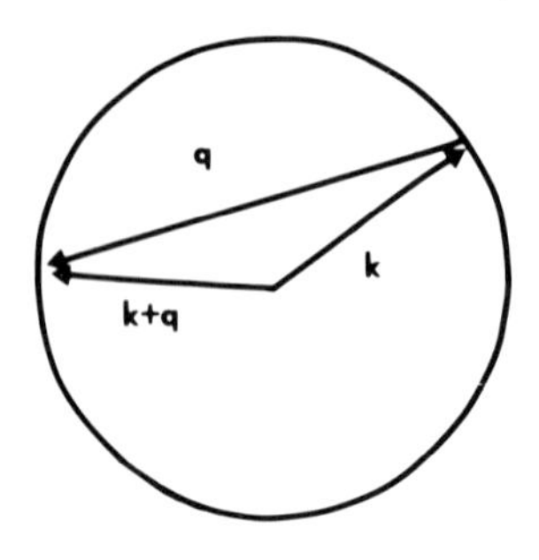

Fig. 11. In the on-Fermi-sphere approximation the pseudopotential is made local by a particular choice of $\mathbf{k}, \mathbf{k}+\mathbf{q}$. For $q < 2k_F$, $\mathbf{k}$ and $\mathbf{k}+\mathbf{q}$ are chosen to be on the Fermi sphere, as illustrated here, while for $q \geq 2k_F$ they are chosen to satisfy $|\mathbf{k}| = k_F, \mathbf{k} \parallel -(\mathbf{k}+\mathbf{q})$.

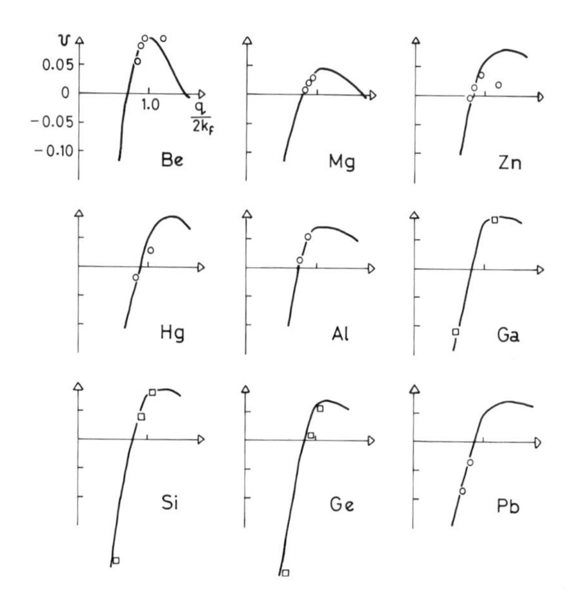

Fig. 12. A comparison of a theoretical[41] pseudopotential with fitted pseudopotential parameters for Be,[45] Mg,[46] Zn,[47] Hg,[48] Al,[49] Ga,[42] Si,[42] Ge,[42] and Pb.[51] Circles correspond to the fitting of Fermi surface data, squares to the fitting of optical spectra (of Ga compounds in the case of Ga). For more comprehensive comparisons of this kind see Cohen and Heine.[51]

3.7. The Cellular Method

The cellular method is of little direct practical importance today. Its significance is historical (as a precursor of the APW and related methods) and conceptual (as a convenient framework for the understanding of the cohesion of solids).

Using Bloch's theorem (Section 2), we can reduce the band structure problem to the solution of Schrödinger's equation in a single Wigner–Seitz cell, with a boundary condition which depends on $\mathbf{k}$. Specifically, the wave function and its gradient must be equal on opposite faces of the cell, apart from a phase shift. The various versions of the cellular method attempt to solve this problem by integrating Schrödinger's equation within the cell and fitting the boundary condition. The classic calculation of Wigner and Seitz[52] concentrated on the bottom of the conduction band (for Na) and made a spherical approximation in which the Wigner–Seitz cell was replaced by a sphere of equal volume. The problem of finding a wave function of s symmetry in this potential which would have zero derivative at the boundary was then relatively simple. Having thus obtained the state at $\mathbf{k} = 0$, they simply assumed that the rest of the band structure was free-electron-like,

$$E = E_0 + \tfrac{1}{2}k^2 \tag{24}$$

since the object of the exercise was to calculate the cohesive energy rather than details of band structure. This was, as is now realized (Section 6.2), a very good approximation for this particular application. If one wishes to investigate the details of band structure rather than just states at $\mathbf{k} = 0$, the spherical approximation to the boundary conditions must be abandoned in favor of a matching of the boundary condition over the true Wigner–Seitz cell. This approach, however, encountered severe mathematical difficulties[53] and has, with a few exceptions,[54] long since been abandoned. It has been suggested[55,56] of late that the cellular point of view is peculiarly appropriate to amorphous semiconductors. Again, this is true in a conceptual sense but may not have significant practical consequences.

3.8. Orthogonalized Plane Wave, Augmented Plane Wave, and Related Methods

Historically speaking, the orthogonalized plane wave (OPW) method should have been discussed before the pseudopotential method (Section 3.6). The work of Phillips and Kleinman,[57] which is the starting point of the development of pseudopotential theory by Cohen, Heine, and others (see Ref. 37, pp. 10–13), was based on a rearrangement of terms in OPW theory, originally formulated by Herring.[58] This was not done here, in order to emphasize the generality of the pseudopotential idea, which need not be tied to OPW theory at all.

As was remarked in earlier sections, the use of plane waves as basis functions is inadequate because of the strong potential near the ion core. This induces rapid oscillations in the wave function which cannot be described by any reasonable number of plane waves. One way of looking at these oscillations is to consider them as necessitated by the condition that the wave function must be orthogonal to all the bound states which lie lower in energy.

Plane waves made orthogonal to these bound states by the addition of suitable linear combinations of the bound states are thus suitable basis functions. They are still plane waves outside the core, but now have the right character in the core to provide a reasonably convergent expansion of the true wave function.

The use of orthogonalized plane waves, which has been reviewed by Woodruff,[59] is not very widespread today, partly because equivalent pseudopotentials provide a more physically transparent picture and partly because the method is somewhat more cumbersome than the APW and related methods.

The augmented plane wave (APW) method is a natural development of the cellular method. Rather than carving the solid up into Wigner–Seitz or atomic cells of complicated geometry, it is here conceived as made up of nonoverlapping spheres surrounding the atoms, and interstitial regions between them. The surfaces on which matching conditions are to be analyzed are now merely spheres—it remains to specify solutions of Schrödinger's equation inside and outside the spheres. At this point two approximations, neither of which is strictly necessary, produce a great simplification of the problem. It is assumed that (a) the potential inside each sphere is spherically symmetric and (b) the potential outside the spheres is flat. The resulting type of potential is called a *muffin tin* potential. It clearly cannot be a good approximation in every case. Any structure which is not more or less closely packed presents difficulties. It has nevertheless enjoyed enormous success, and is even being extended to applications to molecules.[60] Given such a potential, one can proceed in various ways to solve the matching problem on the spheres and this leads to a distinction between the APW and Korringa–Kohn–

Rostoker (KKR) methods, both of which will be briefly sketched below. They have been reviewed by Dimmock[61] and Segall and Ham.[62]

Both methods in practice involve rather simple matrices in their final form, but the mathematics involved in deriving them is fairly complicated.

In the APW method one is confronted at the outset with a choice of various options in the precise definition of the basis functions. Suppose that we are interested in calculating a conduction band eigenvalue E corresponding to a state of crystal momentum $\mathbf{k}$. The basis functions are to be plane waves $\mathbf{k} + \mathbf{g}$ outside the atomic spheres. For each basis function we must choose appropriate combinations of wave functions of principal quantum number n and angular momentum $l = 0, 1, \ldots$ inside each atomic sphere, to match the magnitude of the plane wave $\mathbf{k} + \mathbf{g}$ all over the sphere. For what energy are we to integrate Schrödinger's equation to define u_{nl}? The choice is arbitrary and this energy may be set equal to a constant or $\frac{1}{2}(\mathbf{k} + \mathbf{g})^2$, but the best choice (from the point of view of convergence) would appear to be E, that is, the eigenvalue that is to be calculated. This, of course, can be done, but it results in a secular determinant in which E occurs in off-diagonal components, not just in the diagonal. Matrix diagonalization techniques are then inappropriate, and roots of the determinant must be sought directly. The functions u_{nl} occur in the final secular determinant only as logarithmic derivatives at the radius of the matching spheres. This is also true in the KKR method—the only information that one needs about the potential inside the spheres is the logarithmic derivatives of the eigenfunctions within it, evaluated at the boundary. The fact that all other details of the potential within the spheres are irrelevant is one justification for pseudopotential theory (Section 3.6).

The essential difference between the APW method and the KKR method is that the latter uses an integral equation, equivalent to Schrödinger's equation, as its starting point. This change of tack results in a quite different determinantal equation although based on much the same physical ideas, indeed the same potential and the same basis functions! In this case the determinant is much smaller, having a size dictated by the number of basis functions (u_{nl} in the above notation) necessary for a description of the wave function inside the spheres (typically $1s + 3p + 5d = 9$). The price of this compactness is to be paid in the form of more complicated elements in the determinant. However, these elements break up neatly into parts which depend on the potential (or rather logarithmic derivatives, as before) and purely geometric quantities, the *KKR structure constants*. The latter may be tabulated once and for all for a given structure. However, they are functions of **k** and E as well as structure, so in practice they are available only for symmetry points, for a few particular structures, for certain ranges of energy.[63]

What can be said about the relative merits of these two methods? Ziman[3] and Dimmock[61] have considered this question in some detail. There is probably not much to choose between the two methods when performed *ab initio*; however, in applications for which KKR structure constants are already available KKR has a clear advantage over the APW method.

4. Approximations, Interpolations, Perturbations

4.1. Introduction

The methods of the preceding sections involve, at the outset, few necessary approximations. Even the muffin tin approximation can be dispensed with in the APW method. In practice, of course, all kinds of approximations are involved in the practical application of such schemes, especially when they are used in a semiempirical manner. This is not simply for reasons of convenience but also because one may find that a theory which has all the elements of a first principles calculation has too many free parameters in semiempirical work. For instance, if one chooses to use the parameters $v(g)$ in pseudopotential theory as fitting parameters, one is forced to neglect (or fix the values of) the higher components $g > 2k_F$ entirely (Ref. 51, pp. 83–86), otherwise the number of fitting parameters would be too great and they would be undetermined by the available data.[64]

Any of the existing band structure methods can be adapted for use as a semiempirical scheme, or an interpolative scheme to facilitate the calculation of quantities which depend on interband integrals and the like. Tight binding theory, reduced to its bare essentials, with the overlap parameters used to fit experimental data or as an interpolation scheme in band structure calculation is generally referred to as Slater–Koster[65] theory. Pseudopotential theory used in this way has been dubbed the empirical pseudopotential method (EPM) and has been the subject of a recent comprehensive review.[51] Some comparisons of parameters $v(g)$, which have been fitted to experiment, with theoretical calculations have already been shown in Figure 12.

Which of these (or other) interpolation schemes is to be used depends on the context. For example, in a recent calculation Kunz[66] used Slater–Koster interpolation for the valence band of LiCl and pseudopotential interpolation for the conduction band. In the case of transition metals a mixed tight binding and pseudopotential scheme is needed because the *d* bands, which have essentially tight binding character, lie in the middle of *s–p* bands of essentially free-electron character. Such a scheme, originally set up in a rather intuitive manner by Mueller,[67] has been put on firmer theoretical grounds by further studies.[68,69]

Some further approximative or interpolative schemes are outlined in the following sections. The special area of approximations in disordered systems has been postponed to Section 7, although the discussion of moments in Section 4.2 is highly relevant.

4.2. Moment Methods

The use of moment methods is particularly suited to the use of simple Hamiltonians of tight binding type, such as that given by Eq. (12), although they need not be quite so elementary as that example.

The mth moment μ_m of the electronic density of states $n(E)$ is defined by

$$\mu_m = \int_{-\infty}^{\infty} dE n(E) E^m \tag{25}$$

If $n(E)$ derives from a Hamiltonian of the type mentioned, with a finite number of basis functions per atom, (1) should in general converge. It is related to the Hamiltonian by

$$\mu_m = \mathrm{Tr}\, H^m \tag{26}$$

where Tr denotes the trace of the operator H^m in the function space in which it is defined.

For Hamiltonians of the type specified above the evaluation of (26) for given m reduces to the enumeration of closed paths of m steps from atom to atom in the structure,[70] and the summation of contributions from each path which are traces of matrices whose dimension is the number of basis functions per atom (trivial if this number is one). The beauty of this approach is its extreme simplicity for the first few moments ($m \lesssim 10$). On the other hand, it becomes prohibitively cumbersome for larger m. Therefore, although it may furnish complete information about $n(E)$ in principle, it is somewhat limited in practice.

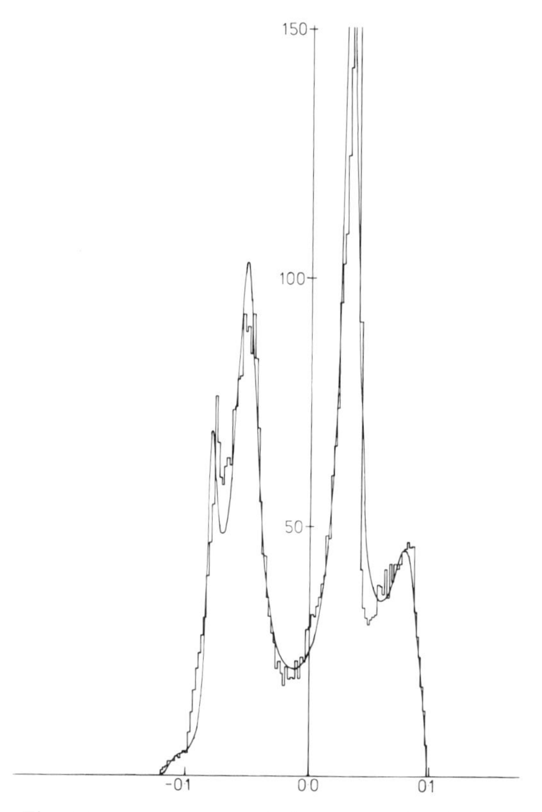

Fig. 13. Comparison of the conventional band structure calculation of the density of states of a transition metal (histogram) with an approximation based on moments.[72]

The question is whether a few moments suffice to reconstruct a spectrum accurate enough for the purpose at hand. Ducastelle and Cyrot-Lackman[70,71] have been remarkably successful in applying the method to calculate the total energy of occupied states in transition metals and hence predict their stable structures, stacking fault energies, etc. They were, of course, concerned here with an integral of $n(E)$ and the smoothed version of the spectrum which was obtained from its first six moments sufficed for this purpose without being a very good approximation to $n(E)$ itself at most points. This work has stimulated a good deal of work on the refinement of the method by the approximation of higher moments.[72] Figure 13 shows a recent calculation of the *d*-band density of states of a bcc transition metal by Haydock *et al.*,[72] which is compared with a histogram based on a conventional band structure calculation by Pettifor.[73] The comparison shows that the moment method, suitably refined, gives a very good picture of the density of states in this case.

At least for tight binding systems, the moment method is one of the principal ways in which one may hope to escape in future from the use of conventional band structure calculations, the difficulty of which increases drastically with the size of the unit cell because of the necessity to solve secular equations with large numbers of variables. The moment method should not suffer from this disadvantage.

4.3. Nearly-Free-Electron Perturbation Theory

In the perturbation series for the energy $E(\mathbf{k})$ in powers of the pseudopotential $v(q)$ the first-order perturbation term may be set equal to zero for a local pseudopotential unless one is interested in absolute energies. However if it is recognized that the pseudopotential is energy dependent and nonlocal, this term must be included and thus even the inclusion of the zeroth-order term changes the band structure from $E = \frac{1}{2}k^2$ to something more complicated (but still isotropic). It is convenient to define an effective mass m^* to include this effect, so that $E = (1/2m^*)k^2$ is the new dispersion law. If this is done by fitting dE/dk at the Fermi surface, the resulting formulas are

$$m^* = \mu^{(E)}\mu^{(k)} \tag{27}$$

$$\mu^{(E)} = 1 - (\partial/\partial E)\langle \mathbf{k}_{\mathrm{F}}|v|\mathbf{k}_{\mathrm{F}}\rangle \tag{28}$$

$$\mu^{(k)} = [1 + (1/k_{\mathrm{F}})(\partial/\partial k_{\mathrm{F}})\langle \mathbf{k}_{\mathrm{F}}|v|\mathbf{k}_{\mathrm{F}}\rangle]^{-1} \tag{29}$$

Here $|\mathbf{k}_{\mathrm{F}}\rangle$ denotes a plane wave with wave vector of magnitude k_{F}, and $\mu^{(E)}$ and $\mu^{(k)}$ are contributions to the effective mass from energy dependence and nonlocality. Such effective masses have been calculated by Weaire[74] and others (Ref. 51, pp. 64–69; Ref. 75). The total effective mass m^* is almost always close to unity, which to some extent justifies simple local pseudopotential theories—on the other hand, any property which involves $\mu^{(E)}$ and $\mu^{(k)}$ in some other combination than $\mu^{(E)}\mu^{(k)}$ may not be accurately treated without their inclusion in the theory (Ref. 51, pp. 64–69).

We have already discussed the next order of perturbation theory for local potentials (Section 3.5). For the complications of nonlocality and energy dependence here see Heine and Weaire (Ref. 11, pp. 319–331) and Harrison (Ref. 39, pp. 259–294).

Some further remarks are in order on the subject of truncation of the pseudopotential. Most current semiempirical studies involve quite large secular determinants (say 50×50) but set $v(g)$ equal to zero for $g \gtrsim 2k_F$. However, a somewhat cruder procedure, that of truncating the basis set at $g = 2k_F$, resulting in a smaller secular determinant, has also been widely used. This procedure may be put on a formal basis by the use of Löwdin perturbation theory,[76] by which a larger secular determinant is reexpressed as a smaller one, with correction terms. For a local pseudopotential the correction terms are given by (Ref. 51, pp. 78–83)

$$\langle \mathbf{k} + \mathbf{g}|\bar{V}|\mathbf{k}\rangle = \langle \mathbf{k} + \mathbf{g}|V|\mathbf{k}\rangle + \sum_{g' > \text{cutoff}} \langle \mathbf{k} + \mathbf{g}|V|\mathbf{k} + \mathbf{g}'\rangle \langle \mathbf{k} + \mathbf{g}'|V|\mathbf{k}\rangle / [\tfrac{1}{2}k^2 - \tfrac{1}{2}(\mathbf{k} + \mathbf{g}')^2] \tag{30}$$

The corrections are indeed small for typical potentials.[77] It is unfortunate that there does not seem to be any such formal procedure for discussion of the errors implicit in the other type of truncation mentioned above.

4.4. The $\mathbf{k} \cdot \mathbf{p}$ Method

The $\mathbf{k} \cdot \mathbf{p}$ method is based on the rewriting of the Schrödinger equation

$$H\psi = E\psi \tag{31}$$

as

$$H_{\mathbf{k}} u_{\mathbf{k}} = E u_{\mathbf{k}} \tag{32}$$

where

$$\psi_{\mathbf{k}} = [\exp(i\mathbf{k} \cdot \mathbf{r})] u_{\mathbf{k}}(r) \tag{33}$$

the function u being periodic, by Bloch's theorem, and

$$H_{\mathbf{k}} = [\exp(-i\mathbf{k} \cdot \mathbf{r})] H [\exp(i\mathbf{k} \cdot \mathbf{r})] \tag{34}$$

The basic idea is to use the known solution of (32) for some $\mathbf{k} = \mathbf{k}_0$ as a first approximation for other values of $\mathbf{k}$.

This can be done either by perturbation theory or by solving a secular determinant, arising from (30), using the $\mathbf{k} = \mathbf{k}_0$ solutions as basis functions for other values of $\mathbf{k}$. The former procedure is called $\mathbf{k} \cdot \mathbf{p}$ perturbation theory. An excellent review has been given by Kane.[78] The other scheme is called $\mathbf{k} \cdot \mathbf{p}$ interpolation theory and has received close attention in a recent article by Van Dyke.[79]

The generality of these methods should be stressed. Viewed in general terms, as above, they do not depend on a particular choice of $\mathbf{k}_0$, although $\mathbf{k}_0 = 0$ is often the choice which is of interest. Second, they may be applied to any Hamiltonian, even for, say, the study of phonon dispersion—indeed $\mathbf{k} \cdot \mathbf{p}$ theory, or its equivalent, has had important applications in making local interpolations of phonon frequencies to calculate total densities of states.[80] The name, however, derives from the usual choice of Hamiltonian for the electronic problem

$$H = -\tfrac{1}{2}\nabla^2 + V(\mathbf{r}) \tag{35}$$

on which case (for $k_0 = 0$)

$$H_{\mathbf{k}} = -\tfrac{1}{2}\nabla^2 + V - i\mathbf{k} \cdot \nabla + \tfrac{1}{2}k^2 \tag{36}$$

so that (apart from the $\frac{1}{2}k^2$ term) the extra term in $H_{\mathbf{k}}$ is $\mathbf{k} \cdot \mathbf{p}$.

The application of perturbation theory to a nondegenerate state (for $\mathbf{k}_0 = 0$) will involve no linear term in $\mathbf{k}$ (by symmetry) and the first significant term, which gives the *curvature* of the band, is given by second order and conventionally expressed as an effective (inverse) mass tensor. If the unperturbed state is denoted by $|0\rangle$ and all other $\mathbf{k} = 0$ eigenstates by $|n\rangle$, the result for this is

$$(1/m^*)_{ij} = \delta_{ij} + 2 \sum_n [\langle 0|p_i|n\rangle \langle n|p_i|0\rangle/(E_n - E_0)] \tag{37}$$

There is clearly no difficulty in generalizing such a formula to the case where there is degeneracy. Equation (37) and its generalizations for degenerate cases have been of great importance in the understanding of the finer points of band structures, especially in semiconductors,[81] where the effective mass parameters are important experimental observables.

The use of $\mathbf{k} \cdot \mathbf{p}$ interpolation theory is motivated by the hope that a very accurate calculation of wave functions and energies at (say) $\mathbf{k} = 0$ can be extended throughout the Brillouin zone within an acceptable loss of accuracy, so that integrated quantities (optical densities of states, etc.) might be obtained with a minimum of computational effort, since the basis set involved in the extrapolation is restricted to the wave functions at $\mathbf{k} = 0$ in the range of energies of interest. One serious drawback of the method is that for such a finite basis set there is nothing in the theory to make bands obey the right symmetry properties at Brillouin zone planes, so that the correct Van

Hove singularities in the density of states are not obtained except at $\mathbf{k} = 0$.

It is difficult to make generalizations about the accuracy of the method.[79] One puzzling aspect of the method which has emerged from Van Dyke's work[79] is that the augmentation of the basis set with lower-lying (core) states and higher states does not seem to improve its accuracy. Since the inclusion of *all* $\mathbf{k} = 0$ states in the basis set ought to give exact results, this seems to imply a significant contribution from very high-energy states. For the moment, anyway, it seems to constitute a rather serious criticism.

4.5. Small-k Expansions for KKR Theory

Segall and Juras[82,83] have shown that it is possible to expand the KKR equations (Section 3.8) for small **k**, obtaining effective mass formulas analogous to those of $\mathbf{k} \cdot \mathbf{p}$ theory. Given the wide utility of KKR, this would appear to be a promising development. The formulas have simple closed forms.[82,83] The results for effective masses are shown in Figure 14 for Cu, compared with a complete KKR calculation. Clearly, if a method could be devised for connecting the bands

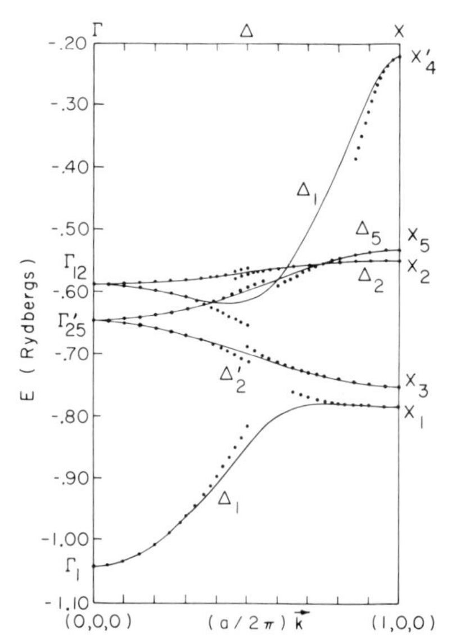

Fig. 14. The full curve is the result of a KKR calculation for Cu. The dotted lines are given by fairly simple closed-form expressions derived by Juras.[83]

given the curvatures at the various symmetry points, an extremely economical method of accurate band structure calculation, involving little or no solutions of secular equations, is within sight!

5. Some Relevant Experiments

5.1. Introduction

Descriptive accounts of energy bands have not usually included much material on the relevant experiments, yet the fact is that many of the important developments in this subject have been stimulated more or less directly by experimental results. Moreover, our judgment of the approximations which we use is based on the experimental comparisons that we make. Thus, for example, the local, energy-independent pseudopotential approximation may have gained a little too much popularity from the concentration on the study of Fermi surfaces, which restricts attention to a single energy and small regions of **k** space. It seems desirable, therefore, to include a survey of the important experiments which have been the yardsticks for band structure calculations. One important class of experimental techniques may be identified as involving an electronic transition. However, with the possible collusion of photons, phonons, and other electrons, the range of possibilities is very great. A recent review by Nagel and Baun[84] is a useful source for the sorting out of some of these techniques and assignment of relative importance. Here we will concentrate on those experiments involving straightforward transitions of single electrons, namely optical absorption and photoemission (valence band → conduction band), and soft X-ray emission and absorption (valence band → unfilled core state, core state → conduction band). Figure 15 is a reminder of this terminology. This selection reflects the historical importance of these experiments, as well as their simplicity, but it does leave aside a number of currently interesting techniques[84] (radiative Auger effect, Compton scattering, etc.).

In either optical absorption or soft X-ray emission and absorption the results are generally interpreted in terms of a simple band picture in which the spectrum is assumed to be given by an integral of a transition probability over occupied (i) and unoccupied (j) states of the same **k** vector. For the case of optical absorption this may be written as

$$\varepsilon_2(\omega) = \pi^{-1} \sum_{ij} \int [f_{ij}(\mathbf{k})/E_{ij}|\nabla_{\mathbf{k}} E_{ij}|]\, dS \tag{38}$$

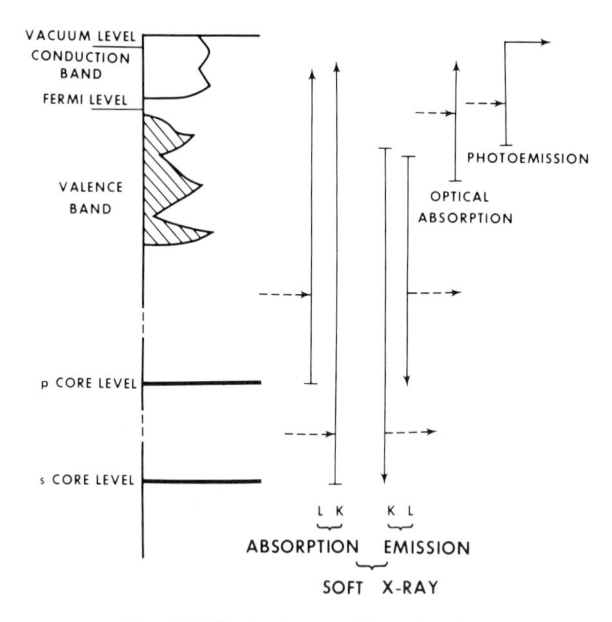

Fig. 15. The basic transitions involved in the four experimental techniques mentioned in the text. Full vertical lines denote electronic transitions. Horizontal dashed lines denote emission or absorption of photons; the horizontal full line denotes emission of an electron.

where the integral is over the surface S in $\mathbf{k}$ space defined by demanding that the states i and j have an energy difference

$$E_{ij} = E_j - E_i \tag{39}$$

exactly equal to $h\omega$, and

$$f_{ij}(\mathbf{k}) = \tfrac{2}{3}|\langle \mathbf{k}, i|\mathbf{p}|\mathbf{k}, j\rangle|^2/E_{ij} \tag{40}$$

It is conceptually convenient and usually a reasonable first approximation to consider the spectrum to be primarily determined by the *joint density of states* of the occupied and unoccupied bands.

$$n_{\text{joint}}(E) = \int dE' n_{\text{occ}}(E') n_{\text{unocc}}(E' + E) \tag{41}$$

modified by the effects of the matrix element and energy denominator in the full formula. Of course, in the case of X rays, one of the two density of states functions entering (41) is just a delta function and the observed spectrum relates simply to the other density of states function involved (the valence band in emission, the conduction band in absorption).

A second class of experiments relates to effects which take place at or about the Fermi level, particularly in metals. Some of these are reviewed in Section 5.4. There are recent reviews of such Fermi surface studies by Lee[85] and Cracknell.[86]

5.2. Soft X-Ray Emission and Absorption

As with many other branches of solid state physics, the technique of soft X-ray emission and absorption spectroscopy, after an exciting *debut* in the 1930's,[87] made only slow progress until recently, when important advances in experimental sophistication have led to a renewed interest. One development which is particularly worthy of note is the use of synchrotron radiation as a source for high-resolution absorption spectroscopy. Some of the advantages of such sources which have been cited[88,89] are: (a) the strength of the source, (b) its smooth profile, (c) its high degree of polarization, (d) its very nearly parallel radiation—all in marked contrast to the properties of the gas discharge and other sources formerly used. With this and other experimental advances the quantity and quality of X-ray data for solids are growing rapidly. The basic interpretation of much of this has not changed since the early work of Skinner.[87] The main features of the spectrum are attributable to features in the density of states, although these may be masked to some extent by the effects of the matrix element in Eq. (40). (Whenever both K and L spectra are available this is not so serious. Various many-body effects can alter the spectrum at its extreme edges—Auger tailing, plasmon satellites, exciton structure, edge singularities in metals (Ref. 9, p. 156). All of these are beyond our scope in this review—suffice it to say that they are not usually sufficiently large effects to seriously detract from the usefulness of soft X-ray spectra as indicators of the position and shape of the broad features of energy bands.

An illustrative example, taken from the work of Weich,[90] is shown in Figure 16. Both K and L emission spectra for crystalline and amorphous Si are shown. Because of the matrix element in Eq. (40), the K spectrum measures (roughly speaking) the density of electronic states weighted with the fractional p-like character of the states. Similarly, the L spectrum is weighted by the s-like character. Wiech has shown that by adding the two spectra for crystalline Si, appropriately

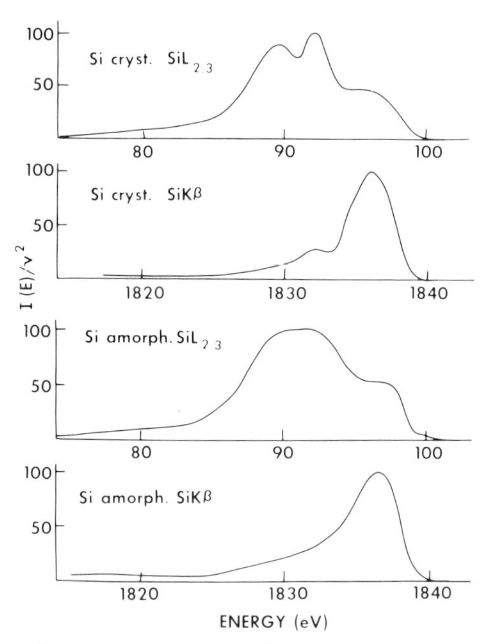

Fig. 16. Soft X-ray emission spectra for crystalline and amorphous Si (after Ref. 90).

weighted, one can obtain a spectrum which is in rough agreement with the theoretical density of states. (The theory of group IV semiconductors is the subject of Section 6.4.) The comparison of amorphous and crystalline Si presented by Wiech illustrates some of the strength and the weakness of the soft X-ray emission technique when put to such purposes. On the one hand, the consequences of the experiment are clear—one needs no deconvolution or other analytic device to infer that two peaks in the density of states of crystalline Si have coalesced on going over to the amorphous case. On the other hand, the resolution of the experiment is not really satisfactory. The peak at the top of the valence band of crystalline Si has a good deal of substructure (see Section 6.4), but this is not resolved here. In absorption the typical resolution is rather better but still does not compare with the extraordinarily precise information available from (modulated) optical spectroscopy, where it is applicable.

In the case of compounds and alloys the use of soft X-ray spectroscopy has the advantage that it yields separate spectra associated with the core levels on the different elements (provided these are not too close in energy). Indeed, the more complicated the system, the more this technique has to offer in comparison with optical spectroscopy.

X-ray photoemission (or ESCA) yields rather similar information to that given by X-ray emission—this technique is currently in a period of rapid development (see, e.g., Ref. 91).

5.3. Optical Spectroscopy

Optical spectroscopy, like X-ray spectroscopy, is going through a period of rapid refinement of techniques. Among recent innovations, the use of synchrotron radiation has opened up the study of systems with large band gaps such as alkali halides,[92] since most previous studies were subject to a cutoff at the largest available band gap (for windows) which was provided by LiF, at about 12 eV. Another important advance is the introduction of *modulated* techniques.

The optical absorption (Figure 15) is given as a function of the frequency ω of the incident radiation by the function $\varepsilon_2(\omega)$, which is the complex part of the dielectric constant:

$$\varepsilon = \varepsilon_1 + i\varepsilon_2 \tag{42}$$

This is in turn related to the complex index of refraction n by

$$\varepsilon = n^2 \tag{43}$$

Thanks to the existence of Kramers–Kronig relations between $\varepsilon_1(\omega)$ and $\varepsilon_2(\omega)$, a complete knowledge of either function serves to determine the other. In practice this means that a measurement of the reflectivity of a clean surface is sufficient to determine $\varepsilon_2(\omega)$. For a review of the basic theory see Phillips.[93]

The function $\varepsilon_2(\omega)$ is again given by the integral formula (38) where j is here an unoccupied state above the Fermi level and i is an occupied state below it. In practice, many-body effects again intrude and exciton structure indeed plays a prominent part in $\varepsilon_2(\omega)$ for many systems, especially those with large band gaps.[92] All of this will, for present purposes, be swept under the proverbial rug.

Even with so much simplification by omission, we are left with a fairly complicated formula for $\varepsilon_2(\omega)$. In this case, unless the electronic density has very prominent peaks corresponding to essentially flat bands, the simple interpretation in terms of a joint density of states may not be reliable. This is because the existence of the **k**-selection rule which is built into the full formula (38). Only initial and final states of the same **k** vector can contribute. In addition, since (38) involves a dipole matrix element, transitions between states of, say, predominantly s and p character will be enhanced. However, **k** selection is the principal effect. So-called *indirect* (non-**k**-conserving) transitions can contribute to $\varepsilon_2(\omega)$ only in cooperation with the emission or absorption of phonons, or in the presence of impurities.

Given such a relatively complicated formula for $\varepsilon_2(\omega)$, how should one proceed with the task of bringing theory and experiment together? On the one hand, there is what is unkindly called the brute force approach, in which a theoretical prediction of $\varepsilon_2(\omega)$ proceeds from essentially first principles to a full numerical calculation based on (38). This involves quite difficult numerical integrals over the Brillouin zone, although the practice of such integration has recently reached the status of a fine art.[94] Examples of such complete calculations are to be found in the work of Dresselhaus and Dresselhaus[95] and Brust.[96] While such calculations are indispensable, they should ideally clear the ground for something more elementary and transparent. How are we to understand such spectra in simpler terms? One approach, associated largely with the work of Phillips,[93] is to regard the Van Hove singularities (Section 2) as the essential elements of the spectrum. These arise from regions in **k** space where occupied and unoccupied bands are parallel—usually, but not always, at symmetry points. This leads to the expectation that an understanding of the band structure at important symmetry points will suffice to explain the spectrum. Conversely, after assignment of Van Hove singularities to particular parts of the band structure, variations with alloying, pressure, etc. could be interpreted without appeal to extensive calculations. This point of view enjoyed great success in the analysis of the group IV and related semiconductors by Phillips[93] and a subsequent semiempirical school of thought involving, *inter alia*, the work of Cohen and Bergstresser.[42]

However, the distinguishing and assignment of critical points in $\varepsilon_2(\omega)$ is not always as straightforward a business as one would like. This point of view would therefore have remained somewhat speculative were it not for the timely arrival of modulation techniques. An excellent review of these has been given by Seraphin.[97]

As with other modulation techniques, the idea is to measure the reflectance while some external parameter is varied periodically. The modulation of the reflected beam can be separated from the unmodulated background, and the resulting spectrum represents some kind of derivative of the original spectrum, showing up its singularities in fine detail. The external parameter may be electric field, pressure or other stress, temperature, or even rotation of the sample.[97] Of these, the first has proven most useful. Electroreflectance is therefore currently the best method for picking out the singularities in an optical spectrum. Further refinements of this approach involve modulating one external parameter, such as electric field, for various static values of another, such as stress.[97]

With the help of these experimental techniques the study of semiconductors in recent years has been a highly successful marriage of theory and experiment. The trends of the important parameters of band structure from one element or compound to the next were found to follow rules which are remarkably simple when one considers the potpourri of effects and approximations involved in band structure calculations. This led Phillips to construct a much simpler semiempirical picture based on bonds rather than bands—but that is another story,[98] told elsewhere in this volume (Chapter 1).

The simple metals have also received considerable theoretical attention in recent years. Again, brute force calculations are reasonably successful.[96] In this case the simple picture on which present interpretations are based is somewhat different. The interband part of the spectrum generally consists of easily identifiable peaks, with a sharp edge on the low-energy side, which may be attributed to the contributions of *parallel* bands.[99–101] To understand how these arise, it should first be recalled that the nearly-free-electron approximation (Section 3.2) is valid for these metals. Because the coefficients $V(\mathbf{g})$ of the pseudopotential are small (Section 3.6) they may be treated independently. In this approximation each coefficient produces a discontinuity in the band structure at the appropriate Brillouin zone plane of magnitude $|2V(\mathbf{g})|$. The point is that this discontinuity is essentially constant over the Brillouin zone plane, as in the simple model of Section 3.5, and there are hence large numbers of states on either side of the plane, separated by about $|2V(\mathbf{g})|$ in energy and having the same (*reduced*) $\mathbf{k}$ value. If those on the low-energy side are occupied and those on the high-energy side unoccupied, a large contribution to $\varepsilon_2(\omega)$ results for $\hbar\omega$ equal to or slightly greater than $|2V(\mathbf{g})|$. The relationship between the optical spectrum, the band structure, and indeed the pseudopotential[101] is thus rather simple in cases for which these parallel band contributions dominate the spectrum. An example is aluminum.[100]

The above remarks do not apply to the alkali metals, for which no Brillouin zone plane intersects the Fermi surface. One would therefore expect a comparatively weak and featureless interband absorption spectrum,[102] which is at variance with the experimental findings of Mayer and El Naby.[103] This was thought for some time to constitute a serious challenge to conventional band theory[93,104] and a variety of tentative explanations were advanced, varying from the stretching of existing theories to the invention of radically new ones.[105] However, current opinion (Ref. 85, p. 115) attributes the discrepancy to experimental artifacts rather than theoretical inadequacies.

In photoemission, electrons excited from below the Fermi level to above the vacuum level (Figure 15) escape from the solid. If the exciting radiation is monochromatic, the resulting spectrum of photoemitted electrons extends from a threshold corresponding to the vacuum level up to an energy given by adding that of the top of the valence band (in an insulator) or Fermi level (in a metal) to the energy $h\nu$ of the exciting radiation. The process can be envisaged as consisting of three steps[106]: excitation of an electron, transport to the surface, and escape. Of these, the first is, for our purposes, the important part. Clearly it may be expected to be describable by formulas similar to those discussed above in the context of optical spectroscopy, although in the case of photoemission the **k**-conservation rule has been questioned.[107] The matter is not entirely settled but it appears to have been shown that at least part of some observed spectra is associated with **k**-conserving transitions.[106] Further uncertainty surrounds the question of the degree to which photoemission measures bulk or surface properties. In recent work by Eastman and Grobman[108] on group IV semiconductors they estimated electron scattering lengths of 25 atom layers for electrons excited by 8.5-eV photons, 11 layers at 10 eV, and four layers at 12 eV. Clearly for some photon energies surface states must make a significant contribution. To see such states was precisely the object of this and other[109] recent work with photoemission. Lastly, there is uncertainty concerning the treatment of the escape probability.[106] A further disadvantage of the method is its inability to probe the density of states just above the Fermi level, since this would be below the vacuum level.

Having noted these drawbacks, most of which may be lessened by future work, it should be mentioned that the method has at least one clear advantage over optical spectroscopy. Since the frequency of the incident radiation $h\nu$ can be changed, this adds an extra dimension to the data, and features in the spectrum can be identified as associated with the initial or final states, depending on how they move around as $h\nu$ is changed.[110]

Perhaps the greatest contributions of photoemission to the understanding of band structure have been for the transition metals, which have such prominent features in the density of states (due to d bands). In particular, Eastman[111] has obtained data for a large number of transition metals and made a comparison with the results of band calculations which is quite satisfactory.

5.4. Fermi Surface Analysis

The information gained from current investigations of Fermi surfaces is extraordinarily precise. It is therefore particularly valuable (quite apart from its direct value in the interpretation of transport properties, etc.) as a test of band structure theories, or as a reliable basis for semiempirical band theory.

The principal technique for Fermi surface analysis is the use of the de Haas–van Alphen effect. The oscillations of the magnetization of a metal in a strong magnetic field are separated from the background and analyzed into component periodicities ΔH. Then each periodicity so detected is directly related to an extremal cross-sectional area A of the Fermi surface, perpendicular to the applied field. The effect may be explained in terms of the quantization of the energy associated with the motion of an electron in the applied field.[112,113]

By varying the direction of the applied field, the extremal cross sections can be determined as functions of this direction. Clearly for a simple centrosymmetric Fermi surface, such as that of an alkali metal

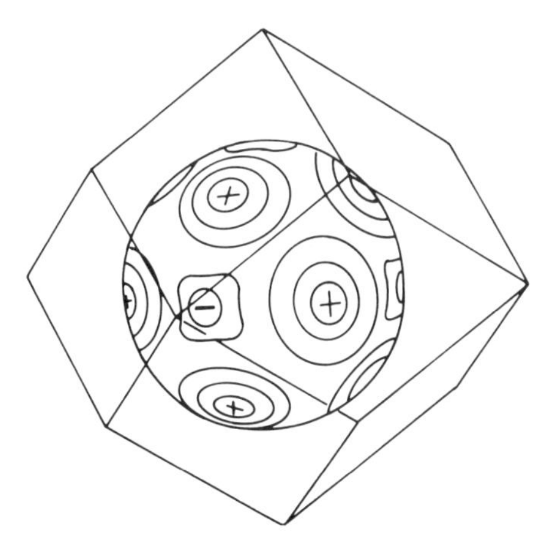

Fig. 17. The Fermi surface of an alkali metal is distorted only slightly from a sphere, being pulled out in the directions of the Brillouin zone faces (reproduced with permission from Ref. 85).

(Figure 17) one would expect such information to be sufficient for a determination of the surface itself. Nevertheless, some uncertainty as to how to proceed with the analysis persisted until Mueller[114] proposed a straightforward inversion scheme. For more complicated Fermi surfaces (an example is shown in Figure 18) the problem is, in a mathematical sense, formidable. Indeed, in Pippard's excellent early account of the subject[116] there is much preoccupation with this question. However, the problem is not really so serious because in

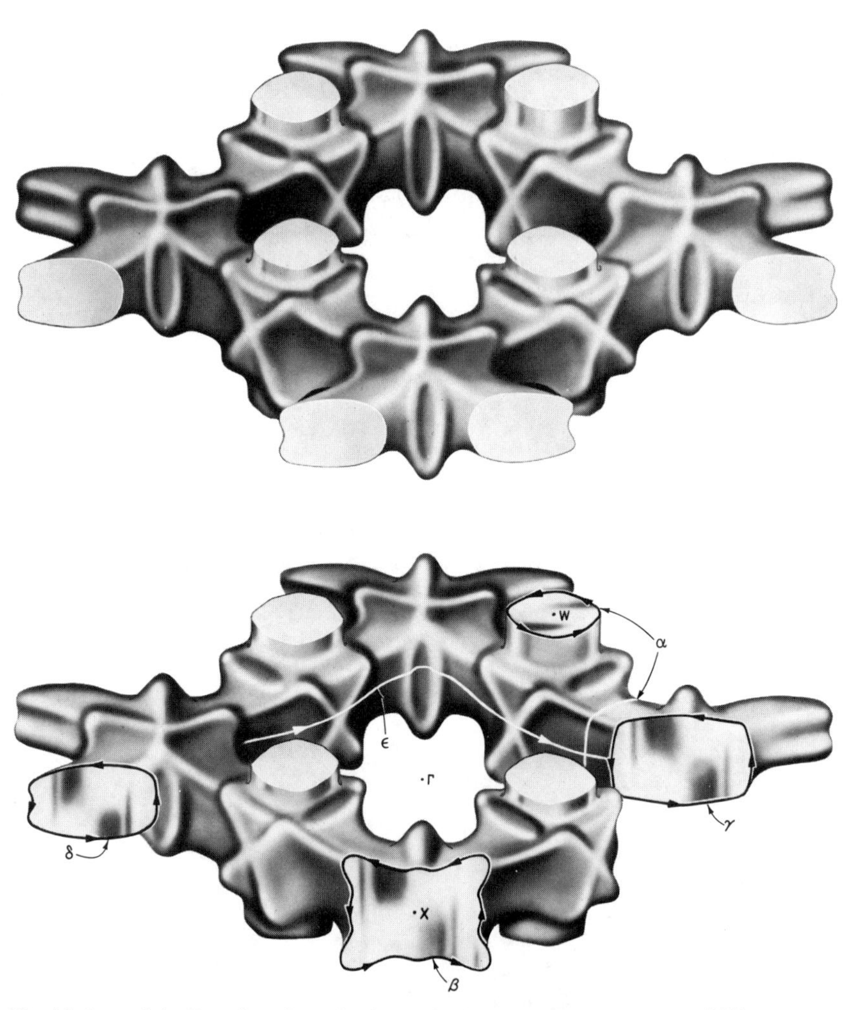

Fig. 18. Part of the Fermi surface of Pd, as calculated by Windmiller *et al.*[115] The lower picture shows various expected extremal orbits.

almost every case theoretical band structure calculations have given good first approximations to the Fermi surface, enabling an identification of the extremal cross sections to be made without any inversion procedure. Indeed, for some simple metals no band structure calculation was necessary, since the construction of Fermi surfaces in the free-electron approximation, combined with a little intuition, was sufficient to interpret the data for these systems.

The de Haas–van Alphen method is not always applicable. In particular, the effect is not detectable in alloys of concentration more than a fraction of a per cent. (The scattering time must be greater than the time necessary to complete an orbit in the magnetic field.) Other techniques, much less accurate, but applicable to cases where the de Haas–van Alphen effect fails, are of considerable current interest. Two examples are the use of positron annihilation(117) and the identification of Kohn anomalies in phonon dispersion relations.(118,119)

Examples of Fermi surfaces have already been given (Figures 17 and 18). Cracknell(86) has compiled a comprehensive review of the strange denizens of **k** space which have been identified in recent years—cigars, needles, crossed pancakes, and the like. In considering this exotic collection it is hard to accept the fact that only a decade or so ago the Fermi surface was a rather abstract concept whose meaningfulness was regarded as debatable.(120)

6. Typical Band Structures

6.1. Introduction

What follows is not intended to be an encyclopaedia of band structures. It is offered as an introduction to some of the major classes of simple solids whose band structure has been studied, with illustrations taken from the recent literature. For more comprehensive bibliographies the reader is referred to Slater,(1) Dimmock(61) (for recent applications of the APW and related methods), and Cohen and Heine(51) (for recent applications of semiempirical pseudopotential methods).

Not many compounds or alloys have been included, partly because the possible combinations of elements become too numerous and partly because much remains to be done in that area. The calculated band structures of these systems often depend critically on assumptions concerning charge transfer (see Section 6.4). It remains to be seen whether the very great successes of band theory for the pure elements will be easily extended to systems containing several elements.

6.2. Simple Metals

The simple metals, whose conduction bands correspond to *s* and *p* shells in isolated atoms, include the alkali metals, the divalent metals Be, Mg, Zn, Cd, and Hg, the trivalent metals Al, Ga, In, and Tl, and the tetravalent metals (white) Sn and Pb. Almost all of their properties which are related to electronic band structure are explicable by nearly-free-electron theory using pseudopotentials (Sections 3.5 and 3.6). The extent to which they conform in detail to this generalization varies from one case to another. For all the metals cited simple pseudopotential theory is fairly successful in predicting or fitting Fermi surface properties. This will be evident from a consideration of the comparisons of theoretical and fitted pseudopotential parameters already shown in Figure 12. However, the use of perturbation theory is not very critical in this context [i.e., the contribution of screening to the values of $v(q)$ which are of interest is not large]. In other contexts the validity of perturbation theory is more critical, and indeed the use of pseudopotential-perturbation theory is then not always so successful. An example is the calculation of phonon dispersion relations by such methods, which has enjoyed remarkable success for Na, Mg, and Al[121,122] but runs into difficulties for the heavier metals and those which have some degree of covalency (e.g., white Sn[123]). Perhaps the latter term should be defined, but it is difficult to do so, and we shall be content to define it somewhat circularly as the breakdown of lowest-order perturbation theory.

The simplest of the simple metals, in many respects, are the alkali metals. Not only are the bands of nearly-free-electron form, but since there is only one electron per atom, this means that the first Brillouin zone is only half filled and all of the gaps in the band structure lie above the Fermi level, resulting in a relatively undistorted Fermi surface, as illustrated in Figure 17. Lee[85] has recently reviewed both theory and experiment for the alkali metals (particularly Fermi surface data). He concludes that, with few exceptions, the experimental evidence is consistent with the straightforward NFE band picture, rather than the spin-density-wave or charge-density-wave models which have been advanced to explain supposed discrepancies.

The polyvalent metals are more complicated in that Brillouin zone planes intersect the Fermi surface and dissect it into various pieces. Rather than review the various minor differences among them, we again refer the reader to Cohen and Heine[51] and Dimmock[61] for detailed reviews.

6.3. Alkali Halides

The alkali halides are exemplary of the class of solids for which tight binding theory is the natural description for the valence band. The rare gas solids might be regarded as the prototypes for such solids, but the alkali halides, which are somewhat similar as far as band structure is concerned, are of greater current interest. Indeed, in the last few years there has been a considerable resurgence of interest in these systems, which, at the crudest qualitative level, were among the first to be understood with regard to such properties as cohesive energy and elastic constants. As for the details of band structure, satisfactory calculations of these have only just begun to emerge as the practical complexities of the tight binding method are mitigated by suitable approximations. Mention should particularly be made of a long series of papers by Kunz and collaborators.[124] Some of the essential features of this work are as follows. The approach is based on Hartree–Fock theory, with various many-body corrections, and rather

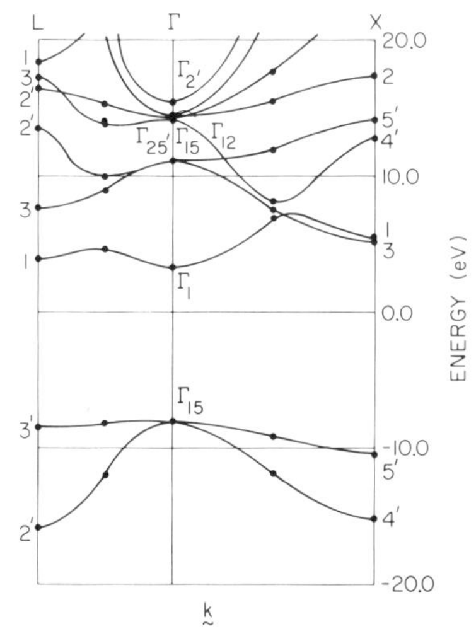

Fig. 19. Self-consistent Hartree–Fock energy bands for LiCl, as calculated by Kunz.[66]

than using *atomic* wave functions as basis functions, uses other localized functions based on the theory of Adams and Gilbert,[125] which are calculated at the outset. This basis set is augmented with (~ 100) plane waves. This might seem strange in view of the above remarks regarding the suitability of tight binding theory. However, the conduction band has much more free-electron character than the valence band, so a calculation which aspires to the description of excited states must take this into account. Indeed, Kunz finds pseudopotential rather than Slater–Koster theory appropriate for interpolating the conduction band. His results[66] for LiCl are shown in Figure 19. A subsequent calculation of the soft X-ray absorption spectrum was rather successful.[126]

6.4. Group IV Semiconductors

The tetrahedrally bonded group IV elements (C, Si, Ge, gray Sn) are intermediate between the NFE and TB extremes of band structure theory (Section 3.2). If one takes the NFE point of view, the first coefficient V_{110} of the pseudopotential is too large for simple perturbation theory (see Figure 10 for the example of gray Sn). This is why the first coefficient fails to agree well with the theoretical curve. Viewed in pseudopotential terms, this is the essence of "covalency." On the other hand, when tight binding theory is used, accurate results cannot be obtained without including interactions between quite distant atoms, and it is only recently that satisfactory results have been obtained by this method.[127]

Pseudopotential theory has so far proved to be the more useful of the two approaches. Indeed, the semiempirical studies of Brust,[96] Cohen and Bergstresser,[42] and others have been remarkable in synthesizing and systematizing a wide range of experimental data with a minimum of computational effort. Nevertheless, there are some aspects of the band structure of these elements which are best studied in the tight binding approximation. These include the general nature of the wave functions within various bands and the question of the correspondence between the electronic properties of crystalline and amorphous Si and Ge. The simplest model Hamiltonian[128] of tight binding form for these elements is

$$H = V_1 \sum_{i,j \neq j'} |\phi_{ij}\rangle\langle\phi_{ij'}| + V_2 \sum_{j,i \neq i'} |\phi_{ij}\rangle\langle\phi_{i'j}| \qquad (44)$$

where ϕ_{ij} denotes an sp^3-directed orbital on the ith atom and the jth bond. This gives the band structure shown in Figure 20, which

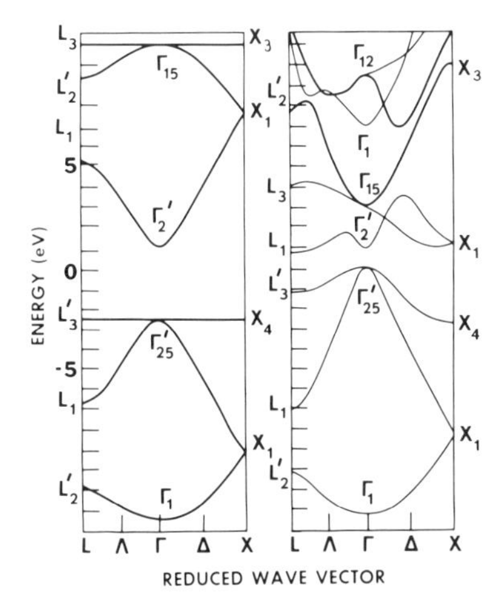

Fig. 20. Band structure of diamond cubic Ge, as calculated using the Hamiltonian (44), compared with the pseudopotential calculation of Herman *et al.*[129]

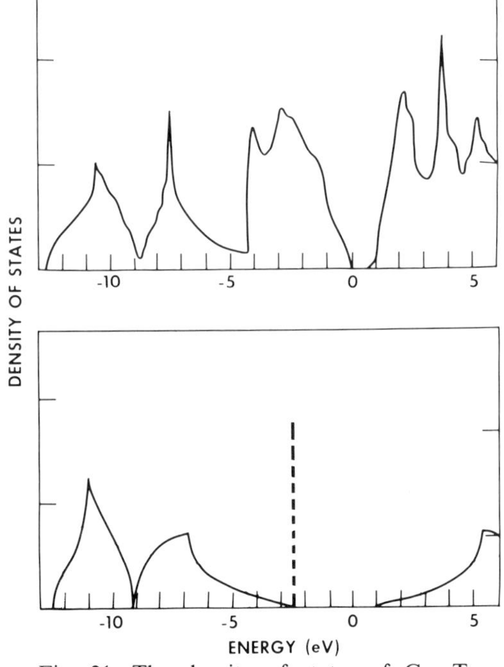

Fig. 21. The density of states of Ge. Top: Herman's calculation.[129] Bottom: the result of a calculation using the Hamiltonian (44). The vertical dashed line denotes a delta function.

may be compared with a more realistic pseudopotential calculation,[129] also shown, for the case of Ge. The same comparison for the density of states is shown in Figure 21. This may be compared with the soft X-ray emission spectrum which was shown in Figure 16. In this simple tight binding model the top of the valence band consists of a flat band of purely *p*-like states, while the continuous bands are made up of a mixture of *s* and *p* basis states. The gap is $2|V_2| - 4|V_1|$. If $|V_1/V_2|$ is increased to the point where the gap vanishes, the flat band becomes detached from the lower half of the valence band and is attached instead to the bottom of the conduction band. For $|V_1/V_2| \geq \frac{1}{2}$ the gap is then zero, so the band structure is that of a semimetal. This is the situation in gray Sn, in which the band structure has this inverted (or "Groves–Paul"[130]) form.

6.5. The III–V and II–VI Semiconductors

The families of compounds which are made up of elements in columns of the periodic table lying on either side of the group IV elements have the same (zinc blende) or related (wurtzite) crystal structure and a similar band structure. (In simple tight binding theory these two structures have an identical density of states, and the same is approximately true for more realistic Hamiltonians.) The zincblende structure is simply the diamond cubic structure with two different atoms occupying the unit cell, e.g., Ga and As. It is convenient to consider the potential (or pseudopotential) to consist of a part which is symmetric about the center of the bond joining the two atoms and a part which is antisymmetric. The first part is then much the same as the potential for a group IV element (in this case Ge) and we are concerned with the effect of adding the antisymmetric part. In pseudopotential theory this picture is particularly useful (Ref. 51, pp. 119–122)—the pseudopotential of the appropriate group IV element is a very good first approximation to the symmetric part of the total pseudopotential, and this was indeed used in much of Cohen and Bergstresser's work.[42] The pseudopotential coefficients corresponding to the antisymmetric components were adjusted to fit observed features of the band structure. The result was a rather successful and simple semiempirical description of this entire family of compounds.

Experimentally, energy differences E_g between states near the top of the valence band and bottom of the conduction band generally obey the law

$$E_g^{\,2} = E_h^{\,2} + C^2 \tag{45}$$

where E_h is the homopolar contribution, associated with the symmetric potential, and C is the contribution from the antisymmetric potential. This does not follow precisely either from pseudopotential theory[131] or tight binding theory, although in both cases one can derive rather similar formulas in which the antisymmetric component is seen to increase the band gap. The fact that in this case the experimental results yield the simple generalization[45] led Phillips to set up a semi-empirical theory,[98] based on the bond concept rather than band theory, in which C in Eq. (45) is the electronegativity difference of the two elements in the compound. This theory, as refined by Phillips and Van Vechten,[132] has enjoyed considerable success. It does not compete with band theory in the description of the finer details of band structure, but it does provide a much more manageable scheme for predicting, by a process of extrapolation and interpolation, many properties that do not depend on these details, e.g., cohesive energies.[133]

According to (45), the principal qualitative difference between group IV elements and these compounds is that the gaps are much larger in the latter. This may be seen in the example shown in Figures 22 and 23, to be compared with Figures 20 and 21 of the previous

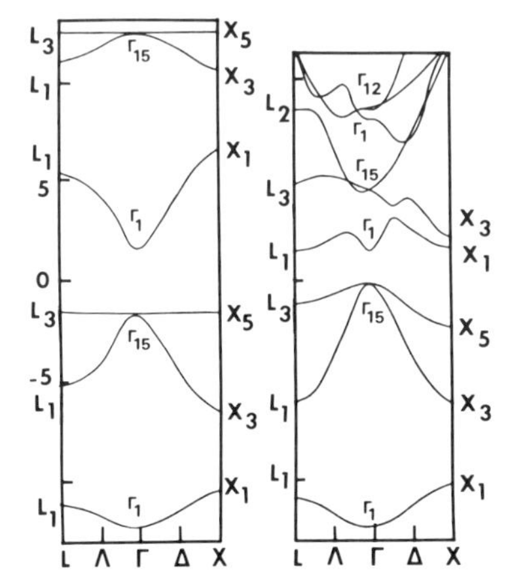

Fig. 22. The band structure of GaAs. Left: the results of a simple model tight binding Hamiltonian. Right: the pseudopotential calculation of Herman and Shay.[134]

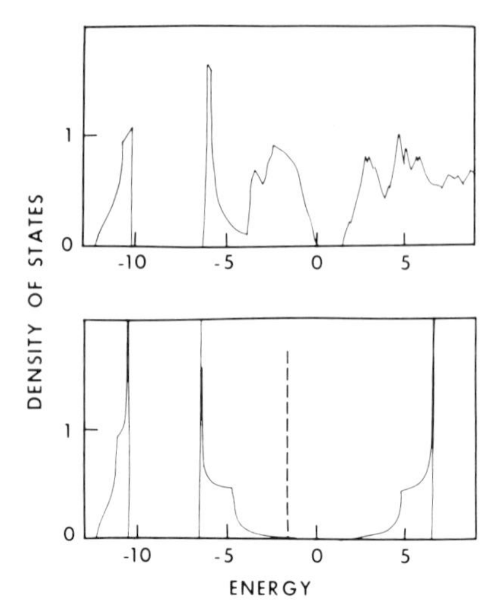

Fig. 23. The density of states of GaAs. Top: the pseudopotential calculation of Herman and Shay.[134] Bottom: the results of a simple model tight binding Hamiltonian. The vertical dashed line denotes a delta function.

section. The incorporation of an asymmetric term in the potential also splits the valence band in the middle. As in the group IV elements, the gap decreases on going downward in the period table, until one again gets semimetallic band structures similar to that of gray Sn. The fact that these compounds can be alloyed (unlike Ge and Sn) offers the possibility of tuning the band gap to very small values.[135]

6.6. Silicon Dioxide

If an oxygen atom is placed in the center of each bond of diamond cubic silicon, the result is the crystal structure of the high crystoballite form of quartz (silicon dioxide). A simple model Hamiltonian of tight binding type, analogous to (44) for Si (and probably much more realistic in this case) can be set up as follows. The oxygen orbitals in the range of energy of interest are $2s$ and $2p$, separated by about 15 eV. Of the three $2p$ orbitals, two may be chosen to be perpendicular to the bond. We will neglect the interaction of these "lone-pair" orbitals with other orbitals. There remain one $2p$ orbital (along the bond) and one $2s$ orbital, both of which mix with the sp^3 Si orbitals on the same bond. With suitable weighting parameters for these interactions and

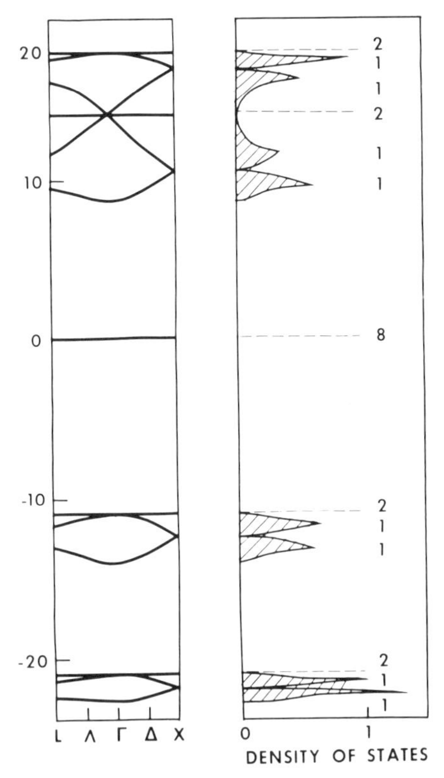

Fig. 24. The band structure and density of states of (high crystoballite) SiO_2, from a simple tight binding Hamiltonian. Again dashed lines indicate delta functions. The total number of states in each band is indicated.

the remaining interaction of sp^3 orbitals on a Si atom [V_1 in the Hamiltonian (44)], roughly estimated from atomic wave functions,[136] the band structure of Figure 24 is obtained.[26,137] The bottom band is similar to the valence band of Si—the wave functions include (mainly) oxygen s orbitals on the bonds. Similarly the next band corresponds to the bonding oxygen $2p$ orbitals. The topmost occupied bands are the lone-pair bands, degenerate in this model, but in reality having a width of several eV.[138] Higher in energy are the (unoccupied) antibonding bands. The broad features of this band structure, which are consistent with spectroscopic data,[138] may thus be understood in terms of a simple molecular description based on the Si–O–Si unit.[139] The splittings produced by the interactions of the orbitals in such a unit are much greater than those which arise from putting the units together, which give rise to the widths of the bands and their fine structure. Not much of this fine structure has yet been resolved,[138] nor has any difference between the different crystalline, indeed amorphous, phases of quartz been detected.

6.7. Transition Metals

The band structure of the transition metals consists of rather narrow *d* bands, which are described rather well by the tight binding method, falling on top of (and therefore hybridizing with) much wider bands of nearly-free-electron character with wave functions of predominantly *s* and *p* character.[140,141] In the noble metals, such as copper (the band structure of which was shown in Figure 13), the Fermi level lies above the *d* bands although their effect (via hybridization) is still felt at the Fermi level, giving rather large band gaps and consequently much greater distortion of the Fermi surface than is found in the alkali metals, which have the same number of *s*–*p* electrons. In the transition metals the Fermi level falls in the midst of the rather complicated *d* bands. They consequently have quite complicated Fermi surfaces (e.g., Figure 8) and many properties depend rather sensitively on the details of the band structure. However, the band structures of all the transition metals are quite similar[141] (apart from differences due to differences of crystal structure). There has therefore been a good deal of success recently in accounting for the trends in properties of transition metals with a single band structure, varying only the number of occupied states and hence the Fermi level. This "rigid band" picture has always been popular in qualitative interpretation of experimental results, but it is quite useful for many quantitative considerations as well (see, however, Section 7.3).

A second simplification is to neglect the influence of the free-electron-like part of the band structure (except inasmuch as it takes up some of the available electrons). Relatively simple moment techniques are then applicable to the remaining (tight binding) Hamiltonian. The results of such a description were shown in Figure 13. Moreover, in the same approximation, transition metal alloys can be quite well described by the coherent potential approximation (Section 7.3).

6.8. Transition Metal Compounds

The transition metal compounds have always been of great interest, both scientific and technological. Even among the simplest of these, the (mono) carbides, nitrides, and oxides, which generally have the rocksalt structure, there are unusually high melting points, high superconducting transition temperatures, and intriguing magnetic and metal–insulator transitions. The metal–insulator transitions

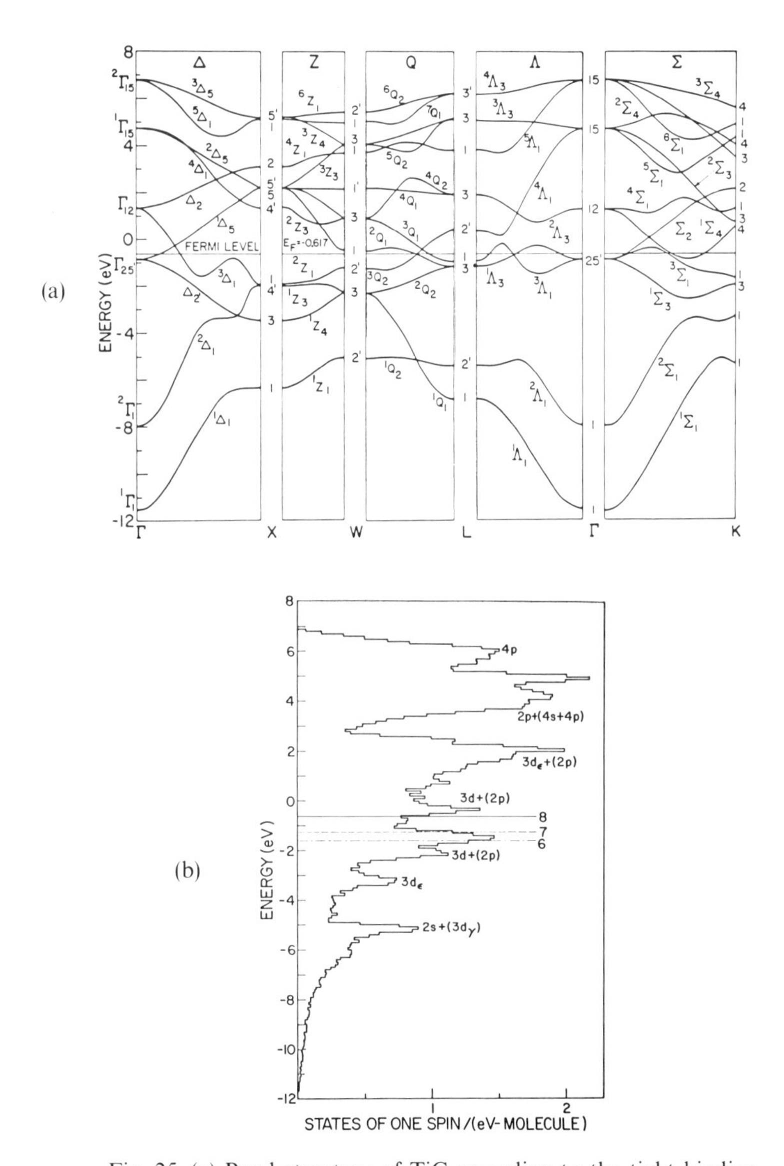

Fig. 25. (a) Band structure of TiC according to the tight binding calculation of Lye and Logothetis.[143] (b) Density of states of TiC, corresponding to the band structure of (a). The $3d$ and $4s$, p states are associated with Ti, and $2s$, p with C.

observed in some of the oxides call into question the applicability of band theory since they have been attributed by some workers to transitions to localized states rather than extended states (Mott transitions). The question is complicated, however, by simultaneous magnetic and/or structural transitions. (Chapter 4 of Volume 2 of this treatise discusses oxides and their electronic structure in some detail.) It seems clear, however, that even in cases where the band picture fails it is useful to have the best possible band calculations, if only to scrutinize their failures!

The properties of all of these compounds are dominated by the transition metal *d* bands. An empirical generalization of long standing states that the metallic compounds behave like the transition metals themselves, with extra electrons contributed by the metalloid atoms. Lye,(142) in analyzing the properties of the carbides, has found this to be consistent with his tight-binding band-structure calculations for TiC, shown in Figure 25, since these involved a charge transfer from C to Ti. This however, does not seem to be the case in the corresponding APW calculations of Ern and Swittendick,(143) and the matter is unsettled. Indeed, the whole question of the definition and implications of charge transfer requires further work. Mattheiss(144) has recently published extensive APW calculations for the transition metal oxides, and has remarked that in these (and other compounds as well) a major difference between the transition metal and the corresponding compound is the elevation of the free-electron-like part of the band structure above the *d* bands (and hence the Fermi level) in the compound. Indeed, it is further conjectured(144) that it is the movement of this part of the band structure that is the determining factor in causing the transition to an insulating state (in VO).

7. Disordered Solids

7.1. Introduction

All of the examples of the preceding sections involved perfectly periodic solids. While a great deal remains to be done in the application of the existing methods to more complicated (periodic) solids, it would seem that there are no very serious obstacles to progress in that direction, given accurate structure determinations. The same can hardly be said for disordered solids, be they alloys, amorphous semiconductors, or whatever. The formulation of a theory of electronic structure without Bloch's theorem has proved to be a difficult task.

While the studies which we shall review have to a large extent been motivated by experimental findings, the connections between theoretical models and experimental data are particularly remote and fanciful in this field. At the outset, therefore, the problems which arise will be defined and discussed as purely mathematical questions, and all mention of relevant experiments is postponed to Section 7.6.

7.2. Definition of Problems

Most studies of disordered solids have been based on simple tight binding Hamiltonians of the kind described in Section 3.3. While this approach is of limited validity, it is at least susceptible to a certain amount of rigorous mathematical analysis. Other Hamiltonians, such as pseudopotential Hamiltonians, which might be more desirable in a given context, pose many more difficulties in a disordered system unless simple lowest-order perturbation theory happens to be adequate, as in the case of the Ziman theory of liquid metals,[145] which is quite successful for the simple metals.

For a tight binding Hamiltonian we can make a distinction between *quantitative* (or cellular) and *topological* disorder. The Hamiltonian is defined by: (a) a given set of nearest-neighbor relationships, which may not form a periodic network (this is *topological* disorder), and (b) matrix elements fluctuating from site to site (this is

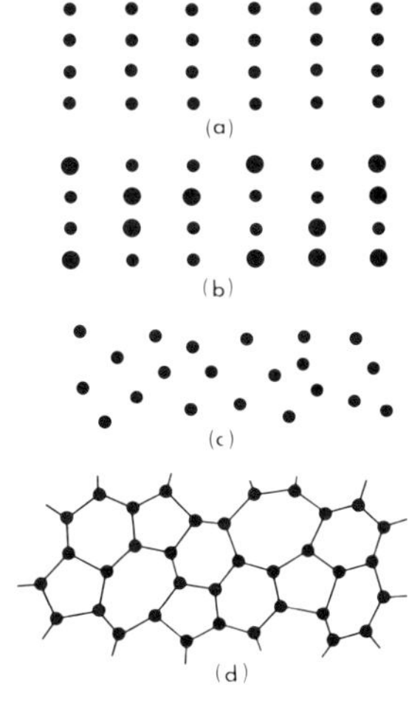

Fig. 26. Various types of disorder encountered in the study of the electronic properties of solids are schematically illustrated: (a) perfect order, (b) quantitative disorder, (c) positional disorder, (d) topological disorder.

called *quantitative* disorder). For example, a disordered substitutional alloy has quantitative disorder only. An amorphous semiconductor has both. Figure 26 illustrates these definitions.

It is obviously convenient to separate these two kinds of disorder. The more important type is quantitative disorder and the study of quantitatively disordered (i.e., alloy) Hamiltonians has been widely pursued. At the simplest level, the question at issue is: How are we to understand the density of states and related properties outside the regions in which simple perturbation theories work? This is the subject of the next section. However, a great deal of recent work has focused on a much more subtle question: What can we say about the extended or localized nature of wave functions for disordered Hamiltonians? This is the "Anderson problem."

One might ask precisely the same question for the case of (purely) topological disorder. Interest in this problem is, however, comparatively recent, and only the first of the two questions posed above has been examined.

7.3. The Density of States in an Alloy

The problem of the determination of the density of states and related properties for Hamiltonians appropriate to alloys is much simpler in one dimension than in three. This statement applies to both numerical and analytical work. In one dimension the notation of a **k** vector (in this case a scalar) generalizes readily to disordered systems. The number of nodes n of a wave function is, for periodic systems, an equivalent concept and is clearly definable for aperiodic systems as well. The wave functions can then be labeled by n in a one-dimensional system—however, no such procedure is possible in three dimensions. There is a considerable history of elegant theorems and precise calculations for one-dimensional systems.[146–149] For instance, it would appear that for a random, one-dimensional array of (delta functions) scatterers *all* eigenstates are localized.[146] However, the methods used to prove this cannot be generalized to higher dimensions and indeed the result is not thought to hold for two or three dimensions. Most recent work on three-dimensional systems has centered on the study of simple tight binding Hamiltonians of the following form:

$$H = \sum_i \varepsilon_i |\phi_i\rangle\langle\phi_i| + \sum_{i \neq j} t_{ij} |\phi_i\rangle\langle\phi_j| \tag{46}$$

The basis functions ϕ_i are localized at the atomic sites (one per site) and only the diagonal elements ε_i vary from site to site. The off-diagonal terms therefore define, by themselves, a periodic system. It remains to assign values to ε_i. For present purposes let us suppose that it can take two values, ε_A for A atoms, ε_B for B atoms, with probabilities x and $(1 - x)$, respectively. The parameter x is therefore the fractional concentration of A atoms. It is assumed that there is no correlation between sites.

Even this idealized problem does not admit of an exact solution for the density of states $n(E)$. One of the few exact statements one can make is that $n(E)$ must be zero in the range defined by $n(E)$ for pure A and pure B systems.[150] It is conjectured but not quite proven that all of this range is occupied by finite density of states. What can one say about the broad features of the energy bands which lie in this range?

The most elementary assumption which one can make is that adding A and B does not change $n(E)$ at all. This is the "rigid band" approximation, often used in the interpretation of experimental results. It is, in the present case, equivalent to the slightly more sophisticated model defined by the virtual crystal approximation, in which the Hamiltonian is averaged over the ensemble defined by the above probability distribution, to define a periodic Hamiltonian of the form (46), with

$$\varepsilon_i = x\varepsilon_A + (1 - x)\varepsilon_B \tag{47}$$

at all sites. In this approximation, the density of states shifts rigidly on change of concentration.

If ε_A and ε_B are far apart in energy, the rigid band model becomes untenable. If we add a few strongly scattering impurities to a periodic system, we expect them to split off impurity states rather than simply shift the band. Indeed, for Hamiltonians as simple as those under discussion the single-impurity problem can be solved exactly and this splitting off of an impurity state demonstrated.[150,151] On the other hand, if ε_A and ε_B are close together, the virtual crystal approximation should be a good one since, when it is subtracted from the true Hamiltonian, the difference is small and might be treated by perturbation theory. It is often convenient to define a dimensionless energy difference δ as $\varepsilon_A - \varepsilon_B$ divided by half the bandwidth of the pure A or B band.

In the last few years an approximation has been developed which has the desirable property of reducing to the virtual crystal

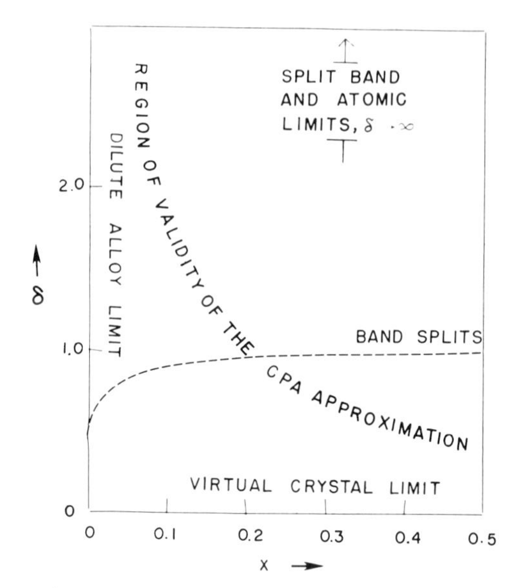

Fig. 27. Regions of validity of various alloy theories. The line denoting the splitting of the band (in CPA theory) is only approximate. (Adapted from Velicky *et al.*[150]).

approximation in the limit in which $\varepsilon_A - \varepsilon_B$ goes to zero, and reducing to the Slater–Koster theory of a single impurity in the dilute limit (x or $1 - x \approx 0$). This coherent potential approximation[150,152,153] (CPA) may therefore be thought of as interpolating between the two types of theory, as in Figure 27. However, when none of $\varepsilon_A - \varepsilon_B$, x, or $1 - x$ is small, the method runs into difficulties.[150] Nevertheless, for the present it provides the best available answer to the question that has been posed.

The CPA is based on the same general idea as the virtual crystal approximation, namely that a periodic Hamiltonian is chosen to represent the aperiodic system. One can always average the Green's function over the ensemble without affecting physical properties of interest, but one cannot strictly average the Hamiltonian, as in the virtual crystal approximation. In the CPA, according to the elegant derivation of Velicky *et al.*,[150] a criterion is chosen for the choice of the best *periodic* Hamiltonian to approximate the ensemble-averaged aperiodic system. The result is a *non-Hermitian* Hamiltonian, which is why one does not simply get the virtual crystal model out of this procedure.

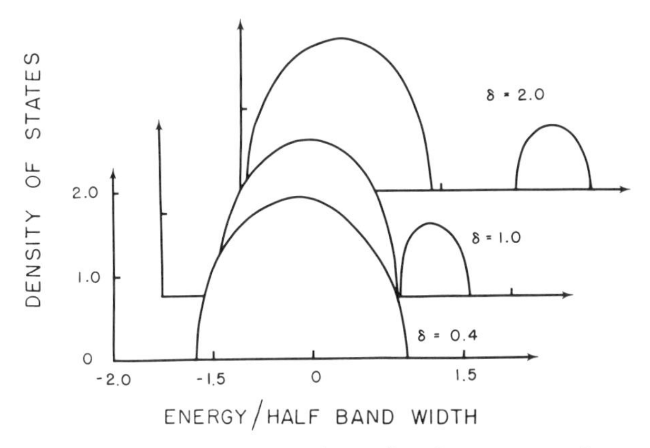

Fig. 28. Density of states of an alloy for concentration $x = 0.15$ and $\delta = 0.4, 1.0, 2.0$ as given by the CPA calculations of Velicky *et al.*[150]

For the simple Hamiltonian (46) the CPA is particularly tractable, and the CPA density of states $n(E)$ for an alloy of given composition can be generated directly from the density of states of the pure A system. Figure 28 gives examples of the calculations of Velicky *et al.*[150] performed with the simple density-of-states function

$$n(E) = (2/\pi\omega^2)(\omega^2 - E^2)^{1/2}, \qquad |E| \leq \omega$$
$$= 0, \qquad |E| \geq \omega \tag{48}$$

for the pure A or B system.

One failing of the approximation is clearly the lack of the expected tails on the various bands, which should extend out to the limits mentioned earlier. These would be very small but are a subject of much current interest, as discussed in the following sections.

A related quantitatively disordered problem has been studied by Lloyd.[154] In Lloyd's problem ε_i has a probability distribution which is the same for all sites (again uncorrelated) and is given by a Lorentzian function. Somewhat surprisingly, the density of states of such a system can be found exactly; indeed, it is simply a broadened version of the density of states generated by the Hamiltonian without the disordered diagonal term. For this case, at least, one can describe the tails, but they are of a particularly nasty kind [all moments of $n(E)$ give divergent integrals] so the model is not quite as attractive as it at first appears.

The next obvious step is to discuss off-diagonal disorder, and various attempts have been made,[155] but the CPA loses much of its appealing simplicity for this case, although it is, in a formal sense, just as valid.

Clearly the studies we have mentioned are model calculations aimed chiefly at qualitative understanding rather than accurate description of real systems. Nevertheless, the CPA has been applied in some detail to transition metal alloys (Section 7.6).

7.4. The Anderson Problem

For a disordered Hamiltonian it is natural to ask what the nature of the wave functions is. In particular, do they extend throughout the solid or are they localized? It is important to note that this is a meaningful question even in the independent-electron approximation and indeed all of this section is in that spirit. What do we mean by localization? Thouless[156] has listed six definitions which have been used, with the statement that they are probably equivalent. Three of these are given below.

(a) For a localized wave function the quantity

$$\int |\mathbf{r} - \mathbf{R}_\alpha|^2 \psi_\alpha(\mathbf{r})|^2 \, d\mathbf{r} \tag{49}$$

is finite for some value of $\mathbf{R}_\alpha$. This corresponds to one's intuitive ideas of the meaning of localization.

(b) The wave functions at the Fermi level are localized if the dc conductivity vanishes (for a static lattice) and the ac conductivity $\sigma(\omega)$ is of order ω^2. This is the most important practical consequence of localization (Section 7.6).

(c) A change of boundary conditions shifts levels corresponding to localized states by an amount of order $\exp(-N^{1/3})$ rather than N^{-1}. This is a very useful condition in numerical studies. In practice, the boundary conditions are changed from periodic to antiperiodic.

Anderson[157] studied a problem of this kind in 1958. (He had in mind disordered spin systems, but the mathematics is equivalent.) The Anderson Hamiltonian is of the general form given by Eq. (46), where ε_i is uniformly distributed between $\pm\frac{1}{2}W$ and the off-diagonal terms are confined to nearest-neighbor interactions of constant magnitude J. In the limits $W/J = 0$ and ∞ it is clear that the eigenstates are extended and localized, respectively. The question is what happens at intermediate values. Anderson confined his attention to the center of the band, and estimated the critical ratio at which there would be a change in the nature of the eigenstates. Anderson's estimate is based on rather difficult mathematical arguments and a definition of localization more subtle than (a)–(c) above. Subsequent numerical studies[156,158] have challenged the approximations although not the general spirit of his calculation, suggesting that his estimate is too large, and the matter is far from settled.[156,159]

Even if W/J is less than its critical value for the center of the band, one expects that states far out in the tails of the band will be localized. Again we can ask what happens at intermediate values. Again the answer can really only be conjectured. Mott has advanced the idea of precise energies at which the states change their character. This is hardly firmly established but is widely accepted. Finally, one can ask what happens to the mobility at this critical energy. Mott has suggested a sharp cutoff, Cohen a square root singularity, and Thouless (guided by numerical calculations) has concluded[156] that probably neither is right, but rather the mobility drops smoothly to zero.

Recent numerical work by Kirkpatrick and Eggarter[162] has shown that in lattices with two interpenetrating sublattices (square, simple cubic, bcc, etc.) localized states can occur at the center of the band while extended states still survive elsewhere, for random binary alloy Hamiltonians. This is in contradiction to some of the conjectured descriptions of localization that have been given, and is indicative of the rather confused state in which this field remains.

Finally, mention should be made of the studies of localization made by Economou and Cohen.[163] While their work is based to some extent on questionable assumptions, these authors have derived a number of interesting results regarding the onset of localization and the position of mobility edges.

7.5. Topological Disorder

The simplest tight binding Hamiltonian for which the effects of topological disorder can be investigated is that given by Eq. (14), in which the matrix elements between basis functions associated with neighboring sites all have the value V. The properties of this Hamiltonian have been investigated for fourfold-coordinated lattices. The immediate interest of such studies derives from their applicability to amorphous Si and Ge. This is not obvious, since the simplest reasonable Hamiltonian for the valence and conduction bands of these materials is that mentioned in Section 6.4, with *four* basis functions per atom. However, for any given structure there is a so-called "one band–two band" transformation[26,164] by which the two Hamiltonians are related. The spectrum of the more complicated Hamiltonian (44) consists of: (a) delta functions at the top of valence and conduction bands and (b) continuous parts which are given by analytic transformation of the spectrum of (12). It is then sufficient to study (12). The spectrum of (12) for the case of the diamond cubic structure was shown in Figure 7. The two peaks correspond to the two peaks at the bottom of the valence band of diamond cubic Si and Ge (see Figure 16). What spectrum is associated with a topologically disordered (fourfold-coordinated) lattice? First, it is easily proven (by Frobenius's theorem) that the spectrum is always bounded by the limits ($\pm 4V$) of the spectrum for diamond. (This translates into the statement that the gap between conduction and valence bands survives in amorphous Si and Ge for such a Hamiltonian.[26]) As for its shape, Weaire and Thorpe[26] have claimed that the two peaks in Figure 7 are to be understood as arising from the sixfold rings of bonds of the diamond cubic structure. This may be demonstrated by studying the Bethe lattice and related exactly soluble models.[26] The Bethe lattice is an infinitely branching fourfold-coordinated "tree" structure which has no closed rings. "Husimi cacti" are made up

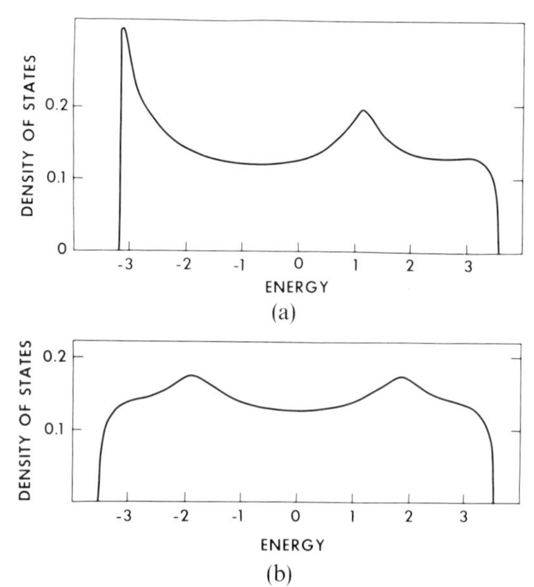

Fig. 29. Densities of states for the Hamiltonian (12) for: (a) a fivefold Husimi cactus and (b) a sixfold Husimi cactus as functions of E/V.

entirely of closed rings of a given number of bonds. Figure 29 shows the exact spectrum of the Hamiltonian (12) for these structures. The Bethe lattice gives a rather flat spectrum, while the introduction of rings results in peaks at twice the eigenvalues of a single ring. (In the case of sixfold rings these are not nearly as large as in the diamond cubic case; however, there are six times fewer sixfold rings through each atom.) The conclusion of this investigation was that for a disordered structure, like that currently proposed for amorphous Si and Ge, the mixture of fivefold, and sixfold, and sevenfold rings would result in a featureless spectrum, to be contrasted with that of the diamond cubic structure.

7.6. Applications

Since the use of the CPA is reasonably tractable only for fairly simple tight binding Hamiltonians with purely site-diagonal disorder, its practical applications are somewhat limited. Nevertheless, there is one important class of alloys for which it is of practical use, namely the transition metals. These have *d* bands which are describable in tight binding terms and which dominate the density of states of the conduction band. Moreover, they have a similar form for different transition metals, particularly those which are close together in the periodic table, so it is indeed reasonable to simply associate a different diagonal term in the Hamiltonian for different elements. When this is done it is found that the difference in this diagonal term from one element to another is quite considerable, so that the virtual crystal approximation is not a very good approximation even in the most favorable cases (i.e., neighboring elements). Thus the CPA finds an ideal testing ground in this context.

Stocks *et al.*[155] have applied the CPA to Cu–Ni alloys and compared the results with experiment in considerable detail. The results are very satisfactory and show the utility of the CPA in the split-band regime. Indeed, these calculations are an excellent demonstration of the state of the art of band theory today. The authors proceed from a first principles KKR calculation (Section 3.8) to an interpolation Hamiltonian of the mixed NFE–TB type (Section 4.1), hence to a Brillouin zone integration for the density of states of the pure metal corresponding to the majority constituent, and finally this becomes the input for the CPA. Some of their results are shown in Figure 30. Considering the fact that the experimental comparison which is being made is with photoemission data (the uncertainties

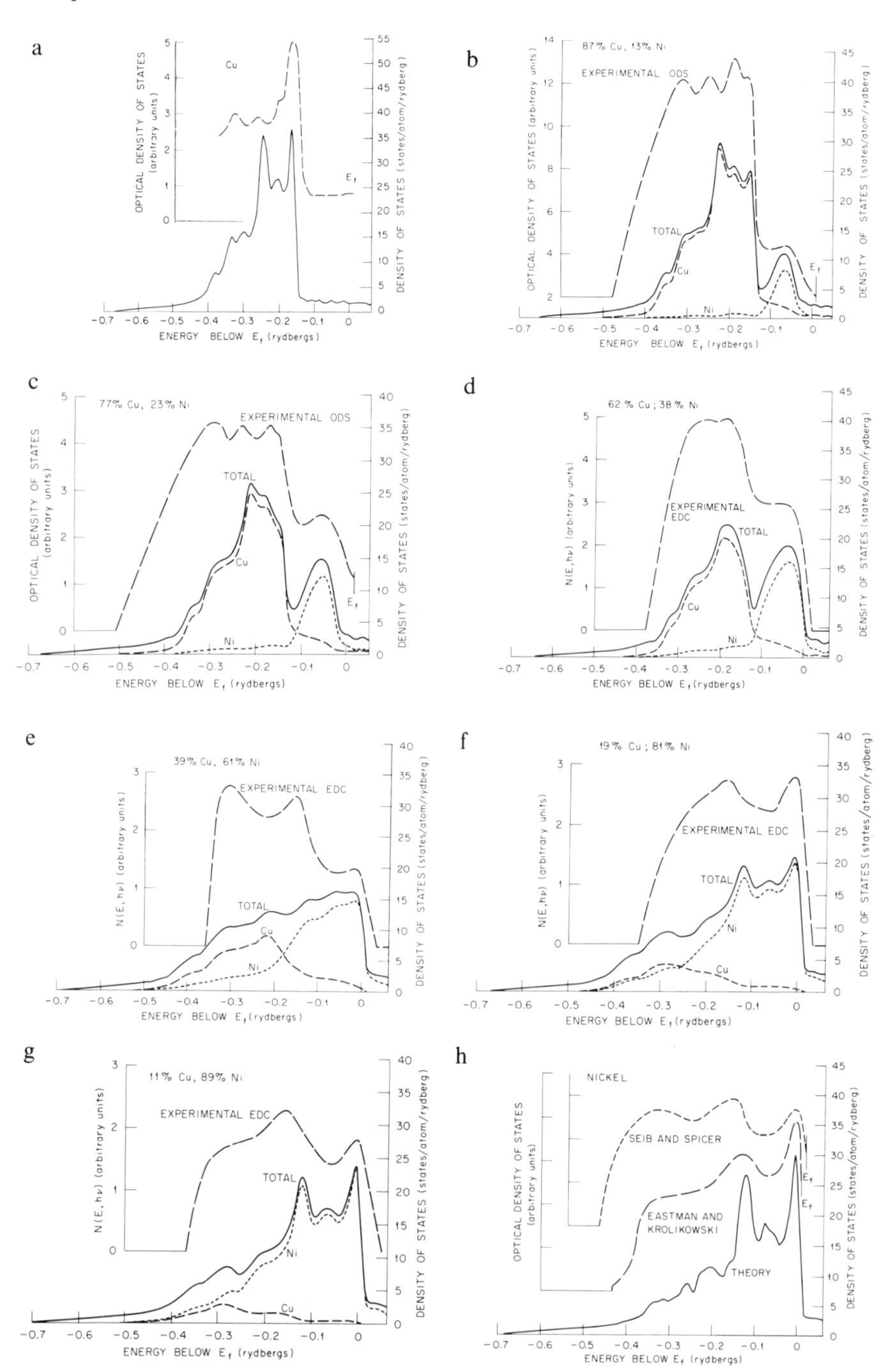

Fig. 30. Comparison of CPA density of states with experiment for Cu–Ni alloys (after Stocks *et al.*[165]) EDC refers to the observed photoemission energy distribution, ODS to a density of states derived from this without consideration of selection rules. Neither can be expected to correspond exactly to the electronic density of states but the general features agree quite well.

of the interpretation of which were mentioned in Section 5.3), the agreement is remarkable.

The study of the nature of states in band tails (the Anderson problem, Section 7.4) has not yet progressed to the point where detailed quantitative comparison with experiment are meaningful. So far the emphasis has been on the extraction of qualitative and semiquantitative consequences of the localization of such states. Mott[166] concluded that, if conduction was due to phonon-assisted hopping among such localized states, it should vary with temperature according to

$$\sigma(T) \propto \exp[-(T_0/T)^{1/4}] \tag{50}$$

This law appears to be obeyed in many systems over large ranges (up to ten orders of magnitude in σ), particularly amorphous semiconductors.[170] Indeed, the agreement of this formula with experiment has been sufficiently exact to cause the very argument[166] which gave it to fall under suspicion, since it involved numerous approximations. As for the question of the shape of the mobility edge at the boundary (in energy) between localized and nonlocalized states, this is as yet experimentally unresolved, although numerical calculations seem to be pointing to a rather undramatic gradual increase of mobility.[158]

The discussion of topological disorder given in Section 7.5 led to the conclusion that the two peaks in the lower half of the valence band of Si and Ge in the diamond cubic phase would merge in the amorphous phase while the top peak survives. This seems to be in accord with the soft X-ray emission data of Wiech and Zöpf[90] (Section 5.2, Figure 16) and also with X-ray[91] and UV[167] photoemission. Weaire and Thorpe have further noted that topological disorder is of considerable importance in discussing the density of states or the nature of the wave functions in any region of energy which is such that wave functions alternate in sign from one atom to the next, since the introduction of randomly distributed odd-membered rings of bonds would have a considerable effect on such states.

Various other approaches to the description of energy bands in amorphous materials (particularly Si, Ge) have been proposed, including the cellular method,[55,56] the use of a smeared version of the density of states of a crystalline phase,[168] the study of finite clusters,[169] and the isotropic Penn model.[170] Probably there is something to be learned from all of these approaches to what remains a difficult problem. Much of the current interest in these materials has centered on the existence of a band gap, which seems to be implied by experimental results for suitably prepared amorphous Si and Ge.

Although Weaire[26,171] has shown that for simple Hamiltonians topological disorder does not destroy the gap, the survival of the gap in spite of the quantitative disorder necessarily implied by bond distortion remains a puzzle. Phillips[172] has suggested that the answer must be sought in the fact that the amorphous solid relaxes to a configuration which gives a (local) minimum of free energy, and such special configurations are associated with the existence of a gap. This is a difficult idea to analyze theoretically. This brings us back to one of the assertions made in the introduction to this chapter, namely that in those contexts where the atomic arrangements are not completely specified this is a serious obstacle to the fundamental description of electronic properties. Nowhere is this more true than in the study of amorphous solids.

8. Conclusion

In a recent review, Ziman[3] has remarked that the stage of formal development of band structure theory appears to be at an end. It seems unlikely that any very new basic methods will be invented in the foreseeable future. That is not to say that the application of the existing methods is yet a matter of routine. The calculation of properties like optical absorption, involving integrals over bands, is an extremely messy business, and in the whole field of approximation, interpolation, and integration of band structures there is still room for improvement. Doubtless there will be continued advances in numerical and analytical techniques for exploitation of band structure calculations, and we may soon reach the stage where the results of the Hartree approximation may be so reliable that many-body effects in spectra can at last be properly isolated and analyzed. The interpretation of some physical properties in terms of band structures, especially those related to the cohesive energy, should remain a challenging problem for many years to come. More intuitive or empirical theories,[98] based on the bond rather than the band concept, may be expected to hold their own for a long time in the description of such properties, and will not easily be displaced by more fundamental schemes.

Acknowledgment

The author wishes to thank J. Faulkner, V. Heine, C. Herring, G. Juras, J. B. Ketterson, S. Kirkpatrick, A. B. Kunz, M. Lee, R. Lye, L. Mattheiss, R. W. Shaw Jr., G. Wiech, and A. R. Williams for their assistance.

Section 7 is partly based on research supported by NSF.

References

1. J. C. Slater, *Quantum Theory of Molecules and Solids*, McGraw-Hill, New York (1965).
2. J. Callaway, *Energy Band Theory*, Academic, New York (1964).
3. J. M. Ziman, The calculation of Bloch functions, *Solid State Phys.* **26**, 1–101 (1971).
4. E. P. Wigner, Effects of the electron interaction on the energy levels of electrons in a metal, *Trans. Faraday Soc.* **34**:678–85 (1938).
5. P. Hohenberg and W. Kohn, Inhomogeneous electron gas, *Phys. Rev.* **136**, 864–67 (1964).
6. W. Kohn and L. J. Sham, Self-consistent equations including exchange and correlation effects, *Phys. Rev.* **140**, 1133–8 (1965).
7. L. J. Sham and W. Kohn, One-particle properties of an inhomogeneous interacting electron gas, *Phys. Rev.* **145**, 561–7 (1966).
8. J. C. Slater, A simplification of the Hartree–Fock method, *Phys. Rev.* **81**, 385–90 (1951).
9. L. Hedin and S. Lundqvist, Effects of electron–electron and electron–phonon interactions on the one-electron states of solids, *Solid State Phys.* **23**, 1–181 (1969).
10. J. Hubbard, The description of collective motion in terms of many-body perturbation theory. II, *Proc. Roy. Soc.* **A243**, 336–52 (1958).
11. V. Heine and D. Weaire, Pseudopotential theory of cohesion and structure, *Solid State Phys.* **24**, 249–463 (1970).
12. K. S. Singwi, M. P. Tosi, R. H. Land, and A. Sjolander, Electron correlations at metallic densities, *Phys. Rev.* **176**, 589–99 (1968).
13. R. W. Shaw, Jr. and W. W. Warren, Jr., Enhancement of the Korringa constant in alkali metals by electron–electron interaction, *Phys. Rev.* **3**, 1562–68 (1971).
14. L. Kleinman, New approximation for screened exchange and the dielectric constant of metals, *Phys. Rev.* **160**, 585–90 (1967).
15. C. Kittel, *Introduction to Solid State Physics*, 4th ed., Wiley, New York (1971).
16. C. Kittel, *Quantum Theory of Solids*, Wiley, New York (1963); N. F. Mott and H. Jones, *The Theory of the Properties of Metals and Alloys*, Clarendon Press, Oxford (1936).
17. L. Van Hove, The occurrence of singularities in the elastic frequency distribution of a crystal, *Phys. Rev.* **89**, 1189–93 (1953).
18. E. O. Kane, Critical point structure in photoelectric emission energy distributions, *Phys. Rev.* **175**, 1039–48 (1968).
19. R. S. Mulliken, C. A. Rieke, D. Orloff, and H. Orloff, Formulas and numerical tables for overlap integrals, *J. Chem. Phys.* **17**, 1248–67 (1949).
20. R. Hoffmann, An extended Hückel theory. I. Hydrocarbons, *J. Chem. Phys.* **39**, 1397–1412 (1963).
21. T. L. Gilbert, in *Sigma Molecular Orbital Theory* (O. Sinanoglu, K. B. Wiberg, eds.), pp. 249–55, Benjamin, New York (1969).
22. M. F. Thorpe, Two-magnon bound state in fcc ferromagnets, *Phys. Rev.* **4**, 1608–13 (1971).
23. T. Wolfram and J. Callaway, Spin wave impurity states in ferromagnets, *Phys. Rev.* **130**, 2207–17 (1963).

24. E. Frikkee, Calculations on magnon impurity modes of a pair defect in a face-centered cubic ferromagnet, *J. Phys.* C **2**, 345–55 (1969).
25. M. F. Thorpe and D. Weaire, Electronic properties of an amorphous solid. II. Further aspects of the theory, *Phys. Rev.* **4**, 3518–27 (1971).
26. D. Weaire and M. F. Thorpe, in *Computational Methods for Large Molecules and Localized States in Solids* (F. Herman, A. D. McLean, and R. K. Nesbet, eds.), pp. 295–315, Plenum, New York (1973).
27. F. Ducastelle and F. Cyrot-Lackmann, Moments developments and their application to the electronic charge distribution of *d*-bands, *J. Phys. Chem. Solids* **31**, 1295–1306 (1970).
28. M. F. Thorpe, Random walks in polytype structures, *J. Math. Phys.* 294–9 (1972).
29. G. H. Wannier, The structure of electronic excitation levels in insulating crystals, *Phys. Rev.* **52**, 191–7 (1937).
30. W. Kohn, Analytic properties of bloch waves and Wannier functions, *Phys. Rev.* **115**, 809–21 (1959).
31. G. Ferreira and N. J. Parada, Wannier functions and the phases of bloch functions, *Phys. Rev.* **2**, 1614–18 (1970).
32. E. I. Blount, Formalisms of band theory, *Solid State Phys.* **13**, 305–73 (1962).
33. P. W. Anderson, Self-consistent pseudopotentials and ultralocalized functions for energy bands, *Phys. Rev. Letters* **21**, 13–16 (1968).
34. W. Kohn, in *Computational Methods for Large Molecules and Localized States in Solids* (F. Herman, A. D. McLean, and R. K. Nesbet, eds.), pp. 245–9, Plenum, New York (1973).
35. N. W. Ashcroft, in *Computational Methods in Band Theory* (P. M. Marcus, J. F. Janak, and A. R. Williams, eds.), Plenum, New York (1971) pp. 368–72.
36. A. R. Williams and D. Weaire, Validity of perturbation theory. I, *J. Phys.* 3 **3**, 387–97 (1970).
37. V. Heine, The pseudopotential concept, *Solid State Phys.* **24**, 1–36 (1970).
38. J. B. Pendry, The cancellation theorem in pseudopotential theory, *J. Phys.* C **4**, 427–34 (1971).
39. W. A. Harrison, *Pseudopotentials in the Theory of Metals*, Benjamin, New York (1966).
40. J. M. Ziman, The *T* matrix, the *K* matrix, *d*-bands and *l*-dependent pseudopotentials in the theory of metals, *Proc. Phys. Soc.* **86**, 337–53 (1965).
41. A. O. E. Animalu and V. Heine, The screened model potential for 25 elements, *Phil. Mag.* **12**, 1249–70 (1965).
42. M. L. Cohen and T. K. Bergstresser, Band structures and pseudopotential form factors for 14 semiconductors of the diamond and zinc-blende structures, *Phys. Rev.* **141**, 789–96 (1966).
43. M. D. Stafleu and A. R. de Vroomen, Fermi surface and pseudopotential coefficients in white tin, *Phys. Stat. Sol.* **23**, 683–96 (1967).
44. M. A. C. Devillers and A. R. de Vroomen, Comments on pseudopotential form factors for white Sn, *Phys. Rev.* **4**, 4631–2 (1971).
45. J. H. Tripp, P. M. Everett, W. L. Gordon, and R. W. Stark, Fermi surface of Be and its pressure dependence, *Phys. Rev.* **180**, 669–78 (1969).
46. D. Jones and A. H. Lettington, The Optical properties and electronic structure of magnesium, *Proc. Phys. Soc.* **92**, 948–55 (1967).

47. R. Stark and L. Falicov, Band structure and Fermi surface of zinc and cadmium, *Phys. Rev. Letters* **19**, 795–98 (1967).
48. J. M. Dishman and J. A. Rayne, Magnetoresistance and Fermi surface topology of crystalline mercury, *Phys. Rev.* **166**, 728–45 (1968).
49. N. W. Ashcroft, The Fermi surface of aluminum, *Phil. Mag.* **8**, 2055–83 (1963).
50. J. R. Anderson and A. V. Gold, Fermi surface, pseudopotential coefficients, and spin–orbit coupling in lead, *Phys. Rev.* **139**, 1459–81 (1965).
51. M. L. Cohen and V. Heine, The fitting of pseudopotentials to experimental data and their subsequent application, *Solid State Phys.* **24**, 37–248 (1970).
52. E. Wigner and F. Seitz, On the constitution of metallic sodium. I, II, *Phys. Rev.* **43**, 804–10 (1933); **46**, 509–24 (1934).
53. F. S. Ham, Energy bands of alkali metals. I, *Phys. Rev.* **128**, 82–97 (1962).
54. S. L. Altmann, in *Soft X-Ray Band Spectra* (D. J. Fabian, ed.), Academic, London (1968).
55. A. I. Gubanov, Cellular method for amorphous semiconductors, *Sov. Phys.—Semiconductors* **5**, 463–66 (1971).
56. D. Weaire, Some properties of random tetrahedrally coordinated structures, *J. Noncryst Solids* **6**, 181–86 (1971).
57. J. C. Phillips and L. Kleinman, New method for calculating wave functions in crystals and molecules, *Phys. Rev.* **116**, 287–94 (1959).
58. C. Herring, A new method for calculating wave functions in crystals, *Phys. Rev.* **57**, 1169–77 (1940).
59. T. O. Woodruff, The orthogonalized plane wave method, *Solid State Phys.* **4**, 367–411 (1957).
60. K. Johnson, in *Computational Methods for Large Molecules and Localized States in Solids* (F. Herman and A. D. McLean, eds.), Plenum Press, New York (1973).
61. J. O. Dimmock, The calculation of electronic energy bands by the augmented plane wave method, *Solid State Phys.* **26**, 103–274 (1971).
62. B. Segall and F. S. Ham, The Green's function method of Korringa, Kohn, and Rostoker for the calculation of the energy band structure of solids, *Methods in Comp. Phys.* **8**, 251–94 (1968).
63. B. Segall and F. S. Ham, Tables of structure constants for energy band calculations with the Green's function method, Unpublished.
64. E. O. Kane, Band structure of silicon from an adjusted Heine Abarenkov calculation, *Phys. Rev.* **146**, 558–67 (1966).
65. J. C. Slater and G. F. Koster, Simplified LCAO method for the periodic potential problem, *Phys. Rev.* **94**, 1498–1524 (1954).
66. A. B. Kunz, Energy bands and optical properties of LiCl, *Phys. Rev.* **2**, 5015–24 (1970).
67. F. M. Mueller, Combined interpolation scheme for transition and noble metals, *Phys. Rev.* **153**, 659–69 (1967).
68. J. Hubbard, The approximate calculation of electronic band structure, *Proc. Phys. Soc.* **92**, 921–37 (1967).
69. R. L. Jacobs, The theory of transition metal band structures, *J. Phys.* C **1**, 492–506 (1968).
70. F. Cyrot-Lackmann and F. Ducastelle, Moments developments. I, *J. Phys. Chem. Solids* **31**, 1295–1306 (1970).

71. F. Ducastelle and F. Cyrot-Lackmann, Moments developments. II, *J. Phys. Chem. Solids* **32**, 285–301 (1971).
72. R. Haydock, V. Heine, and M. J. Kelly, Electronic structure based on the local atomic environment for tight-binding *d*-bands, *J. Phys. C* **5**, 2845–58 (1972).
73. D. G. Pettifor, Theory of crystal structures of transition metals, *J. Phys.* C **3**, 367–77 (1970).
74. D. Weaire, Band effective masses for nineteen elements, *Proc. Phys. Soc.* **92** 956–61 (1967).
75. R. W. Shaw, Jr., Effective masses and perturbation theory in the theory of simple metals, *J. Phys.* C **2**, 2350–65 (1969).
76. P. Löwdin, A note on the quantum mechanical perturbation theory, *J. Chem. Phys.* **19**, 1396–1401 (1951).
77. A. O. E. Animalu, Nonlocal dielectric screening in metals, *Phil. Mag.* **11**, 379–88 (1965).
78. E. O. Kane, The $k \cdot p$ Method, *Semiconductors and Semimetals* **1**, 75–100 (1966).
79. J. P. Van Dyke, First principles full-zone $k \cdot p$ extrapolations critically evaluated, *Phys. Rev.* **4**, 3375–82 (1971).
80. G. Gilat and L. J. Raubenheimer, Accurate numerical method for calculating frequency distribution functions in solids, *Phys. Rev.* **144**, 390–95 (1966).
81. E. O. Kane, Need for a nonlocal correlation potential in silicon, *Phys. Rev.* **4**, 1910–16 (1971).
82. B. Segall and G. Juras, Effective mass parameters for electronic energy bands, *Phys. Rev.* **4**, 3277–80 (1971).
83. G. Juras, to be published.
84. D. J. Nagel and W. L. Baun, in *X-Ray Spectroscopy* (L. V. Azaroff, ed.), Chapter 9, McGraw-Hill, New York (1973).
85. M. J. G. Lee, The Fermi surfaces of the alkali metals, *Crit. Rev. Solid State Sci.* **2**, 85–120 (1971).
86. A. P. Cracknell, The Fermi surface. I, II, *Adv. Phys.* **18**, 681–818 (1969); **20**, 1–141 (1971).
87. W. B. Skinner, The soft X-ray spectroscopy of solids. I, *Phil. Trans. Roy. Soc.* **A239**, 95–134 (1940).
88. C. Gähwiller, F. C. Brown, and H. Fujita, Extreme ultraviolet spectroscopy with the use of a storage ring light. source, *Rev. Sci. Instr.* **41**, 1275–81 (1970).
89. T. Sagawa, in *Soft X-Ray Band Spectra* (D. J. Fabian, ed.), pp. 29–43, Academic, London (1968).
90. G. Wiech and E. Zöpf, Presented at Int. Conf. on Band Structure Spectroscopy of Metals and Alloys, Strathclyde, 1971, to be published; G. Wiech, in *Soft X-Ray Band Spectra* (D. J. Fabian, ed.), pp. 59–70, Academic, London (1968).
91. L. Ley, S. Kowalczyk, R. Pollak, and D. A. Shirley, X-ray photoemission spectra of crystalline and amorphous Si and Ge valence bands, *Phys. Rev. Letters* **29**, 1088–92 (1972).
92. G. W. Rubloff, Far-ultraviolet reflectance spectra and the electronic structure of ionic crystals, *Phys. Rev.* **5**, 662–84 (1972).
93. J. C. Phillips, The fundamental optical spectra of solids, *Solid State Phys.* **18**, 55–164 (1972).
94. F. M. Mueller, Interpolation and k-space integration; A review, in *Computational*

Methods in Band Theory (P. M. Marcus, J. F. Janak, and A. R. Williams, eds.), Plenum, New York (1971).

95. G. Dresselhaus and M. S. Dresselhaus, Fourier expansion for the electronic band structure in silicon and germanium, *Phys. Rev.* **160**, 649–79 (1967).
96. D. Brust, Electronic spectra of crystalline germanium and silicon, *Phys. Rev.* **134**, 1337–53 (1964).
97. B. O. Seraphin, in *Optical Properties of Solids* (E. D. Haidemenakis, ed.), pp. 213–52, Gordon and Breach, New York (1970).
98. J. C. Phillips, *Covalent Bonding in Crystals, Molecules and Polymers*, Univ. of Chicago Press, Chicago, Ill. (1969).
99. A. I. Golashkin, A. I. Kopeliovich, and G. P. Motulevich, Determination of the pseudopotential Fourier components on the basis of interband transitions in the optical range, *Soviet Phys.—JETP* **26**, 1161–66 (1968).
100. N. W. Ashcroft and K. Sturm, Interband absorption and the optical properties of polyvalent metals, *Phys. Rev.* **3**, 1898–1910 (1971).
101. W. A. Harrison, Band structure of aluminum, *Phys. Rev.* **118**, 1182–89 (1960); Electronic structure of polyvalent metals, *Phys. Rev.* **118**, 1190–1208 (1960).
102. P. N. Butcher, The absorption of light by alkali metals, *Proc. Phys. Soc.* **A64**, 765–80 (1951).
103. H. Mayer and M. H. El Naby, Zum inneren lichtelektrischen Effect (Quantensprungabsorption) im Alkalimetall Kalium, *Z. Physik* **174**, 289–95 (1963).
104. F. Abeles, in *Soft X-Ray Band Spectra* (D. J. Fabian, ed.), pp. 191–214, Academic, London (1968).
105. A. W. Overhauser, Spin-density wave antiferromagnetism in potassium, *Phys. Rev. Letters* **13**, 190–93 (1964).
106. N. V. Smith, Photoemission properties of metals, *Crit. Rev. Solid State Sci.* **2**, 45–83 (1971).
107. W. E. Spicer, Possible non-one-electron effects in the fundamental optical excitation spectra of certain crystalline solids and their effect on photoemission, *Phys. Rev.* **154**, 385–94 (1967).
108. D. E. Eastman and W. D. Grobman, Photoemission densities of intrinsic surface states for Si, Ge, and GaAs, *Phys. Rev. Letters* **28**, 1378–81 (1972).
109. L. F. Wagner and W. E. Spicer, Observation of a band of silicon surface states containing one electron per surface atom, *Phys. Rev. Letters* **28**, 1381–4 (1972).
110. T. M. Donovan and W. E. Spicer, Changes in the density of states of germanium on disordering as observed by photoemission, *Phys. Rev. Letters* **21**, 1572–75 (1968).
111. D. E. Eastman, Photoemission studies of *d*-band structure in Sc, Y, Gd, Ti, Zr, Hf, V, Nb, Cr, and Mo, *Solid State Commun.* **7**, 1697–99 (1969).
112. L. Onsager, Interpretation of the de Haas–van Alphen effect, *Phil. Mag.* **43**, 1006–8 (1952).
113. I. M. Lifshitz and A. M. Kosevitch, Theory of magnetic susceptibility in metals at low temperatures, *Soviet Phys.—JETP* **2**, 636–45 (1956).
114. F. M. Mueller, New inversion scheme for obtaining Fermi surface radii from de Haas–van Alphen areas, *Phys. Rev.* **148**, 636–7 (1966).
115. L. R. Windmiller, J. B. Ketterson, and S. Hörnfeldt, De Haas–van Alphen effect in palladium, *Phys. Rev.* **3**, 4213–31 (1971).

116. A. B. Pippard, *The Dynamics of Conduction Electrons*, Blackie and Son, London (1965).
117. L. J. Rouse and P. G. Varlashkin, Angular correlation studies of positron annihilation in copper–nickel alloys, *Phys. Rev.* **4**, 2377–97 (1971).
118. R. I. Sharp, The lattice dynamics of niobium. I, II, *J. Phys.* C **2**, 421–31; 432–43 (1969).
119. R. Stedman and G. Nilsson, Observations on the Fermi surface of aluminum by neutron spectrometry, *Phys. Rev. Letters* **15**, 634–37 (1965).
120. L. M. Falicov and V. Heine, The Many-body theory of electrons in metal or has a metal really got a Fermi surface? *Adv. Phys.* **10**, 57–105 (1961).
121. R. W. Shaw, Jr. and R. Pynn, Optimized model potential; Exchange and correlation corrections and calculation of magnesium phonon spectrum, *J. Phys.* C **2**, 2071–88 (1969).
122. M. A. Coulthard, Pressure dependence of phonon dispersion curves in simple metals, *J. Phys.* C **3**, 820–34 (1970).
123. E. G. Brovman and Yu. Kagan, The phonon spectrum of metals, *Soviet Phys.—JETP*, **25**, 365–82 (1967).
124. A. B. Kunz and N. O. Lipari, Electronic structure of NaBr, *Phys. Rev.* **4**, 1374–81 (1971).
125. W. H. Adams, On the solution of the Hartree–Fock equation in terms of localized orbitals, *J. Chem. Phys.* **34**, 89–102 (1961); T. L. Gilbert, in *Molecular Orbitals in Chemistry, Physics, and Biology* (P. O. Löwdin, ed.), Academic, New York, (1964).
126. F. C. Brown, C. Gähwiller, A. B. Kunz, and N. O. Lipari, Soft X-ray spectra of the lithium halides and their interpretation, *Phys. Rev. Letters* **25**, 927–30 (1970).
127. R. C. Chaney, C. C. Lin, and E. E. Lafon, Application of the method of tight binding to the calculation of the energy band structures of diamond, silicon, and sodium crystals, *Phys. Rev.* **3**, 459–72 (1971).
128. G. Leman and J. Friedel, On the description of covalent bonds in diamond lattice structures by a simplified tight binding approximation, *J. Appl. Phys.* **33**, 281–85 (1962).
129. F. Herman, R. L. Kortum, C. D. Kuglin, and J. L. Shay, in *Proc. Int. Conf. on II–VI Semiconducting Compounds* (D. G. Thomas, ed.), Benjamin, New York, (1967).
130. S. Groves and W. Paul, Band structure of gray tin, *Phys. Rev. Letters* **11**, 194–96 (1963).
131. V. Heine and R. O. Jones, Electronic band structure and covalency in diamond-type semiconductors, *J. Phys.* C **2**, 719–32 (1969).
132. J. A. Van Vechten, Quantum dielectric theory of electronegativity in covalent systems. I, II, III, *Phys. Rev.* **182**, 891–905 (1969); **187**, 1007–20 (1969); and to be published,
133. J. C. Phillips and J. A. Van Vechten, Spectroscopic analysis of cohesive energies and heats of formation of tetrahedrally coordinated semiconductors, *Phys. Rev.* **2**, 2147–60 (1970).
134. F. Herman and J. L. Shay, unpublished.
135. D. L. Carter and R. T. Bate (eds.), *The Physics of Semimetals and Narrow-Gap Semiconductors*, Pergamon, Oxford (1971).
136. F. Herman and S. Skillman, *Atomic Structure Calculations*, Prentice-Hall, Englewood Cliffs, N.J. (1963).

137. M. F. Thorpe and D. Weaire, to be published.
138. T. H. DiStefano and D. E. Eastman, Photoemission measurements of the valence levels of amorphous SiO_2, *Phys. Rev. Letters* **27**, 1560–62 (1971).
139. M. H. Reilly, Temperature dependence of the short-wavelength transmittance limit of vacuum-ultraviolet window materials. II, *J. Phys. Chem. Solids* **31**, 1041–56 (1970).
140. L. F. Matteiss, Energy bands for the iron transition series, *Phys. Rev.* **134**, 970–73 (1964).
141. E. C. Snow and J. T. Waber, The APW energy bands for the body-centered and face-centered modifications of the 3*d* transition metals, *Acta Met.* **17**, 623–35 (1969).
142. R. G. Lye and E. M. Logothetis, Optical properties and band structure of TiC, *Phys. Rev.* **147**, 622–35 (1966); R. G. Lye, A simple model for the stability of transition metal carbides, in *Proc. NBS 5th Materials Research Symp.* to be published.
143. V. Ern and A. C. Swittendick, Electronic Band Structure of TiC, TiN, and TiO, *Phys. Rev.* **137**, 1927–36 (1965).
144. L. F. Mattheiss, Electronic structure of the 3*d* transition metal monoxides, *Phys. Rev.* **5**, 290–306; 307–315 (1972); V. Heine and L. F. Mattheiss, Metal–insulator transition in transition metal oxides, *J. Phys.* C **4**, L191–94 (1971).
145. N. H. March, *Liquid Metals*, Pergamon, Oxford (1968).
146. R. E. Borland, The nature of the electronic states in disordered one-dimensional systems, *Proc. Roy. Soc.* **274**, 529–459 (1963).
147. B. I. Halperin, Properties of a particle in a one-dimensional random potential, *Adv. Chem. Phys.* **13**, 123–77 (1967).
148. H. L. Frisch and S. P. Lloyd, Electron levels in a one-dimensional random lattice, *Phys. Rev.* **120**, 1175–89 (1960).
149. J. Hori, *Spectral Properties of Disordered Chains and Lattices*, Pergamon, Oxford (1968).
150. B. Velicky, S. Kirkpatrick, and E. H. Ehrenreich, Single-site approximations in the electronic theory of simple binary alloys, *Phys. Rev.* **175**, 747–66 (1968).
151. G. F. Koster and J. C. Slater, Simplified impurity calculation, *Phys. Rev.* **96**, 1208–23 (1954).
152. P. Soven, Coherent-potential model of substitutional disordered alloys, *Phys. Rev.* **156**, 809–13 (1967).
153. M. Lax, Multiple scattering of waves, *Rev. Mod. Phys.* **23**, 287–310 (1951).
154. P. Lloyd, Exactly soluble model of electronic states in a three-dimensional disordered Hamiltonian: Nonexistence of localized states, *J. Phys.* C **2**, 1717–25 (1969).
155. J. A. Blackman, D. M. Esterling, and N. F. Berk, Generalized locator-coherent-potential approach to binary alloys, *Phys. Rev.* **4**, 2412–28 (1971).
156. D. J. Thouless, The Anderson model, *J. Noncryst. Solids* **8–10**, 461–69 (1972).
157. P. W. Anderson, Absence of diffusion in certain random lattices, *Phys. Rev.* **109**, 1492–1505 (1958).
158. J. T. Edwards and D. J. Thouless, Numerical studies of localization in disordered systems, *J. Phys.* C (*Solid State Physics*) **5**, 807–20 (1972).
159. J. M. Ziman, Localization of electrons in ordered and disordered systems. I, *J. Phys.* C **1**, 1532–38 (1968).

160. N. F. Mott and E. A. Davis, *Electronic Processes in Noncrystalline Materials*, Oxford Univ. Press (1971).
161. M. H. Cohen, Review of the theory of amorphous semiconductors, *J. Noncryst. Solids* **4**, 391–409 (1970).
162. S. Kirkpatrick and T. P. Eggarter, On the localized states of a binary alloy, *Phys. Rev.* **6**, 3598–609 (1972).
163. E. N. Economou and M. H. Cohen, Localization in disordered materials: Existence of mobility edges, *Phys. Rev. Letters* **25**, 1445–48 (1970).
164. M. Hulin, LCAO energies and wave functions in a covalent semiconductor with topological disorder, *Phys. Stat. Sol.* **52**, 119–25 (1972).
165. G. M. Stocks, R. W. Williams, and J. S. Faulkner, Densities of states in paramagnetic Cu–Ni alloys, *Phys. Rev.* **4**, 4390–4405 (1971).
166. N. F. Mott, Conduction in noncrystalline materials. III, *Phil Mag.* **19**, 835–52 (1969).
167. D. E. Eastman and W. D. Grobman, Photoemission studies of Si, Ge and GaAs using synchrotron radiation in the 7–25 eV range, *Proc. 11th International Conference on the Physics of Semiconductors, Warsaw* (1972), pp. 889–95.
168. D. Brust, Electronic spectrum, *k* conservation, and photoemission in amorphous germanium, *Phys. Rev. Letters* **23**, 1232–34 (1969).
169. J. Keller and J. M. Ziman, Long range order, short range order and energy gaps, *J. Noncryst. Solids*, **8–10**, 111–21 (1972).
170. J. C. Phillips, Electronic structure and optical spectra of amorphous semiconductors, *Physica Status Solidi* **44**, 1–4 (1971).
171. D. Weaire, Existence of a gap in the electronic density of states of a tetrahedrally bonded solid of arbitrary structure, *Phys. Rev. Letters* **26**, 1541–43 (1971).
172. J. C. Phillips, Covalent networks and amorphous semiconductors, *Comments in Solid State Phys.* **4**, 9–11 (1970).

3

Factors Controlling the Formation and Structure of Phases

W. B. Pearson
Faculty of Science, University of Waterloo
Waterloo, Ontario, Canada

1. Introduction

The problems which confront anyone who wishes to study the factors that control the formation and structure of phases or compounds—what combinations of what elements will form stable phases under what conditions and with what crystal structures—can best be expressed by quoting from the proceedings of a conference on phase stability in metals and alloys held in 1966.[1] The remarks, somewhat rearranged in order, are by W. M. Lomer.*

> "... The reason for having this conference [is] the state of depression of physicists with the rate of progress on the major problem of calculating the total energy of any solid system. Ideally one wants to calculate the heat of formation, and the entropy of pure phases and of solid solutions. The idealized calculation that is in mind is complete in the sense that electronegativity factors and size factors would emerge, simply as properties of the atomic potentials involved. If one knew the appropriate potentials, and how to make allowance for electron–electron correlation, there would be no need to introduce other factors. As time goes on, one is learning very much more about both the atomic potentials and the correlation energy estimation, and this formal pathway offers the prospect of a continually improving approach to the desired goal.
>
> "Nevertheless, the formal path is difficult, and the chemical/empirical thinkers try to improve things by injecting "commonsense" variables like

* This extended quotation is reproduced from *Phase Stability in Metals and Alloys*, pp. 569–570, with permission of the McGraw-Hill Book Co., New York.

size, electronegativity, and electronic configuration into the discussion, and most physicists keep trying out little shortcuts. All parties are setting out from a low-lying plain to scale a high mountain; the formal path lies up a tricky shoulder ridge, and is hard going from the start; the chemists' valley goes far into the mountains, giving many perspectives of the peak but never providing a real good climbing base; the physicists have not so much a valley as a set of little terraces where they play with one factor or another in isolation, and they greatly enjoy the game and get to know the boundaries of each terrace rather well. The formal path may still turn out to be the only one which runs from bottom to top; it is not often easy to join part-way up.

"The difficult formal path should perhaps be mapped roughly as the following stages:

1. Assume structure.
2. Evaluate self-consistent fields for the component-free atoms.
3. Superpose these potentials to get band structure, and use the resulting wave function to get self-consistent solution for solid.
4. Evaluate correlation energy correction.
5. Evaluate total energy.
6. Repeat for all suspected rival structures.

"On the empirical side one sets out to collect and correlate data, with a view to determining patterns for predictive purposes, or to identify situations of general theoretical interest. Now the question arises: what shall one use as variables to codify the data—atomic radius, atomic volume, electronegativity, free-atom configuration, type of bond, space-filling ability; the list is long.... One must learn from one's biological friends. The technique of multivariate analysis is well known, and it is well known, too, that no clear measure of the effect of one factor can be obtained if one is at the same time averaging over other significant factors....

"On the physics side, the procedure is very different. It is often considered desirable to investigate closely some particular term in the energy expression on the grounds that it is varying rapidly as a function of some structural parameter. The effects discussed usually depend on the outer electrons, not infrequently on the details of the Fermi surface (e.g., Hume–Rothery rules, magnetic structures, etc.). Then one quickly sees that discussion of distortion of structures, or their local stability can often be managed, but usually comparison between distinct structures is less trustworthy because it rests to an embarrassing extent on assumptions of rigid atomic cores, etc."

Our inability to follow the formal path to determine what phases two given elements may form forces us to rely on experience and intuition, and we have made some progress in being able to say what crystal structure a given phase or compound may adopt. Such experience results from analysis of factors that control crystal structure and of those that lead to distortions of crystal structures. These analyses generally depend on the variation of one property parameter thought to be significant to the stability of a particular crystal structure or series of crystal structures, while assuming other conditions to remain constant, but such assumptions may be unwarranted, as Lomer (Ref. 1, p. 569) indicates and as Kaufman and Bernstein (Ref. 2,

p. 31) later emphasized. Nevertheless, I believe that systems can be chosen where this type of analysis can be performed with ultimate significance. In other cases phenomenological studies can be made that reveal general trends with regard to structure type that are also significantly useful. The danger with such studies is, however, reading more into the conclusions than is warranted by the crudity of the phenomenological model.

The stable phase in a system is that which possesses the lowest Gibbs free energy G, where $G = H - TS$. However, at some compositions in binary systems no single phase is stable, because tangents connecting free-energy surfaces of neighboring phases as a function of composition and temperature or pressure lie at a lower free energy than the surface for any single phase, so that a two-phase region obtains (Figure 1). The properties of the atoms themselves—and their differences—exemplified in electrochemical, chemical bond, geometric, energy band, and size effects, correlate with the enthalpy H, whereas the environmental factors, temperature and pressure, correlate particularly with the entropy S.

This complex dependence of the total free energy of a phase on many factors, some of which are the result of the crystal potential and some of which depend on environmental conditions, limits progress in studies of the relative stability of crystal structures. Nearest-neighbor interaction or bonding makes by far the largest contribution to G, whereas second and third near-neighbors make contributions that are generally at least an order of magnitude smaller. Thus, whenever a complete change of crystal structure occurs in which there is a rearrangement of near-neighbor coordination and distances, the whole basis on which the value of G depends alters drastically—even though the numerical value of the contribution to G from bonding interactions may not be greatly different in the new structure, particularly since the atomic volume will not be greatly different. It is for this reason that studies of change of crystal structure with, for example, electron concentration, which generally makes a relatively small contribution to the overall value of G, are generally meaningless if there are gross changes of structural arrangement. With one or two exceptions, the only occasions on which studies of changing crystal structure with changing electron concentration are meaningful are those cases where structural change can occur without change of near-neighbor coordination. Ordered close-packed AB_3 alloys provide such an example where change of structure occurs as a function of changing electron concentration through change of second and third near-neighbor arrangements, the nearest-neighbor arrangements remaining unaltered.

Where specific types of chemical bonds (e.g., sp^3 bonds) are involved in compounds with given crystal structures, it is not hard to conclude the basis for the stabilities of the particular structures, but we have not made too much progress generally in understanding cohesion in metals. I refer particularly to structures that are held to form for geometric reasons. Large coordination numbers with ligands at different interatomic distances do not occur in order to satisfy formal valence requirements, because these can generally be satisfied in much simpler structures with low coordination number; nevertheless, satisfaction of high coordination in many of these structures is clearly related to cohesion. The close packing, whether tetrahedral, or tetrahedral and octahedral, of many metallic structures is undoubtedly related to the itinerant nature of the valence electrons and results in a low, structure-independent portion of the crystal energy, but more than that probably cannot be said. It is quite clear, for example, that the nearly 500 phases that adopt the Laves structures (Ref. 3, pp. 654–658) strive to achieve 12–16 coordination and good tetrahedral close packing in so doing, but exactly why this should result in a low value of G, giving structural stability in preference to some other structure, has not been determined. Neither is it possible to say definitively why some phases favor tetrahedral close packing whereas others favor tetrahedral and octahedral close packing. At best we can only demonstrate that effective achievement of the particular high coordination is essential to the stability of the structure type for the particular phase. On the other hand, we can well recognize the role of temperature and positional entropy in the development of ordered and disordered close-packed structures and of melting.

In the following sections I make some attempt to show what empirical factors have been recognized as controlling structure type and structural distortions, and therefore the stability of crystal structures. I avoid descriptions of theoretical calculations of cohesion and relative structural stability on the basis that these offer less than the required formal path outlined above. The present stage of development of phenomenological understanding of structural stability allows us to proceed profitably along three main lines: (1) recognition and prescription of the structural types and constraints associated with such features as ionic, saturated covalent, or metallic bonding, (2) recognition of changes of structure resulting from changes in certain properties such as electron concentration or relative atomic size in a series of alloys when other factors can reasonably be expected to remain constant, and (3) recognition and prescription of structural distortions associated with particular interactions between atoms. However, such avenues do not go far in the way of real progress toward the ultimate goal of *a priori* prediction of phase stability and structure type; in fact, with but two exceptions, the subject has advanced very little in the last twenty years. No really certain answer can be given to the question: "Will two components form an intermediate phase and if so, what will its structure be?" We are still concerned with *a posteriori* explanations, and by the time that these give understanding, all possible phases are likely to have been found, so that our understanding cannot be tested by prediction.

I first review several methods of predicting the existence of simple phases and structures and then examine the structural constraints that arise from chemical bonding, whether ionic, saturated covalent, or metallic, and from atomic size. Next, I examine in detail several of the more important factors that control the stability of phases and structures, mainly of metals, and finally I discuss several of the more important atomic interactions that cause distortions of crystal structures.

2. Practical Prediction of Phase Stability

2.1. Metals: Use of Thermodynamic Data

In view of the difficulties of quantum mechanical calculations of phase stability, Kaufman has developed over a number of years thermodynamic methods of calculating and/or estimating the lattice stability of the face-centered cubic (fcc), hexagonal close-packed (hcp), and body-centered cubic (bcc) forms of metals (whether stable or unstable), and the relative stability of the liquid phase (Volume 5, Chapter 4). These methods of determining the relative free energies of the three metallic structures and their subsequent temperature and pressure dependence, as well as the appropriate free energies of mixing when binary systems are considered, rely on such empirical data as are available and various assumptions when they are not. They have been used to calculate binary phase diagrams for transition metal systems with a tolerable amount of success when the only phases observed are fcc, hcp, and/or bcc. However, complete calculation of phase stability as a function of composition and temperature or pressure in a binary system also requires complete thermodynamic

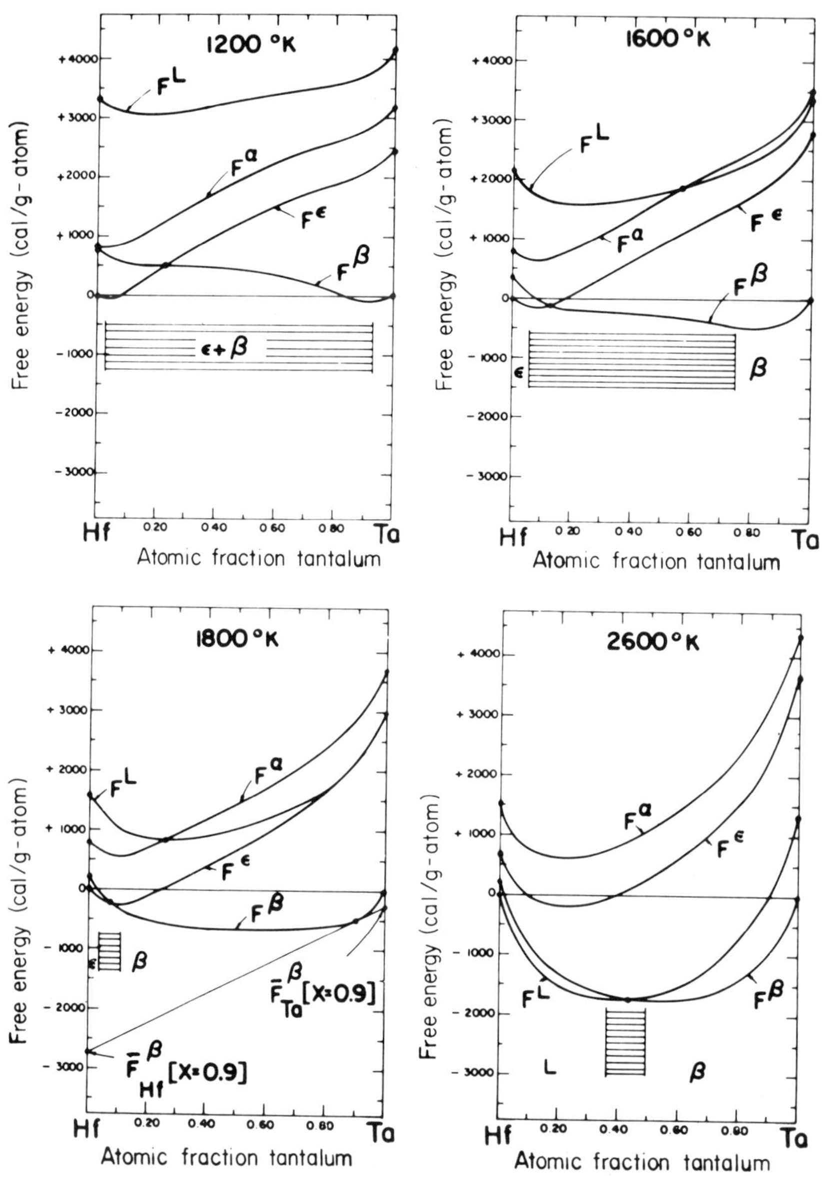

Fig. 1. Calculated free-energy–composition relations in the Hf–Ta system at various temperatures. Phase boundaries are indicated. Reproduced from Kaufman and Bernstein[2] with permission of the publisher.

data on all possible intermediate phases with different structures, which is quite impossible at present.

Complete characterization of the relative stability of the possible crystallographic forms of a pure metal involves the temperature and pressure dependence of their enthalpy, entropy, and volume differences, and the relative stability of the liquid is obtained from the temperature of melting and entropy of fusion of the stable solid form of the metal. In binary systems knowledge of the free energy of mixing would allow the generation of the free-energy curves of the different structural forms as a function of composition at various temperatures, as shown in Figure 1, taken from the work of Kaufman and Bernstein.[2] The binary phase diagram then follows from such curves.

Another method of estimating regions of stability of these phases in binary metallic (particularly transition metal) systems is due to Brewer.[4,5] It arises from the correlation that Engel[6] observed between the outer s, p valence electron concentration and crystal structure: The bcc structure corresponds to one outer s electron (and extends to 1.5 electrons), the hcp structure to two (sp) electrons, and the fcc structure corresponds to three (sp^2) outer electrons. Whereas the first two are well enough in agreement with energy band considerations generally, the fcc structure of the noble metals does not agree with the three outer electrons proposed.

Brewer developed the correlation further, showing that the s, sp, or sp^2 state corresponding to the observed stable bcc, hcp, or fcc elemental structure was indeed generally that which had the largest atomization enthalpy ΔH°_{298} calculated from the promotion energy of the state from the free-atom ground state and the bonding energies of the state, and furthermore that the comparison could be extended to nonintegral states, e.g., $sp^{0.5}$. Applying these rules to binary systems of the transition metals for states $d^{n-1}s$, $d^{n-2}sp$, and $d^{n-3}sp^2$, respectively, and introducing known intermediate phases, such as the σ phase, which occur generally in a certain range of electron concentration, he could predict the phase equilibria. The type of diagram obtained is shown in Figure 2, phase boundaries being derived from the relative values of the atomization enthalpies for the states corresponding to bcc, hcp, and fcc structures in the various alloys, and the curved lines representing constant overall electron concentration on the diagrams. However, in view of the difficulty of reconciling the treatment of the fcc noble metals as having $d^{n-2}sp^2$ states with the results of Fermi surface studies, the significance of his work in advancing an *understanding* of phase stability is uncertain.

In a later paper Brewer[6] develops his earlier work into a bonding description of the structural and thermodynamic properties of the transition metals and their alloys. These depend on the relative distribution of the valence electrons between the d and p orbitals,

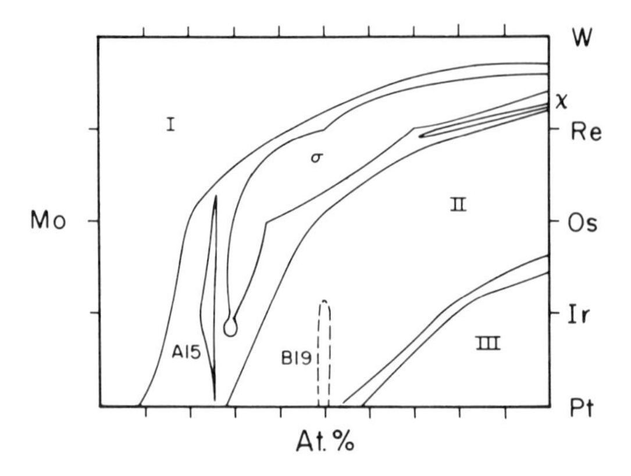

Fig. 2. Representation of phase diagrams of Mo with third-long-period metals from W to Pt. The boundaries of the single-phase regions represent the maximum extent of the phases regardless of temperature. Reproduced from Brewer[4] with permission of the publisher.

since all important configurations of transition metal atoms in the solid state have one *s* electron (s^2 is nonbonding and so unfavored). It is the electrons in *s* and *p* orbitals, that extend beyond nearest neighbors, which influence the long-range order and determine the crystal structure. Structures change with increasing *p* electron concentration from bcc (zero), through hcp (one), to fcc (two). In alloys formed between the transition metals, other structural types may appear intermediately in a definite order of increasing *p*-electron concentration. It is the energy of bonding between electrons in *d* orbitals which determines particularly the relative stability of a series of phases formed between transition metals, or the thermodynamic properties of a series of their mutual solid solutions, and it is changes in the relative amount or effectiveness of *d* bonding with changes of pressure or on alloying that alters the relative stabilities of the bcc, hcp, and fcc structures. The effectiveness of the *d* bonding depends on the size of the *d* orbitals (number of *d* electrons, 3*d*, 4*d*, or 5*d* series) and the number of *d* electrons able to form bonds, which in Brewer's treatment is determined by the available orbitals and by the group number of the elements and their *d*–*s*–*p* configuration consistent with Engel's structural correlation.

2.2. Valence Compounds: Use of Crystal Chemical Knowledge

Knowledge of valence rules alone has of course been one of the prime methods of predicting and finding new compounds, but crystal symmetry and knowledge of valence rules *together* have been used with good effect to predict both the existence of covalent compounds *and their crystal structures* using the derivative structure method that Buerger[7] has discussed systematically. Speaking generally, a set of equipoints* in a crystal structure may be subdivided into two or more subgroups to give a new *derivative* structure of lower symmetry and perhaps larger cell size. Conversely, two sets of equipoints may in certain circumstances be combined to give a new *degenerate* structure of higher symmetry (and possible smaller cell size). Thus, for example, sphalerite (ZnS) is a derivative structure of diamond and a degenerate structure of chalcopyrite ($CuFeS_2$). Where such structural relationships are seen to exist in the structures of valence compounds, prediction only requires the replacement of the cation, say, by two other cations (generally they have to be of different valences) so that their average valence is the same as that of the cation to be replaced, and a new compound with the same general structural arrangement of the atoms is to be expected, the cations occupying the two sets of equipoints in ordered fashion. In some instances, such as the $CdAl_2S_4$ structure (Ref. 3, p. 370) derived from that of sphalerite, one of the subgroups of equipoints in the derived structure is left vacant. The formula could thus be represented $Cd\square Al_2S_4$.

The derivative structure method, either explicitly or implicitly, has been responsible for the discovery of a number of new phases and structures of valence compounds and of metallic phases generally, the group of phases with the $CdAl_2S_4$ structure providing a good example.

Knowledge of certain crystal chemical features associated with a group of covalent compounds has often proved useful in choosing a model in the solution of crystal structures, although it is not particularly useful in predicting the occurrence of compounds. For example, in the case of nonionic valence compounds formed between elements to the right of the Zintl border which have more bonding electrons than valence orbitals, the frequent occurrence of chain, layer, and columnar structural arrangements is well recognized, and in the so-called polyanionic compounds[8,9,10] the anions generally form arrays

* A set of equipoints or a siteset is a set of crystallographically equivalent atomic positions in the unit cell of a crystal structure.

similar to those found in the (8–N) elements lying to the right of the Zintl border, particularly if they occupy only one siteset.

Predictions of the existence of ionic compounds and the type of structure that they may adopt are based on valence, electronegativity difference, and radius ratio of the component atoms.

3. General Structural Consequences of Bonding Types

In structures which are largely ionic, the bonding between the atoms is essentially nondirectional, whereas in compounds with saturated covalent bonding, the bonding between the atoms is very definitely directional and the crystal structures adopted must satisfy the symmetry requirements of the directional bonding. In metals where there are frequently vacant orbitals in the valence shells of the atoms, the covalent bonding is unsaturated and there are more near-neighbors than available valence electrons. The bonding tends to be largely nondirectional, with the result that many of the structures suitable for ionic compounds are also suitable for metals. However, because of the atomic charge in ionic compounds, relative size restrictions must inevitably exist since like ions cannot come in direct contact, because the ionic repulsions that would ensue would render the structures unstable. Such relative size restrictions do not play any noticeable role in the stability of compounds with saturated covalent bonding, although of course they may give rise to stearic effects, particularly in organic compounds. Neither do they play a noticeable role in the structures of binary metallic phases except when the sizes of the two components are very disparate, as, for example, in the Hägg interstitial phases.

3.1. Ionic Crystals

The potential energy V of an ionic crystal can be expressed approximately as the sum of the Coulomb interaction and the Born repulsive potential:

$$V = -Ae^2Z^2/R + Be^2/R^n$$

where e is the electron charge, Z the valence of the ions, R the interatomic distance, A the Madelung constant, B the repulsive coefficient, and n the Born exponent. From this it follows that the crystal energy U_0, defined as $-NV_0$ (N is Avogadro's number), is given by

$$U_0 = (NAe^2Z^2/R_0)(1 - 1/n)$$

The Madelung constant can be defined as

$$A/R = \sum_j{}' [(\pm)/r_j]$$

where R is the nearest-neighbor distance, r_j is the distance of the jth atom from the reference ion, and $(\pm)$ represents the sign and number of the ions in a shell a distance r_j from the reference ion. Thus, other things being equal, an ionic structure with the larger near-neighbor coordination sphere will have the larger Madelung constant.

Because of the Coulomb repulsion between like ions, the prime condition controlling the crystal structure adopted by an ionic substance is that like ions cannot come in contact. Since anions are generally larger than cations, the limiting factor is the number of anions that may surround the smaller cation in its coordination shell. The limiting value of cation–anion radius ratio R_C/R_A at which the anions would come in contact can be calculated for any coordination polyhedron. Such values are listed in Table 1. Thus ionic substances

TABLE 1
Lower Limiting Values of Radius Ratios for Stability of Ionic Coordination Polyhedra

Coordination about cation	Distribution of anions	Minimum R_C/R_A for stability
9	Corners of right triangular prism and outside rectangular faces thereof; all edges equal	0.732
8	Corners of cube	0.732
6	Corners of octahedron	0.414
4	Corners of square	0.414
4	Corners of tetrahedron	0.225

do not adopt structures in which the cations are octahedrally surrounded by anions if R_C/R_A is less than 0.414. A second consequence of Coulombic repulsions between like ions is that in the structures adopted by strongly ionic substances the anion coordination polyhedra ideally only share corners. Structures in which the anion coordination polyhedra share edges or faces, so that direct cation–cation interaction is possible, are adopted by weakly ionic and covalent substances. Pauling (Ref. 11, p. 561) observes that if the polyhedra in the structure of a strongly ionic compound do share an edge, the edge is shortened compared to other edges of the polyhedron, whereas shared edges or the edges surrounding a shared face are generally lengthened relative to other edges in the structure of a compound with covalent bonding. This is expressed as one of Pauling's rules (Ref. 11, pp. 543–562) that the presence of shared edges and especially of shared faces in an ionic structure decreases its stability, the effect being large for cations of high valence and low coordination.

In any series of structures for ions of the same type (say, M^+X^-), the Madelung constant is generally larger, the greater the number of anions in the coordination polyhedron about the cations. The expression for the crystal energy indicates that the larger the Madelung constant of an ionic structure, the more stable it will be. For this reason it is to be expected that a compound would adopt a structure with the largest coordination of the anions about the cations permitted by the radius ratio. Although this is the general rule, it is not invariably followed and several M^+X^- compounds with R_C/R_A greater than 0.732 take the NaCl (CN 6) rather than the CsCl (CN 8) structure. One reason for this may come from the increase of ionic diameter with increase of coordination of the ions. Thus in the expression for the crystal energy given above, although A increases with change of structure from the NaCl to the CsCl type, so does the interionic separation R_0. Depending on the relative sizes of these changes, the crystal energy might only improve slightly or not at all if the compound adopted the CsCl instead of the NaCl structure.

A consequence of the spherical symmetry of the electron distribution of an ion and the nondirectional nature of the interaction between ions is that a relatively few crystal structure types are required, and in fact used by simple ionic substances, in contrast to the much greater number adopted by simple covalently bonded substances and metals.

Over 40 years ago Pauling[11] also gave several other rules governing the formation of ionic structures. His electrostatic valence rule says essentially that in a stable ionic structure the anion charge (valence V) shall be exactly, or very nearly, satisfied by the electrostatic bonds from the cations surrounding it, i.e.,

$$V = \sum_i (Z_i/C_i)$$

where Z is the cation charge or valence and C is the coordination number of the cations by the anions. His cation distribution rule essentially states that in structures with several different cations those with high valence (and low CN) tend to be as far from each other as possible.

3.2. Compounds with Saturated Covalent Bonds

In the structures of compounds with saturated covalent bonds the directional requirements of the chemical bonds must be satisfied by the structural arrangement of the atoms, and tetrahedra and octahedra are the commonest coordinations found. The method of linking these polyhedra by corners, edges, or faces is of some interest both in complex minerals such as silicates and sulfosalts and in the structures of simpler compounds, because of the direct interactions that take place between atoms centering the polyhedra when they share either edges or faces. As noted on p. 124, the relative sizes of the atoms rarely appear to impose restrictions on the structural arrangements, and this is examined further in Section 4.

The structures of these compounds are generally built up of a three-dimensional network of bonds, but in the structures of the group V to group VII (nontransition metal) elements and compounds formed between them and also with group IV atoms, the atoms have more valence electrons than bonding orbitals, and molecular, chain, layer, or columnar structural arrangements are generally found. The chains, layers, etc. are separated essentially by van der Waals distances because of the filled valence shells displayed in these directions. The structures therefore represent a very poor use of all the valence electrons available.

The formation of compounds with saturated covalent bonds is subject to the restriction of bond dehybridization. Two light elements may form a compound with saturated covalent bonds, but two heavier elements from the same groups and therefore having the same maximum valences may not, because of the phenomenon known as dehybridization or metallization. The tendency to dehybridization increases with the average principal quantum number $\bar{n}$ of the two components, and this can be taken as a rough measure of dehybridization. When $\bar{n} = 4$ dehybridization may be important, but when $\bar{n} = 6$ it is complete. Considering the most general case of bonds derived from atomic s and p orbitals (e.g., hybridized sp^3 bonds), dehybridization results from the increasing energy separation of the ns and np atomic levels as n increases. When $n = 6$ a lower overall electron energy is achieved by a $6s^2\,6p^2$ configuration, although when

$n = 2$, sp^3 gives a lower overall electron energy than a $2s^2\,2p^2$ configuration.

3.3. Metallic Phases

Normal valence rules are generally not satisfied in phases that are metallic, and the number of ligands of an atom usually greatly exceeds the number of valence electrons available from forming bonds of unit strength. The bonds of atoms in a metallic structure are generally nondirectional when the number of electrons in partly filled sheets of the Fermi surface is not more than one or two electrons per atom. When the number of valence electrons is more—say, two to four electrons per atom—structures are formed which show the influence of directed chemical bonds.

All metals have a Fermi surface which may or may not interact strongly with Brillouin zone planes (planes of energy discontinuity in reciprocal space). Therefore electron concentration, which controls the size of the Fermi surface; structure, which specifies the symmetry and sequence of Brillouin zone planes; and the component atoms, which give rise to the atomic potentials and therefore control the size and distribution of the energy band gaps across the Brillouin zone planes; all have their effects on phase stability. Nevertheless, these effects are generally relatively small compared to the contribution to the enthalpy that arises from the nearest-neighbor interactions (bonding or cohesion).

Relative atomic size does not generally influence the structures of metals except in cases where the sizes of the atoms are very disparate, say 20% or more different, although it influences the distribution of the component atoms on the different crystallographic sites in structures. Thus the structures of alloys in which very high coordination numbers occur (CN 18–26) show some dependence on the relative sizes of the atoms. Hägg interstitial alloy phases formed between transition metals and the metalloids, boron, carbon, and nitrogen (and hydrogen) also depend on relative atomic size. When the ratio of the radius of the metalloid to that of the transition metal is less than about 0.59, interstitial structures occur in which the metal forms a close-packed cubic or hexagonal, or a simple hexagonal, array. When it is greater, more complex structures occur in which there is evidence of directed chemical bonds. The explanation of this rule is probably that when the metalloids are small enough the transition metal band structure still obtains, but when they are larger so that transition metals are too far separated, it breaks down and structures with directed bonds from the metalloids occur.

The structures of metals are often closely packed, and intermetallic phases generally have smaller atomic volumes than predicted by Zen's law.[12] Two types of close packing are very common: normal close packing, in which each atom has 12 neighbors and space is filled with tetrahedra and octahedra in the ratio two to one, and tetrahedral close packing, in which space is filled with distorted tetrahedra—regular tetrahedra cannot pack together to fill space. In such structures the coordination of the atoms is generally 12, 14, 15, or 16.

3.4. *A Priori* Separation of Structure Types

Confronted with the *a priori* question of whether two particular elements are likely to form compounds and with what structures, the first thing to establish is whether they can form compounds in which valence rules are satisfied and if so, whether they are likely to form ionic compounds, or whether they are only likely to form metallic alloys. The electronegativity difference of the two components furnishes a rough means of determining whether ionic compounds are likely to be formed. As a rough rule of thumb, when $\Delta x > 1.0$* they are likely to form compounds with a fairly ionic character if chemical valences match, and if relative size restrictions do not interfere. When $0.5 < \Delta x < 1.0$ saturated covalent compounds may be formed if the normal or possible chemical valences of the two components match, and if dehybridization is not likely to be important. When $\Delta x < 0.5$ and the chemical valences do not match and are both less than four, metallic compounds are likely to be formed. Phenomenological discussions such as these suggest that a general separation of ionic, saturated covalent, and metallic substances could be effected in a three-dimensional space defined by axes Δx, $\bar{n}$, and R_C/R_A, and that this could delineate in a very crude manner where certain types of crystal structures might be found.

The first real quantitative success in separating structure types according to these general principles was achieved by Phillips and van Vechten† (Chapter 1), who derived ionicities from spectroscopic

* Thermochemical values of, e.g., Pauling (Ref. 11, p. 93) and Haissinskij.[13]

† See references given by Phillips.[14]

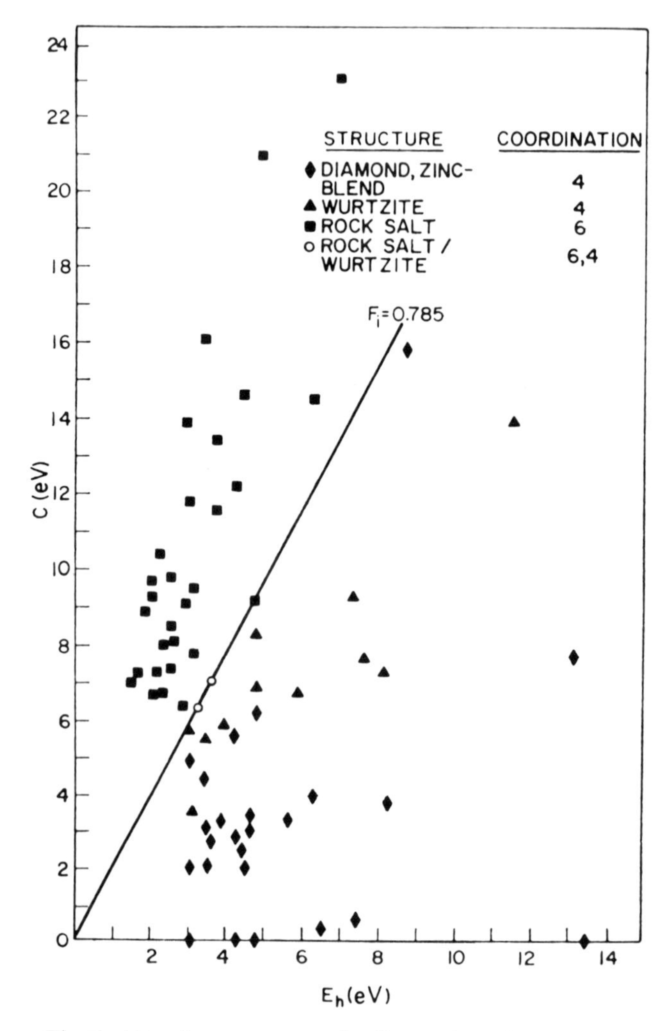

Fig. 3. AB valence compounds: Separation of compounds with the rocksalt structure (CN 6–6) from those with the diamond, sphalerite or wurtzite structures (CN 4–4) on a diagram of the spectroscopically defined covalent and ionic energy band gaps, E_h and C. After Phillips and Van Vechten (see Ref. 14).

measurements on some 70 AB-type valence compounds with the sphalerite or rocksalt structures, excluding those of the rare earths and transition metals. Any partially ionic substance must have a covalent as well as an ionic component to its bonding, and this was recognized as giving rise to a covalent and an ionic contribution to the spectroscopically derived energy band gap. In their relationship $E_g^2 = E_h^2 + C^2$, E_g is the energy band gap between the bonding and antibonding states, E_h ($\propto a^{-2.5}$, where a is the cubic cell edge for the AB compound with either the rocksalt or sphalerite structure) is the part of the energy band gap due to covalent bonding, and C is that part due to ionic charge transfer. The values of these quantities were derived from absorption spectra of the AB crystals. The quantity C amounts to a spectroscopically defined electronegativity difference for the compound AB, but it does not express the electronegativities of the individual atoms A and B. The spectroscopic ionicity of the compound AB is defined as $f_i = C^2/E_g^2$, and the covalent fraction of the bond as $f_c = E_h^2/E_g^2$.

Phillips and Van Vechten showed that there is a unique separation of AB compounds with the rocksalt structure (6–6 coordination) from those with tetrahedral (4–4) coordination in the diamond, sphalerite, or wurtzite structures in a diagram of C versus E_h (Figure 3). The straight line separating the two groups of structures passes through $E_h = C = 0$, and its slope corresponds to an ionicity value $f_i = 0.785 \pm 0.010$. The misfortune accompanying this analysis is that all possible AB compounds of the class are now known. If they were not, it is clear that valence considerations would establish a new compound as belonging to the class, and the results of absorption spectra measurements would prescribe the type of crystal structure that it would adopt with such certainty that an X-ray study would be unnecessary, unless the compound lay very close to the line separating the two coordination groups.

Although such studies have been performed only for this group of AB compounds, a forerunner was the $\bar{n}$ versus Δx diagram of Mooser and Pearson,[15] which was intended to separate valence compounds according to their tendency to dehybridization on the one hand, and ionicity on the other. The third parameter R_C/R_A was conveniently neglected since its influence would only be apparent in extreme cases of size disparity (using the covalent radius scale). These diagrams separated compounds with structures having high coordination numbers from those with low coordination numbers, the former being generally ionic or metallic with nondirectional bonds, and the latter being generally saturated covalent structures in which the bonds were directional. Such phenomenological studies (see, e.g., Figure 4) lacked the quantitative strength of Phillips and Van Vechten's analysis, but had the advantage that they could be applied readily to transition metal and nontransition metal compounds and to

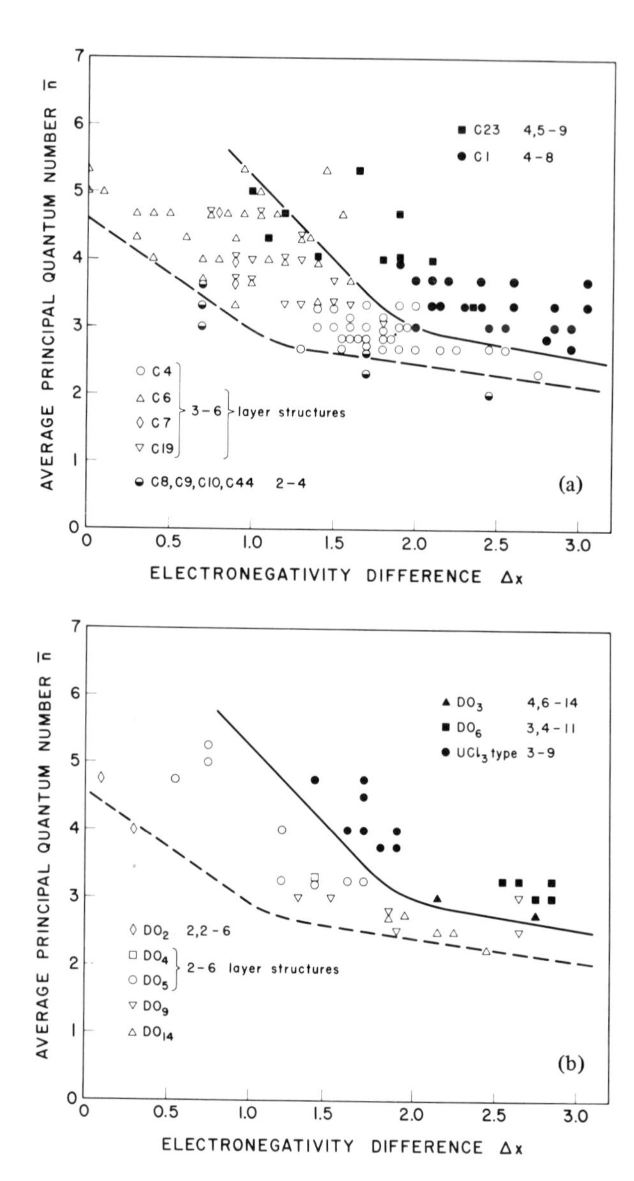

Fig. 4. (a) AB_2 and (b) AB_3 valence compounds shown on an $\bar{n}$ versus Δx diagram. Symbols indicate the different crystal structures given as *Strukturbericht* types, together with the near-neighbor coordination. After Mooser and Pearson.[15]

valence compounds of any compositions: AB, AB_2, A_2B_3, $A'A_2B_4$, etc. At the time when they were produced they did show without doubt that structural features were linked to the relative ionicities and the extent of dehybridization of the chemical bonds that valence compounds contained. This in itself was an advance in understanding structural stability because until that time the only general principles that had been firmly established (apart from the connection between electron concentration and structure in some alloys and some general geometric principles in the structures of metals) were the importance of directed chemical bonds, and of radius ratios and Madelung constants in controlling ionic structures.

4. Atomic Size and Structural Constraint

The previous sections suggest that a thorough understanding of atomic size and the roles that it plays in relation to structural stability is most important. It is not so much the numerical value of the atomic diameter that is generally of importance, but the state of the atom that gives rise to its apparent size. This is because the state of the atom prescribes how the atom will react in any particular structural constraint and whether or not its apparent size will exert any influence.

Thus atoms which have achieved filled valence subshells either by electron sharing or by ionic transfer are prevented from approaching each other closely because overlap of the tails of the electron wave functions causes the development of repulsive forces that keep the atoms apart. Uncharged atoms with filled valence subshells are held together by fluctuating dipolar or higher-order forces and their apparent diameters are known as van der Waals diameters. Charged ions similarly have filled valence subshells, but in this case the attraction of opposite ionic charges results in the ions approaching each other somewhat more closely than the sum of the van der Waals radii of the atoms. Conversely, neighboring ions of like charge encounter added repulsion which imposes the structural constraint that they cannot come in contact. Such atoms therefore behave much like rigid billiard balls and size is a feature that is very meaningful in their crystal structures since it restricts the range of radius ratios of oppositely charged ions beyond which a given coordination polyhedron cannot be stable.

Atoms which have incompletely filled valence subshells normally achieve chemical bonding by the overlap of their atomic wave functions, as in covalent compounds and metals. Atoms in this state approach each other much more closely than they could if they had filled valence subshells, so that covalent or metallic radii are smaller than van der Waals or negative-ion radii. In substances where valence rules can be satisfied by the formation of unit or multiple strength covalent bonds and appropriate coordination to satisfy the valence requirements can be attained, relative atomic sizes do not introduce constraints in the selection of structure type as they do in ionic crystals. In these compounds electron sharing to satisfy valence rules generally results in the atoms attaining filled valence subshells, so that in the nonbonded directions they do not approach each other more closely than their van der Waals diameters permit. Not infrequently in such compounds the formation of covalent bonds also results in some charge separation, so that in addition to the attraction of unlike atoms by the shared electron-pair bonds there is an additional ionic attraction between them. This results in even closer atomic approaches than those expected from the covalent bonding alone.

In metals the atoms frequently have vacant atomic orbitals in addition to those containing electrons, and furthermore, they are generally surrounded by far more neighbors than can be held by shared-pair bonds of unit strength. Therefore the atoms are generally separated by distances considerably greater than those for unit strength bonds, and in the direction of any particular ligand the atomic wave functions overlap to a considerably lesser extent than if the two atoms were to form a covalent bond of unit strength. Thus there is considerable latitude in how closely any particular pair of atoms may approach each other without generating a strongly repulsive interaction. It is for this reason that many different interatomic distances are frequently found in the structures of metals and alloys. The metallic atom behaves as a rather soft rubber ball that can be variously dented in the directions of different ligands, apparently without difficulty. Thus in truly metallic structures it is the average atomic volume, rather than the individual interatomic distances, which appears to be important.[16] When atoms approach each other more closely than the bond strength for the particular coordination would demand, it is not necessarily to be interpreted as the result of increased bonding in these directions, since it may arise trivially in order that effective contacts between ligands in other directions can occur. This behavior is by no means uncommon; for example, it occurs in most of the 500 AB_2 phases known to take a form of the Laves structures (see p. 142). Such compression of the atoms can occur provided that the energy penalty paid for it is not larger than the energy advantage of effectively achieving the large number of contacts (high coordination). Similar effects are also found in some structures whose stability derives from the satisfaction of particular chemical bonds.

Thus it can be envisaged that the cell parameters (dimensions) are the result not so much of the *a priori* diameters of the atoms as of the overall atomic volume and satisfaction of geometric or bonding arrangements and contacts in metallic or nonionic valence structures. Apparent atomic diameter is in large measure the result, rather than the cause of factors that give rise to structural stability, and certain types of metallic structures thus introduce a new concept of atomic size. Indeed, such concepts suggest that atomic volume would be a much more satisfactory parameter than interatomic distances in metals. Use of atomic volume, however, runs into trouble when considering alloys formed with metalloids lying to the right of the Zintl border which attain a filled valence subshell by electron sharing and/or transfer (Ref. 3, p. 143). Consider, for example, selenium. Its elemental volume results from an atom which has two ligands at single-bond distances and four more neighbors at distances approaching those of the van der Waals diameter, whereas the atomic volume of Se in $PtSe_2$, with the CdI_2 structure, results from Se that has three unit strength bonds and three Se neighbors approximately at van der Waals distances. In $NiSe_2$, with the pyrite structure, the atomic volume of Se would represent an atom that has four unit strength bonds and no Se neighbors at van der Waals distances. The atomic volume of Se in each of these three situations differs greatly, decreasing as the atom forms more unit strength bonds, directing its filled valence subshell to fewer and fewer Se neighbors.

Arguments involving exact details of what should be a metallic atom's size appear to be unimportant in questions of structural stability. Any reasonable self-consistent set of radii is sufficient for discussion, and Pauling's $R_{(1)} - R_{(n)} = 0.3 \log n$ rule[17] can be used to convert radii appropriate for one coordination number to another. Regardless of the failings of any particular set of metallic radii, Pauling's rule appears to be very reliable.

One feature of atomic size that must be recognized is the effect of crystal field stabilization energy on the diameters of the transition metal atoms,[18–21] and to a much lesser extent on the sizes of the rare earths. This may cause as much as 25% reduction in radius compared to the radius expected from normal Coulombic contraction on proceeding across a transition metal period. The size of the crystal field stabilization energy depends on the symmetry of the anion coordination polyhedron surrounding the transition metal atom, on the number of d electrons, and on their configuration in high- or low-spin states if applicable (d^0, high-spin d^5, and d^{10} states do not experience this stabilization energy). Crystal field stabilization energy is found not only in ionic compounds such as the transition metal monoxides with the rocksalt structure, but also in sulfides, selenides, arsenides, and antimonides that are not strongly ionic and which are either semiconductors or metals.[21] Thus the consequences of crystal field stabilization energy on transition metal atom sizes appear not to be confined only to nonmetallic alloys. Furthermore, the effects are so diverse, depending on the *d*-electron number (element valence), spin state, and coordination polyhedron, that it is impossible to give any simple system of transition metal sizes that would be appropriate to all such compounds.

The critical diameter D_c for change from localized to collective behavior of the *d* electrons in transition metal compounds is also an important parameter that affects properties and structure, as discussed by Goodenough (Ref. 22, pp. 26–28, 265–266, 295–297).

The effect of relative atomic size on alloy structures was discussed on p. 134. It also influences the extent of terminal solid solubility in metallic alloys as expressed in the Hume-Rothery 15% rule[23,24] and its extension by Darken and Gurry[25] to also take account of the effect of electronegativity difference in limiting solid solubility. The Hume-Rothery rule has now come to be stated that if the sizes of the solute and solvent atoms differ by less than 15%, extended terminal solid solutions may form, whereas if they differ by more, it is very unlikely that extended solid solutions will form. The rule is only permissive for the formation of extended solid solutions at radii differing by less than 15%, since other factors such as, for example, large electronegativity difference may still prevent their forming.

5. Factors Influencing the Stability of Crystal Structures

5.1. Electrochemical Factor

When there is a large difference in the electronegativity of two components of a binary alloy there is a strong tendency for the formation of very stable compounds in which valence rules are satisfied. This electrochemical factor is the strongest influence controlling alloy structure, dominating other influences such as energy band and geometric factors.

The effect of a strong electrochemical factor in an alloy system depends on the electron concentration, as Hume-Rothery has pointed out.[26] When the electron concentration is high it leads to the formation of valence compounds of great stability and narrow range of homogeneity, but when the electron concentration is low as, for example, in Au–Mg or Au–Cd alloys, an increase in electrochemical factor leads to an increase in both stability and range of homogeneity of intermediate phases. In either case, however, a large electrochemical factor leads to restricted terminal solid solubility; that is to say, in general the stability of random atomic arrangements is low. This influence is apparent in the Darken–Gurry examination of the Hume-Rothery 15% rule (p. 136). The partial breakdown of the Hume-Rothery electron concentration rules in alloys of Cu, Ag, and Au with the group VB elements As, Sb, and Bi, where the electrochemical factor is much larger than in the alloys with group IIB to IVB elements —where the rules hold well—is evidence of its domination over energy band effects.

5.2. Geometric Effects

For many years the importance of close packing, in which space is filled with tetrahedra and octahedra in the ratio of two to one (tto) has been recognized in the structures of metals. Such arrangements are generated by the stacking one above the other of triangular (3^6)* nets of atoms at the appropriate spacing. Much more recently there has been an increasing awareness of the importance of tetrahedral close packing (tcp), which, if ideally realizable, would be a closer packing than tto. However, space cannot be filled with regular tetrahedra, and most tcp structures that are known are generated by interpenetrating icosahedra and CN 14, 15, or 16 polyhedra which have triangulated surfaces (Figure 5). All atoms comprising an icosahedron have surface coordination number (SCN) five. The CN 14, 15, and 16 polyhedra have respectively two, three, and four atoms with SCN six, the remaining atoms having SCN five. Many of these tcp structures are built up of pentagon–triangle, pentagon–hexagon–triangle, or hexagon–triangle layers of atoms interleaved by triangular, triangle–square, or square layers of atoms which lie over the centers of the pentagons or hexagons of the main layers. Indeed, most of the structures of metals can be described according to the stacking of various layer nets of atoms. This is no doubt because it is a means of generating the simplest coordination polyhedra: tetrahedra, octahedra, triangular prisms, cubes, anticubes or Archimedian antiprisms, pentagonal

* Schläffli notation; see Ref. 27.

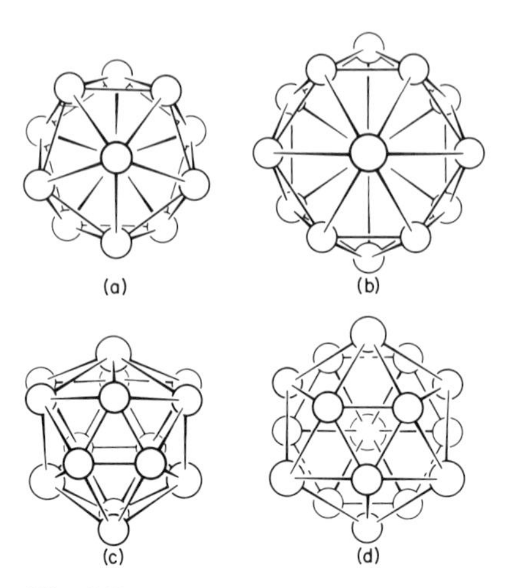

Fig. 5. Representation of (a) the icosahedron and (b–d) CN 14–16 polyhedra with triangulated surfaces.

prisms and antiprisms, and hexagonal prisms and antiprisms. The relative sizes of the atoms generally play an important role in controlling site occupation in tcp structures.

Apart from any advantages accruing from high coordination and close packing, very little is known of the basis of geometric arrangement as a source of structural stability. Close-packed structures generally occur when the concentration of electrons in partly filled sheets of Fermi surface is quite low, probably less than two electrons per atom. When it is higher—from two to four electrons per atom—directed chemical bonds play a role in the structural arrangement and the structures are no longer close packed. Many of these structures, particularly those of valence compounds (semiconductors and insulators) are generated by inserting further triangular layers of atoms at one-fourth, one-half, and/or three-fourths of the spacing between "close-packed" layers.

One of the major difficulties in assessing the importance of geometric effects on a quantitative basis is the lack of precise knowledge of the appropriate sizes of the atoms in a metallic alloy and the observation that the ligands of an atom in a complex metallic structure generally surround it at several different distances. Certainly any model regarding metallic atoms as incompressible billiard balls is quite inappropriate!

The precisely known parameters respecting any crystal structure which has been accurately determined are the arrangement of the atoms within the structure—the geometry of the structure—and the unit cell size and atomic parameters (if any) of phases which take the particular structure. Known much less certainly are the absolute and relative sizes of the atoms of phases that take the structure, data which are needed to decide whether or not the atoms come in contact and to determine how the unit-cell parameters depend on the relative sizes of the atoms. Nevertheless, such are the facts that must be assessed in any analysis of a geometric basis for structural stability.

The geometry of any particular structure adopted by a binary phase A_aB_b can be represented as a diagram composed of straight lines, one for every independent interatomic distance in the structure.[28] The diagram is drawn with a reduced strain parameter $(D_A - d_A)/D_B$ as ordinate and the diameter ratio D_A/D_B of the two component atoms as abscissa. Here D_A and D_B are the diameters of the two components for the (average) coordination in the structure, and d_A is the distance between an arbitrarily chosen pair of A atoms in the structure. Actual phases with the structure can be represented on the diagram by individual points by determining the strain parameter values from the measured cell dimensions (and atomic parameters) of the phases. From the distribution of these points on the diagram, factors governing the stability of the structure and controlling the unit-cell dimensions can be deduced. The assumed sizes of the component atoms enter these considerations in establishing the extent to which an atomic contact may have been made; however, insofar as values of the strain parameter and radius ratio are coupled, an error in the assumed radius of one component may not greatly influence the conclusions to be drawn from the distribution of points for actual phases on the diagram. The derivation of these diagrams and the effects of variable atomic parameters or unit-cell axial ratios are discussed in the Appendix, p. 171.

When the point for a phase lies on a particular contact line on the diagram the atoms just come in contact according to the sum of their assumed radii. When it lies below the line contact has not been made, and when it lies above the line that particular contact is compressed according to the assumed sizes of the atoms (i.e., they are closer than the sum of their radii for the appropriate coordination). Thus, assuming the radius ratio for any pair of atoms, it is clear that the cell dimensions of a phase formed by them are free parameters that determine which contacts are not made, just made, or compressed in a particular structure.

The characteristic feature of a structure that is controlled by a particular chemical bond A–B is that the points for all known phases with the structure lie along the line for that particular bond contact. Thus, all known phases with the sphalerite structure lie along the line for 4–4 A–B contacts (Figure 6) so that the A–B contacts control the unit-cell dimensions.[28] Indeed, it is well known that phases with this structure owe their stability to the adamantine bond distribution. Less well known is that phases with the $AsNa_3$ structure (Ref. 3, pp. 395–398) owe their stability to 2 + 6 As–Na bonds as shown by the structure diagram (Figure 7), so that As is surrounded by eight Na atoms which satisfy the requirements of two sets of pivotally resonating sp^3 bonds.[28] Such bonding is consistent with the semiconducting properties of the compounds (attainment of a filled valence subshell by As)

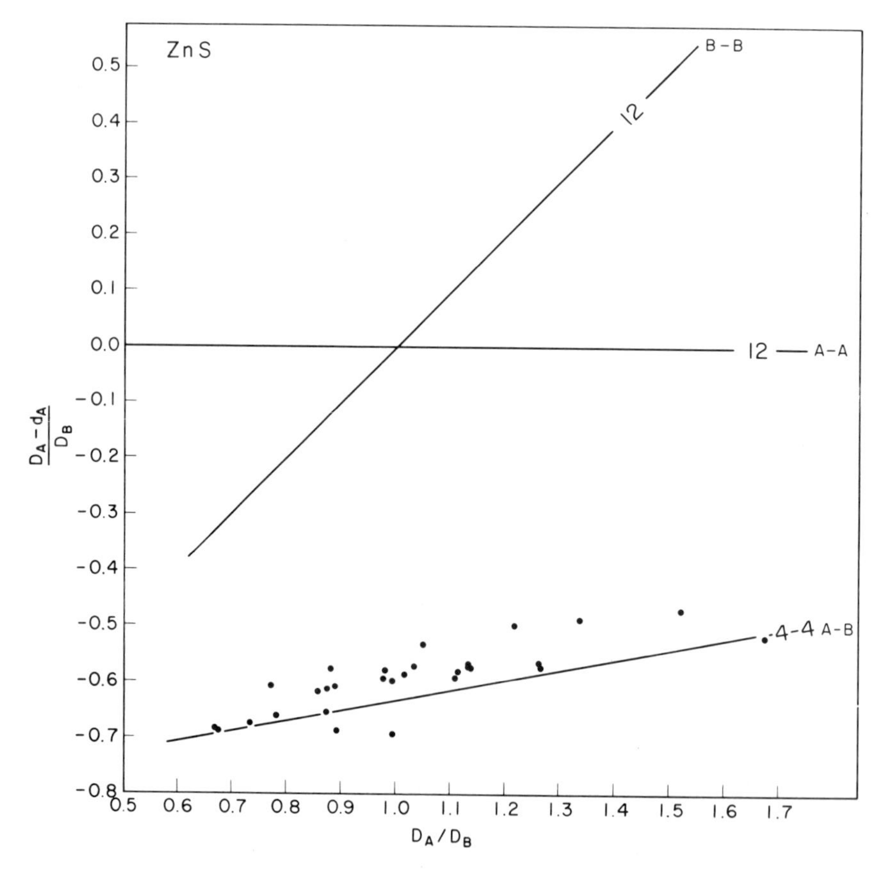

Fig. 6. Structure diagram for phases with the sphalerite structure. After Pearson.[28]

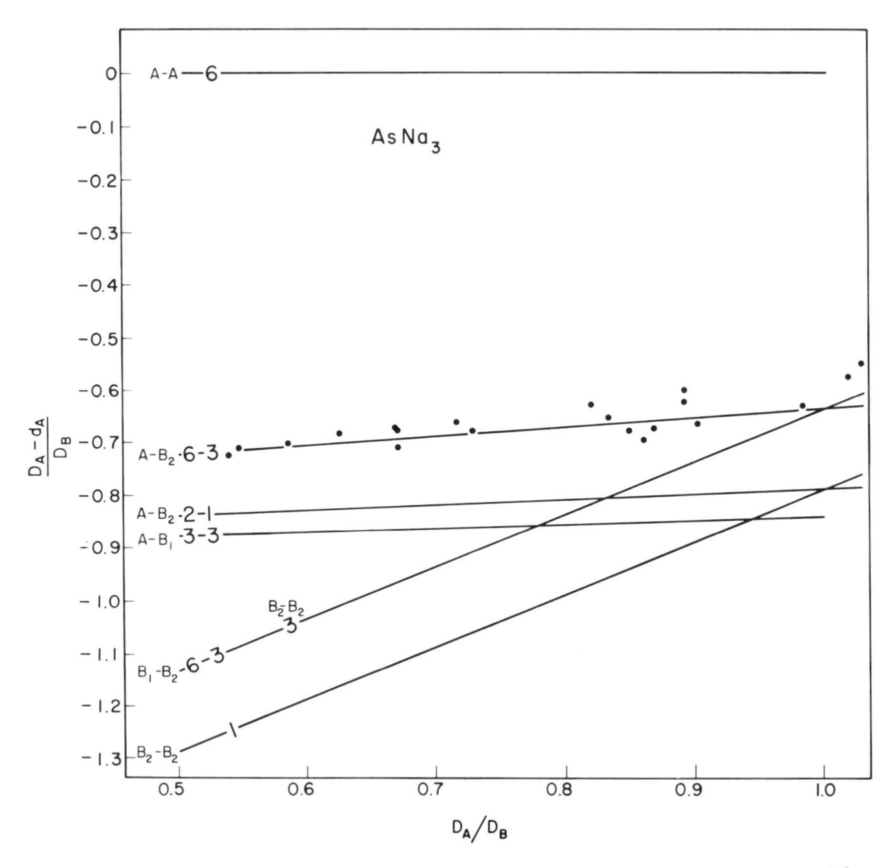

Fig. 7. Structure diagram for phases with the $AsNa_3$ (AB_3) structure. Constructed for axial ratio $c/a = 1.78$ and atomic parameter $x_{Na} = \frac{5}{12}$. Numbers indicate number of A–B, etc. contacts. After Pearson.[28]

and with the exceedingly constant axial ratio (1.79 ± 0.01) of some 20 substances that take the structure. This constancy results from the angular rigidity of the sp^3 bond (as in the wurtzite structure with tetrahedral coordination, compared, say, to the NiAs structure with octahedral coordination of As about Ni). In achieving these contacts, however, three A–B contacts and several B–B contacts are considerably compressed. Nevertheless, the distribution of these phases, despite the different radius ratios of the component atoms, leaves little doubt that it is the six As–Na bonds, particularly, which control the unit cell dimensions of phases with this structure. The compressed contacts are not to be regarded as evidence for strong bonding; rather, they are a trivial consequence of making the other contacts. If this were not so, it is probable that substances with this structure would be metallic rather than semiconductors.

In the case of a structure said to form for geometric reasons it is expected that phases would be distributed on the structure diagram in such a way as best to satisfy the required (high) coordination. Thus in the complex tcp structure of σ phases (Ref. 3, pp. 673–676), whose stability depends on the effective achievement of the 12, 14, and 15 CN of the atoms, the structure diagram consists of many lines which cross at radius ratios between 1.00 and 1.10 (Figure 8). Points for phases with the structure lie in this region of the diagram, showing that the coordination is indeed satisfied to the best extent possible (Ref. 3, p. 55); there are very few σ phases formed by elements whose radius ratios lie outside these limits, so that effective attainment of the coordination places limits on the relative sizes of the component atoms that can form phases adopting the structure. Not all tcp structures have structure diagrams that are as complex as that of the σ phase. The structure diagrams for the $MgCu_2$ and ideal $MgZn_2$ Laves phases (AB_2), for example, only contain three important contact lines.

The Laves phase structures are built up of interpenetrating icosahedra and Friauf polyhedra (CN 12 and 16).[3] The phases were always believed to form for geometric reasons since at a radius ratio

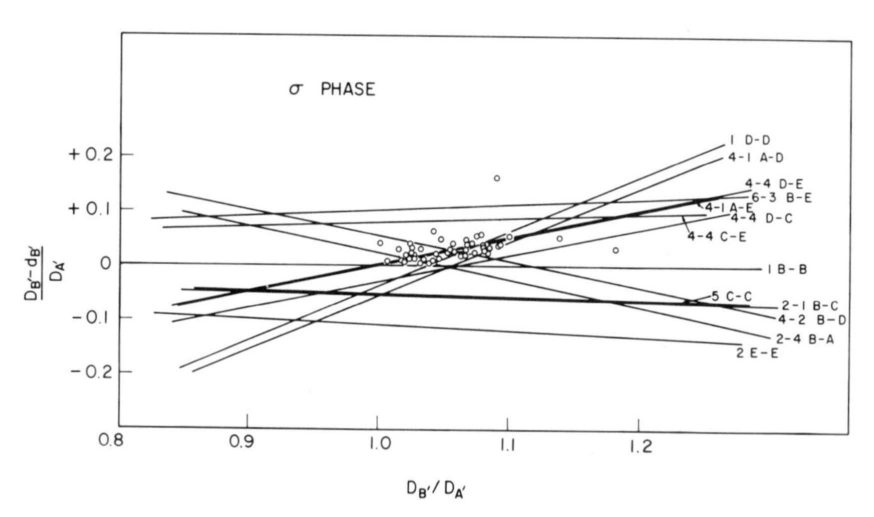

Fig. 8. Structure diagram for phases with the σ phase structure. After Pearson.[3] A to E represent the five different sitesets found in the structure. A refers to positions 2(*a*), B to 4(*g*), C to 8(*i*), D to 8(*i*), and E to 8(*j*). A′ and B′ designate the component atoms. Constructed for axial ratio and atomic parameters of σ-FeCr. Numbers indicate the number of D–D, A–D, etc. contacts.

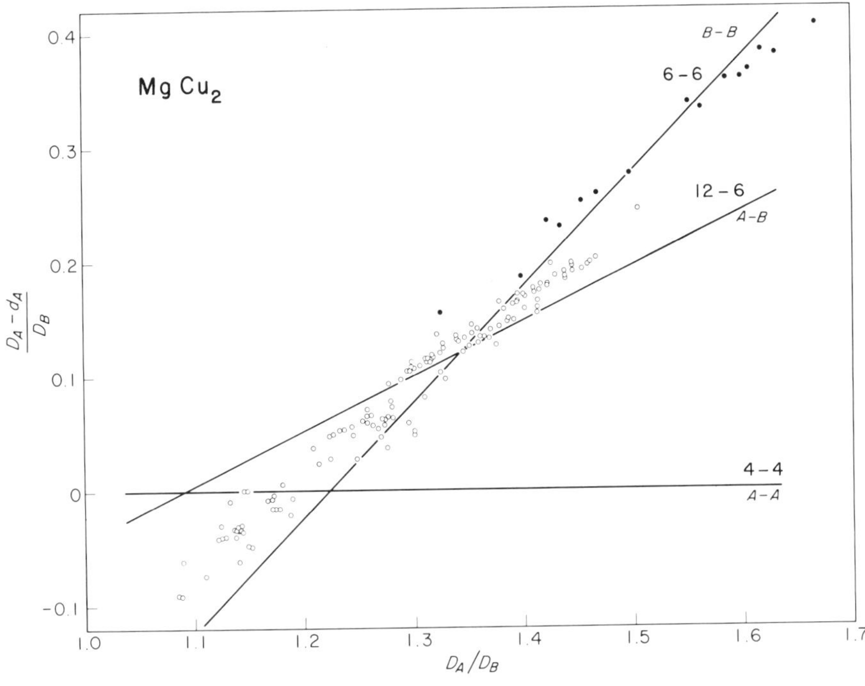

Fig. 9. Structure diagram for phases with the $MgCu_2$ (AB_2) structure. After Pearson.[29] Numbers indicate the number of A–A, etc. contacts.

$R_A/R_B = 1.225$, the volume of two B atoms equals that of one A atom and hence a good space filling could be obtained. In fact, Laves phases with the $MgCu_2$ structure are known to occur at radius ratios varying from 1.10 to 1.67, so it appears that their stability rests on more than this simple fact. The 16–12 coordination of the structure is made up of 4–4 A–A contacts, 12–6 A–B contacts, and 6–6 B–B contacts. The structure diagram (Figure 9) shows that at radius ratios greater than 1.225, the 4–4 A–A contacts have to be compressed in order that the 12–6 A–B and 6–6 B–B contacts can be achieved.[28,29] Indeed, at high radius ratios the compression of the A–A contacts is considerable, but the energy penalty for this compression is clearly exceeded by that gained through achieving the 16–12 coordination, and so structures formed by components with radius ratios as high as 1.67 are known.

The asymmetry of distribution of phases about the so-called ideal radius ratio of 1.225 is apparent from Figure 9. Far fewer phases occur at lower radius ratios and furthermore, the atoms are not compressed sufficiently for contacts in the 16–12 coordination to be formed effectively. This is because at radius ratios less than 1.225 it is the six B–B rather than the four A–A contacts that have to be compressed in order to achieve 16–12 coordination. However, there are twice as many B as A atoms in the structures, and six rather than four contacts have to be compressed. The energy penalty paid for compressing two-thirds of the atoms in the structure is not recovered by the achievement of 16–12 coordination, and at radius ratios a little less than 1.225 the four A–A and the 12–6 A–B contacts are not fully made, and below a radius ratio of 1.10 no further phases take the $MgCu_2$ structure.

These structure diagrams permit a good description of the occurrence and cell dimensions of the Laves phases to be given, and they show in semiquantitative fashion that a geometric effect giving rise to structural stability so controls the unit-cell dimensions of phases with the structure that all of the contacts required by the particular coordination are effectively made. Furthermore, they show that this process can lead to structural stability over a very wide range of radius ratios of the component atoms, provided that the energy penalty paid for compressing any contacts that may be necessary to achieve the coordination is not too high. Similar behavior is found for

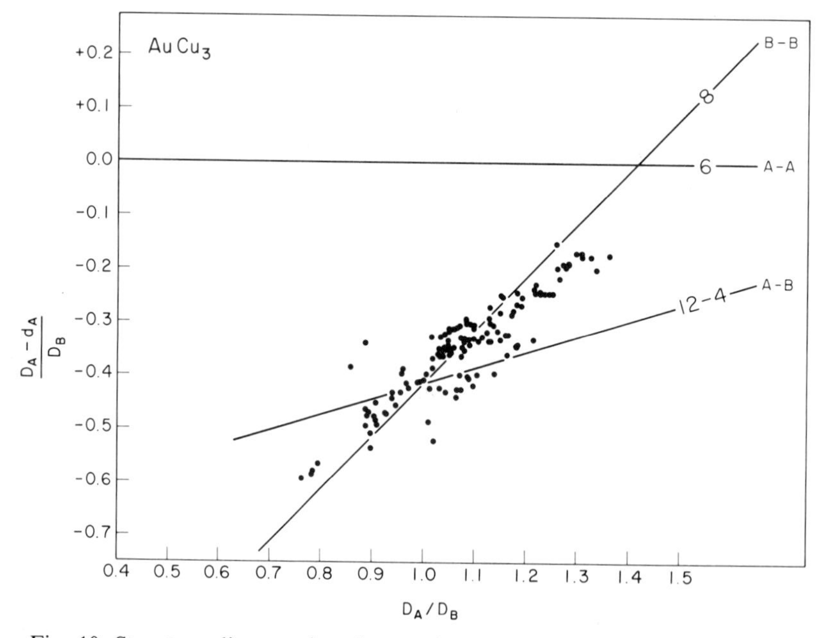

Fig. 10. Structure diagram for phases with the $AuCu_3$ (AB_3) structure. After Pearson.[28] Numbers indicate the number of A–A, etc. contacts.

tto close-packed structures, as indicated in Figure 10, which gives the structure diagram for phases with the $AuCu_3$ superstructure.

I have already emphasized our rather weak basis for understanding why so-called geometric effects in structures should lead to low values of the free energy and therefore to structural stability, except insofar as they are manifestations of tcp or tto close packing, which are the closest possible packings of atoms and therefore give low values for the structure-independent contribution to the crystal energy. However, several facts do stand out clearly. In fairly complex coordination structures, phases are generally only formed from atoms whose relative sizes allow effective achievement of the high coordination. Not all structures in which a high coordination is possible are necessarily formed for this reason. The α-$ThSi_2$ structure (Ref. 3,

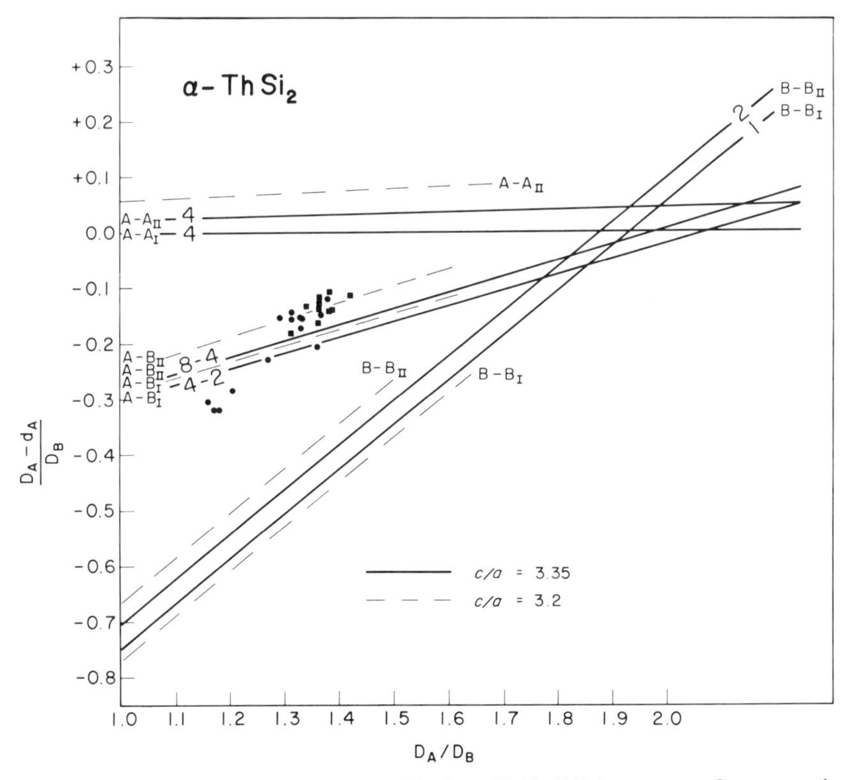

Fig. 11. Structure diagram for phases with the α-$ThSi_2$ (AB_2) structure. Constructed for the axial ratios indicated and the ideal value of the atomic parameter $x_{Si} = \frac{5}{12}$. Numbers indicate number of A–B, etc. contacts: Squares indicate phases with defect $ThSi_2$ structures. After Pearson.[28]

pp. 518–520), for example, if formed from components with radius ratios from 1.85 to 2.10 would, with the correct cell dimensions, give 20–9 coordination, but no phases with the structure have such radius ratios.[28] Instead, they are found to have radius ratios from about 1.15 to 1.40, and it is clear from the structure diagram (Figure 11) that A–B chemical bonding controls the overall structural dimensions and structural stability.[28] One reason why the high coordination does not control this structure is probably because it does not result from either tcp or tto close packing. On the other hand, in simpler tcp or tto close-packed structures where the maintenance of high coordination is possible over a wide range of radius ratios by adjustment of cell dimensions, sufficient lowering of the free energy can be achieved for the phases to be stable.

5.3. Energy Band Effects

Since changes in electron concentration result in changes in the numbers of electrons about the Fermi surface, they are expected in the general case only to influence the energies of a small fraction of the outer or valence electrons, while leaving the energies of the main body of electrons, which are responsible for cohesion, largely unchanged. It is for this reason that electron concentration effects are generally expected to have only a small influence on the free energy of a phase or structure and therefore its stability, compared to cohesion or bonding, which is expected to have an effect that is at least an order of magnitude larger. Thus the interactions of Fermi surface and planes of energy discontinuity are, with the exception of the case noted below, not going to have a very big influence on structural free energies compared to that resulting from the overall near-neighbor coordination, even though electron concentration is a factor entering into the stability of any phase with a Fermi surface, i.e., any metal.

Realization of these facts prescribes the conditions under which structural change as a function of electron concentration might be studied. It is essential to find a range of structures in which the near-neighbor coordination does not change from structure to structure. Only in this way can the main contribution to the free energy be kept essentially constant as the influence of the interaction of various planes of energy discontinuity with the Fermi surface is examined when crystal structures change with changing electron concentration. Two such systems of structures have been recognized and studied. In one the introduction of antiphase domain boundaries "of the first kind" which do not change the overall near-neighbor concentration in ordered AB or AB_3 structures generates planes of energy discontinuity in reciprocal space which can be so adjusted as to follow a Fermi surface expanding or contracting with increasing or decreasing electron concentration. In the other the regular introduction of planar stacking faults in ordered close-packed AB_3 structures, where the close-packed planes individually have AB_3 stoichiometry, creates families of polytypic structures in all of which the near-neighbor coordination remains unchanged. Such structures involving changes in second or third nearest neighbors introduce changes in the planes of energy discontinuity about the Fermi surface which allow quantitative correlation of structural change with electron concentration.

The effect of a plane of energy discontinuity at or close to the Fermi surface is to lower the electron energies relative to those expected from the free-electron model. Figure 12 shows the parabolic E versus **k** relationship for free electrons. Introduction of a plane of energy discontinuity lowers the electron energies at **k** vectors less than that of the plane and raises them at **k** vectors higher, as shown by the broken line in Figure 12. If therefore the Fermi surface extends nearly to, or makes contact with, the plane of energy discontinuity, the electron energies are lowered, and if the Fermi surface contains

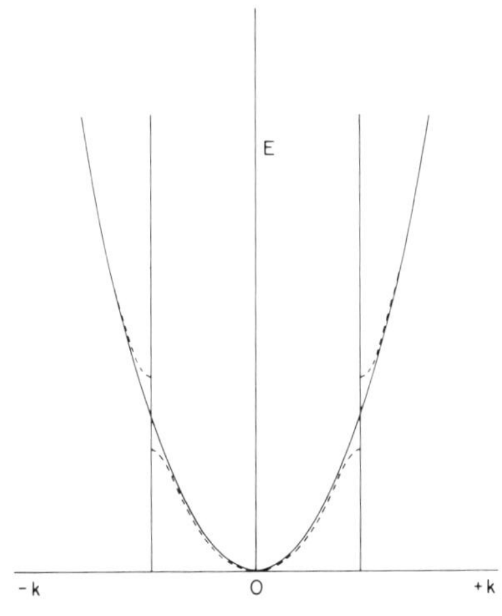

Fig. 12. Parabolic E versus **k** relationship for free electrons (full line), showing the effect of planes of energy discontinuity (vertical lines) on the electron energies (broken line).

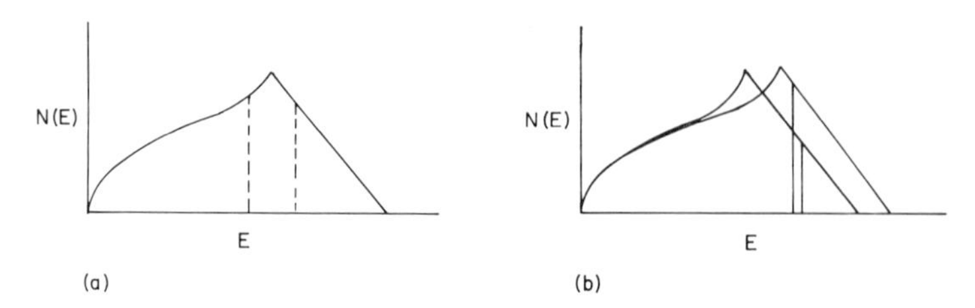

Fig. 13. (a) Representative form of an $N(E)$ curve. The peak corresponds to the Brillouin zone planes touching the Fermi surface. (b) Representative $N(E)$ curves for two different structures showing the different energies of the Fermi level for a phase with the same electron concentration in each structure.

large flat areas which are surrounded or touched by planes of energy discontinuity, the electron energies are lowered along **k** vectors covering a large fraction of the solid angle, the total effect being to give considerably added stability to the structure. To be effective, such energy discontinuity planes need not constitute a specific Brillouin zone, but assuming that they do, the degree of filling of the zone might correspond to Fermi energies located between the two broken lines in Figure 13(a) if the maximum stabilizing effect is to be achieved. When the Brillouin zone is filled up further, after the Fermi surface has made contact with the zone planes, fewer and fewer states are available and the density of states must decrease if the energy band gaps across the planes are high enough to prevent overlap into the succeeding zone so that the electrons have to assume increasingly higher energies as the electron concentration is increased. Eventually some other structure with a higher density of states becomes more stable, since it can accommodate the valence electrons at a lower total energy, as indicated in Figure 13(b). The succession of Brillouin zones (first, second, third, etc.) for any structure is prescribed by the crystal symmetry, there being only 14 different families of Brillouin zones corresponding to the 14 Bravais lattices in real space. However, the atoms which occupy the structure determine both the electron concentration and the atomic potentials that control the size and distribution of the energy band gaps across the Brillouin zone planes, so that structural stability depends directly on the properties of the atoms in the structure.

Thus energy bands influence relative structural stability in two ways: (i) by the lowering of electron energies by planes of energy discontinuity close to or just touching the Fermi surface, stability is increased, and (ii) by raising the electron energies as a Brillouin zone becomes more and more filled, without overlap taking place into the succeeding zone, stability is decreased relative to other structures. Nevertheless, it is emphasized that the influence of both of these effects on the free energy G of the phase is generally small compared to the influence of the near-neighbor "bonding."

5.3.1. Lowering Electron Energies by Planes of Energy Discontinuity

It was Sato and Toth[30,31] who showed that when low-energy imperfections such as antiphase domain boundaries were introduced into an ordered structure without changing the near-neighbor coordination, the positions of (some of) the Brillouin zone boundaries were altered so that they followed an expanding Fermi surface and maintained structural stability despite the increase in electron concentration due to alloying. Consider, for example, the ordered AuCu I structure. The reduced Brillouin zone is made up of {100} planes and the second extended zone is made up of (002) and {110}-type planes. Mapped back in the reduced zone, the second zone has a square cross section normal to the k_z axis. With two electrons per primitive cell, the Fermi surface overlaps the {001} planes of the first zone and touches the {110} planes of the second zone.

Brillouin zone planes can be drawn in suitably scaled reciprocal space by joining the appropriate reciprocal lattice points to the origin and creating the plane normal to this line through its midpoint. Thus projections of {110} Brillouin zone planes are obtained in a section through the $k_x k_y$ plane in Figure 14. Creation of a one-dimensional long-period superstructure by introducing an out-of-step shift of $\frac{1}{2}(\mathbf{a} + \mathbf{c})$ in the (010) plane of every M AuCu I pseudocells along the b direction would result in splitting the {110}-type reciprocal lattice points in direction b of the superperiod. The degree of the splitting would be $x = 1/2M$ in terms of $1/a$, the unit reciprocal lattice parameter. Thus each of the reciprocal lattice {110} energy discontinuity planes in Figure 14 is replaced by two new planes, one passing nearer to the origin, the other farther away. These are derived from the two new inner reciprocal lattice points arising from each {110} point, and the plane's traces are obtained by exactly the same construction as used to draw the trace of the {110} planes (Figure 14). The four outer planes can accommodate a larger Fermi surface than the original {110} planes as the Fermi surface expands with increasing electron concentration of the alloys. By decreasing the period M of

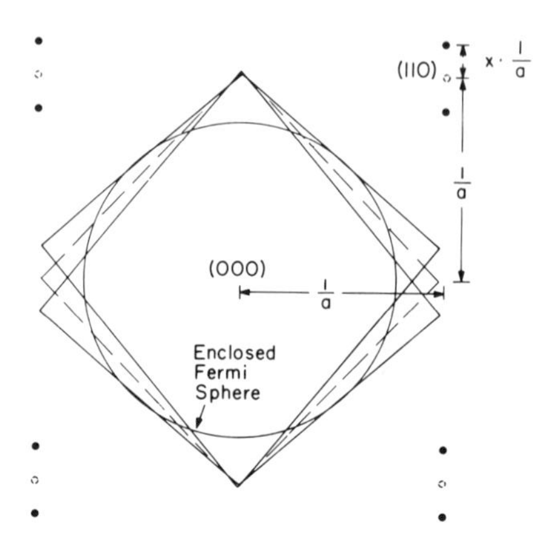

Fig. 14. Reciprocal lattice for the CuAu II structure in a plane through the origin parallel to (001), showing the Fermi sphere and planes of energy discontinuity arising from the splitting of the superstructure reflections. After Sato and Toth.[32] Broken lines indicate the Brillouin zone for CuAu I.

the antiphase domains along the *b* direction, these planes can be further opened up to accommodate a greater expansion of the Fermi surface. Conversely, if the effective electron concentration is decreased below one electron per atom by alloying with certain transition metal elements, the inner set of four planes can be made to follow the contracting Fermi surface by decrease of the antiphase domain period *M*.

Since the electron concentration e/a controls the size of the Fermi sphere, it is possible to relate the splitting x of the reciprocal lattice $\{110\}$ points exactly to the electron concentration which gives a Fermi sphere that would just touch the inner or outer set of $\{110\}$-type planes.[32] The relationship is $e/a = (\pi/t^2 12)(2 \pm 2x + x^2)^{3/2}$, where it has been found necessary to introduce an empirical factor t representing the degree of distortion in the direction of the $\{110\}$ planes that the Fermi surface has from a free-electron sphere. Taking a value of about 0.95 for this factor, it is found that the electron concentration as a function of the domain period *M* follows the relationship rather closely for a series of AuCu alloys substituted with metals of higher valence such as Zn and Al, as shown in Figure 15.[32] Similar results are found for one-dimensional, long-period superstructures in $AuCu_3$ alloys for which, it so happens, an identical relationship relates electron concentration and antiphase domain period. Figure 16

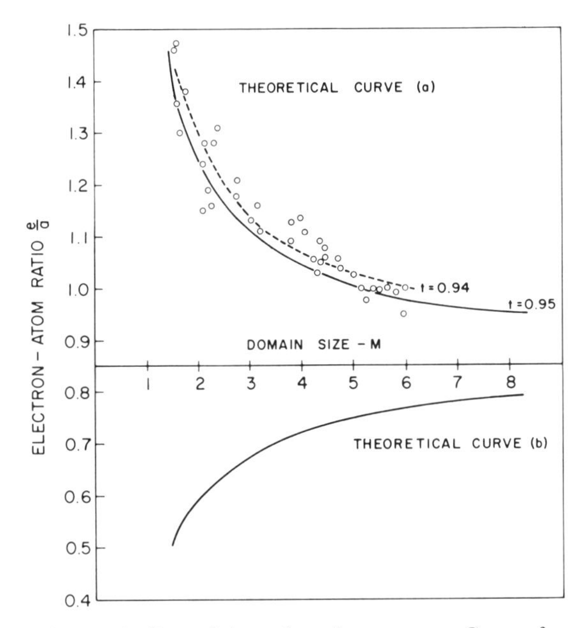

Fig. 15. AuCu antiphase domain structure: Curves for electron atom ratio versus domain size M are derived from the equation given in the text. Circles show experimental data determined by electron diffraction from thin films of alloys. After Sato and Toth.[32]

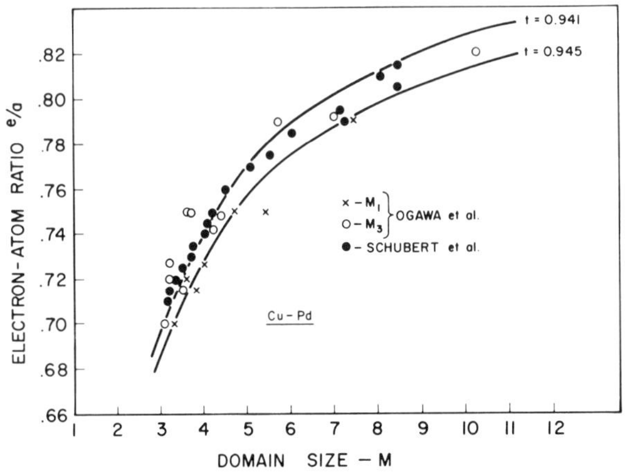

Fig. 16. $AuCu_3$ antiphase domain structure in Cu_3Pd alloys. Curves for electron–atom ratio versus domain size M are derived from the equation given in the text. Points show experimental data determined for Cu–Pd alloys. The electron concentration is calculated on the assumption that Pd contributes no electrons. After Sato and Toth.[33]

shows data for Cu_3Pd alloys as the Cu–Pd ratio is varied[33]; the electron concentration is calculated on the assumption that Pd contributes zero to the effective valence electron concentration. Measurements of the antiphase domain period in such alloys were obtained by electron diffraction from evaporated thin films.

These experiments undoubtedly show that structural stability is maintained by adjusting the period of the antiphase domain boundaries so as to maintain planes of energy discontinuity about the expanding or contracting Fermi surface in a series of alloys. Furthermore, the experimental conditions that were prescribed (p. 146) for such a demonstration to be possible have indeed been maintained—the near-neighbor coordination remained constant throughout the series of structural changes. In the case of AuCu I only two sublattices are required to describe the domain boundaries, and this condition is satisfied in the one-dimensional, long-period superstructure; however, in the $AuCu_3$ structures four sublattices are required, and two-dimensional superstructures characterized by two periods M_1 and M_2 are also possible. The four sublattices result in two kinds of one-dimensional superstructure. That known as the "first kind" is characterized by a step shift $\frac{1}{2}(\mathbf{a}_1 + \mathbf{a}_2)$ which lies in the plane of the antiphase domain boundary, and boundaries of the first kind do not change the relative numbers of near-neighbors of the atoms. Boundaries of the "second kind" are characterized by vectors $\frac{1}{2}(\mathbf{a}_2 + \mathbf{a}_3)$ and $\frac{1}{2}(\mathbf{a}_3 + \mathbf{a}_1)$ which are equivalent, but which do not lie in the plane of the antiphase boundary. In order to create the domain boundary by shifting the crystal along these vectors, it is necessary to "remove a plane of atoms," thus changing the near-neighbor coordination, which requires a larger energy than formation of boundaries of the first kind. For this reason only boundaries of the first kind are found in one-dimensional, long-period superstructures of phases with the $AuCu_3$ structure.

Formation of one-dimensional, long-period antiphase domain structures slightly changes the ratio of the pseudocell edges in the direction of the superperiod ($<1\%$), so that the AuCu antiphase domain structure changes from tetragonal to orthorhombic, and the $AuCu_3$ structures are tetragonal instead of cubic. The observed changes of pseudocell edge a_1'/a_2' are predictably greater or less than unity as the electron concentration is greater or less than unity, and these predictions are also confirmed by experimental observations.[30]

When the value of the antiphase domain period M is small, say one or two, it tends not to respond to changes of electron concentration on alloying. However, structural stability may still be maintained with changing electron concentration by altering the planes of energy discontinuity about the Fermi surface through polytypic structure change, as Sato and Toth have shown.[30] This is brought about by the introduction of planar stacking faults at regular intervals in the sequence of close-packed layers. Thus, for example, the introduction of a stacking fault on every fourth and fifth layer of an ABC sequence of close-packed layers, creates a hexagonal polytype (41 in Ždanov-Beck notation) of the original cubic structure. The near-neighbor coordination in hexagonal or rhombohedral polytypes so created is the same as in the cubic prototype AB_3 structure; change of structure results from changes in second or third nearest neighbors. The different planes of energy discontinuity in reciprocal space resulting from the change of crystal structure may give a better surrounding of an expanding or contracting Fermi surface, so as to keep the electron energies low and maintain phase stability.

Any close-packed arrangement of atoms can be described on hexagonal or orthorhombic axes with the close-packed layers parallel to the basal plane and the c axis perpendicular to them. In the orthorhombic reciprocal lattice the separation of the closest spot to the origin in the c direction [(111) of a fcc array] and the origin itself gives the reciprocal distance between neighboring close-packed planes. In a periodically modulated structure the number of reflections occurring between the origin and the fcc (111) reflection gives the number of close-packed layers (or groups of similarly related layers) L in the repeat sequence (corresponding to the period M of the antiphase domain structures).[34] Even situations, $L = 2, 4, 6, \ldots$, correspond to hexagonal structures and odd $L = 1, 2, 5, \ldots$, to structures with rhombohedral symmetry. The period L may represent a single close-packed layer or it may represent groups of similarly oriented layers. Thus in rhombohedral structures it represents groups of three layers, and in hexagonal structures it may represent one or several layers. For example, the $L = 1$ structure of Sato *et al.*,[34] called $1R^s$, is a six-layer structure, 6R in Ramsdell's notation.[35]

Analysis by Sato and co-workers[34] of reciprocal lattice electron diffraction reflections for modulated structures derived from the $M = 1$ structure (Al_3Ti type) indicates that the volumes included within their Brillouin zones increase in the order of structures $1R^s$, $5H^s$, $3R^s$, and $6H^s$ (6R, 10H, 18R, and 6H in Ramsdell's notation). This is exactly the order of occurrence of modulated structures in Au–Mn alloys with 22–28 at. % Mn. At 22 % Mn the $M = 1$ antiphase domain structure occurs as expected for an electron concentration of 1.45 electrons per atom, assuming Mn is trivalent. As the Mn content and therefore the electron concentration are increased, the modulated structures occur in the order indicated above, although more than one of the polytypes may coexist in any given alloy.

Alloys about Au_3Cd have the $M = 2$ antiphase domain structure (Au_3Zr type). With increasing Cd content (increasing electron concentration) modulated structures (polytypes of the $M = 2$ structure) are obtained in the order $1R^s$, $4H^s$, $6H^s$, and $3R^s$ (12R, 4H, 6H, and 36R in Ramsdell's notation), in agreement with estimates from the electron diffraction patterns of the order of increasing volumes included within their Brillouin zones.[36]

Thus there seems to be little doubt that, provided the near-neighbor coordination can be kept constant, structural stability can be maintained (as electron concentration is increased or decreased by alloying) by the introduction of low-energy defects, such as antiphase domain boundaries of the first kind or planar stacking faults, so as to maintain planes of energy discontinuity about the expanding or contracting Fermi surface and keep the lowered electron energies that result therefrom. The structural changes resulting from the introduction of these imperfections only alter the second or third nearest-neighbor coordination, nearest-neighbor coordination being preserved, so that the major contribution to the free energy of the structures remains essentially constant. In these conditions it is possible to observe systematic structural changes resulting from the relatively small effects on the overall free energy that altering planes of energy discontinuity about the Fermi surface induce. In addition to the semi-quantitative investigations discussed here, many qualitative observations have been made of the change of stacking sequence in ordered AB_3 polytypes (increase of percentage of hexagonally surrounded layers) and of the change of ordering arrangement within the AB_3 layers as the electron concentration changes. Some of these results are summarized in diagrams by Pearson (Ref. 3, pp. 87–90).

The Laves phases can also be considered as a polytypic family of structures with the cubic Cu_2Mg as the prototype. In the appropriate direction, Laves phases can be regarded as made up of four-layer groups composed of three triangular network layers (3^6 in Schläfli symbols) and a Kagomé network layer (3636). These four-layer groups can be successively stacked along a direction normal to them in three positions *A*, *B*, *C* corresponding to the *A*, *B*, *C* sites of conventional close packing. Near-neighbor coordination is identical in all AB_2

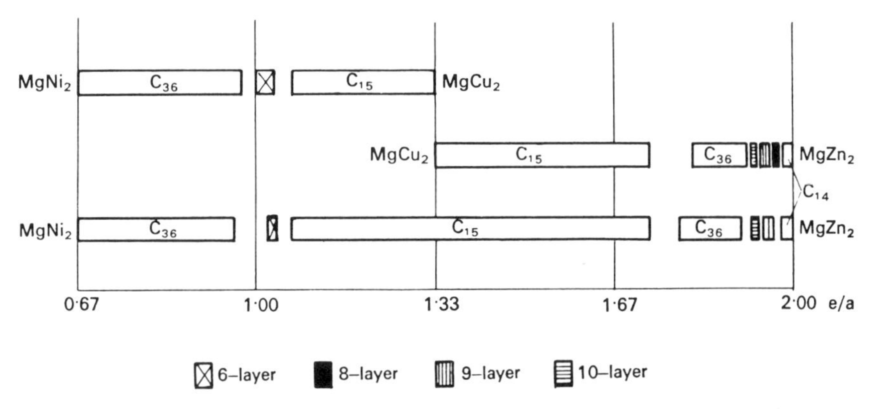

Fig. 17. Diagram indicating the $MgCu_2$ polytypic structures found in the $Mg(Ni, Cu)_2$, $Mg(Cu, Zn)_2$, and $Mg(Ni, Zn)_2$ systems. The layer numbers indicate the number of four-layer groups in the repeat sequence of the polytype. The $MgZn_2$ structure (C 14) is the two-layer type, $MgCu_2$ (C 15) is three-layer, and $MgNi_2$ (C 36) is four-layer. Electron concentrations are based on the assumption that Ni contributes zero valence electrons; however, the appropriate number might possibly be 0.6 in these alloys. After Komura *et al.*[38]

Laves phases (A–4A, A–12B, B–6A, and B–6B), so that they conform to the condition required for the observation of structural change as a function of electron concentration. Indeed, such observations have been made in alloy series with Laves phase structures[37,38] and Figure 17 shows the various polytypes occurring in $Mg(Ni, Cu, Zn)_2$ alloys as a function of electron concentration.[38]

The γ-brass superstructure probably derives part of its stability from similar considerations, as Lomer (Ref. 1, p. 280) has pointed out. It is generally described as a superstructure of the bcc structure formed as a block of $3 \times 3 \times 3 = 27$ bcc pseudocells, with atoms at the cell corner and body center omitted to give 52 atoms in the cubic supercell. Gamma brasses are the Hume-Rothery 21/13 electron compounds following the β-brasses, whose upper limit of stability is at an electron concentration of 3/2. In the second Brillouin zone the Fermi surface of ordered β'-brasses with the CsCl structure contains large, rather flat electron surfaces parallel to the cubic zone faces. A superlattice vector one-third of the β' reciprocal lattice vector would couple two such areas across the zone boundaries and stabilize a supercell which has a tripled cell edge in real space, through lowering of the electron energies. Omission of two atoms from the supercell should give a good value of the necessary matrix element (Ref. 1, p. 280). Thus the γ-brass superstructure occurs, and the most favorable electron concentration for achieving stability appears to be about the electron–atom ratio 21/13. Actually, the γ-brass structure is also for the most part a good approximation to tetrahedral close packing and it also undoubtedly derives much stability from the geometric arrangement of the atoms.[39]

5.3.2. Structural Changes Because of Brillouin Zone Filling

In the light of this discussion, the Hume-Rothery α, β, γ, and ε phases, which were the first crystal structure types discovered whose stability was influenced by electron concentration, require some explanation, since the structural changes from one to the other involve changes of nearest-neighbor coordination. The question naturally arises how the structural changes depend on electron concentration since they involve complete structural rearrangement. Hume-Rothery originally found that in alloys of Cu, Ag, and Au with subsequent group II–IVB elements, the α fcc phase existed up to an electron concentration of about 1.4 e/a; the ordered or disordered β phases occurred next and were stable up to an electron concentration of 1.5 e/a. These were followed by γ phases about an electron–atom ratio of about 21/13, and thereafter by close-packed hexagonal ε phases about an electron–atom ratio of 7/4. Excluding the β–γ phase relationship discussed above, the reason for these rules appears, in the rigid band framework, to be the decreasing $N(E)$ curves as the appropriate Brillouin zones for the structures become filled up but not overlapped, and ultimately the greater stability of another structure with a better $N(E)$ relationship. Second, owing to the low electron concentration, all of the outer valence electrons occupy sheets of Fermi surface in partly filled Brillouin zones, so that the differences of interaction of Fermi surface and Brillouin zone boundaries in one structure or the other have an unusually large influence on the cohesive electrons and therefore the enthalpy and free energy of the phases, thus exerting a uniform controlling influence on the structures, despite changes of near-neighbor coordination. Third, the comparative similarity of the group I–IVB metals severally leads to similar and not too pronounced changes (as a function of electron concentration) of the enthalpy arising from near-neighbor interactions as the composition of the alloys changes on proceeding across the phase fields in the various alloy systems. Thus the controlling influence of electron concentration on structure through the energy band effect can be observed. The essential correctness of these observations can be appreciated from the partial collapse of Hume-Rothery's rules in alloys of Cu, Ag, or Au with the group VB elements where a significant electrochemical difference is introduced. This increases the importance of near-neighbor interactions and the chemical bond factor, so that the energy band effect is much less of a dominant influence on the stability of the structures.

5.3.3. Pseudopotential Considerations

Pseudopotential theory provides a method of assessing the energy band contribution to structural stability (see, e.g., Ref. 40). Although the band structure energy U_{bs} is generally a small fraction of the total energy U of a crystal and although the present results of the application of the theory are largely interpretational rather than strictly quantitative, it offers some promise since it is the pseudopotentials of the *pure* metals that are involved in deriving data for alloys. However, the real limitation in considering, say, the rearrangement potential of the atoms (see below) may be the sufficient satisfaction of the constant-volume condition in alloy phases.

The total energy of a crystal in atomic units per atom can be written as

$$U = U_0 + U_E + U_{bs}$$

where U_0, the energy relative to the separated ions and electrons, is the difference of the total ionization energy and the bonding energy. The energy U_0, which is the most important term in the above expressions, is independent of structure and includes the main part of the Ewald electrostatic energy of point ions in the electron gas. The term U_E is the structure-dependent part of the Ewald energy, which favors simple structures since its value for these is small, and it increases rapidly with structural distortion. The term U_{bs} is the band gap energy, which is proportional to the square of the band gap ($U_{bs} \propto -[V(g)]^2$). It may be important when the Fermi surface of a metal approaches or touches planes of energy discontinuity in reciprocal space. Figure 12 indicates the nature of the electron-energy lowering relative to the free-electron energy, which results from the presence of planes of energy discontinuity.

The smallest reciprocal lattice vector **g** giving planes of energy discontinuity for the bcc structure is a set of 12 equivalent (110) vectors. For the fcc structure, there is a set of eight (111) vectors and there is a further set of six (200) vectors at a slightly larger **g** value. The

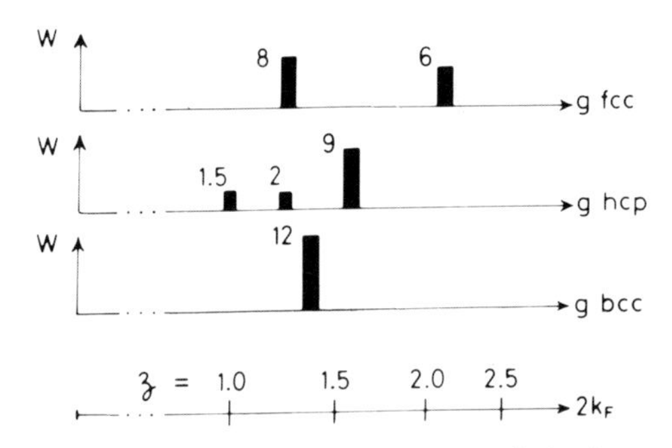

Fig. 18. Magnitudes of reciprocal lattice vectors **g** with structural weights W relative to $2\mathbf{k}_F$ for different electron–atom ratios z for the fcc, hcp, and bcc structures. Reproduced from Heine and Weaire(40) with permission of the publisher.

hcp structure has three sets of reciprocal lattice vectors with small **g** values as indicated in Figure 18. If in a particular structure of an element the smallest reciprocal lattice vectors **g** lie close to q_0 where the value of the pseudopotential $V(q)$ passes through zero (Figure 19), the band gap energy will be small. If, however, the reciprocal lattice vector lies at a greater value than q_0, the band gap energy is larger, and a contraction of the unit cell of the crystal in real space would increase **g** further, giving a still larger band gap and hence a greater band gap lowering of the total energy of the crystal. If **g** lies at a lower value than q_0 (Figure 19), the band gap will also be larger and an ex-

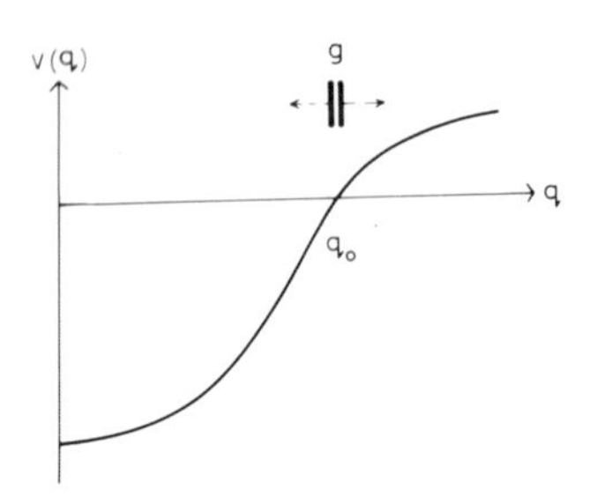

Fig. 19. Representative pseudopotential $V(q)$ with reciprocal lattice vectors **g** falling near q_0. Reproduced from Heine and Weaire(40) with permission of the publisher.

pansion of the crystal in real space would give a still greater band gap lowering of the crystal energy. If the smallest **g** for a close-packed structural form of metal were to lie close to q_0, giving a small band gap energy, U_{bs} could be increased by distortion of the structure so as to split the reciprocal lattice vectors, moving some to lower and some to higher **g** values. Such is believed to be the cause of the distorted crystal forms of Ga, In, and Hg.[40]

The reciprocal lattice vectors **g** and their structural weights in relation to the $2\mathbf{k}_F$ value (**k** value corresponding to the Fermi level) expressed in terms of electron per atom ratio for the fcc, hcp, and bcc structures (Figure 18) do, in a naïve way, show how the hcp or fcc structure might be expected at an electron concentration of one per atom, the bcc structure at about 1.5 e/a (as the β phases), the hcp structure at (1.6–2) e/a, and the fcc structure at (2.3–3) e/a.[40]

Structural stability can be further examined in terms of an interatomic potential $\Phi(R)$, where R refers to real space, which is a rearrangement potential of the atoms at constant volume. The potential $\Phi(R)$ is not a complete potential; it does not contain the purely volume-dependent terms in U_0, being composed of the Coulomb repulsion Z^2/R and the band structure energy expressed as a spherically symmetric interatomic potential $\Phi_{bs}(R)$.[40] The form of $\Phi(R)$ which is shown in Figure 20, and particularly that of the first minimum, is essentially determined by the form of the pseudopotential $V(q)$ of the metal or alloy. It is characteristic of the form of $\Phi(R)$ that the nearest-neighbor distances falling in the range of strong Coulomb repulsion, or in the region of the first maximum, would be a cause for structural instability. Thus Figure 20 shows that gallium in an assumed fcc structure would have the 12 closest neighbors at an R value lying on the rapidly rising function well above R_m. Distortion of the structure to give both closer and considerably larger interatomic distances (Figure 20) could therefore be expected to lower the structural energy. Nevertheless, the structural distortion must be subject to the rearrangement condition of constant volume resulting from the volume-dependent energy terms in U_0. This condition appears to be well satisfied by the observed structure of Ga, which places seven near-neighbors in the region of R_m and moves the other five to larger distances in the second minimum of $\Phi(R)$.[41]

Similar considerations may account for the axial ratios found in the group II hcp metals and for the structures of α- and β-Hg.[42] For example, it is found that in Mg with nearly ideal axial ratio the nearest neighbors lie close to R_m in $\Phi(R)$, but for Cd they would lie at

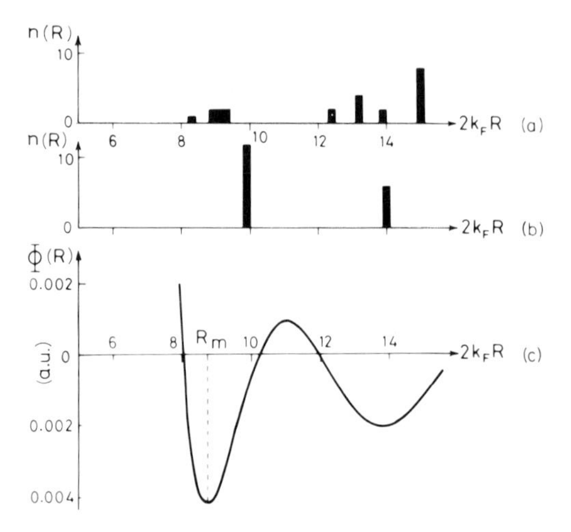

Fig. 20. Number of neighbors $n(R)$ at distances R in (a) the Ga structure and (b) in Ga with hypothetical fcc structure, together with the interaction potential $\Phi(R)$ for Ga. Reproduced from Heine and Weaire[40] with permission of the publisher.

larger R values on the steeply rising curve, suggesting that a splitting into six plus six neighbors with a considerable increase of c/a could give a structure of greater stability. Harrison[43] has shown that the Ewald energy of the hcp structure has a minimum close to the ideal axial ratio, but the imposition of the effects of the different $\Phi(R)$ functions resulting from the different q_0 values in the pseudopotentials of these metals, which increase in order from Be to Hg, could result in a total U which has lower values at either higher or lower c/a values than the ideal. Weaire[42] has calculated this total energy for the group II metals as a function of the axial ratio, with the results shown in Figure 21. This indicates that c/a should be smaller than ideal for Be and considerably larger than ideal for Zn and Cd, and it confirms the observed instability of Hg in the hcp structure. One of the pleasing features of Weaire's discussion is his account of the rapid increase of c/a in Mg–Cd alloys containing more than 60 at. % Cd in terms of the pseudopotentials of Mg and Cd. This increase of c/a was originally held to be a good example of energy band effects on relative lattice dimensions until Fermi surface studies showed that the particular Brillouin zone face concerned in the proposed overlap in alloys with about 60 and more at. % Cd had already occurred in Mg.

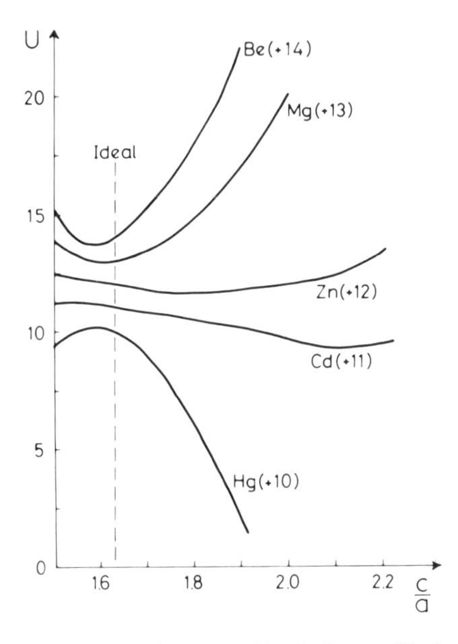

Fig. 21. Total energy U relative to ideal c/a as a function of c/a for hexagonal metals calculated using the model potential. Units are 10^{-3} a.u. per ion and the numbers on the diagram represent the amount of displacement of the curves upward from zero. Reproduced from Heine and Weaire[40] with permission of the publisher.

If the application of pseudopotential theory to problems of structural stability is to be of further value at this stage, it is hoped that quantitative predictions of structural change and of the existence or nonexistence of specific phases with specific structural arrangements can be made and verified; otherwise it only becomes one more theoretical manipulation struggling to account for facts which are already known.

5.4. Environmental Factors

5.4.1. Effect of Temperature on Phase Stability

The phase which is stable at the lowest temperatures is generally that with the greatest binding energy, because the entropy term TS is small. However, as the temperature increases, another phase with higher entropy due to uncertainty of positional or electronic parameters may achieve a lower free energy. Positional disorder arises from thermal vibration, leading to uncertainty in the location of the atoms about their equilibrium positions, or from the wrong occupation of sites in ordered binary or multicomponent alloys, and also from the random introduction of vacant sites. Vibrational entropy tends to be larger in structures where the atoms are loosely bound and separated by greater distances. Thus it might be held responsible for the transformation from α- to β-Sn at 292°K.[44] α-Sn, which has interatomic distances 2.81 Å and is stable at low temperatures, is held to have the largest binding energy, whereas β-Sn, with interatomic distances 3.02 Å and 3.18 Å, has a greater vibrational entropy. Above 297°K, $T\,\Delta S$ for $\alpha \rightarrow \beta$ exceeds ΔH and the transformation to the metallic form occurs.

Superstructure ordering of metallic phases may occur at low temperatures because of A-likes-B interactions or energy band effects, or because of the influence of large ion cores, where the stable structure is that in which atoms with large cores are surrounded by those with smaller cores. As the temperature increases, any positional disorder that is induced decreases the ordering force and cooperatively creates more disorder, until the influence of the $-TS$ term results in the disordered structure becoming the stable form. A similar situation occurs regarding the disordering of magnetically aligned spins in transition metal or rare earth compounds. Antiferromagnetic spin ordering at low temperatures may result in a change of crystal structure if the compound with disordered spins has cubic symmetry, as, for example, in CoO and NiO. As the temperature of the antiferromagnetic form is increased, thermal vibrations disorder some spins and this decreases the coupling force, leading cooperatively to further disordering with further temperature increase until all long-range ordering of the spins is destroyed and the structure reverts to the cubic form.

The increasing vibrations of the atoms about their equilibrium positions with increasing temperature ultimately result in the melting of a solid. Most alloys melt to monatomic liquids, but some structures, such as that of hexagonal Se, melt to a liquid containing randomly arranged chains of atoms. This is because the forces holding the chains of Se atoms together in the crystalline solid are much weaker than those along the chains. Thermal vibrations which are strong enough to destroy the binding between the chains are not so strong as to break down the bonds along the chains; hence the solid melts to a liquid made up of a random arrangement of Se chains.

Thus the influence of the entropy term in the expression for the free energy of a phase can be clearly understood with regard to positional and spin disorder, as well as melting, but understanding the factors that contribute to the enthalpy is a much more difficult part of accounting for phase and structural stability.

5.4.2. Effect of Pressure on Phase Stability

Pressure alters the relative dispositions of the free-energy-composition surfaces in an analogous manner to temperature, so that the type of phase changes occurring under the influence of pressure are similar to those that take place under the influence of temperature (Volume 5, Chapter 9).

Pressure induces transformations to structures with smaller atomic volume and therefore higher coordination and/or denser packing. The types of transformation that are observed are to structures with higher coordination, as in the transformation of KCl from the NaCl to the CsCl type structure, or to distorted structures with the same overall coordination, as in the transformation of diamond Si to Si II, or of SnTe from the NaCl to the GeS structure. In the case of the close-packed metallic elements, transformations are frequently observed from one close-packed structure to another polytypic form, as in the case of the rare earths, Fe, Zn, and Cd. The reason for this type of change is not at present understood, unless it is only a matter of the new polytype giving a slightly smaller atomic volume. Magnetic free-energy differences are involved in the $\gamma \rightleftarrows \varepsilon$ transformation in iron (Ref. 2, pp. 16–24).

5.4.3. Rapid Quenching: Metastable Phases

The very rapid quenching speeds (10^8 °C/sec) which can be obtained by the splat-cooling method have been shown to be capable of retaining metastable alloy phases when a molten alloy is quenched (Volume 5, Chapter 10). Such phases are particularly found in alloys between the group II to group VB metals, which generally do not form stable intermediate phases. For an intermediate metastable phase to be obtained, its free-energy curve must be below that of the liquid state, yet above the tangents to the free-energy curves of the terminal solid solutions or intermediate phases in the alloy systems.

6. Distortions of Crystal Structures

Various types of atomic interactions lead to distortions of simple crystal structures. Sometimes these distortions occur at specific transformation temperatures and sometimes they are present in the structure of the phase when it is formed from the melt. Since experience gained in understanding the causes of these distortions provides a modest ability to predict distortions in the structures of undiscovered or unexamined substances, a number of examples are discussed here. Among recognized interactions causing structural distortions are covalent bonding between certain cations, Jahn–Teller condensation, spin–orbit coupling, magnetic exchange forces, and mechanical instability of structures to shear. With the exception of the last of these, the compounds of interest are mainly transition metal or rare earth oxides, chalcogenides, and pnictides. In the transition metal compounds the interactions depend on the numbers of *d* electrons and the degeneracy of their orbitals, which in turn depend on the anion coordination polyhedra about the transition metals and on their relative configurations, since these determine the possible extent of cation–cation (*C–C*) and cation–anion–cation (*C–A–C*) interactions. Structural arrangement and electrical and magnetic properties are thus closely connected and information on one may be a source of information on the other. Whether or not the *d* electrons are localized or collective has important consequences for the properties and, sometimes, structures of these compounds. Since the size of an electron shell decreases with the number of electrons in it, collective electron behavior is likely to be found when the number of *d* electrons is small, and localized behavior when it is large (say 6–10). However, the main principle that is found to control the structures of these compounds is the endeavor of the transition metal atoms to achieve a filled valence subshell of *d* electrons, the t_{2g}^6 subshell being the most stable.

Electrostatic repulsion of the *d* electrons by surrounding negative ions (whether oxide, chalcogenide, or pnictide) causes splitting of the energies of the *d* levels into a group of three and a group of two levels for octahedrally or tetrahedrally coordinated cations. The splitting energy Δ is in conflict with exchange forces, which require as many electrons as possible to have parallel spins. A large value of Δ means that with four to seven *d* electrons, spin-paired configurations ("low-spin") are obtained rather than spin-parallel ones ("high spin"). Both high- and low-spin states are found in transition metals of the first long period, but only low-spin states are known for the transition metals of the second and third long periods, where the values of Δ are relatively larger.

The sizes of the transition metal atoms and their d-electron configurations depend on the relative energies of crystal field, spin–orbit coupling, and Coulombic interactions. In the first long period, Coulombic interaction energies are generally greatest, with crystal field interactions dominating those of spin–orbit coupling, so that high-spin states may be realized for electron concentrations between d^4 and d^7. However, in the second and third long periods the energies of the three effects are all comparable, and because of the larger crystal field energies, only low-spin configurations occur.[45] The relative importance of these three energies also depends on the symmetry and arrangement of the anion coordination polyhedra as well as on the number of d electrons.

6.1. Distortions Arising from Cation–Cation Bonds

Direct cation–cation interactions can occur if anion octahedra surrounding the cations share either edges or faces. Competing with these are possible cation–anion–cation interactions, which are strongest when the C–A–C angle is 180°. Goodenough (Ref. 22, pp. 165–168; Ref. 46) has given rules governing the relative strengths and consequences of these interactions. Generally C–C interactions are strongest with three or fewer d electrons and C–A–C interactions are strongest from d^4 to d^8, although at d^4 Jahn–Teller distortion or other effects may dominate. Nevertheless, when the C–C distance is greater than the critical distance for collective electron behavior C–C interactions are of necessity weak.

Antiferromagnetic interactions are common between cations in neighboring octahedral sites with half or less filled t_{2g} orbitals, whereas the interactions are generally ferromagnetic in interacting t_{2g} orbitals that are degenerate and more than half-filled. Alternately, and of interest here, t_{2g} orbitals on neighboring transition metal atoms with five or fewer electrons may form shared-pair covalent bonds with loss of spin moment. Thus metallic d^3 VO with the rocksalt structure changes structure at 114°K on cooling and becomes an insulator, either because of shared-pair covalent bonding giving effectively t_{2g}^6 subshells or because of magnetic separation of a d^3 half-band with aligned spins (Ref. 22, p. 269; Ref. 47). In the corundum structure (Ref. 3, p. 450) cations occupy two-thirds of the octahedra in a close-packed

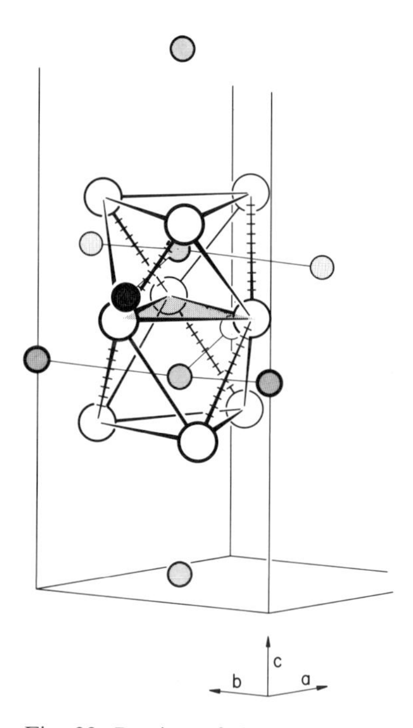

Fig. 22. Portion of the corundum (Al_2O_3) structure, showing transition metal–transition metal neighbors along the c axis and in the basal plane. Two octahedra sharing a face parallel to (001) are shown. Edges marked by bars lie between neighbors in the basal plane.

array of anions in such a way that pairs of occupied octahedra share a face perpendicular to the c axis of the structure (Figure 22). Formation of covalent bonds between these pairs of cations along [001] occurs noncooperatively in Ti_2O_3 and V_2O_3, causing a change of axial ratio rather than of crystal structure. The anion octahedra also share edges across the basal plane (Figure 22) so that weaker C–C interactions in the basal plane can also occur. These lead to an abrupt cooperative transition with change of crystal structure in V_2O_3 at a lower temperature than the [001] interaction (Ref. 22, pp. 256–266; Ref. 48). The compound Ti_2O_3, with one d electron, pairs this through the [001] noncooperative transition, becoming an insulator so that it does not form d bonds in the basal plane. Whereas d^2 V_2O_3 is still metallic following the upper noncooperating transition, it becomes an insulator following the pairing of the remaining

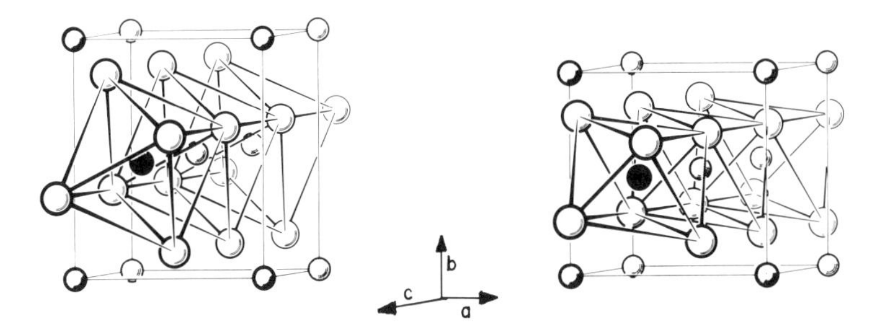

Fig. 23. Comparison of the marcasite, FeS_2, (left) and rutile, TiO_2, (right) structures, showing the lines of octahedra along [001] which share edges.

d electron in basal plane bonds. The transformation is accompanied by change of structure to a monoclinic form. Both Cr_2O_3 and α-Fe_2O_3 are semiconductor-insulators with antiferromagnetic spin ordering.[22,48]

The rutile (Ref. 3, p. 415) and marcasite (Ref. 3, pp. 411–413) structures resemble each other in that the transition metals are surrounded octahedrally by the anions, the octahedra sharing edges normal to the [001] direction (only) (Figure 23). This allows direct *C–C* interactions in this direction, and d^1 VO_2, which is metallic at high temperatures, becomes monoclinic and a semiconductor at 340°K, the V atoms moving together in pairs alternately along [001] as the *d* electrons form shared-pair bonds, resulting in loss of magnetic moment (Ref. 49; also see Ref. 22, pp. 270–275). The compound TiO_2, having no *d* electrons, is a semiconductor as expected and undergoes no such transition; neither does CrO_2 (metal) or MnO_2 (semiconductor), but MoO_2, WO_2, TcO_2, and ReO_2 have a monoclinic structure similar to that of VO_2, and interatomic distances suggest multiple *C–C* bonds along [001] of the rutile structure. Similar behavior is found in the d^5 transition metal pnictides, which have a monoclinically distorted form of the marcasite structure, in which the transition metals come together in pairs alternately along [001] of the marcasite structure (Ref. 3, p. 414). Such compounds are semiconductors and have essentially no magnetic moment (if the pairing is sufficient), the inference being that the distortion results from covalent electron sharing between pairs of atoms so that each attains a stable, filled t_{2g}^6 subshell. The d^4 pnictides have the marcasite structure, but compared to that of d^6 marcasites, the transition metal atoms are crowded together along the [001] direction, suggesting that they might form two shared-pair covalent bonds, each atom thus obtaining a filled t_{2g}^6 subshell, in keeping with the semiconducting properties and essentially zero magnetic moment of the compounds. It has also been suggested that this compression along [001] is the result of Jahn–Teller condensation (Ref. 10, pp. 367–369), but this is put in question by the recent discovery that in four of the compounds which were examined the cation is not at the center of the distorted anion octahedron.[50] The characteristic of this group of structural distortions is the formation of covalent bonds between cations centering octahedra. Generally they lead to loss of magnetic moment because of covalent pairing of the *d* electrons, and they may lead to semiconductor-insulator properties if the electron sharing gives a filled *d* subshell, attainment of which is generally the reason for the distortion.

6.2. Jahn–Teller Distortions

The degeneracy of the ground state of a cation (not a Kramer's doublet) may be removed and the crystal energy lowered by distortion of the anion coordination polyhedron to lower symmetry.* Whether or not such condensation will occur in the d^1, d^2, d^4, d^5, d^6, d^7, and d^9 high- and/or low-spin transition metal states where degeneracy can occur in octahedral coordination depends on the influence of other effects such as spin–orbit coupling (which must be quenched) and lattice vibrations (temperature). It is only in the high-spin d^4 ($t_{2g}^3 e_g{}^1$) and d^9 ($t_{2g}^6 e_g{}^3$) configurations that it is well recognized, since the e_g levels react most strongly with the anions. Tendency to Jahn–Teller distortion of polyhedra surrounding cations tetrahedrally is less, since the *d* orbitals do not point directly toward any of the anions. In Jahn–Teller distortion the cation remains at the center of the distorted anion polyhedron[22,48,51] and this allows distinction between the covalent bond distortions discussed above.

Jahn–Teller distortion of high-spin d^4 compounds is recognized in CrS, MnF_3, MMn_2O_4, and $HMnO_2$, for example, and in d^9 compounds such as CuO, CuF_2, $CuCl_2$, and $CuBr_2$. However, there is no Jahn–Teller distortion of d^4 MnAs, MnSb, and MnBi, and complex magnetic interaction energies must be comparable to those that could be derived from Jahn–Teller distortion (Ref. 22, pp. 279–282). The weak (MnP type) distortion of the NiAs structure of MnAs that is

* See, e.g., Ref. 51; Ref. 22, pp. 64–67, 202–215; Ref. 48, pp. 215–223; also see Chapter 4 and Volume 2, Chapter 1.

found between about 130°C and 40°C on cooling[52] is not of the Jahn–Teller type, since the metal atoms move from the center of the distorted anion octahedra. Below 40°C MnAs undergoes a first-order structural change back to the NiAs structure in which it is ferromagnetic.[53] Differences such as these emphasize the difficulty of predicting certainly whether Jahn–Teller distortion will occur.

Jahn–Teller distortions of the anion polyhedra are observed in the spinel structure (Ref. 3, p. 402) of several $M'M_2X_4$ compounds (Ref. 22, pp. 193–215; Ref. 54). In octahedral sites d^4 Mn^{3+} and d^9 Cu^{2+} give Jahn–Teller distortions, whereas d^8 Ni^{2+} and d^9 Cu^{2+} give distortions of tetrahedral sites of some spinels. Whether the spinels are normal, $MM_2'X_4$, or inverse, $M'(M, M')X_4$, depends on the relative crystal field stabilization energies of the transition metal ions in octahedral or tetrahedral coordinations.[55] For example, Ni^{2+} has a large octahedral field stabilization energy compared to that for tetrahedral field stabilization. Thus Ni^{2+} spinels are inverse except for that of Cr^{3+}, which has an even larger excess octahedral field stabilization energy. The d^5 Fe^{3+} ion, which has no crystal field stabilization because it is spherically symmetric, forms inverse spinels because of the excess crystal field stabilization energy of octahedrally compared to tetrahedrally coordinated M^{2+} ions (excluding the spherically symmetric d^5 Mn^{2+}).

6.3. Spin–Orbit Coupling Distortions

Spin–orbit coupling, as discussed by Goodenough* stabilizes a degenerate state of the d electrons, in contrast to Jahn–Teller distortion, which stabilizes nondegenerate states without contribution to angular momentum. It may be important when the orbital angular momentum is not quenched by the crystal field in the undistorted crystal. Crystal distortion resulting from spin–orbit coupling is in the opposite sense to that expected from Jahn–Teller distortion; hence the two can be distinguished. For example, Goodenough attributes the rhombohedral and tetragonal distortions to antiferromagnetic FeO and CoO below their Néel points to spin–orbit coupling effects.

* Ref. 56. Also see Ref. 22, pp. 64–67, 190–193, 213–215.

6.4. Magnetic Exchange Energies

Ferromagnetic spin ordering occurs at a second-order transformation and thus does not normally lead to a change of crystal structure, except in special cases such as MnAs, referred to above, where the structural change can be attributed to special effects. Antiferromagnetic spin ordering, on the other hand, frequently creates new crystal symmetry, converting cubic crystals to uniaxial crystals as in the examples of FeO and CoO referred to above. Hexagonal crystals normally adjust to the strains of antiferromagnetic ordering by a change of axial ratio. In some cases antiferromagnetic ordering leads to mechanical instability of the crystal so that it undergoes a shear (martensitic) transformation, as in the ordering of some fcc Mn–Cu alloys.

6.5. Mechanical Instability

In certain circumstances a crystal structure may become mechanically unstable relative to shear and subsequent deformation (at temperatures below those where diffusion processes can occur), so that it undergoes martensitic (diffusionless) transformation to some new structure type. Such transformations are, for example, found in Li, Na, Fe containing a little carbon, Nb_3Sn, V_3Si, and Mn–Cu alloys and in β-brass in certain two-phase ($\alpha + \beta$) alloys. Causes for such mechanical instabilities may be strains introduced as a result of magnetic ordering, interaction between large atomic cores in alloys which have been prevented from ordering,[57] as, e.g., in some β-brasses, strains resulting from interactions of Fermi surface and Brillouin zone boundaries, or complex electronic interactions as suggested in the case of Nb_3Sn and V_3Si.[58] Prediction of the occurrence of martensitic transformations is generally impossible, although knowledge of the appropriate elastic constants and their temperature dependence would give an indication of impending mechanical instability. Thus the fcc lattice would become mechanically unstable as the value of $\frac{1}{2}(C_{11} - C_{12})$ goes to zero, and experiments on In–Tl alloys do indeed show that this is the reason for the change from fcc to fc tetragonal structure by martensitic transformation involving $\{110\}$ $\langle 1\bar{1}0 \rangle$ shears. A rapidly increasing value of the anisotropy ratio $2C_{44}/(C_{11} - C_{12})$ of a cubic crystal with composition or temperature indicates impending mechanical instability and it is not surprising to find that β-brass, which under certain conditions undergoes martensitic transformation, has a large value of the anisotropy ratio.

7. Epilogue

Our acknowledged inability to calculate theoretically what phases with what structures may be formed even in a binary system, and the examples discussed here of some attempts to determine the factors which give phase stability or structural distortion, reveal how slightly the subject has progressed in terms of truly quantitative results. For example, even an energy band calculation and total energy evaluation of a metallic phase with a particular structure which could be considered exact and in agreement with Fermi surface studies and optical properties would provide but a small fraction of the work required to show why that phase was the stable one, since similar data should be obtained for the phase in all other possible crystal structures and for all other possible competing phases. A realistic appraisal of the situation in these circumstances is that, although the real goal lies much beyond our abilities, chemical and structural association is still a useful practical means of predicting the occurrence of phases and structures, and that we can obtain meaningful data concerning the control of phase stability by specific factors if we are careful enough to arrange experiments so that competing influences remain unchanged.

Appendix—Structure Diagrams

Let the structure of a binary alloy A_xB_y be compressed until, successively, according to the geometry, various A–A, A–B, and B–B contacts are made. These are considered to occur when the interatomic distances (d_A, d_{AB}, d_B) have the values $d_A = D_A$, $d_{AB} = \frac{1}{2}(D_A + D_B)$, and $d_B = D_B$, where D_A and D_B are the atomic diameters of the two components for the average CN in the alloy. The distances of all close interatomic contacts in the structure can be expressed in terms of the unit-cell and atomic parameters and d_A, the distance between an arbitrarily selected set of A–A contacts. Each of the interatomic distances in the structure can thus be represented on a structure diagram in terms of a reduced strain parameter $(D_A - d_A)/D_B$ (ordinate) and the diameter ratio D_A/D_B (abscissa), where they appear as straight lines (Figure 6). This diagram represents the geometry of the structure. Selecting arbitrarily a different A–A contact only moves the whole diagram relative to the scale of reduced strain parameter, and selecting a different strain parameter [e.g., $(D_B - d_B)/D_B$] only rotates the whole diagram; the relative positions of the lines remain fixed and no new information is introduced.

The d_A values for actual phases that take the structure can be calculated from the unit-cell (and atomic) parameters of the phases. Each phase can therefore be represented by a point on the diagram since its reduced strain parameter and radius ratio can be determined from d_A, D_A, and D_B. From the positions of the points representing a number of phases with the structure it is possible to determine whether chemical bond or geometric factors control the stability of the structure, as explained on pp. 139–146.

When a structure has variable axial ratios, interaxial angles, or atomic parameters, lines on the structure diagram representing interatomic contacts may change their relative positions as the structural parameters change. Generally it is possible to construct the structure diagram for representative parameter values (as, e.g., for the σ phase or the $AsNa_3$ structure) or for several representative sets of structure parameters (as, e.g., for the CdI_2 structure with c/a values of 1.33, 1.63, and 1.75). If points for observed phases follow a particular contact line when, for example, the axial ratio changes as in the case of the CdI_2 structure, then there is very strong evidence that that contact (A–B in the case of the CdI_2 structure) does indeed control the stability and dimensions of the structure.

Acknowledgments

I am grateful to McGraw-Hill for permission to quote widely from pp. 569 and 570 of *Phase Stability in Metals and Alloys.* I am also indebted to the following publishers and authors for permission to reproduce figures: Figure 1, from *Computer Calculation of Phase Diagrams* by L. Kaufman and H. Bernstein, Academic Press, New York. Figure 2, from an article by L. Brewer in *High-Strength Materials*, John Wiley & Sons, New York; permission granted by United States Atomic Energy Commission. Figures 18–21, from an article by V. Heine and D. Weaire in *Solid State Physics*, Vol. 24, Academic Press, New York.

References

1. W. M. Lomer, in *Phase Stability in Metals and Alloys* (P. S. Rudman, J. Stringer, and R. I. Jaffee, eds.), McGraw-Hill, New York (1967).
2. L. Kaufman and H. Bernstein, *Computer Calculation of Phase Diagrams*, Academic, New York (1970).
3. W. B. Pearson, *The Crystal Chemistry and Physics of Metals and Alloys*, Wiley—Interscience, New York (1972).

4. L. Brewer, in *High-Strength Materials* (V. F. Zackay, ed.), pp. 12–103, Wiley, New York (1965).
5. L. Brewer, in *Phase Stability in Metals and Alloys* (P. S. Rudman, J. Stringer, and R. I. Jaffee, eds.), pp. 39–61, McGraw-Hill, New York (1967); L. Brewer and P. R. Wengert, *Met. Trans. AIME* **4**, 83–104 (1973).
6. N. Engel, *Ingenioren* **1939**, N101; **1940**, M1; *Haandbogi Metällare: Selskabet for Metalforskning*, Copenhagen (1945); *Kemisk Maandesblad*, **30**(5), 53; (6), 75; (8), 97; (9), 105; (10), 114 (1949); *Powder Met. Bull.* **7**, 8 (1954); *ASM Trans. Quart.* **57**, 619 (1964).
7. M. J. Buerger, Private communication (1969).
8. E. Busmann, *Z. anorg. Chem.* **313**, 90–106 (1961).
9. W. B. Pearson, *Acta Cryst.* **17**, 1–15 (1964).
10. F. Hulliger and E. Mooser, in *Progress in Solid State Chemistry* (H. Reiss, ed.), Vol. 2, pp. 330–377, Pergamon, Oxford (1965).
11. L. Pauling, *The Nature of the Chemical Bond*, 3rd ed., Cornell Univ. Press, Ithaca, N.Y. (1960).
12. E-an Zen, *Am. Min.* **41**, 523 (1956).
13. M. Haissinskij, *J. Phys. Radium* **7**, 7 (1946).
14. J. C. Phillips, *Rev. Mod. Phys.* **42**, 317–356 (1970).
15. E. Mooser and W. B. Pearson, *Acta Cryst.* **12**, 1015–1022 (1959).
16. H. W. King, in *Alloying Behavior and Effects in Concentrated Solid Solutions* (T. B. Massalski, ed.), pp. 85–104, Gordon and Breach, New York (1965); *J. Mat. Sci.* **1**, 79–90 (1966).
17. L. Pauling, *J. Am. Chem. Soc.* **69**, 542–553 (1947).
18. J. H. van Santen and J. S. van Wieringen, *Rec. Trav. Chim. Pays-Bas*, **71**, 420 (1952).
19. N. Hush and M. H. L. Pryce, *J. Chem. Phys.* **28**, 244–249 (1958); **26**, 143–144 (1956).
20. G. Blasse, *J. Inorg. Nucl. Chem.* **27**, 748–750 (1965).
21. W. B. Pearson, *Z. Kristallogr.* **126**, 362–375 (1968).
22. J. B. Goodenough, *Magnetism and the Chemical Bond*, Interscience, New York (1963).
23. W. Hume-Rothery, G. W. Mabbott, and K. M. Channel-Evans, *Phil. Trans. Roy. Soc.* A**233**, 1 (1934).
24. W. Hume-Rothery and G. V. Raynor, *Structure of Metals and Alloys*, pp. 100–104, Institute of Metals, London (1954).
25. L. Darken and R. W. Gurry, *Physical Chemistry of Metals*, pp. 86–89, McGraw-Hill, New York (1953).
26. W. Hume-Rothery, in *Phase Stability in Metals and Alloys* (P. S. Rudman, J. Stringer, and R. I. Jaffee, eds.), pp. 3–23, McGraw-Hill, New York (1967).
27. H. M. Cundy and A. P. Rollett, *Mathematical Models*, p. 56, Clarendon Press, Oxford (1954).
28. W. B. Pearson, *Acta Cryst.* B**24**, 1415–1422 (1968).
29. W. B. Pearson, *Acta Cryst.* B**24**, 7–9 (1968).
30. H. Sato and R. S. Toth, *Bull. Soc. Fr. Minér. Crist.* **91**, 557–574 (1968).
31. H. Sato and R. S. Toth, in *Alloying Behavior and Effects in Concentrated Solid Solutions* (T. B. Massalski, ed.), pp. 295–419, Gordon and Breach, New York (1965).

32. H. Sato and R. S. Toth, *Phys. Rev.* **124**, 1833–1847 (1961).
33. H. Sato and R. S. Toth, *Phys. Rev.* **127**, 469–484 (1962).
34. H. Sato, R. S. Toth, and G. Honjo, *J. Phys. Chem. Solids* **28**, 137–160 (1967); H. Sato, R. S. Toth, G. Shirane, and D. E. Cox, *J. Phys. Chem. Solids* **27**, 413–422 (1966).
35. L. S. Ramsdell, *Am. Min.* **32**, 64–82 (1947).
36. M. Hirabayashi, N. Ino, and K. Hiraga, *J. Phys. Soc. Japan* **22**, 1509 (1967).
37. F. Laves and H. Witte, *Metallwirtschaft* **15**, 840 (1936).
38. Y. Komura, M. Mitarai, I. Nakatani, H. Iba, and T. Shimizu, *Acta Cryst.* **B26**, 666–668 (1970); Y. Komura, M. Mitarai, A. Nakaue, and S. Tsujimoto, *Acta Cryst.* **B28**, 976–978 (1972).
39. J. K. Brandon, R. Brizard, K. McMillan, and W. B. Pearson, to be published.
40. V. Heine and D. Weaire, in *Solid State Physics*, Vol. 24, pp. 249–463, Academic, New York (1970).
41. V. Heine, *J. Phys.* **C1**, 222–231 (1968).
42. D. Weaire, *J. Phys.* **C1**, 210–221 (1968).
43. W. Harrison, *Pseudopotentials in the Theory of Metals*, Benjamin, New York (1966).
44. J. Lumsden, *Thermodynamics of Alloys*, pp. 93–95, The Institute of Metals, London (1952).
45. D. H. Martin, *Magnetism in Solids*, p. 129, Iliffe, London (1967).
46. J. B. Goodenough, *Phys. Rev.* **117**, 1442–1451 (1960); D. G. Wickham and J. B. Goodenough, *Phys. Rev.* **115**, 1156 (1959).
47. F. J. Morin, *Phys. Rev. Letters* **3**, 34–36 (1959).
48. J. B. Goodenough, in *Progress in Solid State Chemistry* (H. Reiss, ed.), Vol. 5, pp. 145–399, Pergamon, Oxford (1971).
49. A. Magnéli and G. Andersson, *Acta Chem. Scand.* **9**, 1378–1381 (1955); B. Marrinder and A. Magnéli, *Acta Chem. Scand.* **11**, 1635–1640 (1957).
50. H. Holseth and A. Kjekshus, *Acta Chem. Scand.* **23**, 3043–3050 (1969); G. Brostigen and A. Kjekshus, *Acta Chem. Scand.* **24**, 1925–1940 (1970).
51. L. E. Orgel, *An Introduction to Transition-Metal Chemistry*, pp. 57–63, Methuen, London (1960).
52. R. H. Wilson and J. S. Kasper, *Acta Cryst.* **17**, 95–101; R. O. Kornelsen, *Can. J. Phys.* **39**, 1728 (1961).
53. C. Guillard, *J. Phys. Radium* **12**, 223–227 (1951); B. T. M. Willis and H. P. Rooksby, *Proc. Phys. Soc.* **B67**, 290–296 (1954); Z. S. Basinski and W. B. Pearson, *Can. J. Phys.* **36**, 1017–1021 (1958).
54. J. D. Dunitz and L. E. Orgel, *J. Phys. Chem. Solids* **3**, 20–29 (1957).
55. J. D. Dunitz and L. E. Orgel, *J. Phys. Chem. Solids* **3**, 318–323 (1957).
56. J. B. Goodenough, *Phys. Rev.* **117**, 1442–1451 (1960).
57. C. Zener, in *Phase Stability in Metals and Alloys* (P. S. Rudman, J. Stringer, and R. I. Jaffee, eds.), pp. 31–33, McGraw-Hill, New York (1967).
58. J. Labbé and J. Friedel, *J. de Phy.* **27**, 153–165, 303–308, 708–716 (1966).

4

Structure and Composition in Relation to Properties

J. H. Wernick
Bell Laboratories, Inc.
Murray Hill, New Jersey

Examination of crystal structure, and thus local coordination, and consideration of the nature of the atoms comprising the solid, i.e., position in the periodic table, have been for many years a crude means of predicting some of the bulk properties of solids. For example, the local coordination and chemical composition are suggestive of the nature of the bonding and have been used to classify materials as metallic, semiconducting, or insulating. A number of generalizations of this sort are discussed in a rigorous manner in Chapter 1 in the context of crystal ionicity. Unifying concepts and trends are always welcome, and it is safe to say the future of solid-state chemistry will rely on how well we utilize them for the discovery of new materials.

There are bulk properties which are sensitive to the details of structure and chemical composition to an extent that *a priori* predictions concerning bulk behavior cannot as yet be made. Thus, the solid-state chemist, whether he or she is engaged in the synthesis of new materials to exhibit a particular behavior, or in altering the properties of a known material, is generally guided to a first approximation by a knowledge of the chemistry and physics of known materials and the effect of structure and composition on the properties of these materials. One's "materials" education, then, should include an awareness of past experience on the role of several aspects of structure and composition in relation to properties and the basic reasons, where possible, for the observed behavior. In this chapter several aspects of the magnetic, superconducting, dielectric, and mechanical behavior of solids will be discussed with the above thought in mind.

1. Magnetic Behavior

1.1. Introduction

Advances during the last ten to fifteen years in elucidating the role of structure or atomic environment on the magnetic behavior of atoms, and thus bulk magnetic behavior, are a direct consequence of combining standard crystallographic and bulk magnetization measurement techniques with those microscopic techniques (NMR, EPR, Mössbauer resonance, and neutron diffraction) which give information regarding the local electronic environment of ions and atoms. In the following few paragraphs several topics relevant to the role of local environment will be touched upon to set the stage for subsequent topics discussed in this section.

Exchange forces, quantum mechanical in origin, are responsible for producing spontaneous collective interatomic behavior leading to ferromagnetic, ferrimagnetic, and antiferromagnetic ordering in the absence of an applied field. The magnetic exchange energy W of two spins S_i and S_j (in the Heisenberg isotropic exchange model)* is

$$W = -2J_{ij}S_i \cdot S_j$$

where $|S_i|$ and $|S_j|$ are one-half the total number of unpaired spins on the i and j sites. For example, for Gd^{3+}, with a half-filled $4f$ shell, $|S_i| = 7/2$. For the Mn^{2+} ion $|S_i| = 5/2$. Here J_{ij} is the exchange energy or exchange integral between atoms i and j and is related to the overlap of the charge distributions or wave functions of the atoms i and j. In general, the magnitude of J falls off rapidly as the neighbor distance increases and inclusion of second nearest neighbors is sufficient for most cases.

The interatomic exchange could be direct, if, for example, the interatomic distances were sufficiently small so that the electron wave functions directly overlap and thus form a band.† The exchange

* Anisotropic exchange and higher-order (biquadratic $S_i^2 \cdot S_j^2$) terms are not considered in this discussion.

† The band model of ferromagnetism in Fe-group metals is the current picture of collective behavior in these metals (see Herring[1] for a discussion of itinerant versus localized-spin model for ferromagnetic metals). For a fuller discussion of the subjects discussed in this section, the reader is referred to Kittel.[3]

TABLE 1
Effective Number p of Bohr Magnetons for Iron-Group Ions

Ion	Electron configuration	Ground-state level	p(calc) $= g[J(J+1)]^{1/2}$	p(calc) $= 2[S(S+1)]^{1/2}$	p(exp)
Ti^{3+}, V^{4+}	$3d^1$	$^2D_{3/2}$	1.55	1.73	1.8
V^{3+}	$3d^2$	$^3F_2$	1.63	2.83	2.8
Cr^{3+}, V^{2+}	$3d^3$	$^4F_{3/2}$	0.77	3.87	3.8
Mn^{3+}, Cr^{2+}	$3d^4$	$^5D_0$	0	4.90	4.9
Fe^{3+}, Mn^{2+}	$3d^5$	$^6S_{5/2}$	5.92	5.92	5.9
Fe^{2+}	$3d^6$	$^5D_4$	6.70	4.90	5.4
Co^{2+}	$3d^7$	$^4F_{9/2}$	6.63	3.87	4.8
Ni^{2+}	$3d^8$	$^3F_4$	5.59	2.83	3.2
Cu^{2+}	$3d^9$	$^2D_{5/2}$	3.55	1.73	1.9

between localized moments may be indirect via a coupling with electrons on neighboring nonmagnetic ions or by polarization of conduction electrons; the former is known as superexchange and it is the mechanism in operation in insulators, such as ferrite, garnets, and perovskites. The exchange interaction between neighboring atoms via polarization of conduction electrons is the predominant mechanism in rare earth metals and alloys. If J_{ij} is positive, ferromagnetic exchange is said to occur; if it is negative, antiferromagnetic interactions occur.

Atoms or ions containing incompletely filled d orbitals are sensitive to the symmetry of the crystal field or electrostatic field of their surroundings because the d wave functions are exposed to the surrounding ions. The crystalline field destroys the spherical symmetry of an isolated transition metal ion and the magnitude of the splitting or removal of the degeneracy of the $3d$ states depends on the type, local symmetry, and interionic distances of the surrounding ions.* One consequence of this is the quenching of orbital angular momentum of the d-shell transition ions. The moment per ion is then often due primarily to the spin angular momentum. The effective moment p

* For example, under the influence of an octahedral crystal field (sixfold coordination), the fivefold degenerate d orbitals are split and some are lowered in energy relative to others. The e_g orbitals (or d_γ or Γ_3) consist of the $d_z{}^2$ and $d_{x^2-y^2}$ orbitals and are directed along the three Cartesian axes. The second group, the t_{2g} (or d_ε or Γ_5) orbitals, consist of d_{xy}, d_{yx}, and d_{zx} orbitals, and are directed between the Cartesian axes. The t_{2g} orbitals are lower in energy than the e_g orbitals.

in units of Bohr magnetons of an ion is

$$p = g[J(J + 1)]^{1/2}$$

where J is the angular momentum quantum number. The spectroscopic splitting factor g is defined for Russell–Saunders coupling by

$$g = 1 + \{[J(J + 1) + S(S + 1) - L(L + 1)]/2J(J + 1)\}$$

where L and S are orbital and angular spin momenta, respectively. Table 1 illustrates the effect of the crystalline field in quenching the orbital angular momentum of some $3d$ ions. Good agreement between experiment and theory is observed when only the spin angular momentum is considered.

The interatomic distances in ionic lattices containing $3d$ transition metal ions, which have a nonspherically symmetric electronic ground state,[2] are shortened by the crystal field relative to that of a similar structure in which the interionic potential is purely of the Madelung type. This is illustrated in Figure 1. If the binding between ions is of the latter type, a nearly monotonic (dotted line) change is expected in the equilibrium interionic separation r_e with atomic number of the cation in isostructural salts with the same anion. The

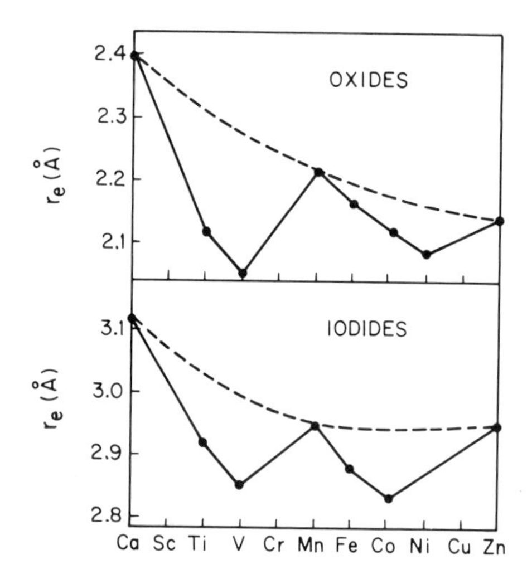

Fig. 1. Interionic distances as a function of atomic number of isostructural oxides (O_h^5) and isostructural iodides (D_{3d}^3). Ca^{2+} ($3d^0$), Mn^{2+} ($3d^5$), and Zn^{2+} ($3d^{10}$) have S or spherically symmetric ground states (after Hush and Pryce[2]).

lattice stabilization energies also show this periodic variation. Without crystal field stabilization the lattice energies would also change monotonically. Crystal field stabilization is zero for Ca^{2+} ($3d^0$), Mn^{2+} ($3d^5$), and Zn^{2+} ($3d^{10}$) ions (*s*-state ions).[(2)]

Pearson,[(40)] by examining the radii of transition metal atoms in a number of compounds of group 5 (pnictides) and group 6 (chalcogenides) elements, which included both metals and semiconductors, has shown that the radii depend on the number and configuration of the *d* electrons and on the symmetry of the anions coordinated to it in a manner similar to that observed for ionic oxides and halides.

The orbital angular momenta are essentially quenched in the magnetically ordered 3*d* transition elements, Fe, Co, and Ni, and their alloys. Because the *d* states are outer electronic states, a band of *d* states with a continuous distribution of electron energies forms. This *d* band is quite narrow compared to the bands in nontransition metals, and this is the reason for their comparatively high electronic specific heat (and high density of states at the Fermi surface) and lack of integral atomic moments. If the *d* electrons were completely localized, they would yield integral moments.

The 4*f* electrons, responsible for the magnetic behavior of the rare earth metals and ions, are localized within the inner shell (4*f*), and the atomic or ionic moment is due to both spin and orbital angular momenta. The radial portion of the 4*f* wave function as a function of distance is shown in Figure 2 to illustrate that the 4*f* shell is shielded by the 5*d* and 6*s* shells. Because of this shielding, the *f* orbitals are relatively insensitive to the symmetry of the crystal field in which they are placed. Table 2 illustrates the agreement between observed and calculated values of *p* when the orbital moment is included.

Crystal field splittings in rare earth salts are known in fairly good detail from magnetic and optical studies and there is evidence from magnetic and thermal studies that crystalline fields affect the magnetic properties of rare earth metals, some metallic rare earth intermetallic compounds, and the behavior of rare earths as impurities in other metals.

Clearly, the magnetic behavior of solids is determined by the kind and number of magnetic ions present, nature of the local environment of the magnetic ions, and nature of the exchange interactions between ions. In the following these points will be illustrated by discussing the magnetic behavior of a number of metallic and nonmetallic systems.

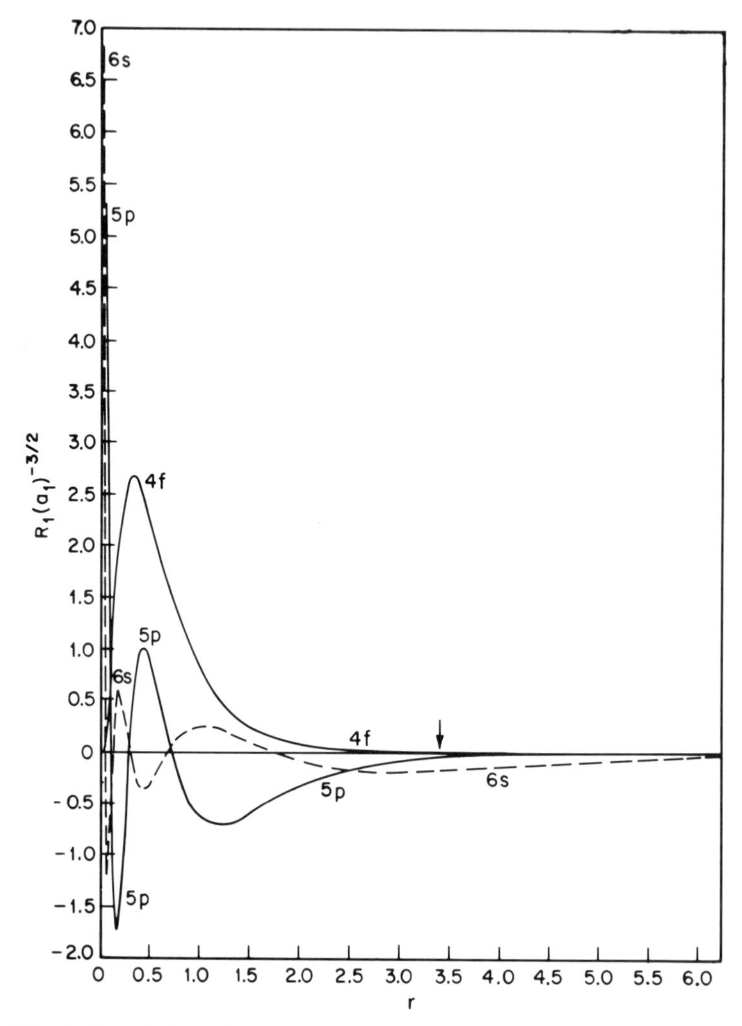

Fig. 2. The $4f$, $5p$, and $6s$ radial wave functions for the free atom of Gd illustrating the localized nature of the $4f$ electrons. The nearest-neighbor interatomic midpoint in hexagonal close-packed Gd is indicated by the arrow (units are the Bohr radius, a_1) (after Weiss[4]).

1.2. The 3*d* Transition Elements

Only five *d* transition element metals exhibit long-range magnetic order. Iron, Ni, and Co are ferromagnetic, while α-Mn ($T_N = 95°K$) and Cr ($T_N = 313°K$) are antiferromagnetic. Alloys and intermetallic compounds that order magnetically (with the exception of $Sc_{0.76}In_{0.24}$[5], Au_4V, and $ZrZn_2$[6]) contain at least one of these elements.

TABLE 2
Effective Number p of Bohr Magnetons for Trivalent Lanthanide Group Ions (near Room Temperature)

Ion	Electron configuration	Ground-state level	p(calc)	p(exp)
Cr^{3+}	$4f^{1}5s^{2}p^{6}$	$^{2}F_{5/2}$	2.54	2.4
Pr^{3+}	$4f^{2}5s^{2}p^{6}$	$^{3}H_{4}$	3.58	3.5
Nd^{3+}	$4f^{3}5s^{2}p^{6}$	$^{4}I_{9/2}$	3.62	3.5
Pm^{3+}	$4f^{4}5s^{2}p^{6}$	$^{5}I_{4}$	2.68	—
Sm^{3+}	$4f^{5}5s^{2}p^{6}$	$^{6}H_{5/2}$	0.84	1.5
Eu^{3+}	$4f^{6}5s^{2}p^{6}$	$^{7}F_{0}$	0	3.4
Gd^{3+}	$4f^{7}5s^{2}p^{6}$	$^{8}S_{7/2}$	7.94	8.0
Tb^{3+}	$4f^{8}5s^{2}p^{6}$	$^{7}F_{6}$	9.72	9.5
Dy^{3+}	$4f^{9}5s^{2}p^{6}$	$^{6}H_{15/2}$	10.63	10.6
Ho^{3+}	$4f^{10}5s^{2}p^{6}$	$^{5}I_{8}$	10.60	10.4
Er^{3+}	$4f^{11}5s^{2}p^{6}$	$^{4}I_{15/2}$	9.59	9.5
Tm^{3+}	$4f^{12}5s^{2}p^{6}$	$^{3}H_{6}$	7.57	7.3
Yb^{3+}	$4f^{13}5s^{2}p^{6}$	$^{2}F_{7/2}$	4.54	4.5

* After Kittel.[2] $p = g[J(J + 1)]^{1/2}$, where g is the spectroscopic splitting factor and $J = |L \pm S|$, with L the total orbital angular momentum and S the total spin angular momentum.

In magnetically ordered Fe (bcc) the atomic moments are aligned along the cube directions even though the nearest-neighbor direction is along $\langle 111 \rangle$ and neighboring moments are coupled parallel to one another by exchange forces. The preference for a specific crystallographic direction is due to spin–orbit coupling with the small residual unquenched orbital moment which gives rise to the magnetocrystalline anisotropy. The unpaired $3d$ electrons have a slight preference for the e_g orbitals which point in the cube direction.

On the other hand, the easy axis of magnetization in Ni (fcc) is along the $\langle 111 \rangle$ direction, due to spin–orbit coupling, because the electrons prefer the t_{2g} orbitals which point in this direction. For cobalt (hcp), the easy axis of magnetization is along the c axis. The magnetocrystalline anisotropy energy is about 10^4 times smaller than the exchange energy in the $3d$ transition elements and relatively low fields (less than several thousand gauss) are required to rotate the magnetization from the easy axis into the hard direction. This is to say the spin–orbit coupling, although present, is relatively weak, and contrasts the behavior of rare earth metals to be discussed below.

The α-Mn structure is cubic with 58 atoms per cell. There are four nonequivalent crystallographic positions and they are designated A_1, A_2, A_3, and A_4. The atomic distribution is two atoms in A_1, eight in A_2, 24 in A_3, and 24 in A_4. The nearest-neighbor interatomic distances vary from 2.24 to 2.96 Å. Even within each of the four crystallographically different sites the interatomic distances are not the same. However, a total coordination number (CN) can be assigned to each site and an average interatomic distance. Thus, atoms A_1 and A_2 have CN 16, whereas A_3 and A_4 have CN 13 and CN 12, respectively.[7] Neutron and X-ray diffraction studies[8] have shown that the low-temperature chemical and magnetic unit cells of antiferromagnetic α-Mn are identical and that the moments on atoms A_1, A_2, A_3, and A_4 are 1.72, 1.46, 1.11, and 0.02 μ_B respectively. The magnetic behavior of Mn atoms notoriously depends on interatomic distance in alloys and intermetallic compounds and this will become apparent in a subsequent section.

Chromium (bcc), when magnetically ordered, is antiferromagnetic. The antiferromagnetism is based on an itinerant electron rather than on a purely localized model whereby the magnitude of each atomic moment varies spatially so that there is a continuous variation with position (spin density waves) and the spin structure or magnetic unit cell is incommensurate with the chemical cell, i.e., it is not a rational number of chemical cells. The magnetic unit cell contains about 20 chemical cells.[9,11–15] Like the classical antiferromagnetic state, the moments at the cube corners are antiparallel (along a cube direction) but the magnitudes are not quite equal to those at the cube centers, thus giving rise to a sinusoidal variation of the spin density. A spin density wave model for Cr is supported, not only by the incommensurate magnetic cell, but also by the absence (from neutron scattering experiments) in the paramagnetic region (above the Néel temperature) of localized moments.[16] This implies that the antiferromagnetic state results, on cooling, from an unusual instability of the conduction electrons. Small amounts of Mo, Tc, Ru, Rh, and Re in solid solution, which increase the electron-to-atom ratio, raise the Néel temperature and magnetic moment of Cr, and at about 1% solute the magnetic structure becomes commensurate and demonstrates that the antiferromagnetic state is critically dependent on the electronic band structure.[17]

1.3. Rare Earth Metals

Lanthanum ($Z = 57$), with the electron configuration xenon core ($4f^0 5d^1 6s^2$), is the first element in the $5d$ transition series and the rare earths, which begins with Ce ($Z = 58$), are often referred to as the

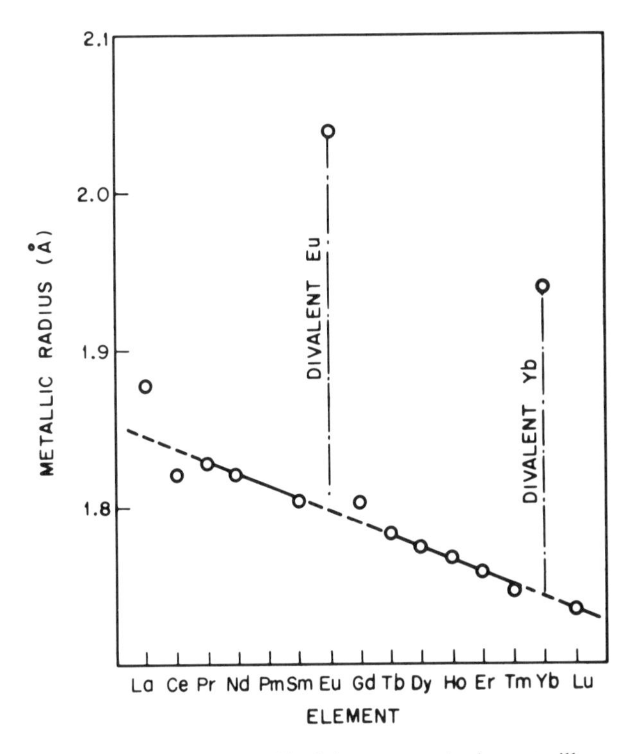

Fig. 3. The metallic radii of the rare earth elements illustrating the lanthanide contraction (after Taylor[10]).

lanthanides. For La the empty 4*f* shell is at a higher energy than the $5d^1$ level, but as electrons are added to the 4*f* shell (and protons to the nucleus), the energy of the 4*f* shell is lowered and the 4*f* states are occupied (Figure 2). Only for Gd (half-filled 4*f* shell) and Lu (filled 4*f* shell) is there one 5*d* electron in the ground-state electron configuration (Hund's rule). Nevertheless, most of the rare earth atoms are trivalent, the "5*d* electron" coming effectively from the 4*f* shell. The valence electrons become conduction electrons in the metals. The experimental *atomic* moments of the metals are generally consistent with a simple *ionic* treatment using Hund's rules (Table 2). Both Sm and Eu are exceptions because the level splitting is such that the first excited states have to be included.*

Since half-filled or filled 4*f*-shell configurations are quite stable (Hund's rule), Eu ($4f^7 5d^0 6s^2$) and Yb ($4f^{14} 5d^0 6s^2$)† are divalent metals

* For less than half-filled shell the total moment is calculated from $J = |L - S|$. For half-filled shell $J = |S|$, and for more than half-filled shell $J = |L + S|$.

† Yb metal is paramagnetic with a temperature-independent susceptibility.

TABLE 3
Room-Temperature Crystalline Structures of the Lanthanides

Element	Structure	Element	Structure
La	d-Hex	Tb	HCP
Ce	FCC	Dy	HCP
	d-Hex	Ho	HCP
Pr	d-Hex	Er	HCP
Nd	d-Hex	Tm	HCP
Sm	Sm type	Yb	FCC
Eu	BCC	Lu	HCP
Gd	HCP		

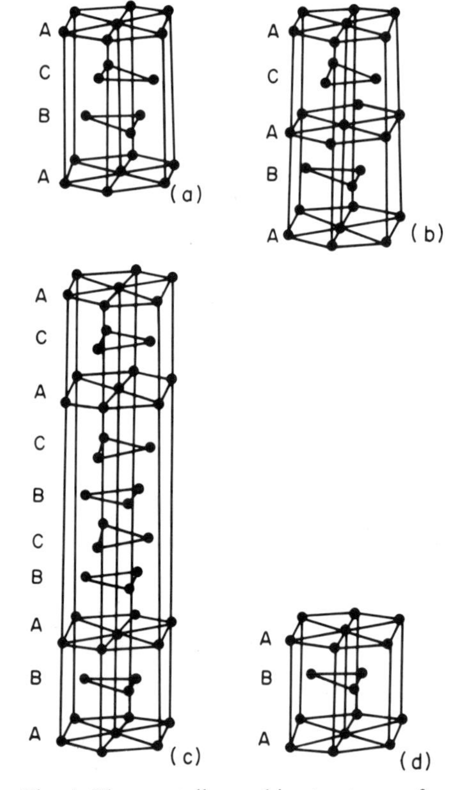

Fig. 4. The crystallographic structures of the rare earth elements in terms of three basic layers. (a) face centered cubic (fcc viewed in the ⟨111⟩ direction), (b) double hexagonal (d-hex), (c) samarium type, (d) hexagonal close packed (hcp) (after Taylor[10]).

and divalent in many metallic alloys and intermetallic compounds. For Ce there is a tendency to lose one or both $4f$ electrons and thus exhibit trivalent and tetravalent behavior in metals and nonmetals. The increase in nuclear charge with increasing atomic number gives rise to the lanthanide contraction (Figure 3). As noted earlier, the magnetic behavior of the rare earth atoms or ions is due essentially to the unpaired $4f$ electrons, and since they are localized, the coupling mechanism necessary for long-range magnetic order is not due to direct overlap of electron wave functions as occurs for the $3d$ metals Fe, Co, and Ni.

The room-temperature crystal structures are tabulated in Table 3. The structures can be discussed in terms of the stacking of three types of layers, *A*, *B*, and *C* (Figure 4). The double hexagonal structure exists at the lower temperatures and is confined to the lighter rare earth elements. Both Eu and Yb exhibit cubic structures.

All of the lanthanide elements, except La ($4f^0$), Yb ($4f^{14}$), and Lu ($4f^{14}$), exhibit magnetic order below room temperature. The

TABLE 4

Ordering Temperatures and Antiferromagnetic Spin Structures of the Magnetic Rare Earth Elements (after Taylor[10])

Element	Néel temperature, °K	Curie temperature, °K	Antiferromagnetic spin structure
Ce	12.5	—	Ferromagnetic in planes with moments along *c* axis but net antiferromagnetic
Pr	25	—	Adjacent layers antiparallel;
Nd	19*	—	sinusoidal modulation in plane
Sm	14.8	—	?
Eu	90	—	Helix
Gd	—	293	—
Tb	229	222	Helix
Dy	179	85	Helix
Ho	131	20	Helix
Er	84	20	Sinusoidal *c* axis 85–53.5°K; helix + *c*-axis sinusoidal 53.5–20°K
Tm	56	25†	—

* Hexagonal sites order at this temperature followed by a change in the ordering at 7.5°K due to ordering of the cubic sites.

† Transition to ferrimagnetic. Metamagnetic–ferrimagnetic coupling can be overcome by the application of fields in excess of 28 kOe.

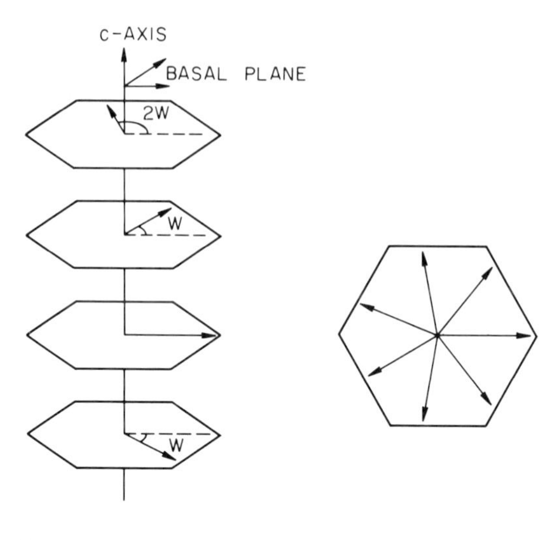

Fig. 5. Illustration of a helical spin structure. Ordering in each plane is ferromagnetic but the moment direction from plane to plane changes through a constant "turn angle" W resulting in overall antiferromagnetic behavior (after Taylor[10]).

magnetic ordering temperatures and the antiferromagnetic spin structures are given in Table 4. A helical spin or magnetic structure is generally exhibited by the antiferromagnetic state. Such a helical spin system is schematically illustrated in Figure 5. The heavy rare earths, beginning with Tb, show both ferromagnetic and antiferromagnetic ordering, the latter occurring at higher temperatures. Thus these materials exhibit complex magnetic structures and magnetization behavior. For example, Dy can be made to behave ferromagnetically up to the Néel point in sufficiently high applied fields, i.e., the antiferromagnetic coupling is overcome by the applied field. A material which exhibits this effect is usually referred to as metamagnetic. The behavior of the magnetization resulting from the destruction of a helical spin structure by an applied field is schematically illustrated in Figure 6.[10]

Because of the localized nature of the $4f$ electrons, the exchange interaction between rare earth atoms in the rare earth metals and between rare earth and $3d$ atoms in alloys leading to long-range order

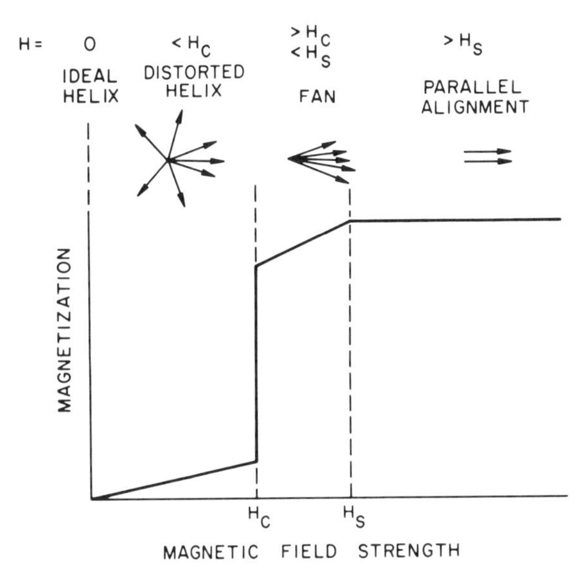

Fig. 6. Illustration of metamagnetic behavior via destruction of ideal helical spin structure by an applied field (after Taylor[10]).

is indirect via conduction electrons. This exchange mechanism is known as the Rudermann–Kittel–Kasuya–Yosida (RKKY) mechanism[18–20] and leads to a polarization of conduction electrons in the neighborhood of an ion which is oscillatory in character, decaying at large distances. The exchange interaction between two spins is also an oscillating function of distance and accounts for the existence of helical spin structures.

1.4. Role of Local Atomic Environment Regarding Development of Atomic Moments and Long-Range Order

Iron atoms exhibit moments only in some hosts.[22] The effect of local atomic environment in random solid solutions in the development of an atomic or localized moment* on Fe solute atoms in nonmagnetic bcc Nb–Mo alloys containing 1 % Fe[21] and in fcc Rh–Pd alloys containing 1 % Co[22] was first emphasized by Jaccarino and Walker.[21] NMR studies indicated that the average moment

---

* Atoms or ions exhibit local moments or are paramagnetic if they exhibit net spin densities which are "long lived." This can result in either long-range magnetic order or a temperature-dependent susceptibility.

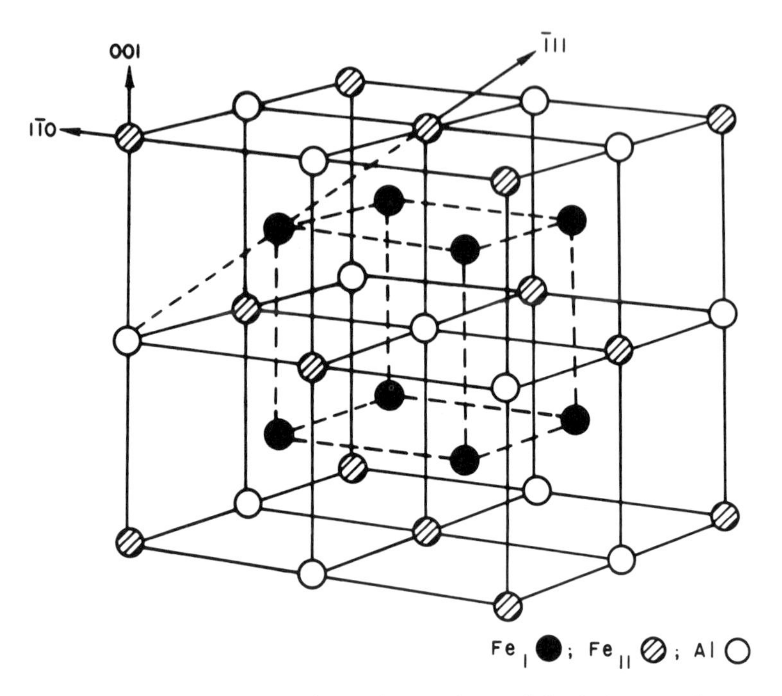

Fig. 7. Crystal structure of Fe_3Al. Fe_I (eight): full circles. Fe_{II} (four): shaded circles. Al: open circles.

associated with 1% Co increases with Pd content of the Rh–Pd solid solutions in a manner such that only those Co atoms with at least two Pd nearest neighbors (NN) exhibit a local moment. The effect of the number of NN and next-nearest-neighbor (NNN) impurities on the hyperfine field of Fe-rich alloys, determined by Mössbauer spectroscopy, has been studied fairly extensively (see, e.g., Ref. 23). In this section we give a few examples to illustrate the effect of local environment on the magnetic behavior of ordered alloys.

The intermetallic compound FeAl exists with the CsCl structure (bcc). Ordered, stoichiometric as possible, FeAl exhibits a very small susceptibility, does not exhibit magnetic order, and Mössbauer studies on annealed FeAl have shown that no magnetic moment is associated with the iron atoms.[24] When some of the Al atoms are replaced by Fe atoms, so that Fe atoms now have Fe NN, local moments on Fe atoms with Fe NN develop. Crushing or plastic deformation also produces Fe NN (see Section 1.5.2). The experiments with nonstoichiometric FeAl indicate that Fe atoms with eight Fe NN have a moment of 1 μ_B and at low temperature, $Fe_{1.1}Al_{0.9}$ orders ferromagnetically.

These results explain the bulk magnetic properties observed in non-stoichiometric Fe-rich and plastically deformed FeAl.

Ordered Fe_3Al (fcc structure, DO_3, Figure 7, four formula units/cell), in contrast to FeAl, forms from disordered Fe–Al solid solutions via an order–disorder reaction and is ferromagnetic with $T_c \simeq 750°K$. In the ordered state, the twelve Fe atoms are situated on two sets of nonequivalent sites, designated as Fe_I and Fe_{II}. Eight Fe atoms (Fe_I) are located in a simple cubic sublattice. Each has four Fe_{II} and four Al atoms as NN. The Fe_{II} atoms have eight Fe_I NN and are situated at the body-centered positions of the small cubic subcells. Neutron diffraction studies[25] at room temperature indicate that Fe_{II} atoms exhibit a spin moment of 2.16 μ_B and a moment of $\sim 1.64\ \mu_B$ is associated with Fe_I atoms. These results are summarized in Table 5 and compared with spin moments in α-Fe and FeAl.[26,27] It should be noted that the Fe_{II} atoms, with eight Fe_I NN, exhibit a moment, within experimental error, identical to that of pure iron and that the second and third NN have little or no effect, similar to that of Fe in FeAl. The reduced moment for Fe_I atoms is attributed to a reduction of Fe NN. The difference in atomic moments exhibited by Fe_I and Fe_{II} atoms was shown from neutron[25] and Mössbauer[28] studies to be due to a difference in $3d$-electron configuration and antiferromagnetic interactions are ruled out.

On the other hand, Pt_3Co, in the ordered state, has the cubic $L1_2$ structure (Cu_3Au prototype), and is ferromagnetic with a Curie temperature of 12°C.[49] Each Co has 12 Pt NN, yet each Co atom exhibits a localized moment and the exchange interaction is such that strong ferromagnetic exchange occurs. When disordered, Co will have Co NN, and the Curie temperature rises to 209°C.[49] Since the Curie temperature is an indication of the strength of the exchange

TABLE 5

Atomic Environments and Spin Moments in Fe_3Al and FeAl at Room Temperature, Corrected for Perfected Atomic Order[26,27]

Phase	α-Fe	Fe_3Al		FeAl
Atom	Fe	Fe_{II}	Fe_I	Fe
Nearest neighbors	8 Fe	8 Fe_I	4 Fe_{II}, 4 Al	8 Al
Second-nearest neighbors	6 Fe	6 Al	6 Fe_I	6 Fe
Third-nearest neighbors	12 Fe	12 Fe_{II}	12 Fe_I	12 Fe
μ_{Fe} at 300°K, μ_B	2.175	2.18 ± 0.1	1.8 ± 0.1	0

interactions, the ferromagnetic coupling becomes even stronger with decrease in the Co–Co distance. Polarization by Co of the Pt *d* electrons plays an important role in the magnetic behavior of this alloy, as well as in the magnetic behavior of dilute solutions of Co and Fe in Pd and Pt.[50] Both Pd and Pt become ferromagnetic at low temperatures at very low concentrations of Fe and Co. "Giant" moments are produced in the neighborhood of the impurities by polarization of host-atom *d* electrons.

In contrast to Pt_3Co, isostructural Pt_3Fe in the ordered state is *antiferromagnetic* with a Néel temperature of 130°K.[51] Thus in this material a negative exchange interaction dominates at large Fe–Fe distances. The moment per Fe atom is 3.0 μ_B. Disordering (by plastic deformation) results in ferromagnetic behavior ($T_c = 425°C$) and the moment per Fe atom is 4.0 μ_B for an alloy containing 24.9 at. % Fe.[51]* Recalling that Fe atoms in bcc Fe exhibit a moment of 2.175 μ_B, it is apparent that polarization of *d* electrons of the intervening Pt atoms plays an important role.

To further illustrate nearest-neighbor and interatomic distance effects, the phases Ni_3Mn, MnAlGe, and Au_4V will be considered. Particularly significant is Au_4V because it is metallic and V metal does not exhibit magnetic order†; Au_4V can be ferromagnetic.

It is well known that the magnetic properties of alloys containing Mn are very dependent on the Mn–Mn distance. For example, Ni_3Mn in the disordered form is fcc, but upon ordering (~800°K) assumes the cubic $L1_2$ structure. Each Mn is coordinated to 12 Ni atoms while each Ni is coordinated to four Mn and eight Ni atoms. In the disordered alloy the magnetization of the alloy is quite small, the Mn–Mn atoms roughly paired, and the Mn–Mn exchange antiferromagnetic. In the ordered state the alloy is ferromagnetic ($T_c \sim 728°K$) and the Mn atoms have only Ni atoms as nearest neighbors. In the ordered state, neutron and bulk magnetization data give $\mu_{Mn} = 3.18 \pm 0.25\ \mu_B$ and $\mu_{Ni} = 0.30 \pm 0.05\ \mu_B$ and the Mn–Ni exchange ferromagnetic[29] as the most likely interaction. From a magnetization study of this alloy as a function of degree of order it appears that Mn will become antiferromagnetically coupled with respect to its neighbors if the NN shell contains three or more Mn atoms; the condition existing in the

* Isostructural ordered Pt_3Cr and Pt_3Mn are ferromagnetic with μ_B of 2.2/Cr and 4.2/Mn.[52]

† V_2O_3 (discussed in Section 1.7.2) in the insulating state is antiferromagnetic. The V–V distance is larger than it is in the metallic state and the V ions exhibit local moments.

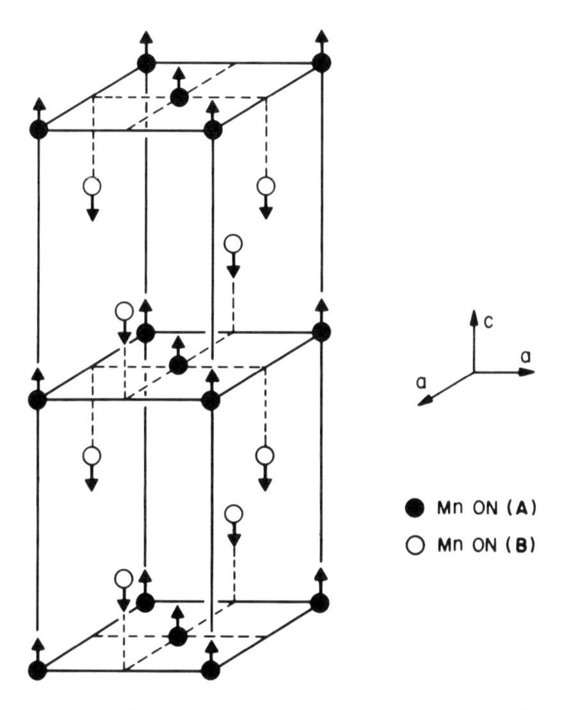

Fig. 8. Crystal and ferrimagnetic structures of Mn_2Sb. The latter shows the antiparallel coupling of Mn moments on the two sets of nonequivalent sites. Two cells are shown and the Sb atoms omitted for clarity.

disordered state, on the average, will have three Mn NN to each Mn atom.[30]

Plastic deformation of ordered Ni_3Mn also produces disorder and Mn–Mn pairs (see Section 1.6.2). As a result, the magnetization will decrease with increasing amounts of deformation.[53]

Neutron diffraction measurements have shown Mn_2Sb* to be ferrimagnetic at room temperature (Figure 8), with the Mn moments in the nonequivalent sites to be unequal and antiparallel along the *c* axis.[31] Replacing one-half the Mn and all of the Sb atoms with Al and Ge to form MnAlGe (Figure 9) results in a ferromagnetic phase whose easy axis of magnetization is the *c* axis. Thus, only long-range

* Mn_2Sb is tetragonal, C_{38} structure. The Mn atoms are distributed on two sets of nonequivalent sites designated *A* and *B*.

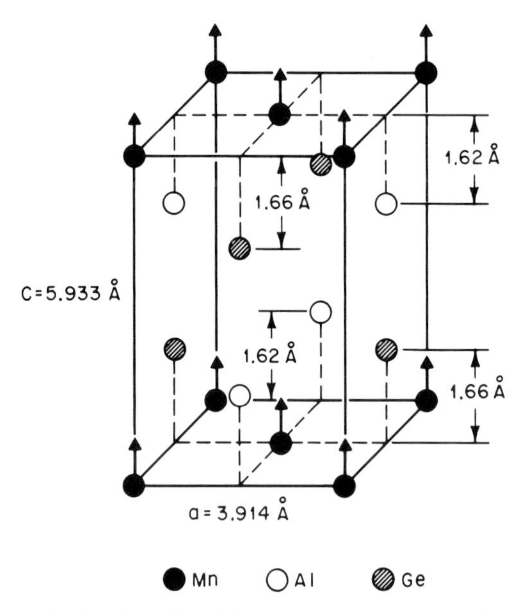

Fig. 9. Crystal and ferromagnetic structure of MnAlGe.(32–34)

ferromagnetic exchange is produced via conduction electrons by increasing the Mn–Mn distance.[32]

As indicated above, Au_4V can be made to be ferromagnetic, a rather surprising result in view of the fact that V metal does not order magnetically at any temperature. We have seen that Fe atoms develop local moment, i.e., exhibit intraatomic order, when they have Fe NN.* In contrast, V atoms appear to develop local moments in Au_4V when they do not have V nearest neighbors. Au_4V is disordered fcc at elevated temperatures and undergoes an ordering below 565°C to the Ni_4Mo body-centered tetragonal structure shown in Figure 10. Disordered Au_4V is paramagnetic at all temperatures, while ordered Au_4V becomes ferromagnetic at $T_c = 43°K$.[35,36] Later work showed $T_c = 67°K$, presumably due to better stoichiometry and order.[37,38] On the basis of magnetic susceptibility measurements[36,41] on Au_4V, NMR studies of dilute V solutions in Au,[42] and Mössbauer

* In the Mn phases discussed above, the Mn atoms exhibit local moments regardless of whether the interatomic exchange is ferromagnetic or antiferromagnetic.

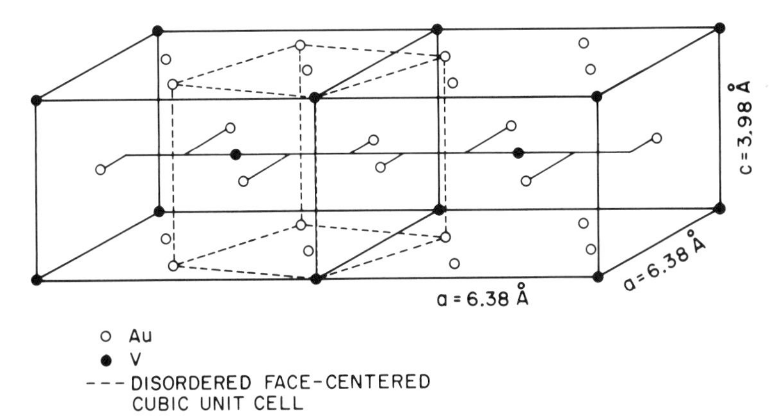

Fig. 10. The body-centered tetragonal Au_4V structure, with the Au atoms placed in ideal positions of 0.2, 0.4, 0· −0.2, −0.4, 0; −0.4, 0.2, 0; and 0.4, −0.2, 0 around each V atom; the exact values have not been determined. The (disordered) fcc unit cell is shown by the dashed lines. In the fcc structure $a = 4.04$ Å with an average volume/atom ratio of 16.5 Å^3. When transformed to the idealized bct structure, one of the *a* axes shrinks to 3.98 Å (the *c* axis) and the other two *a* axes shrinks to 4.03 Å, resulting in a drop of the average volume per atom to 16.2 Å^3 (after Chin *et al.*[38]).

effect studies of Au^{197} in Au_4V,[43–45] in particular the comparison of the hyperfine fields in Au_4V and Au_4Mn,[44,45] the values of the magnetic carrier ratio q_c/q_s being two[36,39,41] it is concluded that the V atoms possess localized moments. Note that the nearest-neighbor V–V distance is 3.98 Å in the *c* direction, and it is considerably greater than that of 2.63 Å in V metal. When disordered, either by heat treatment or plastic deformation, each V atom has, on the average, $0.2 \times 12 = 2.4$ V NN at a considerably smaller distance, 2.95 Å. As the degree of long-range order decreases, the magnetization will decrease in a continuous manner.[38,46–48]

When a coarse-grained sample is order-annealed under a uniaxial stress so that predominantly one kind of the three types of ordered domains results, the specimen exhibits strong uniaxial magnetic properties.[37] The uniaxal properties appear to be due to a large magnetocrystalline anisotropy, the *c* axis being the easy axis of magnetization.[37] A hysteresis loop of Au_4V is shown on Figure 11 and the very large coercivity, ~60,000 Oe, may be a manifestation of the large magnetocrystalline anisotropy.

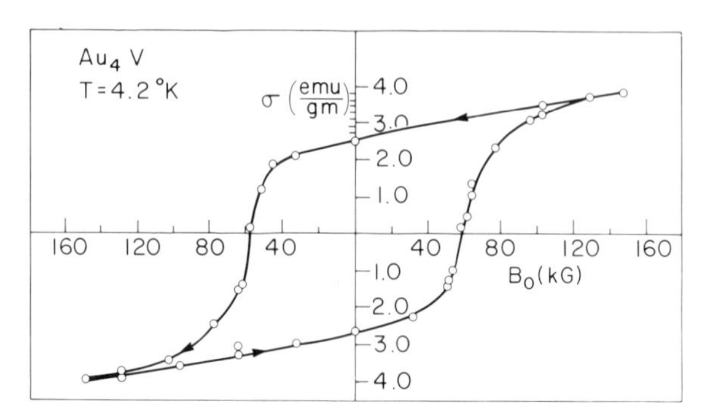

Fig. 11. Magnetic hysteresis loop for ordered Au_4V at 4.2°K (unpublished data of S. Foner).

1.5. Directional Ordering and Magnetic Anisotropy

1.5.1. Introduction

We have already seen that the nature of the local environment about a given atom, arising from ordering or disordering due to changes in stoichiometry or in composition, by crystallographic transformations, or by plastic deformation, may result in pronounced changes in bulk magnetic behavior, such as saturation magnetization, because the central atom may develop local moments which can subsequently lead to long-range magnetic order or because of a change in sign of the dominant exchange interactions. Plastic deformation of the atomically ordered forms of a number of phases, such as Ni_3Mn and Au_4V, results in disorder and drastic changes in saturation. On the other hand, plastic deformation of ordered Ni_3Fe and FeCo results in only relatively slight changes in saturation. However, a magnetic anisotropy energy is introduced which results in the formation of "easy" axes of magnetization which are directions of minimum anisotropy energy. In this section we consider specifically the effect of plastic deformation on magnetic anisotropy. Magnetic anisotropy introduced during annealing and a growth-induced anisotropy found in some solid solutions crystals of rare earth iron garnets is discussed in Volume 2, Chapter 6.

Plastic deformation of metals and alloys occurs by slip and/or twinning on certain crystallographic planes and along certain crystallographic directions and is accomplished by the movement of dislocations (Volume 2, Chapter 7). Plastic deformation results in considerable disorder, leading to phase transformations in some cases. The deformation of Au_4V, for example, leads to the formation of regions fcc in character. Crystal lattice rotation also occurs, giving rise to preferred orientations, or texture, in a polycrystalline material.

1.5.2. Deformation-Induced Magnetic Anisotropy. Ni_3Fe Alloys ($L1_2$ Structure)

The theory of directional ordering,[62,63] originally formulated to account for a related anisotropy which is induced by annealing some materials below the Curie temperature in an applied magnetic field, considers a solid solution of A and B atoms for which the pseudodipolar magnetic coupling energies of AA, AB, and BB atom pairs are different and a uniaxial magnetic anisotropy is expected whenever there is an asymmetric distribution of the atom pairs. Annealing in the presence of a magnetic field tends to align in the field direction those atom pairs with the minimum coupling energy (Volume 2, Chapter 6). Subsequently, Bunge and Müller[64] and Chikazumi *et al.*[65] proposed that the crystallographic slip process also produces an asymmetric distribution of atom pairs and hence directional order. Extensive investigations of Chikazumi and co-workers[65–67] and Chin and co-workers[68–73] have shown conclusively that "slip-induced directional order" theory accounts for the observed anisotropy.

Slip-induced directional order is illustrated in a simple manner in Figure 12 for an ordered two-dimensional alloy. Application of a shear stress causes the atoms to slip over one another to form an antiphase domain boundary. AA and BB atom pairs are created across the boundary only in the vertical direction. If the pseudodipolar magnetic coupling energy is lowest for an AB pair, the vertical direction becomes a high-energy or hard magnetic axis. Thus, the number of induced like-atom pairs will determine the strength of the anisotropy and will depend on the degree of long-range atomic order as well as on the slip character, i.e., the amount of shear on a given plane. The latter is illustrated in Fig. 12(b). By referring to the lower slipped plane, it can be seen that like-atom pairs do not form if an even number of slip steps occur. It is also important to note that the number of atom pairs and the direction of their alignment will also depend on which of several equivalent slip systems operate during the deformation, and the latter depend on initial crystal orientation (or initial texture in a polycrystalline material) and on the imposed shape change or deformation processing.

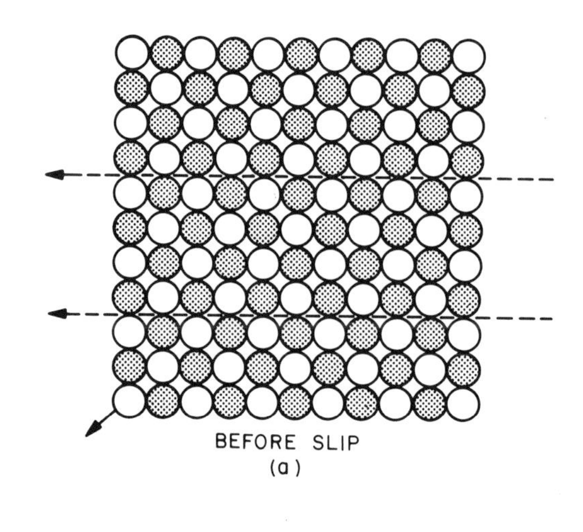

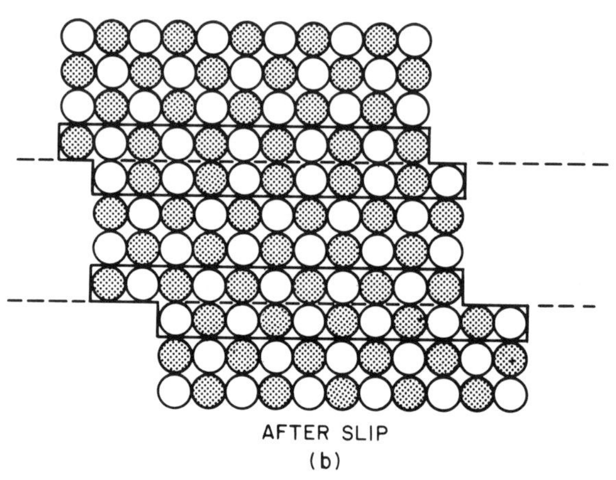

Fig. 12. Schematic illustration of slip-induced directional order in a simple AB ordered alloy. (a) Symmetric distribution of unlike-neighbor pairs prior to slip along potential horizontal planes (dashed lines), (b) atomic distribution after slip. (After Chin.[46])

Directional order theory[62,63] yields an expression for the induced macroscopic anisotropy energy E which results when there is an asymmetric distribution of atom pairs N_{BBi} (or N_{AAi} or N_{ABi})

$$E = l \sum_i N_{BBi} \cos^2 \Phi_i + \text{const}$$

where l is the pseudodipolar energy ($l \equiv l_{AA} + l_{BB} - 2l_{AB}$), N_{BBi} is

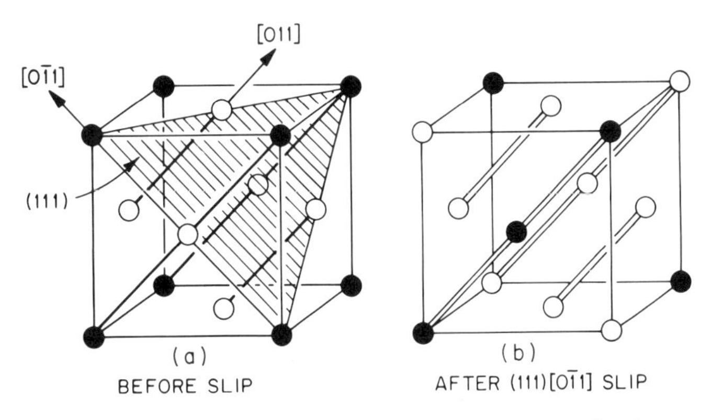

Fig. 13. Illustration of the development of like-atom pairs due to slip on the (111) [0$\bar{1}$1] system in the $L1_2$ structure. (a) Before slip in (111) [0$\bar{1}$1] system. Heavy bars join nearest neighbors in [011], which is perpendicular to the [0$\bar{1}$1] slip direction. (b) After slip on (111) planes. Double bars join like-atom pairs in [001] induced by slip. Pair distribution is unchanged in the other five ⟨110⟩ directions (after Chin[46]).

the number of BB pairs per unit volume created by slip on slip system i, and Φ_i is the angle between the local magnetization and the induced BB pair direction. The sign of l indicates whether like-atom pair or unlike-atom pair directions are hard axes of magnetization. For Fe–Ni alloys $Nl \approx 3 \times 10^8$ ergs/cm^3, while for FeCo, $Nl \approx -3 \times 10^8$ ergs/cm^3 (N is the number of atoms per unit volume).[68,72] In the former the like-atom pair direction is hard, while it is the easy axis in Fe–Co and Fe–Al alloys. The above equation has been developed in detail by Chikazumi *et al.*[66] and extended by Chin[46,68,69] for directional order induced by slip in the commercially important Ni_3Fe structure, taking into account the Bragg and Williams long-range order (LRO) and the Bethe short-range order (SRO) parameters.*

The Ni_3Fe alloys are disordered fcc at elevated temperatures and on cooling, undergo an order–disorder transformation at 503°C. They can be prepared, by proper heat treatment, to exhibit LRO, and the ordered structure is of the cubic Cu_3Au type ($L1_2$, fcc). They can also be heat treated to produce SRO. The development of like-atom pairs in ordered Ni_3Fe for (111) [0$\bar{1}$1] slip is illustrated in Figure 13.

* For a review of directional order theory and its application to a number of other structure types and the effect of plastic deformation on the magnetic properties of metals and alloys, the reader is referred to the paper by Chin.[46]

Like-atom pairs, denoted by double bars, are produced along the [011] direction perpendicular to the [0$\bar{1}$1] slip direction, while the pair distribution is unchanged in the other five ⟨110⟩ directions. For the SRO case, the direction of BB pairs induced by (111)[0$\bar{1}$1] slip are no longer along [011], but equally probable in the three ⟨110⟩ directions connecting the (111) slip planes, i.e., [011], [101], and [110].

For the LRO case[46] the total pair energy E_L is given by

$$E_L = \tfrac{1}{8} nlP_0 P' s^2 \sum_i |S_i| (\alpha_1 \beta_{1i} + \alpha_2 \beta_{2i} + \alpha_3 \beta_{3i})^2$$

where N is the number of atoms per unit volume; s is the long-range order parameter; P_0 is the probability that a unit dislocation is not paired with another; P' is the probability of passing a dislocation per slip plane; S_i is the average number of dislocations passed per slip plane; β_{1i}, β_{2i}, and β_{3i} are the direction cosines of BB pairs induced by slip on system i; and α_1, α_2, and α_3 are the direction cosines of the local magnetization.

For the SRO case[46]

$$E_S = (1/48) nlP' \sigma \sum_i |S_i| (n_{2i} n_{3i} \alpha_2 \alpha_3 + n_{3i} n_{1i} \alpha_3 \alpha_1 + n_{1i} n_{2i} \alpha_1 \alpha_2)$$

where n_{1i}, n_{2i}, and n_{3i} are the direction cosines of the slip plane normal to the ith system and σ is the Bethe short-range order parameter.

From a knowledge of the initial orientation of a crystal or the texture of a polycrystalline material, the nature of the deformation processing, active slip systems, and degree of order, the slip-induced easy axis of magnetization can be determined from the above equations.[46,68,71]

The role of initial atomic ordering on the magnitude of the induced anisotropy is illustrated in Figure 14 for crystals of 2Mo–76Ni–22Fe composition. The induced anisotropy constant is plotted against true strain (or reduction in thickness) for LRO and SRO single crystals and polycrystalline samples. The deformed LRO single crystal exhibits the strongest induced anisotropy; $K_u = 460{,}000$ ergs/cm^3 and it is the largest ever reported for alloys nominally Ni_3Fe. The SRO single crystal and LRO and SRO polycrystalline samples do not exhibit as strong an induced anisotropy or asymmetric distribution of atom pairs and this is due to poor alignment of atom pairs from grain to grain (or from SRO to SRO regions) and to greater disorder and randomization produced by intersecting slip and lattice rotation, particularly in the polycrystalline samples. This, of course,

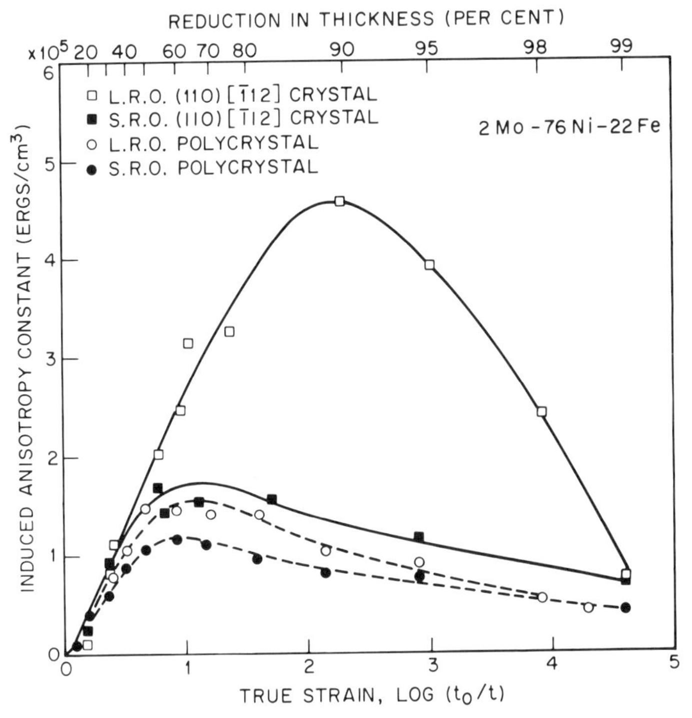

Fig. 14. Slip-induced uniaxial anisotropy constant as a function of thickness strain for single crystals compressed in (110) $[\bar{1}12]$ orientation and for rolled polycrystalline material; 2Mo–76Ni–22Fe alloy (after Chin[46]).

also accounts for the decrease in induced anisotropy at large deformations.

1.6. Magnetic Oxides

1.6.1. Introduction

We confine our discussion to the insulating ferrimagnetic oxides having the cubic spinel and garnet structures because these materials are of great technological importance.* The crystal chemistry of the ferrites and garnets offers wide latitude for the control of magnetic properties and many different compositions have been developed for

* The cubic spinel oxides and garnets containing Fe are usually referred to as ferrites or soft magnetic materials. The hexagonal $PbFe_{12}O_{19}$, $BaFe_{12}O_{19}$, and $SrFe_{12}O_{19}$ phases, useful for some permanent magnet applications, are referred to as hard ferrites.

a wide range of applications. Polycrystalline or ceramic bodies of ferrites are used in microwave devices and computer memories and as core materials for inductors and transformers. Single crystals are used for low-wearing recording heads. Some polycrystalline garnets are used in microwave devices and single crystals of solid solutions of garnets are potentially of great importance for a new class of memory and logic devices called bubble domain devices.

As will be discussed below, the bulk magnetization behavior of these materials depends on the kind, number, and distribution of magnetic ions on the crystallographic sites. However, there are many properties, such as coercive force and permeability, which are structure sensitive, and technologically useful materials can be achieved only by careful control of microstructure. Microstructure includes grain size and texture in polycrystalline ceramics and uniformity of composition in both single crystals and polycrystalline aggregates. The role of defects, both chemical and physical, on magnetic behavior, is discussed in Volume 2, Chapter 6.

1.6.2. Spinel Oxides

The general chemical formula for the *normal* spinel oxides is $A^{II}[B^{III}]_2O_4$, where A^{II} is a divalent ion with ionic radius between 0.6 and 1 Å, and includes Mn, Fe, Co, Ni, Cu, Zn, Mg, and Cd.* The B^{III} ions are usually Fe, Cr, Mn, Al, and Ga. The crystal structure is cubic with eight formula units per cell (space group O_h^7, $Fd3m$). The A and B ions occupy special positions within the lattice, while the coordinates of the positions of the 32 oxygen ions include a parameter, the so-called u parameter, which varies from oxide to oxide. The oxygen atoms are in an almost perfect close packing. If perfect, the u parameter would be 0.375 ($\frac{3}{8}$) and in actual materials u deviates from this value because of the sizes of the cations. The eight divalent cations occupy tetrahedral holes in the close-packed oxygen array, i.e., each is coordinated tetrahedrally to four oxygen ions. The 16 trivalent ions occupy 16 octahedral holes, i.e., coordinated to six oxygen ions.†

* The prototype spinel phase is $MgAl_2O_4$. If tetravalent or group four atoms are involved, the formula will be $A^{IV}B^{II}O_4$.

† There are 64 possible tetrahedral sites and 32 possible octahedral sites, but the metal ions only occupy certain of these sites. The tetrahedral sites are often referred to as the A sublattice and the octahedral sites as the B sublattice.

In the *inverse* cubic spinels, the eight divalent ions occupy one-half of the octahedral sites and eight of the trivalent ions occupy the eight tetrahedral sites. Thus, the chemical formula for the inverse spinel can be written as $B^{III}[A^{II}B^{III}]O_4$. While $ZnFe_2O_4$ is a normal spinel, $NiFe_2O_4$ or $Fe[NiFe]O_4$ is an inverse spinel. Chromites with spinel structure are always normal.

The cation distribution in completely normal and completely inverse spinels represent the two extreme cases and most of the Fe-containing oxides are intermediate between them. For example, a portion of the Mn ions in $MnFe_2O_4$ are on octahedral sites and the chemical formula illustrating this distribution is $Mn_{0.8}Fe_{0.2}\cdot[Mn_{0.2}Fe_{1.8}]O_4$. Similarly, the formula $Mg_{0.1}Fe_{0.9}[Mg_{0.9}Fe_{1.1}]O_4$ depicts the cation distribution in $MgFe_2O_4$. The tetrahedral site is smaller and trivalent ions, which are usually smaller, tend to favor tetrahedral sites and inversion. Even within a pseudobinary system between two spinels an inversion from one distribution to the other can occur. X-ray, neutron, and Mössbauer spectroscopy, together with bulk magnetization measurements for a number of cases, yield site distribution information and magnetic moments associated with the ions.[57,58]

The general cation distribution can be written[55] as $M^{II}_{\delta}Fe^{III}_{1-\delta}\cdot[M^{II}_{1-\delta}Fe^{III}_{1+\delta}]O_4$, where δ is a measure of the degree of inversion and for $\delta = \frac{1}{3}$ a completely random distribution of cations is present. High temperatures favor randomization and it is apparent that the method and temperature of preparation and heat treatment will affect the ion distribution and thus the magnetic and transport properties.

Several AB_2O_4 phases exhibiting tetragonal and orthorhombic symmetry are known and can be derived by distortion of the cubic spinel structure. The distortion can occur via the Jahn–Teller effect (next section) or by an ordering of the cations on the A or B sublattices. An example of the latter is the inverse spinel Fe_3O_4 ($Fe^{III}[Fe^{II}Fe^{III}]O_4$). Below $\sim 120°K$ the Fe^{II} and Fe^{III} ions are ordered on the B sublattice resulting in an orthorhombic structure or superstructure. Now this is not a classical order–disorder, in the sense that ions do not diffuse at this low temperature. The extra valence electrons only order, i.e., the temperature is sufficiently low that charge hopping does not occur (resistivity increases rather abruptly below this temperature). An example of an order–disorder system is $Fe[Li_{0.5}Fe_{1.5}]O_4$. The Li and Fe ions are ordered on the B sublattice below $\sim 1000°K$.[55]

In the above discussion we have used formal charges to describe the oxidation state of the cations and the implication was given that the cations were indeed ions exhibiting oxidation states equal to the formal charges. Although the bonding in these materials is indeed quite ionic, there is a covalent contribution to the bonding. For example, the divalent cation is tetrahedrally coordinated to four oxygens, implying sp^3 hybrid covalent bonds and the trivalent cation is coordinated to six oxygens, implying octahedral hybrid covalent bonds involving d orbitals (d^2sp^3, for example, if perfect octahedral coordination).

Several factors in addition to ionic size enter into the choice of lattice site preferred by a particular cation. For example, divalent Zn and Cd are larger than divalent Ni and trivalent Cr, yet the former show a marked preference for the tetrahedral sites, while the latter prefer an octahedral environment. The s electrons of Zn and Cd can easily form sp^3 hybrid covalent bonds with oxygen, while the crystal field of the octahedral environment lowers the energy of Ni and Cr (crystal field stabilization).* Electrostatic considerations also enter in, i.e., ions of small positive charge will be coordinated to a smaller number of negative ions. Calculations suggest that the inverse structure is favored (lowest lattice energy) when $u < 0.379$ and the normal structure is favored when $u > 0.379$.[(55)]

In the spinel lattice each oxygen is coordinated to three octahedral-site cations and one tetrahedral-site cation, forming a slightly distorted tetrahedron. The net moments on the tetrahedral and octahedral sublattices are in opposition to one another, and magnetic behavior will be affected by cation distribution. The principal magnetic exchange interactions are of the superexchange type via the oxygen ions. The negative A–O–B interaction is the strongest and is responsible for the ferrimagnetic behavior. The A–O–B angle is approximately 125°. The A–O–A and B–O–B superexchange interactions are generally weak, although in the nonoxide spinels the positive B–X–B interactions can be relatively strong.

1.6.3. Jahn–Teller Effect

The Jahn–Teller effect[(54)] refers to the distortion in the NN coordination about $3d$ transition metal ions because of the distribution of electrons in the d orbitals in various crystalline fields (see also

---

* In general, the crystal fields of the oxide spinels is sufficient to quench the orbital angular momenta of the d electrons so that the spin angular momenta only determine the magnetic moments associated with the magnetic ions.

Chapter 3, and Volume 2, Chapter 1). The distortion of the local ligand symmetry, in extreme cases, results in cooperative, macroscopic phase transformations. A well-known example is the cubic-to-tetragonal phase transformation which occurs in some transition metal spinels (e.g., Mn_3O_4, $ZnMn_2O_4$, $CuFeO_4$). For the high-spin ground-state electron configurations $3d^4$ and $3d^9$ in an octahedral crystalline field, tetragonal distortion of the octahedra occur. Under the influence of a tetrahedral field high-spin ions having d^3, d^4, d^6, d^8, and d^9 configuration will suffer Jahn–Teller distortions.*

By way of an example, the spinel oxides will be briefly considered. As indicated above, they can be discussed in terms of two extreme cases, normal and inverse spinels, and they can exist with intermediate ion distributions, and ionic size, electron configurations, stoichiometry (method of preparation), and heat treatment will influence the distribution of the transition metal ions.[55,56]

The sites occupied by a given cation also depend to a great extent upon the nature of the other cations present. The Cr^{3+} ion is an ion which appears always to occupy octahedral sites. It is a d^3 ion and, in the high-spin state, the occupied orbitals are directed away (antibonding orbitals) from the ligands and will not give rise to a Jahn–Teller distortion.

$NiFe_2O_4$ is an inverse spinel and shows no Jahn–Teller distortion (Ni^{2+}, $3d^8$, in high-spin) while $NiCr_2O_4$ is a normal spinel and exhibits a Jahn–Teller distortion.

$CuFe_2O_4$ is an inverse spinel. Since Cu^{2+} ions ($3d^9$) occupy octahedral sites, large tetragonal distortions arise and a cubic-to-tetragonal phase change occurs on cooling, and thus the relative site occupancy of the Cu ions affects the transition temperature and magnetic properties.[55] On the other hand, $CuCr_2O_4$ is a normal spinel and since the Cu^{2+} ions occupy tetrahedral sites, a Jahn–Teller distortion also occurs in this material.[55]

During the past few years cooperative distortions attributable to Jahn–Teller splitting have been observed at low temperatures, in several materials containing rare earth ions. Examples are $DyVO_4$,[59] $TmAsO_4$,[60] and $DyAlO_3$.[61] In $DyVO_4$ and $TmAsO_4$ the crystallographic phase transition is magnetically controllable.

* Whether an ion is in a high-spin or low-spin state will depend on the strength of the crystalline field.

1.6.4. Rare Earth Iron Garnets

The rare earth (R) iron garnets, $R_3Fe_5O_{12}$ or $R_3^{3+}Fe_2^{3+}Fe_3^{3+}O_{12}^{-2}$, are crystallographically cubic and belong to space group *Ia3d*.* There are eight formula units per cell. The 24 R ions occupy the (24*c*) positions and the 40 Fe ions occupy the (16*a*) and (24*d*) positions. The oxygen ions occupy the (96*h*) positions and require three parameters to specify their coordinates.

Of the 40 Fe ions, 24 (Fe_3^{3+}) occupy tetrahedral sites, while 16 (Fe_2^{3+}) occupy octahedral sites. Each R ion is coordinated to eight oxygen ions in a so-called dodecahedral site.

The R-ion garnets are ferrimagnetic and the long-range interactions between localized moments are via superexchange through the oxygen ions. The moments of the Fe^{3+} ions ($3d^5$ configuration) on the tetrahedral sites, although parallel to one another, are antiparallel to the moments of the Fe^{3+} ions on the octahedral sites, and since there are three tetrahedral ions to every two octahedral ions, there is a net unbalance of five unpaired electrons, leading to relatively low net magnetic moments and thus low saturation magnetizations near room temperature. The moment of the rare earth ion sublattice contributes somewhat to the net moment in the neighborhood of room temperature because of partial alignment or polarization due to the net moment of the Fe ions. However, at low temperatures the rare earth ion contribution becomes increasingly important and dominates, leading to compensation points† for the heavy rare earths. By substituting other cations (magnetic and nonmagnetic), the magnetic properties can be altered.

For example, $Y_3Fe_5O_{12}$ is ferrimagnetic but shows a ferromagnetic magnetization–temperature curve ($T_c \approx 500°K$) which reflects the net magnetization of the Fe sublattices (Y^{3+} is diamagnetic). $Gd_3Fe_5O_{12}$ is also ferrimagnetic, but since the Gd^{3+} sublattice moment is antiparallel to the net Fe moment, a compensation temperature ($\sim 300°K$) is exhibited. The compensation temperature and total magnetization behavior can be altered, for example, by substituting Y^{3+} for Gd^{3+}. Similarly, the magnetic behavior of the garnets can be altered by substituting diamagnetic (full-shell) trivalent ions, such as Al^{3+} and Ga^{3+}, for Fe ions. Garnets, nonmagnetic at

* The prototype material is the mineral grossularite, $Ca_3^{2+}Al_2^{3+}Si_3^{4+}O_{12}$.

† The compensation point is the temperature where the moment of the R sublattice is equal, but antiparallel, to the net moment of the Fe sublattices, resulting in zero bulk moment if the crystal is fairly pure and perfect.

room temperature, can be prepared by completely replacing the Fe ions (e.g., $Nd_3Ga_5O_{12}$ and $Y_3Al_5O_{12}$). Such garnets are of importance for nonmagnetic substrates for epitaxial growth of films of the magnetic garnets.

The cationic size as well as the relative size of the cations and thus the lattice constant are important in determining the existence of some garnets and site preferences. For example, the R^{3+} ions prefer the dodecahedral or *c* sites, yet the garnet $Nd_3Fe_5O_{12}$ does not form, presumably because of size considerations.[55a] On the other hand, Nd can be substituted for some of the Y in $Y_3Fe_5O_{12}$. Large divalent ions, such as Ca^{2+}, can be substituted on *c* sites by simultaneous substitution of tetravalent ions on the tetrahedral (*d*) or octahedral (*a*) sites for maintenance of charge neutrality.[55a] Small R ions, such as Yb, Tm, Er, Ho, and Dy, can also enter octahedral sites (e.g., $\{Nd_{3-y}R_y\}[R_xGa_{2-x}(Ga_3)]O_{12}$).[55a,55b] Both Al^{3+} and Ga^{3+} prefer the tetrahedral and octahedral sites. It is apparent that the crystal chemistry of the garnets offers wide latitude for the control (and adjustment) of magnetic behavior. Geller[55a] has recently presented a rather extensive review of the crystal chemistry of the garnets.

1.7. Magnetic Semiconductors

1.7.1. Introduction

Magnetic semiconductors have been receiving increased attention in recent years from scientific and technological viewpoints. Scientific interest centers on elucidating the electronic structure, conduction mechanisms, and magnetic exchange interactions, while application-oriented research is usually concerned with exploiting the interesting magnetoresistance and magnetooptical properties exhibited by many of these materials. Many show semiconductor or insulator-to-metallic transitions at a critical temperature or pressure. Interatomic distances and doping (including stoichiometry) affect the magnetic interactions and transport properties and the electronic structure is affected markedly by magnetic ordering, thus affecting magnetic and transport properties.

In general, most of the magnetic semiconductors are to be found among the oxides and chalcogenides (sulfides, selenides, and tellurides) of transition elements and may exhibit one of the three possible types of magnetic order. For the insulating oxides the conduction electrons belong to bands formed primarily from the *d* orbitals of the cations. The *d* bands are narrow because the ions are relatively far apart. This gives rise to low carrier mobilities (of the order of 0.1 cm^2/V-sec) and conduction at elevated temperatures is most likely due to a thermally activated hopping mechanism, particularly in the highly ionic solids, such as ferrites and garnets. Another consequence of the narrowband or localized nature of the *d* electrons is the postulated existence of quasiparticles called polarons.[74–77] Semiconducting behavior in the high-energy-gap materials, i.e., the oxides, is a result of the lack of stoichiometry or the presence of other ions of different oxidation state. Chalcogenides, such as ferromagnetic $CdCr_2S_4$, $CdCr_2Se_4$, and ferromagnetic europium chalcogenides (EuO and EuS), with energy gaps below 2 eV, exhibit larger mobilities (10–100 cm^2/V-sec) and the conduction mechanism in these materials is still uncertain. For reviews of magnetic semiconductors, polarons, conduction mechanisms in insulators, and metal–insulator transitions, the reader is referred to Refs. 74–77 and to Chapter 4, Volume 2.

In this section only the highlights of a few magnetic semiconductors of simple structure are discussed in the general context of this chapter.

1.7.2. Metal–Insulator Transitions

Crystallographic transitions and magnetic ordering sometimes accompany metal (M)–insulator (I) transitions and changes in composition (and stoichiometry) and pressure may have a pronounced effect on the transition temperature and magnetic and electrical properties. We illustrate some of this behavior for the oxide V_2O_3. The M–I transition in this material was discovered by Föex[110] and has been recently studied in detail largely by McWhan and his associates.[78–82] Above ~155°K, V_2O_3 is metallic and paramagnetic[81] and has the hexagonal corundum (Cr_2O_3) structure. At ~155°K on cooling V_2O_3 transforms abruptly to an insulator and the structure distorts to one of monoclinic symmetry. Associated with this transition to the insulating state, and structural change, is an increase in the nearest-neighbor V–V distance[82] and localization of *d* electrons on the V sites to give localized moments and long-range antiferromagnetic (AF) order. Mott[84] proposed that long-range Coulomb interactions between charge carriers could lead to localization of conduction electrons and nonconducting behavior and the transition would be discontinuous as a function of volume, i.e., would occur at a critical value of the interatomic distance. In fact, the antiferromagnetic phase is suppressed by pressure so that an antiferromagnetic insulator-to-metal transition occurs as a function of volume at low temperatures. A metal–insulator transition involving localization–delocalization of electrons is often referred to as a Mott transition. It is important to note that metal–insulator transitions have been observed which do not require invoking localization–delocalization of conduction electrons. As an example, the broadband semiconductors such as Si, Ge, and some 3–5 compounds, undergo first-order transitions to the metallic state at elevated pressures.[85] The metallic phases are of different crystal structure and possess

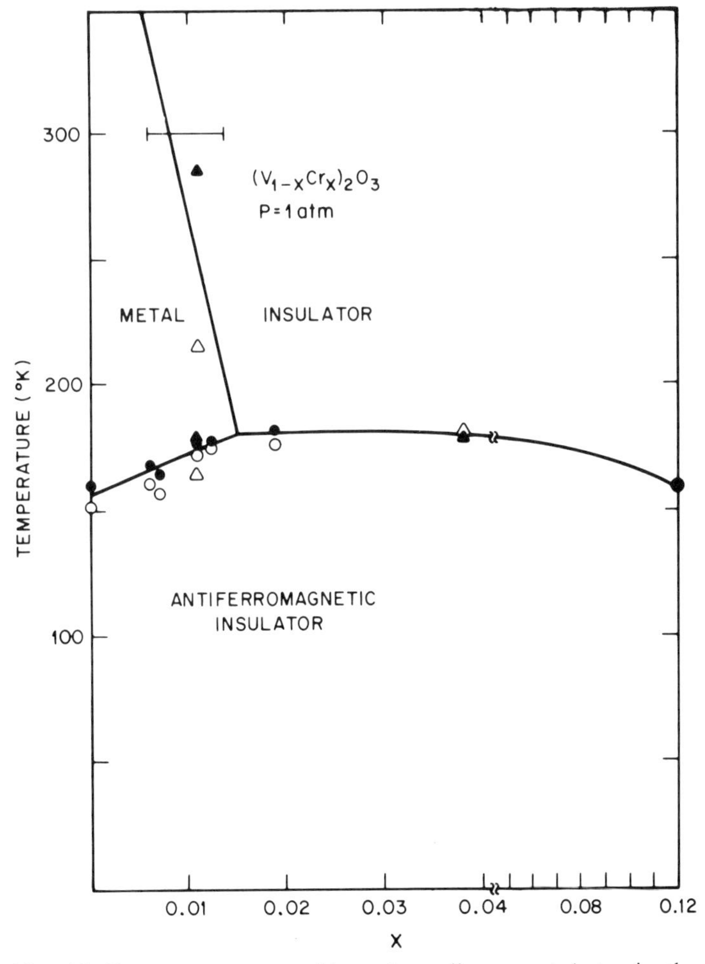

Fig. 15. Temperature–composition phase diagram at 1 atm in the system $(V_{1-x}Cr_x)_2O_3$ (after McWhan and Remeika[78]).

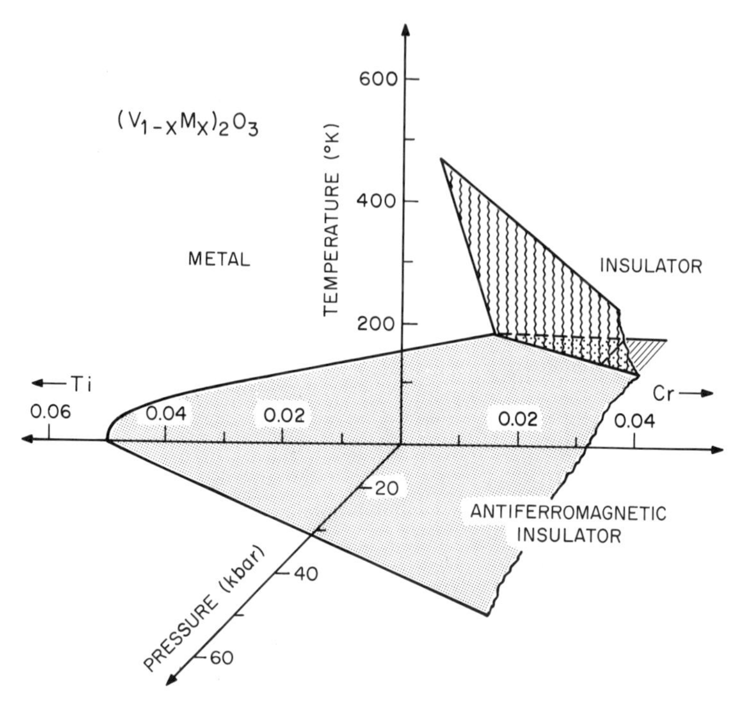

Fig. 16. Temperature–pressure–composition phase diagram showing M–I, M–AF, and I–AF surfaces for $(V_{1-x}M_x)_2O_3$, where M = Ti, Cr (after McWhan *et al.*[111]).

higher atomic coordination and these transitions can be discussed in terms of band theory only.

Because Cr_2O_3 is insulating at room temperature and the c/a ratio behaves anomalously[87] at the V-rich end of the $(V_{1-x}Cr_x)_2O_3$ system, McWhan and his associates investigated the V-rich end of the system via pressure, magnetic, electrical, and structural studies. The results of much of this work are summarized in Figures 15 and 16 in the form of temperature–composition and temperature–pressure–composition phase diagrams. Above about 1.5% Cr in V_2O_3 (Figure 15) the formation of the metallic phase is completely suppressed. On the other hand, application of pressure results in the formation and stabilization of the metallic phase which at 1 atm does not show a M–AF transition. Replacement of V by Ti^{3+}, with one d electron, has the opposite effect of Cr with one more d electron, i.e., suppresses the M–AF transition and for $x \geq 0.05$ in the $(V_{1-x}Ti_x)_2O_3$ system it appears that the AF insulating phase does not exist (Figure 16).[88,111]

An example of a semiconductor-to-metal transition without any change in crystal structure is the recently observed pressure-induced transitions in SmS, SmSe, and SmTe (cubic NaCl structure).[86] For SmS (optical gap $\approx$ 0.2 eV) the transition occurs at a discrete pressure (6.5 kbar), while for SmSe $\Delta E \approx$ 0.46 eV) and SmTe ($\Delta E \approx$ 0.6 eV) the transition occurs continuously with increasing pressure (0–50 kbar for SmSe and 0–60 kbar for SmTe). The transitions appears to be due to the promotion of a $4f$ electron to a $5d$ state with increasing pressure, thus converting Sm^{2+} to smaller Sm^{3+} ions.[86] They are not considered Mott transitions.

1.7.3. Cadmium Chromium Spinel and Europium Chalcogenides

All of the magnetically ordered semiconducting spinels contain Cr^{3+} ions on the octahedral sites. If there are no magnetic ions on the tetrahedral sites, the phases can exhibit ferromagnetic or antiferromagnetic ordering. If there are magnetic ions, such as Fe^{2+}, Co^{2+}, or Mn^{2+} on the tetrahedral sites (e.g., $FeCr_2S_4$), ferrimagnetic ordering is exhibited.

The magnetically ordered behavior of semiconducting chalcogenide spinels containing magnetic ions solely on the octahedral sites appears to result from a competition between antiferromagnetic and ferromagnetic interactions. In general, the sign of θ in the Curie–Weiss relation for the paramagnetic susceptibility, $\chi_m = C_m(T - \theta)^{-1}$, indicates whether ferromagnetic or antiferromagnetic interactions dominate. This information is obtained from measurements above the ordering temperature. If θ is positive, ferromagnetic interactions dominate. Yet, even with large, positive θ, antiferromagnetic interactions can be sufficiently strong to produce an antiferromagnetically ordered ground state at low temperatures.[89] Experimental and theoretical studies[90] indicate that next-nearest-neighbor and more distant neighbor cation–cation interactions must be considered in accounting for the occurrence of long-range magnetic order via superexchange.

Both $CdCr_2S_4$ and $CdCr_2Se_4$ are ferromagnetic,* and magnetic, electrical, and optical energy gap data for them are shown in Table 6. The octahedral crystal field quenches the orbital angular momenta

* $CuCr_2S_4$, $CuCr_2Se_4$, and $CuCr_2Te_4$ are *metallic* ferromagnets with T_c = 420, 460, and 365°K, respectively. On the other hand, replacement of one Se in $CuCr_2Se_4$ with Br ($CuCr_2Se_3Br$) gives rise to a semiconductor with T_c = 274°K.[91]

TABLE 6
Some Structural, Magnetic, Optical, and Electrical Data for Ferromagnetic $CdCr_2S_4$ and $CdCr_2Se_4$[92,93,95]

	Lattice parameter, Å	T_c, °K	θ, °K	ΔE (eV) at 300°K	Electron mobility at 300°K, cm^2/V-sec	Hole mobility at 300°K, cm^2/V-sec
$CrCr_2S_4$	10.244	84.5	156	~ 1.6	<0.5	~ 30
$CdCr_2Se_4$	10.755	129.5	210	(1.2–1.3)	7	~ 30

of the Cr ions. They are in the high-spin d^3 ($S = \frac{3}{2}$) configuration and occupy the t_{2g} states. A simplified room-temperature, zero-field band structure scheme for stoichiometric $CdCr_2Se_4$ is schematically illustrated in Figure 17. The t_{2g} orbitals overlap a Se p band, but still remain localized. This band picture has been useful in interpreting magnetic and transport data. Magnetic ordering markedly affects the band structure and thus the transport and optical properties and an applied magnetic field has a profound effect on the electronic properties.

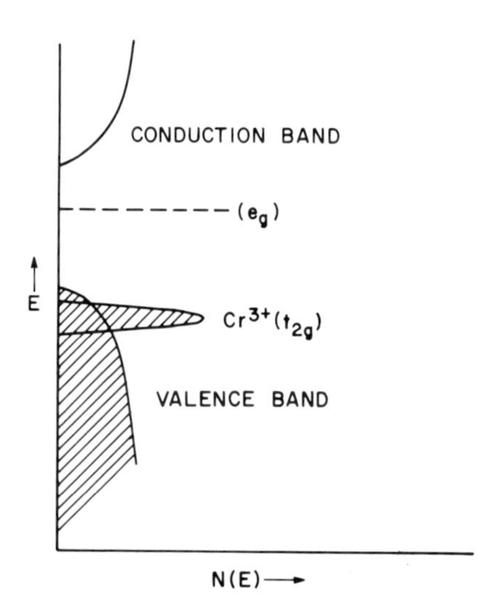

Fig. 17. Simplified band structure scheme at room temperature and zero applied magnetic field proposed for $CdCr_2Se_4$.

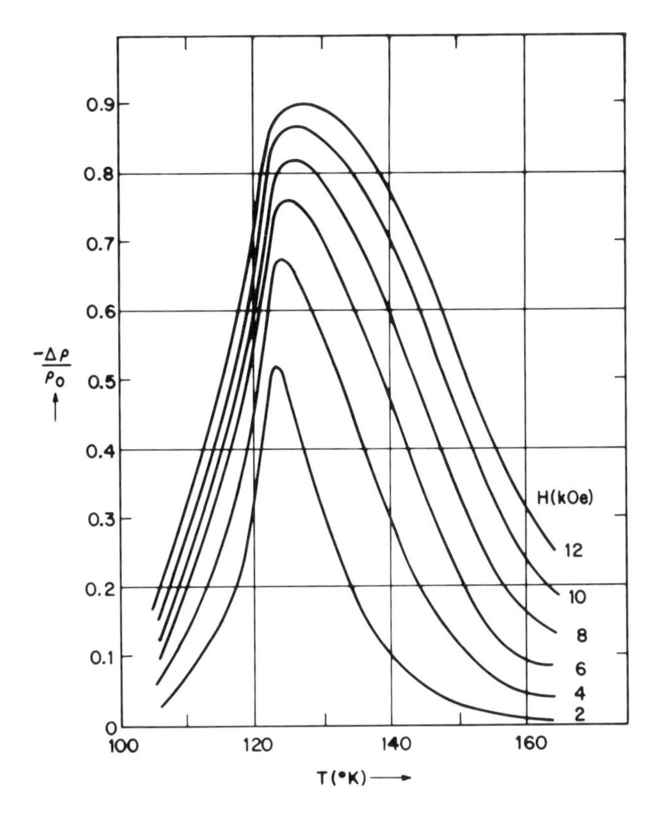

Fig. 18. Observed magnetoresistance for *n*-type $CdCr_2Se_4$ (after Bongers *et al.*[94]).

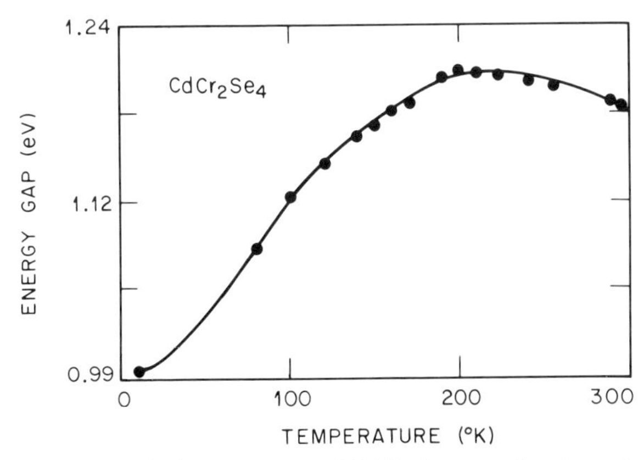

Fig. 19. Optical energy gap of $CdCr_2Se_4$ as a function of temperature (after Shepherd[95]).

Doping with Ag and Au produces *p*-type conductivity, while Ga or In yields *n*-type conductivity. The *n*-type material exhibits a large resistivity peak and "giant" negative magnetoresistivities near the Curie temperature (Figure 18), while *p*-type material does not show a resistivity maximum and only a small magnetoresistance effect.[93,94] The optical gap of $CdCr_2Se_4$ shows the normal increase with decrease in temperature, but reaches a maximum near the Curie temperature and then decreases (Figure 19). A model for the interaction of charge carriers with localized moments has been proposed which results in spin-splitting of the valence and conduction bands and spin-disorder scattering and satisfactorily accounts for the above results.[94]

The europium chalcogenides, EuO, EuS, EuSe, and EuTe, having the cubic NaCl structure, are a group of magnetic semiconductors that have received considerable attention from the standpoint of magnetooptical properties.[96–98,100–102] The Eu ion in these phases is divalent and in an *S* state with $S = \frac{7}{2}$. Each Eu^{2+} ion has 12 Eu^{2+} ions as nearest neighbors and six Eu^{2+} ions as next-nearest neighbors. Some structural, magnetic, and optical energy gap data are shown in Table 7. Both EuO and EuS are ferromagnetic and the magnetic ordering temperatures decrease with increasing volume of the unit cell. The chalcogenide EuSe is weakly antiferromagnetic but can be made ferromagnetic by an applied field (metamagnetic) or by doping with several per cent of a trivalent R atom, such as Gd. The ordering temperature of EuO is also increased by doping.[109] The chalcogenide EuTe is metamagnetic. A plausible band structure scheme, based on optical data, suggests that localized 4*f* states are located within the gap. The empty 5*d* states form a narrow band of empty states located near the bottom of the conduction band. The conduction band is 6*s*

TABLE 7
Some Structural, Magnetic, and Optical Energy Gap Data for Eu Chalcogenides[96–98,100–102]

	Type of ordering	Lattice parameter, Å	T_c, °K	θ, °K	ΔE(eV) at 300°K	Temperature of ΔE(max), °K
EuO	Ferromagnetic	5.143	69.2	~80	1.115	90
EuS	Ferromagnetic	5.968	16.5	19	1.645	36
EuSe	Antiferromagnetic	6.197	$T_N = 4.6$	9	1.780	20
EuTe	Antiferromagnetic	6.603	$T_N = 9$	−6	1.050	—

in character, while the valence band is *p*-like. Magnetic exchange splits the bands.

Two magnetic interatomic exchange interactions are operative: (1) a so-called direct interaction involving a virtual transition of a 4*f* electron to a 5*d* state which strengthens a positive, direct exchange occurring via overlap of 4*f* wave functions, and (2) a negative, indirect interaction via the anions.[99–106] The negative, indirect superexchange via the anions becomes increasingly important[99,106] with increase in atomic number of the anion, even though θ is still positive for EuSe. A somewhat similar situation exists for isostructural MnO, MnS, and MnSe,[108] which are all antiferromagnetic (T_N = 122, 165, and 190°K, respectively) and it appears that indirect superexchange becomes larger (and more dominant) with increasing lattice parameter.

The behavior of the Eu chalcogenides resembles that of the chalcogenide spinels in several respects. The optical energy gap (ΔE) of the Eu chalcogenides increases with decreasing temperature and reaches a maximum value at temperatures somewhat above the ordering temperature (Table 7). An applied magnetic field at low temperatures decreases the energy gap and this is illustrated in Figure 20 for EuS. The Eu chalcogenides exhibit maxima in resistivity and giant negative magnetoresistivities near the ordering temperatures[109] and the increase in T_c due to doping is attributed to enhanced positive exchange via the conduction electrons.[106]

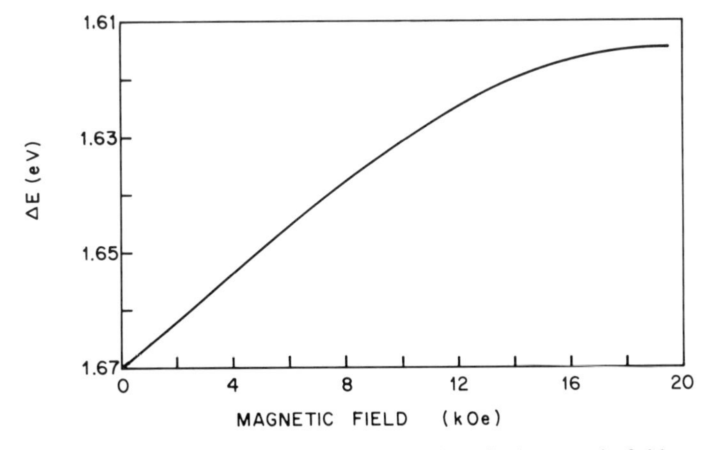

Fig. 20. Optical energy gap as a function of applied magnetic field at 20.1°K for EuS (after Busch and Wachter[96]).

1.8. Linear and Two-Dimensional Magnetic Behavior

Crystals in which the magnetic atoms are located on well-separated linear chains or in two-dimensional layers may be expected to behave magnetically as linear and two-dimensional systems, respectively. The main impetus for the study of such phases has been the desire for the acquisition of information relevant to the theories of magnetic order. A linear chain of interacting magnetic spins will not exhibit long-range order for either Heisenberg exchange or even for the pure Ising case ($J_{ij}^{\perp} = 0$).[112] Long-range order exhibited by real systems results from interchain interactions.[113] An isolated layer will likewise not support[114] long-range magnetic order if the magnetic coupling is isotropic (Heisenberg model),* but even the smallest anisotropy will result in a finite magnetic ordering temperature. For a review of the state of theory and experiment the reader is referred to Ref. 116 and the references contained therein. In this section several examples will be presented to illustrate two-dimensional magnetic behavior.

Both $CrCl_3$ and $CrBr_3$ are isostructural and exhibit a hexagonal layer structure. The layers of Cr ions are separated by two layers of Cl ions at a distance of about 6 Å. In $CrBr_3$ the Cr^{3+} spins are ferromagnetically coupled within the layers and are parallel to the c axis. In $CrCl_3$ the Cr^{3+} spins are also ferromagnetically coupled within the layers but are normal to the c axis. Weak interlayer interactions, due to dipolar forces and/or superexchange, occur in both materials, but the interlayer interactions are ferromagnetic in $CrBr_3$ ($T_c = 37°K$) and antiferromagnetic in $CrCl_3$ ($T_N = 16.8°K$).[117] The exchange constant J within the layers in $CrCl_3$ is small and a field of ~ 160 Oe applied parallel to the spins (normal to the c axis) will produce spin-flip, resulting in metamagnetic behavior and a two-dimensional ferromagnet.[117]

The structure K_2NiF_4 (Figure 21) is an example in which antiferromagnetic coupling exists within the plane[116,118,119] (spins parallel to c-axis). In this structure each Ni^{2+} ion layer is separated by two nonmagnetic KF layers. The magnetic exchange interaction between layers, which will primarily be due to superexchange through the nonmagnetic intermediaries or dipole–dipole, is expected to be several orders of magnitude lower than the intralayer exchange.[116] This fluoride is regarded as a near-Heisenberg antiferromagnet exhibiting anisotropy-induced magnetic ordering[120,121] and recent

* However, such systems might undergo a magnetic phase transition involving no long-range order, according to the theoretical treatment of Stanley and Kaplan.[115]

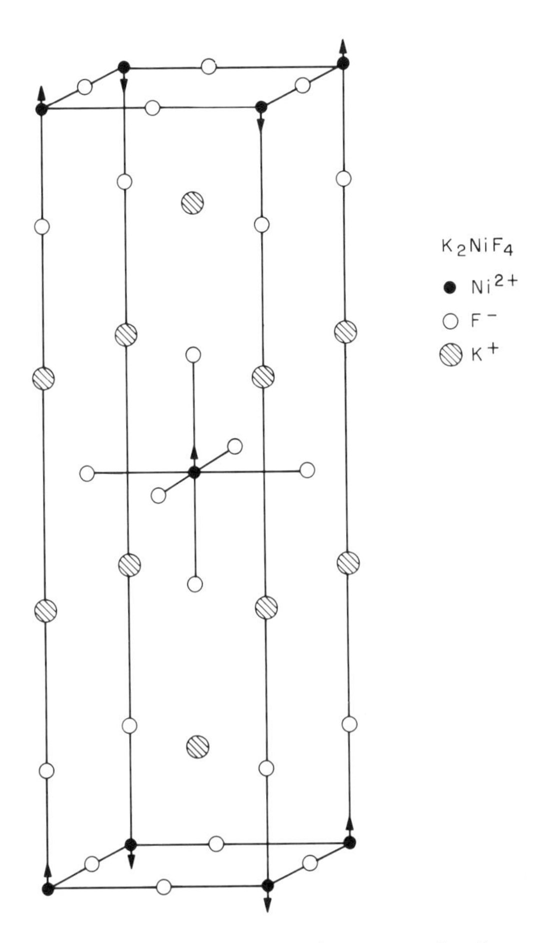

Fig. 21. The K_2NiF_4 crystal structure showing the magnetic ordering of the Ni^{2+} ions. The $NiKF_3$ perovskite layer is apparent in the center of the structure (after Lines[116]).

neutron scattering studies show K_2NiF_4 to be a true two-dimensional antiferromagnet.[119]

On the other hand, Rb_2FeF_4, isostructural to K_2NiF_4, is more anisotropic and this results in the Fe spins having directions lying in the layers.[122] The transition to long-range order in Rb_2FeF_4 is accompanied by a crystallographic distortion.

Increased separation of planes containing magnetic ions will, of course, result in a reduction of interplanar interactions. This has been accomplished by synthesis of compounds of the form

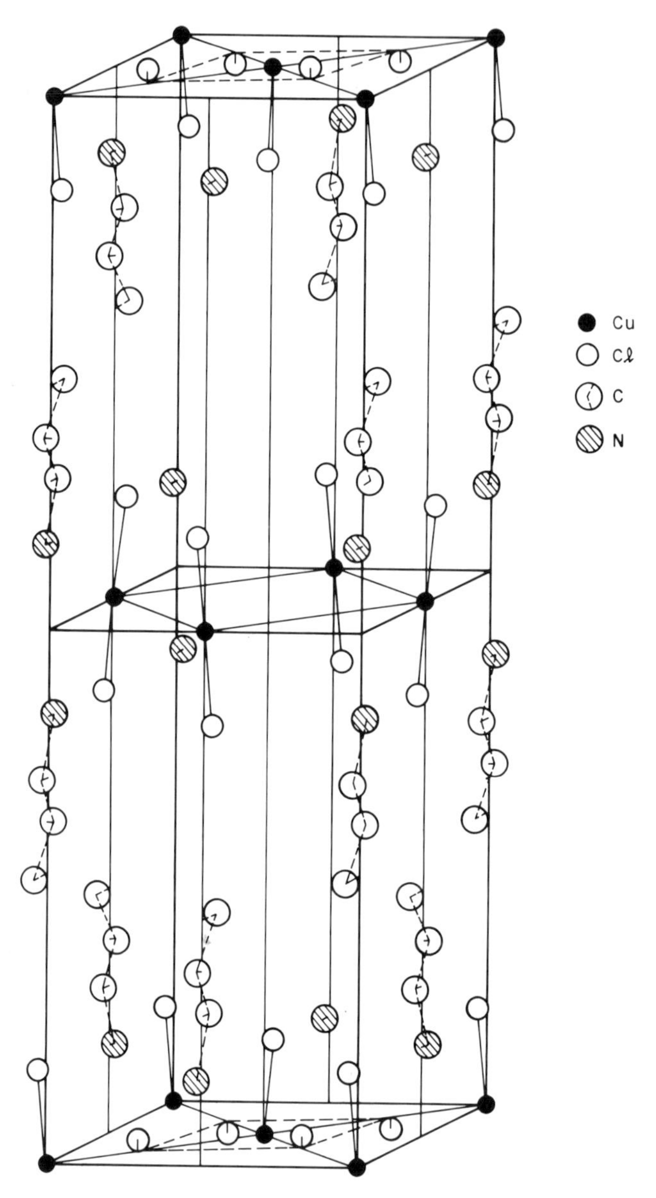

Fig. 22. Crystal structure of $Cu(C_3H_7NH_3)_2Cl_4$. Not all propyl groups shown, for the sake of clarity (after de Jongh *et al.*[123]).

$Cu(C_nH_{2n+1}NH_3)_2Cl_4$.[123] They exhibit magnetic order and approximate a two-dimensional Heisenberg system.[123] These layered materials are orthorhombic and the crystal structure is shown in Figure 22. The $CuCl_2$ layers are separated by two layers of $(C_nH_{2n+1}\cdot$

TABLE 8
Ordering Temperatures and Interionic Distances for Some Chlorides and Bromides of the Form $Cu(C_nH_{2n+1}NH_3)_2Cl_4$ (after de Jongh *et al.*[123])

Compound	T_0, °K	d_1, Å	d_2, Å
$Cu(CH_3NH_3)_2Cl_4$	8.9	5.247	9.99
$Cu(C_2H_5NH_3)_2Cl_4$	10.2	5.240	11.22
$Cu(C_5H_{11}NH_3)_2Cl_4$	7.26	5.265	17.80
$Cu(C_2H_5NH_3)_2Br_4$	10.72	5.541	11.44
$Cu(C_3H_7NH_3)_2Br_4$	10.41	5.548	12.78

NH_3)Cl and the Cu^{2+} ions form a two-dimensional magnetic lattice. Ordering temperatures T_0 for some of these materials, as well as for two isostructural bromides, are given in Table 8. The distances between Cu ions within each layer (d_1) and between nearest-neighbor Cu ions in different layers (d_2) are also shown. Note that d_1 remains nearly constant, while d_2 varies from ~ 10 to 13 Å. Although the susceptibility data on these materials suggest ferromagnetic interactions of the Cu^{2+} ions within the layers, the behavior below the ordering temperature differs for the various compounds. The analogous ferrous complexes have recently been synthesized and susceptibility and Mössbauer studies indicate antiferromagnetic behavior with ordering temperatures between 72 and 96°K.[124]

1.9. Amorphous Magnetic Materials

It is apparent from the above discussion that the lack of long-range atomic order in the presence of a regular crystalline lattice, although affecting the sign and magnitude of the net exchange interaction, is not an impediment to the existence of long-range magnetic order. In fact one need only recall that atomically ordered FeAl does not order magnetically. Only when Fe atoms have Fe nearest neighbors do localized moments and long-range magnetic order develop. The question naturally arises whether long-range magnetic order can exist in the extremely disordered state, i.e. in amorphous material, in which short-range atomic order is confined to perhaps the first-neighbor shell. In 1960 Gubanov[125] theoretically predicted that ferromagnetism could be exhibited by amorphous materials. In comparison with its crystalline counterpart, the amorphous phase could exhibit higher or lower magnetizations and Curie temperatures, depending on whether the exchange integral increases or decreases. Handrick[126] considered a model in which the exchange integral always decreases due to the structural disorder and this leads to lower magnetization and Curie temperatures.

Amorphous ferromagnetic Co–Au films were first produced by vapor deposition on substrates cooled to liquid nitrogen temperature,[127] but stimulus for expanded effort in this direction resulted from the development of techniques for rapidly quenching alloys from the liquid state to produce metastable phases. These techniques are referred to as "splat cooling."* Magnetic and Mössbauer studies[129–131] of $(Pd_{1-x}M_x)_{80}Si_{20}$ alloys, where M = Fe, Co, or Ni, and of Fe-rich Fe–P–C alloys,[132,133] and magnetic[134,135] and X-ray studies[136] of Mn–P–C alloys show that magnetic ordering temperatures are lower than the crystalline counterparts, usually multiphase, and that the magnetic behavior of these amorphous metals is largely due to the presence of atomic clusters rich in magnetic atoms.

Amorphous rare-earth transition-metal alloys have recently been prepared and the net exchange interaction will depend on whether the R-ion is light or heavy, just as occurs in the crystalline R-transition-metal alloys and insulators.[306]

1.10. Summary

The magnetic behavior of solids is strongly dependent on structure and composition and only several topics related to this broad subject have been discussed in this chapter. Saturation magnetization, which depends on the kind and density of magnetic atoms or ions in a crystal, their disposition, and the dominant exchange mechanism in operation, is important in technological considerations and we have mainly concentrated our discussion on factors which bear on saturation. Imperfections, including microstructure, such as the presence of magnetic and nonmagnetic second-phase particles, have a pronounced effect on the magnetization behavior and anisotropy of ferromagnetic and ferrimagnetic materials and thus on magnetic hysteresis. Indeed, most technological applications of magnetic materials also depend on the particular way a magnetically ordered material responds to magnetization and demagnetization and this aspect of the role of structure is discussed in Volume 2, Chapter 6.

---

* See Volume 5, Chapter 10. For recent reviews see Ref. 128.

2. Superconducting Behavior

2.1. Introduction

Stucture and composition in relation to superconducting behavior are discussed in this section mainly insofar as the superconducting critical temperature T_c is concerned. The main thrust will be concerned with high-T_c materials, generally alloys and intermetallic compounds containing transition elements. Very little will be said concerning those microstructural details, including structural defects, which give rise to hard superconducting behavior, such as large current-carrying capacities important for superconducting magnets.

Many of the properties of superconductors have been successfully accounted for by the Bardeen–Cooper–Schrieffer (BCS) microscopic theory of superconductivity.[137] However, the theory cannot predict the occurrence of superconductivity in materials, i.e., it does not tell us what atomic constituents should be put together and what crystal system is necessary in order to obtain materials exhibiting high critical temperatures T_c. However, prior to the advent of the BCS theory a large body of information regarding the occurrence of the superconductive state in elements, alloys, and intermetallic compounds was accummulated, notably by Matthias and his

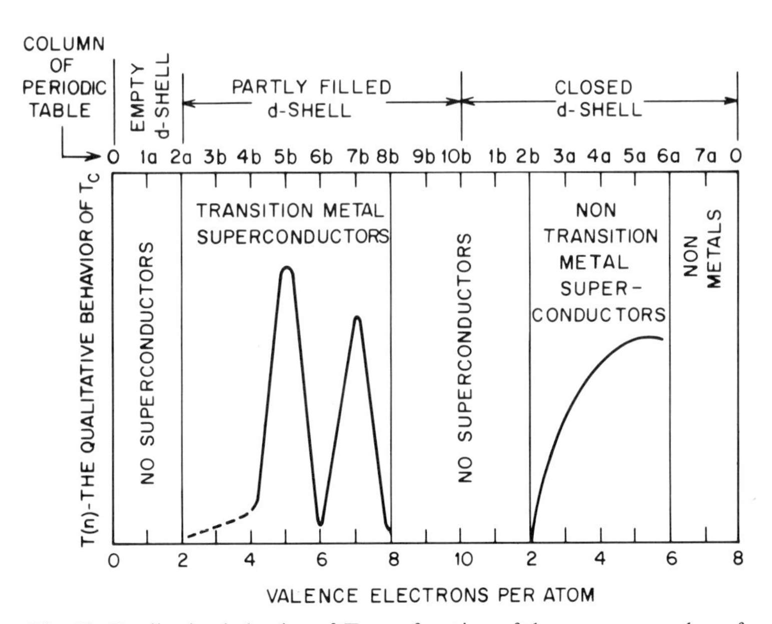

Fig. 23. Qualitative behavior of T_c as a function of the average number of valence electrons per atom as proposed by Matthias.[138,140,141]

associates[138,139] and this information has been, and still is, useful in the search for new superconductors. It was shown (Figure 23) that the occurrence of superconductivity or the maxima in critical temperature depends on the average number of valence electrons per atom. This is computed on the basis of counting all electrons outside a closed shell for each atomic species, weighted as to composition. Since materials which exhibit the highest T_c contain transition metal atoms,* one should note particularly the peaks at electron concentrations of approximately 4.5 and 6.5 for transition metal superconductors. Electron concentration peaks at ~ 4.5 and ~ 6.5 valence electrons/atom are illustrated for Cr_3Si-, α-Mn-, and $MgZn_2$-type phases in Figures 24–26. The tetragonal σ phases ($D8_b$ type; example, $Mo_{0.4}Re_{0.6}$, $T_c \sim 8.4°K$), which exhibit relatively high T_c, also show a T_c peak at ~ 6.7 electrons/atom.†

The BCS theory says that in the presence of an attractive interaction, electrons near the Fermi level having equal and opposite momentum and spin may attract each other to form pairs (known as Cooper pairs), resulting in the superconducting state. This attractive interaction involves the lattice of positive ions, i.e., is phonon induced. The paired superconducting state is at a lower energy and is separated from the normal-state energy by a finite gap.

The basic equation of the BCS theory, known as the BCS equation, is

$$kT_c = 1.14\langle \hbar\omega \rangle_{av} \exp[-1/N(0)V]$$

---

* Ferromagnetic and antiferromagnetic metals do not exhibit superconductivity down to the lowest temperatures accessible for measurement. The effect of localized magnetic moments is discussed in Section 2.5.

† The Cr_3Si-type phases ($A15$), also referred to as β-tungsten phases, are ordered cubic materials containing eight atoms per cell. The larger B atoms (A_3B) occupy the bcc positions (Figure 27) at 000 and $\frac{1}{2}\frac{1}{2}\frac{1}{2}$ and have 12 nearest A neighbors at a distance $a\sqrt{5/4}$ and eight B NN neighbors at $a\sqrt{3/2}$. The smaller A atoms are in pairs on the faces of the unit cube at positions $\frac{1}{4}0\frac{1}{2}$; $\frac{1}{2}\frac{1}{4}0$; $0\frac{1}{2}\frac{1}{4}$; $\frac{3}{4}0\frac{1}{2}$; $\frac{1}{2}\frac{3}{4}0$; $0\frac{1}{2}\frac{3}{4}$. The nearest neighbors to an A atom are two other A atoms at a distance of $a/2$. The σ phase has a complex tetragonal structure with 30 atoms/cell. There are five nonequivalent crystallographic positions for the 30 atoms. Significant features of this phase are broad homogeneity ranges and most decompose eutectoidally at elevated temperatures. The α-Mn-type structure is cubic with 58 atoms/cell distributed among four nonequivalent positions. The $MgZn_2$ type is hexagonal with 12 atoms/cell. The P (orthorhombic, 56 atoms/cell), R (rhombohedral, 53 atoms/cell), and M (rhombohedral, 13 atoms/cell) phases are structurally related to Cr_3Si, σ, α-Mn, and $MgZn_2$ types, but the Cr_3Si and Laves phases are the simplest. All are tightly packed structures and in a given binary system in which the end members are bcc and fcc there is a definite sequence in the appearance of these phases[144]: BCC $\rightarrow$ Cr_3Si $\rightarrow$ σ $\rightarrow$ Laves $\rightarrow$ α-Mn $\rightarrow$ hexagonal variant $\rightarrow$ fcc. It is not necessary, of course, that all phases occur.

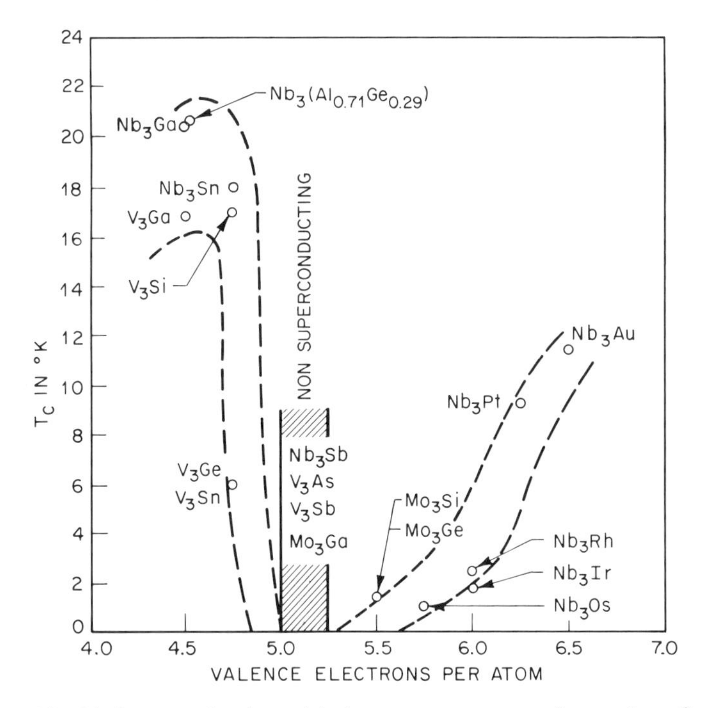

Fig. 24. Superconducting critical temperature versus the number of valence electrons per atom for some A_3B Cr_3Si (*A*15) compounds.

where $\langle \hbar\omega \rangle_{av}$ is the average energy of the phonons that scatter electrons at the Fermi surface; $N(0)$ is the normal-state density of electron states per unit energy range at the Fermi surface, and V is an interaction energy of electrons close to the Fermi surface, and is composed of an attractive electron–electron pairing potential arising from the electron–phonon interaction and a Coulomb repulsive interaction between electrons. Here k is Boltzmann's constant. The above equation is also written in terms of the Debye temperature θ_D by taking $1.14\langle \hbar\omega \rangle_{av}$ proportional to $k\theta_D$:

$$T_c = 0.85\theta_D \exp[-1/N(0)V]$$

The term $N(0)$ can be computed from a knowledge of the coefficient of electronic contribution γ to the heat capacity, determined from low-temperature heat capacity measurements. The Debye temperature θ_D can also be determined from heat capacity data as

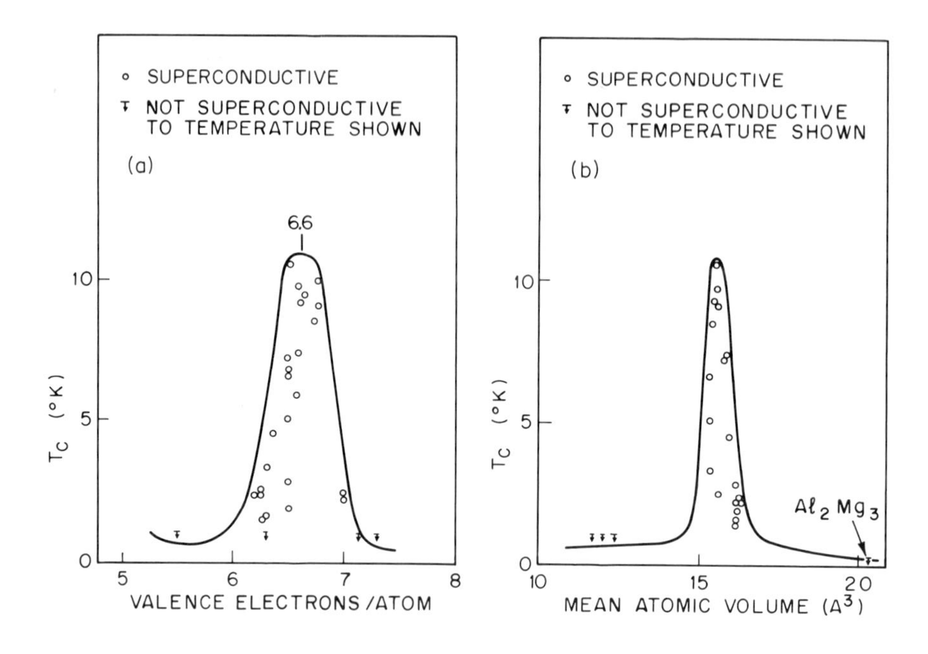

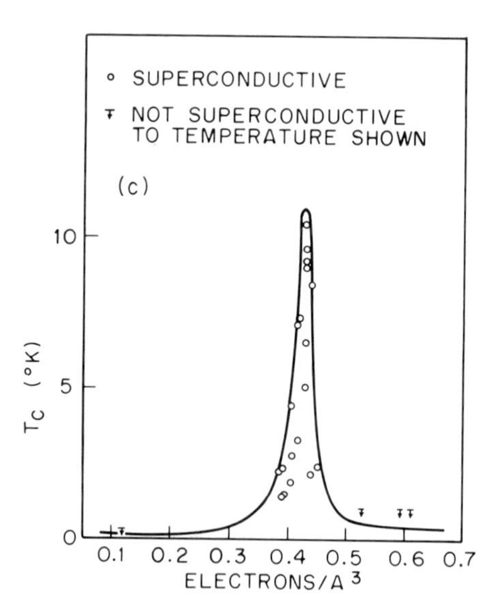

Fig. 25. Superconducting critical temperature versus (a) number of valence electrons per atom, (b) mean atomic volume, and (c) electron density for cubic *A*12 (α-Mn) type compounds (after Roberts[140]).

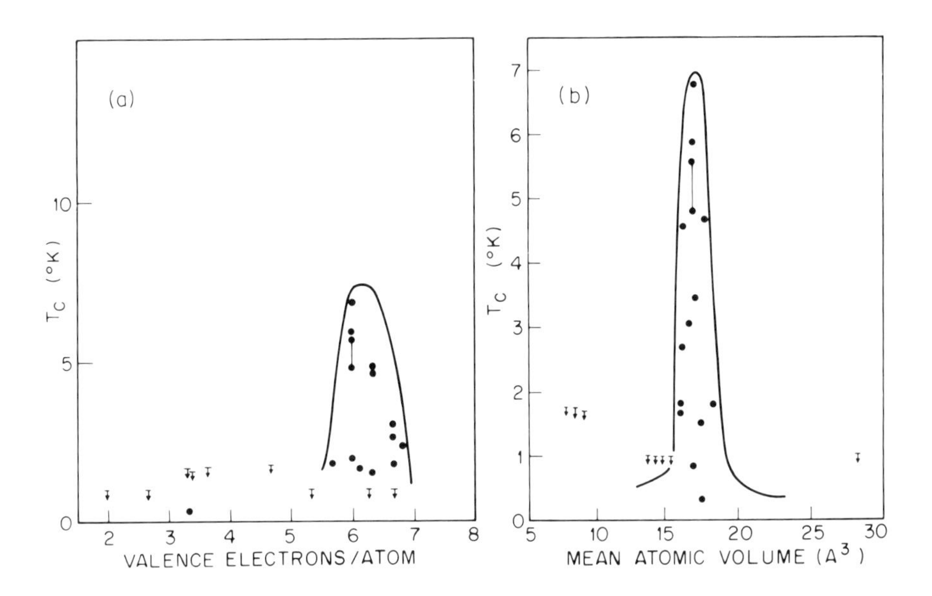

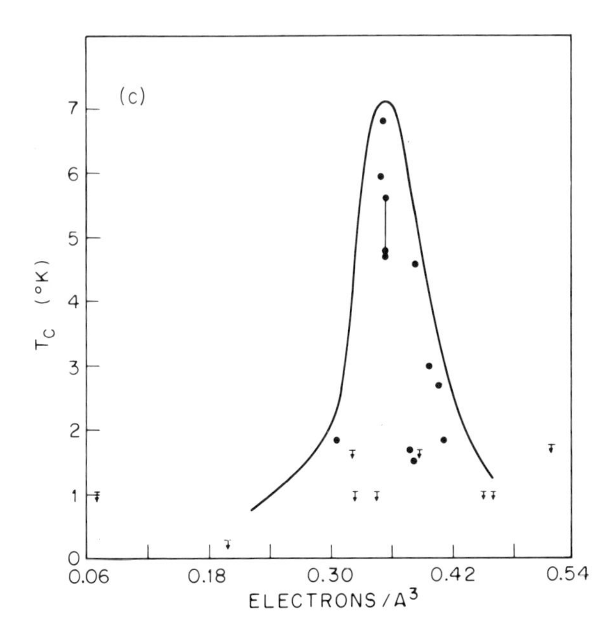

Fig. 26. Superconducting critical temperature versus (a) number of valence electrons per atom, (b) mean volume and (c) electron density for *C*14 ($MgZn_2$) type Laves phase compounds (after Roberts[140]).

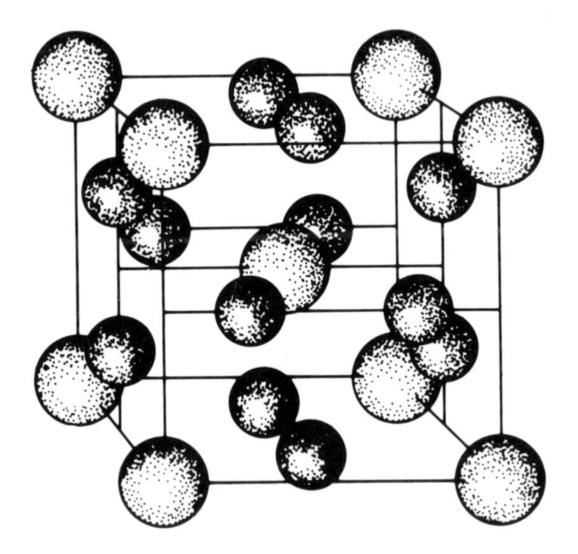

Fig. 27. The Cr_3Si structure showing the chains of transition metal atoms on the faces which are parallel to the cube directions.

well as from ultrasonic measurements.* This information, together with a knowledge of T_c, allows one to determine an approximate value of the interaction energy V. For a number of superconductive alloys V was found to be approximately constant,[138] suggesting the importance of $N(0)$ and the phonon spectrum regarding the occurrence of superconductivity and the realization of high T_c in the search for new superconductors. The peaks in T_c as a function of the number of valence electrons per atom illustrated in Figures 24–26 were suggested as resulting from a similar variation of $N(0)$.[143] A more complete treatment has been given by McMillan,[156] who showed that the

* The total heat capacity at a given temperature is the sum of a lattice or phonon contribution (αT^3) and an electronic contribution (γT), the latter dominating at low temperatures. Here α and γ are constants and γ is referred to as the electronic specific heat coefficient. The electronic heat capacity is proportional to the spin susceptibility and to the density of states at the Fermi surface. The value of γ is less than 20×10^{-4} cal/mole-deg^2 for most elements, including the superconducting elements, in the normal state (nonsuperconducting) at low temperatures. For the superconducting Cr_3Si-type compounds V_3Ga and V_3Si, γ is in excess of 150×10^{-4} cal/mole-deg^2, or ~ 50 per transition metal atom, and is indicative of a high density of states.

important parameter is the phonon spectrum for high-T_c materials (transition metals and alloys).*

The use of BCS theory to quantitatively compute T_c for a given material requires detailed knowledge of both the electronic structure and the phonon spectrum. Information of this sort for complex materials (i.e., alloys and intermetallic compounds) is extremely difficult to obtain and one still relies on past experience and empiricism in the search for new materials.

Clearly, electron concentration or density of states and the phonon spectrum are important factors determining the occurrence of superconductivity and emphasis on these factors will be placed in subsequent sections in discussing superconducting behavior.

2.2. The Cr_3Si (β-W) Transition Metal Nitride and Carbide Phases. Electron Concentration and Lattice Instability

All the materials exhibiting T_c values above about 17°K have either the cubic Cr_3Si† or the cubic NaCl structure at room temperature and contain at least one transaction element as a constituent. An interesting feature of the atomic arrangement in the Cr_3Si-type materials is the unusually small interatomic distance between transition metal atoms. In the superconducting Nb_3B compounds, for example, the Nb–Nb distance is approximately 10% smaller than in metallic Nb and the B atoms are 8–14% smaller. The dense linear chains of transition metal atoms is the basis of one-dimensional-like

* Superconducting tunneling and inelastic neutron scattering are techniques for determining phonon spectra. McMillan's[156] strong electron–phonon coupling theory yields an equation for T_c:

$$T_c = (\theta_D/1.45)\exp\{-1.04(1+\lambda)/[\lambda - \mu^*(1+0.62\lambda)]\}$$

λ is an electron–phonon coupling constant and μ^* is the Coulomb pseudopotential. The density of states obtained from specific heat data in this case is enhanced by a factor $(1+\lambda)$ over the "band structure" or "bare" density of states $N(0)$. According to the arguments of McMillan, T_c is related less to changes in $N(0)$ than to changes in the phonon spectrum which determines λ, and low atomic vibrational frequencies can lead to an enhancement of T_c. Hopfield's theory[179] of high-T_c transition metal materials expresses the electron–phonon interaction in terms of an atomic parameter and mean phonon stiffness and says that the electronic density of states is not a relevant parameter as long as it is large. T_c is correctly calculated for Nb–Ta and La–Y alloys as examples. It is pointed out that the most likely materials for higher T_c are compounds whose atomic volumes are unusually low, and, of these, those with abnormally low interatomic force constants.

† Commonly referred to as the β-W phases.

band structure schemes for these materials.[146,147] Narrow, high-density-of-states d bands are suggested as being associated with the chains.

In the V-based compounds NMR and susceptibility studies[148–151] have shown that those compounds with high T_c show unusual normal-state temperature dependences of the Knight shift, nuclear spin–lattice relaxation times, and magnetic susceptibilities. The nonsuperconductors of that series showed no such anomalies. It is believed that a very narrow d band with high density of states at the Fermi energy is the cause of the above dependences.[149] The situation with high-T_c Nb-based compounds is different.[152–154] The Nb^{93} Knight shift in Nb_3Al ($T_c = 18.7°K$) and Pt^{195} shift in Nb_3Pt ($T_c = 8.5°K$) and magnetic susceptibilities in both these compounds are roughly temperature independent. The Sn^{119} Knight shift in Nb_3Sn ($T_c = 18°K$) was found to be strongly temperature dependent, but its behavior as a function of temperature below $\sim 50°K$ is different from the measured magnetic susceptibility. A recent NMR study of Nb_3X (X = Sn, Al, Au, Pt, Ir) compounds showed the spin components of the magnetic susceptibility and Knight shift to be smaller and less temperature dependent than in the case of the V_3X compounds of comparable T_c.[155]

Alloying additions are commonly made to known superconducting materials in order to raise T_c. By suitable solid solution alloying, the electron concentration and thus the Fermi level (and density of states) may be altered to such an extent so as to result in an increase in T_c. At the same time the phonon spectrum is altered. For atomically ordered phases disorder and volume changes may be partly responsible for the changes in the phonon spectrum or density of states. These and other alloying effects are interrelated and difficult at the present time to explicitly separate from one another.

The substitution of Ge for some of the Al in Nb_3Al results in an increase of T_c.[157–160]* Specifically, $Nb_3(Al_{0.71}Ge_{0.28})$, after proper annealing, exhibits $T_c = 20.5°K$, the highest reported to date.[160] Although the electron to atom ratio e/a is increased from 4.50 to 4.56, the exact reason for the increase is not known.

The results of a detailed study by Testardi and his associates on the effect of alloying V_3X superconductors are summarized in Figure 28.[161] For the moment, neglect the effect on lattice parameter. Note

* Replacement of small amounts of Sn in Nb_3Sn with Al, Ga, and In raise T_c of Nb_3Sn (18.05°K) by several tenths of a degree.[180,181]

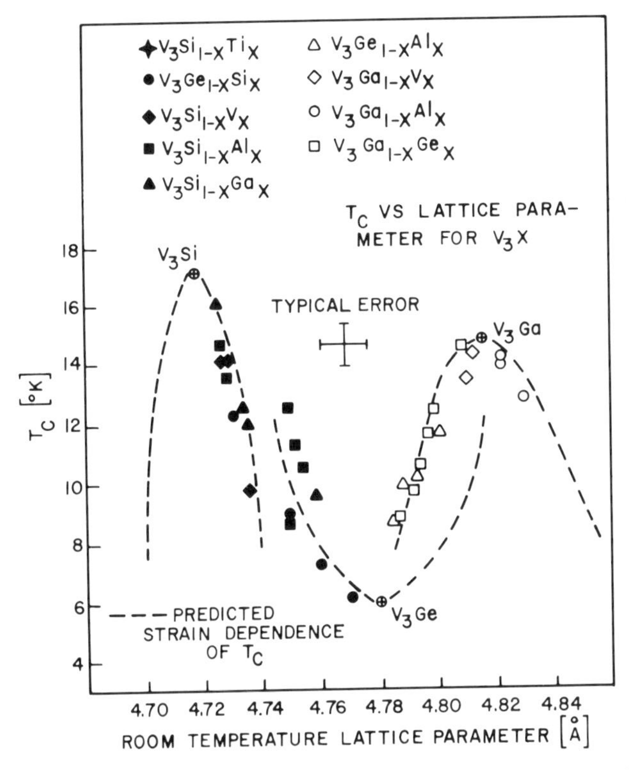

Fig. 28. Plot of T_c versus lattice parameter for V_3X–V_3Y solid solution alloys. The predicted strain dependence for three compounds is shown by the dashed lines (after Testardi *et al.*[161]).

that for V_3Si ($T_c = 17.1$°K) and V_3Ga ($T_c \approx 16.5$°K) alloying to achieve an increase, a decrease, or no change in electron concentration results in a decrease in T_c. On the other hand, for V_3Ge ($T_c \approx 6$°K) T_c is raised by alloying.

It has been shown that several high-T_c Cr_3Si phases undergo a cubic-to-tetragonal martensite-type shear transformation at low temperatures, but just above T_c [V_3Si (~20–27°K),[162] Nb_3Sn (~45°K),[163] $Nb_3Al_{0.75}Ge_{0.25}$ (>24°K),[164] V_3Ga (>50°K)[172]]. For V_3Si the tetragonality can be as much as $(c/a) - 1 \approx 1.0025$ at 4.2°K.[161] Low-temperature X-ray, resistivity, specific heat, ultrasonic, and NMR studies have been used for detecting these transformations. The sublattice distortion has been detected in Nb_3Sn but not V_3Si.[165] NMR studies did not detect this transition in Nb_3Al,[153]

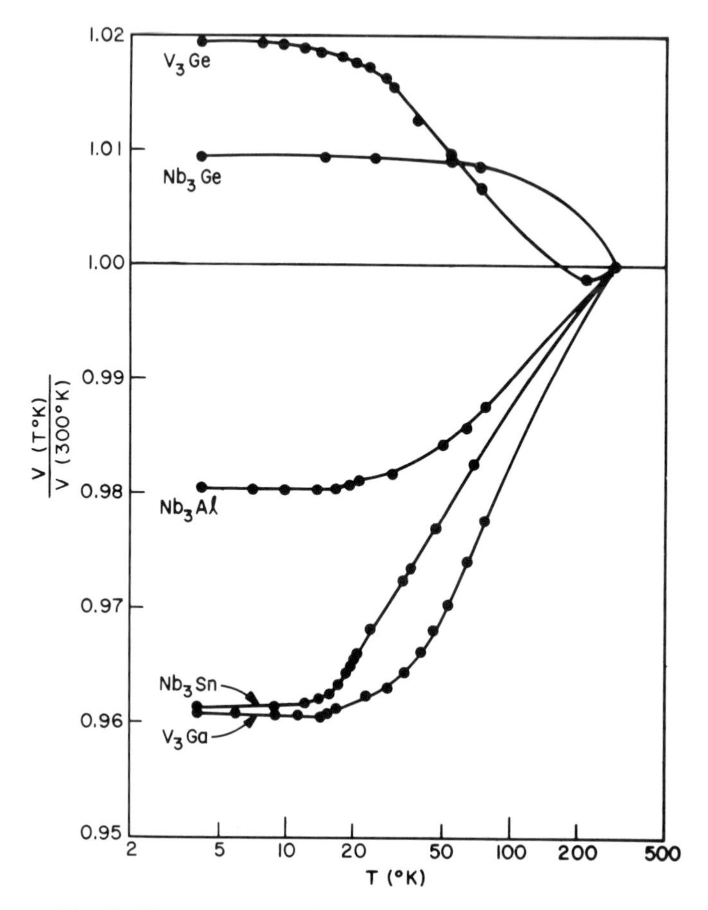

Fig. 29. The temperature dependences of the sound velocities in *A*15 structure compounds. For V_3Ge the data are for shear waves propagating along [001]. All other data are for longitudinal waves in polycrystalline samples (after Testardi *et al.*[166]).

but this does not rule out the occurrence of one, because of sample purity and homogeneity.*

This structural instability is associated with the occurrence of high T_c; i.e., a "softened" phonon spectrum in the normal state strongly influences T_c and is a prelude to the structural transformation.[161,166–169] It is thought that the conditions favorable for high-T_c superconductivity also lead to structural instabilities, and that the structural transformation which results may also lead to a lowering of T_c. In the case of V_3Si a reduction of $\sim 0.38°K$ was calculated.[167]

* Not all samples of V_3Si show a transformation. Impurities and second phase may play an important role.[161]

Prior to the transformation, on cooling, the lattice softens. For example, ultrasonic measurements on Nb_3Sn, Nb_3Al, Nb_3Ga, V_3Si, and V_3Ga show anomalous temperature dependences of the elastic moduli, i.e., an elastic softening (Figure 29).[166] The anomalous temperature dependences disappear in the superconducting state. This unusual behavior does not occur for the isostructural low-T_c compounds V_3Ge and Nb_3Ge ($T_c \approx 5$–$7°K$), although for V_3Ge there is a tendency toward softening at the elevated temperatures. Note that although Nb_3Al is abnormal insofar as other properties are concerned, as discussed above, its elastic behavior is consistent with a high T_c.

Near T_c the sound velocity of V_3Ge and V_3Si exhibits large discontinuities in the temperature derivative at T_c.[167] This is illustrated for V_3Ge in Figure 30, which shows the sound velocity of [001] longitudinal waves as a function of temperature. The discontinuity of the slope at T_c arises from the quadratic strain dependence of T_c.[167] Application of a magnetic field sufficient to destroy the superconducting state gives rise to normal-state behavior below T_c. A thermo-

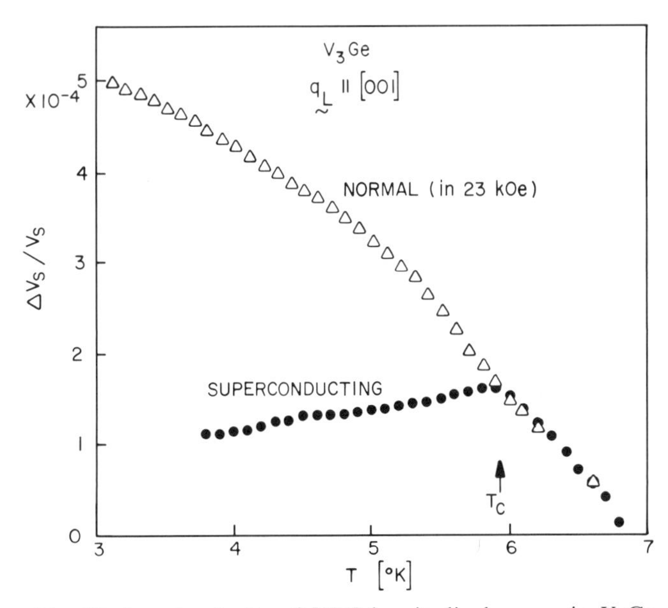

Fig. 30. Sound velocity of [001] longitudinal waves in V_3Ge versus temperature. The discontinuity of the slope at T_c arises from the quadratic strain dependence of T_c. The zero position of $\Delta V/V$ is chosen arbitrarily. Note that the application of a magnetic field sufficient to destroy superconductivity results in normal metallic behavior (after Testardi[167]).

dynamic treatment of sound velocity and specific heat data[167] for a second-order phase transition yields a general dependence of T_c on strain and quantitatively predicts for V_3Si the reduction in T_c which results from the structural transformation, the arrest of this transformation at T_c, the strain dependence of the specific heat discontinuity at T_c and of the structural transformation, and the anisotropic stress dependence of T_c. The strain dependences are very large, mainly quadratic. For strains greater than $\sim 10^{-3}$, T_c will be lowered for cubic V_3Si and will raise T_c for V_3Ge. Now returning to Figure 28, we see that T_c is affected by virtue of the strain induced by alloying (increase or decrease in average cell volume) and its effect on the phonon spectrum, rather than by a chemical change such as a shift in the Fermi level and thus a change in the density of states.[161]*

Further confirmation of a structural transition preceding the onset of superconductivity was obtained for V–Ru alloys[169] and for HfV_2.[170]†

That a phonon or lattice instability precedes high-T_c superconductivity suggested that if instabilities associated with phase transitions other than martensitic, such as eutectoid decompositions, were frozen into the lattice, high T_c's might be achieved. This is important because it suggests a new avenue of approach in the search for new materials. Evidence for the confirmation of this idea was obtained for Mo–Re alloys.[168] The Mo–Re σ phase decomposes eutectoidally at approximately 1150°C (Figure 31). In bulk form alloys having compositions encompassing the σ-phase compositions exhibit T_c values of the order of 8°K. By sputtering $Mo_{0.38}Re_{0.62}$ films onto substrates held near the eutectoid temperature (Figure 31), T_c is raised to ~ 15°K‡ Reducing the Re content caused the T_c peak to broaden and decrease in magnitude. For $Mo_{0.55}Re_{0.45}$, T_c(max) was ~ 13°K. The structure of the films deposited at 1250°C showed X-ray patterns simpler than that of the σ phase and appeared to be either cubic or tetragonal, the latter having an a axis larger and a c axis slightly smaller in comparison to the σ phase.

* T_c of Al films can be increased by three times (1.2°K to 3.6°K) by sputtering Al with Al_2O_3 or Ge. The lattice parameter increases and corresponds to a volume increase sufficient to account for the results.[213]

† For $V_{0.548}Ru_{0.452}$ and HfV_2 the transformation temperatures are 55°K and ~ 120°K, respectively.

‡ Cooling to room temperature after sputtering is rapid so that the high-temperature state is frozen-in.

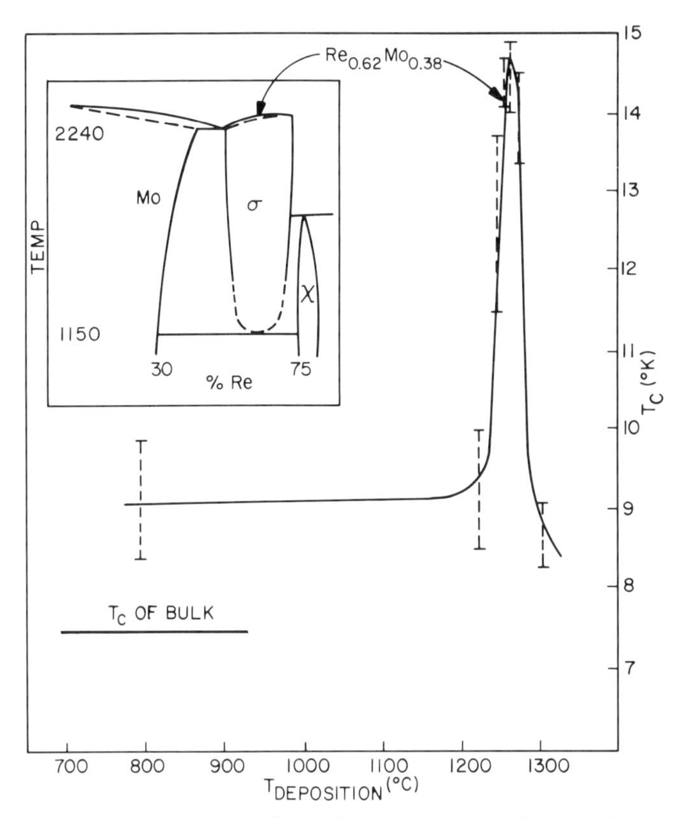

Fig. 31. Plot of T_c versus deposition temperature of $Mo_{0.38}Re_{0.62}$ sputtered films. The width of the superconducting transition is given by the vertical length of the data line. Insert shows part of the Mo–Re phase diagram after Knapton.[171] Note that the maximum in T_c is obtained for substrate temperatures near the eutectoid temperature. (After Testardi *et al.*[168])

As indicated earlier, some of the mononitrides and monocarbides having the cubic rocksalt structure at room temperature exhibit high T_c (NbN, 15°K,[173] NbC, 10.95°K). The temperature T_c can be raised by alloying; for example, $NbN_{0.72}C_{0.28}$ exhibits $T_c = 17.8$°K.[174] Using NbN as a solvent, Hulm and his associates have systematically studied the effect of alloying isostructural carbides and nitrides on T_c, θ_D, and density of states.[175–177]* Their results are summarized in

* These alloys have extremely high melting points and are usually prepared by powder metallurgy techniques. Sound velocity measurements on sintered samples of NbC–NbN alloys could not be performed because of the extreme attenuation due to porosity and grain boundaries.[166] Stoichiometry is difficult to obtain in these materials.

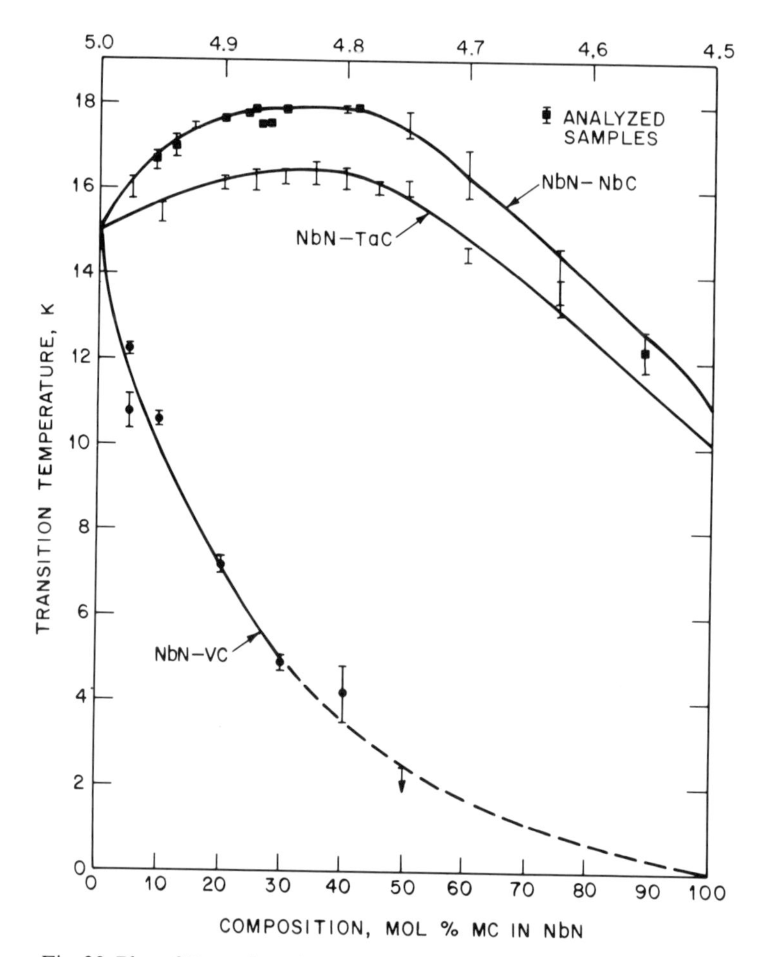

Fig. 32. Plot of T_c as a function of composition and e/a in the NbN–NbC and NbN–TaC systems (after Hulm *et al.*[177]).

Figures 32–35. The addition of NbC to NbN raises T_c, which reaches a flat maximum, while VC lowers T_c (Figure 32). On the other hand, VC raises T_c of VN (Figure 33). The rapid depression of T_c for NbN by VC and VN is not due to the presence of magnetic impurities. For the nitrides only TiN raises T_c of NbN (Figure 34).

From an analysis of C_p and susceptibility data[178] NbN has been shown to exhibit a low density of states and this suggested that it is an *s–p* superconductor* and its high T_c is due mainly to a large

* An *s–p* superconductor is one in which the electrons or states at the Fermi surface are *s* and *p* in character. An *s–d* superconductor has primarily *s* and *d* electrons at the Fermi surface and is usually the case for superconductors containing *d*-electron transition elements.

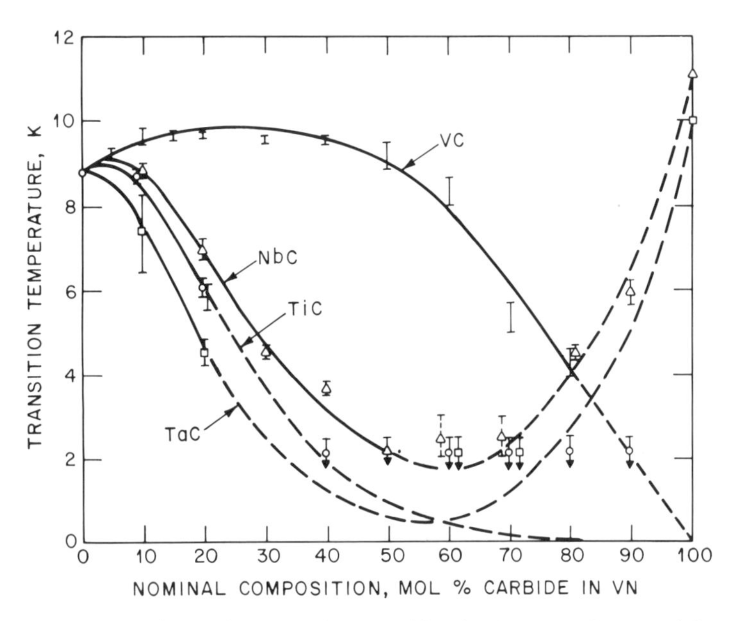

Fig. 33. Plot of T_c as a function of composition for the VN–VC, VN–NbC, VN–TiC, and VN–TaC systems (after Hulm *et al.*[177]).

electron–phonon interaction. The density of states of NbN is about the same as Pb (an *s–p* superconductor) even though T_c of NbN is about twice that of Pb. Recently energy band calculations by Mattheiss and NMR studies[212] have shown, however, that the energy bands near the Fermi energy are predominantly *d*-like. Hulm and associates[177] have determined θ_D and γ for a number of other carbides and nitrides and, using these together with T_c data, employed McMillan's[156] strong coupling theory to analyze the above solid solution results. Using several assumptions, they computed the electron–phonon attraction parameter λ and on a plot of λ versus $N(0)$ suggested that the NaCl compounds can be separated into two groups, one in which λ is characteristic of *s–p* superconductors and the other characteristic of *s–d* superconductors (Figure 35). The Cr_3Si phases, which are considered *s–d* superconductors, are also shown for comparison and it is suggested that TiN and VN behave as *s–d* superconductors. Their analysis further suggests that alloys comprised of two *s–p* superconductors or two *s–d* superconductors as end members should exhibit T_c values which vary essentially linearly, although shallow maxima or minima are possible. On the other hand, alloys comprised of *s–p* and *s–d* superconductors as end members

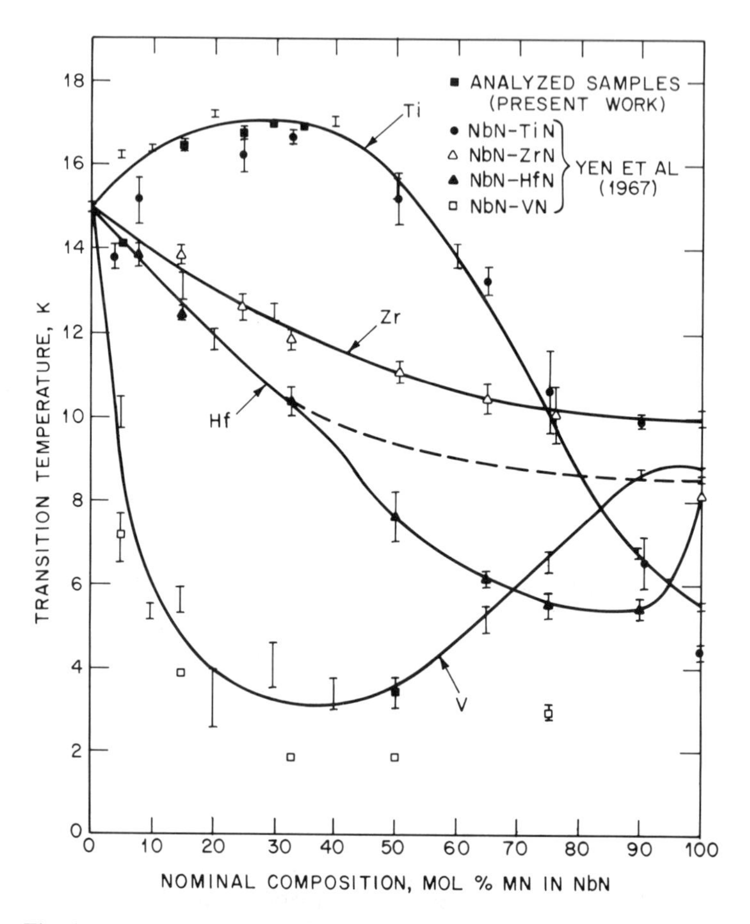

Fig. 34. Plot of T_c as a function of composition for the NbN–TiN NbN–ZrN, NbN–HfN, and NbN–VN systems (after Hulm *et al.*[177] including the data of Yen *et al.*).

will exhibit T_c values as a function of composition which depart greatly from linearity. Usually a deep minimum in T_c is observed, suggested as being due to nonoverlap of *d* and *s–p* bands. It is suggested that their analysis might be useful to account for the depression of T_c of Cr_3Si alloys, but insufficient C_p data are available.[177]

2.3. Role of Stoichiometry and Atomic Order

Implicit in the above discussion is that alloying can either raise or decrease T_c either through affecting the density of states or electron–phonon interaction. These two parameters are related in a

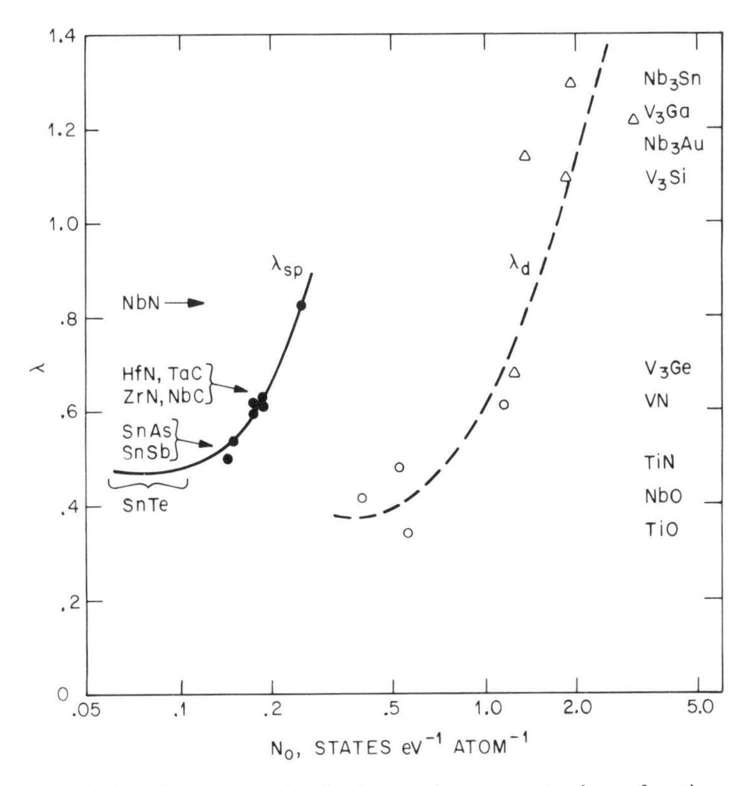

Fig. 35. The superconducting interaction parameter λ as a function of the bare density of states for various NaCl and Cr_3Si structure compounds (after Hulm *et al.*[177]).

complex way to stoichiometry and atomic order. Disorder introduced into an ordered phase via alloying has not always been detrimental to achieving increased T_c. $Nb_3(AlGe)$ is such an example, although for this case most of the disorder is presumably associated with the non-transition element sublattice with little alteration of the transition atom chains. On the other hand, increased T_c values have been obtained in nonstoichiometric Cr_3Si phases by increasing the degree of order by preparing ordered phases and more stoichiometric phases.

For example, Nb_3Ge, when prepared by normal techniques, such as arc-melting, exhibits $T_c \approx 6.9°K$ and is deficient in Ge. When prepared by splat-cooling to obtain a metastable phase, superconductivity begins at 17°K.[182] The transition is fairly broad. Geller[186] was the first to recognize that Nb_3Ge ($T_c = 6.9°K$) exhibited a lattice parameter different from that predicted on the basis of stoichiometric material. Nearly stoichiometric material was obtained only by splat cooling, although disorder was introduced; about one-half of the Ge atoms were on Nb sites. Annealing was unsuccessful in increasing order and T_c[182]; Ge precipitated during annealing.

Similarly, T_c of normal Nb_3Ga (T_c = 14.5°K) can be raised to 20.3°K,[183] presently the highest T_c for a binary phase, when prepared so that it is nearly stoichiometric. High-T_c Nb_3Ga is formed by quenching alloys from above the solidus followed by a low-temperature anneal.[183,185] The lattice constant of near-stoichiometric material is a minimum.

Both Nb_3Au and V_3Ga exist at elevated temperatures in the disordered bcc state and can be retained by quenching to room temperature. These disordered phases are not high-T_c superconductors. By an ordering anneal these phases transform to the Cr_3Si structure with T_c values of 11.5°K and ~16°K, respectively.

Aluminum is soluble in bcc V to the extent of 40 at. % at 500°C and the bcc composition, V_3Al, is not superconducting. Several workers have tried unsuccessfully to induce order by heat treatment. Recently Hartsough and Hammond[184] have shown that V_3Al having the Cr_3Si structure (T_c = 9.6°K) can be prepared by codeposition of V and Al in vacuum onto substrates held at temperatures in the range 350–450°C. The deposition temperature range is quite critical and this phase appears to be unstable at temperatures above ~500°C.

2.4. Metastable Superconducting Phases

Splat-cooling of liquid alloys has been used for preparing new phases in the search for new superconductors. Hexagonal MoC, the stable phase at room temperature, is only superconducting below 4.2°K. The high-temperature form, cubic NaCl type, can be retained at room temperature by splat-cooling[188] and is superconducting at 14.3°K. Similarly, splat-cooled melts of cubic WC are superconducting at 10°K.[188]

Metastable superconducting simple cubic alloys (one atom/cell) have been prepared by splat-cooling.[189,190] Very few crystalline solids exhibit a simple cubic structure and all that are known are superconducting.[189] The simple cubic alloys $Au_{0.25}Sb_{0.75}$, $Pd_{0.165}Sb_{0.835}$, $Ag_{0.25}Te_{0.75}$, and $(Au_{0.55}Pd_{0.45})Te_2$, for example, become superconducting at 6.7, 4.9, 2.6, and 4.5°K, respectively. Note that the component elements are not superconductors and Sb and Te are semimetals.

2.5. Paramagnetic Impurities in Superconductors

Paramagnetic impurities possessing localized magnetic moments drastically lower T_c by depairing Cooper pairs. An example of the effect is shown in Figure 36.[191] Figure 36 shows the effect of Mn, Cr, and Fe on splat-cooled cubic MoC ($T_c = 14.3$°K). The initial rates of decrease of T_c are of the same order of magnitude as with Fe additions to Mo–Re alloys (22°K/at. %)[193] and Cr in NbN (> 15°K/at. %).[194]

The effect of well-localized inner unpaired $4f$ electrons on T_c of $LaAl_2$ ($T_c = 3.24$°K) has been studied.[192] The initial rate of depression of T_c correlates rather well with the de Gennes factor[202] for the heavy rare earths.

Unpaired d and f electrons or impurity atoms can give rise to well-defined (long-lived) local moments if the electronic state lies well below the solvent Fermi level. If interactions between impurity moments are not important, usually the case for dilute solutions, the theory of Abrikosov and Gorkov[195] generally describes the depression of T_c. It predicts a linear decrease with concentration for very dilute solutions and a rate of decrease which increases at higher concentrations. While MoC(Mn) appears to behave according to theory, MoC(Cr) and MoC(Fe) do not. Impurity interactions at high concentrations, clustering of magnetic impurities, and spin–orbit effects will result in a decreasing rate of lowering of T_c.[195–197]

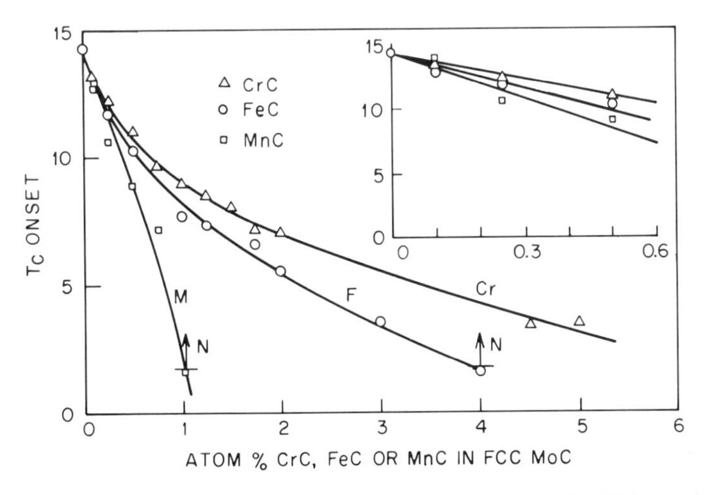

Fig. 36. Variation of T_c of cubic MoC with substitutional additions of paramagnetic $3d$ impurities. Not shown is a 7 atm. % CrC addition which depressed T_c to below 1.7°K (after Willens and Buehler[191]).

Although the Abrikosov–Gorkov theory has been quite successful in explaining many of the effects of well-defined magnetic impurities, some experimental results are not explainable by theory and have been discussed by Sugawara.[201]*

2.6. Ternary Superconducting Chalcogenides

Two new structural groups of ternary superconductors (not ternary solid solutions of binary phases), noncubic at room temperature, have recently been reported.[203,204] They are discussed briefly here because they are the first known chalcogenide phases exhibiting relatively high T_c. The critical temperatures, however, are under 14°K.

The first group consists of phases represented by $Li_xTi_{1.1}S_2$, where $0.1 < x \leq 0.3$, and exhibit T_c from 10 to 13°K ($Li_{0.3}Ti_{1.1}S_2$, $T_c \approx 13$°K).[203] They are hexagonal at room temperature with the Ti_3S_4 structure.[203] They are unstable between 77°K and ~600°K. When they are held for several hours at room temperature their transition temperature drops by a few degrees. Subsequent annealing at ~900°C and quenching restores bulk superconductivity. Thus this is another example of the existence of metastability or a tendency toward structural instability as a precursor to the existence of high-T_c superconductivity.

The materials belonging to this structural group are not intercalated phases in the sense that metal atoms are found within and between the fully filled layers. Phillips[206] has presented a microscopic model for the structure and superconductivity in the $Li_xTi_{1+y}S_2$ phases. The Ti atoms are distributed alternately between filled and partially occupied sheets and as y approaches 0.5, the ideal Ti_3S_4 structure is realized. In the latter the metal layers are alternately filled and half-filled. The y-Ti atoms act as "bridges" (thus referred to as "bridge compounds") connecting filled Ti sheets and thus contribute to the density of states at the Fermi level. He suggests that the Ti bridge atoms are "soft" at low temperatures, below 600°K, and contribute to the structural instability and relatively high T_c.

* Transition metal impurities not exhibiting or nearly exhibiting a localized moment, such as Cr, Mn, and Fe in Al and Ce and U in Th, can also lower T_c.[198,199] This may occur when the d or f electrons are in virtual or bound states near the Fermi level at the temperatures of interest. For this case the depression of T_c is linear at low concentrations and shows a decreasing rate of depression of T_c at higher concentrations. For reviews of theory and experiment concerned with transition metal impurities in superconductors the reader is referred to Refs. 200 and 201.

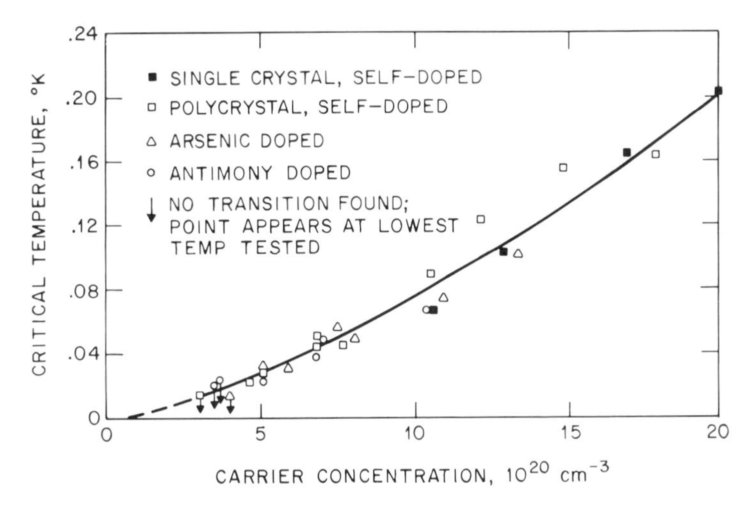

Fig. 37. Plot of T_c of SnTe as a function of carrier concentration (after Hulm *et al.*[209]).

The second structural group is composed of compounds of the form $Mo_{6-x}A_xS_6$, where A = Cu, Zn, Mg, Ag, Cd, Sn, or Pb. These compounds were first synthesized by Cheverel *et al.*[205] and are rhombohedral at room temperature. Here T_c varies from ~2.5°K for Mo_5CdS_6 to ~13°K for $Mo_{5.1}Pb_{0.9}S_6$.[204]

2.7. Superconductivity of Degenerate Semiconductors

Some semimetals or heavily doped semiconductors (carrier densities in excess of $10^{19}/cm^3$) were first shown to be superconducting by Hein and his associates[207] (GeTe, SnTe). Subsequently, nonstoichiometric and doped $SrTiO_3$ were also shown to be superconducting.* The compound $SrTiO_3$ is ferroelectric and highly polar.

Although T_c for these materials is very low (~1°K and less), these materials are of theoretical interest because their superconducting behavior can be studied over a wide range of carrier concentration [and therefore $N(0)$].

The material SnTe (cubic rocksalt structure at room temperature) has been studied in detail by Hulm and his associates.[209,210] Figure 37 shows how T_c varies as a function of carrier concentration down to

* It is interesting to note that GeTe and $SrTiO_3$ undergo low-temperature structural transitions.

$\sim 4 \times 10^{20}/\text{cm}^3$. There is no indication of a cutoff in superconductivity down to the lowest temperature of measurement or to a maximum in T_c with increasing carrier concentration. For the titanates $SrTiO_{3-x}$ and $Ca_ySr_{1-y}TiO_{3-x}$, maxima in T_c occur between $10^{19}/\text{cm}^3$ and $10^{20}/\text{cm}^3$[208,211] and may be due to paramagnetic impurities.[210] In the titanates the superconductive state is attributed to the interaction between electrons and soft optical phonons.[211]

2.8. Summary

The search for new high-T_c materials ($> 17°$K) still relies, for the most part, on the large body of empirical information accumulated during the last 25 years. The presently known high-T_c materials are densely packed intermetallic compounds containing at least one nonmagnetic *d*-electron transition element as a constituent and have either the cubic Cr_3Si or the cubic NaCl structure at room temperature. The transition element is the source of high density of states at the Fermi level, a prime requirement for high T_c. It is thus likely that the as-yet undiscovered high-T_c materials will also contain *d*-electron transition elements. The details of the phonon spectrum are also important because a large phonon-assisted electron–electron attractive interaction also leads to high T_c. Low atomic vibrational frequencies can lead to an enhancement of T_c and it has been recognized that lattice instabilities are associated with, or a precursor to, the occurrence of high T_c. This notion has been a guide in the recent search for new materials. Alloying is used to raise T_c by changing the density of states but the disorder so introduced may adversely affect T_c, particularly if the disorder is associated with the transition metal sublattice. Thus a knowledge of crystal chemistry and the relation of structure to superconducting properties are of prime importance for the materials scientist engaged in the search for new superconducting materials.

3. Dielectric Materials

3.1. Ferroelectrics

3.1.1. Introduction

Ferroelectric crystals are noncentrosymmetric* and exhibit long-range ordering of permanent electric dipole moments below a charac-

* Ferroelectricity is restricted to ten noncentrosymmetric point groups.

teristic temperature (Curie temperature). Even in the absence of an external electric field, there is a spontaneous polarization. In addition, there exist regions in which all the moments are aligned. These are called domains. Above the Curie temperature ferroelectric order does not exist; the materials are no longer polar and this state is referred to as paraelectric. A dielectric anomaly usually occurs in the neighborhood of the Curie temperature.

The paraelectric–ferroelectric transition is usually accompanied by small permanent relative displacements of ions or molecular groups from the symmetry positions in the paraelectric phase. Local electric dipoles result from the ion displacements and these crystals are referred to as displacive ferroelectrics. The structural instability may be associated with highly temperature-dependent low-frequency transverse optical phonons in the paraelectric phase which predominate at the Curie temperature (see, for example, Refs. 214–216 and Volume 2, Chapter 3). The nature and magnitude of the ion displacements determine many of the properties of ferroelectric crystals.

The dielectric constant decreases rapidly with increase in temperature above the Curie temperature in a manner similar to the Curie–Weiss law for the paramagnetic region of a ferromagnet. The ferroelectric crystal, at the Curie temperature, is composed of local electric dipoles, while the magnetic crystal is composed of local atomic moments. A curve of polarization versus applied field below the Curie temperature resembles a magnetic hysteresis loop and ferroelectric crystals and polycrystalline aggregates can exhibit a remanent polarization and possess a coercive field necessary to reverse the sense or direction of polarization. A crystal or polycrystalline aggregate as a whole can appear unpolarized if the volume of domains in opposite senses are equal. The polarization process in an applied field, then, involves the motion of domain walls. A single crystal can also be made single domain by "poling," i.e., cooling the crystal in an applied field from an elevated temperature. Thus, the name ferroelectric is a result of the similarity in response to an applied electric field and temperature to the way a ferromagnetic material responds to an applied magnetic field and temperature.

The existence of an *antiferroelectric* state[217] is possible. The antiferroelectric state occurs, for example, when neighboring lines of ions are displaced in opposite directions, resulting in no net spontaneous polarization below the Curie point, although the ordering may be accompanied by changes in the dielectric constant. Crystals of WO_3, $PbZrO_3$, and $(NH_4)_2H_3IO_6$ with ordering temperatures of 1010, 593, and 245°K are examples of antiferroelectric crystals.

Some ferroelectric crystals are known which do not exhibit a Curie point because they melt before leaving the polarized state. Examples are the $BaMF_4$ compounds, where M can be either a divalent 3*d* transition metal ion (Mn, Fe, Co, or Ni) or a nonmagnetic divalent ion (Mg or Zn).[218] Others, like Rochelle salt, $NaK(C_4H_4O_6)\cdot 4H_2O$, and $SrTeO_3$,[219] exhibit two Curie points, i.e., they are only ferroelectric between an upper and a low temperature.

3.1.2. Classification of Ferroelectrics

Ferroelectrics can be placed into two main classes.[220] The first comprise the displacive ferroelectrics. Typical representatives of this class have the perovskite or a perovskite-like structure. The transition is usually related to the low-frequency optical lattice vibrations in the paraelectric phase.

The second class[220] comprises the hydrogen-bonded crystals, such as KH_2PO_4 and KH_2AsO_4. In these materials the proton can occupy one of two positions equally at elevated temperatures, but at low temperatures the proton system condenses into an ordered configuration. In this sense the transition can be viewed as of the order–disorder type, but is discussed from a quantum mechanical tunneling point of view.[220]

A classification scheme for ferroelectrics has recently been given which is based on the nature of the atomic displacements involved in the polarization reversal.[221,222] In this scheme ferroelectrics can be either one-dimensional, two-dimensional, or three-dimensional. One-dimensional ferroelectrics refers to those in which the displacements are essentially parallel to the polar axis. In two-dimensional ferroelectrics the atomic displacements or vectors connecting positions of identical atoms are confined to parallel planes containing the polar direction. Three-dimensional ferroelectrics are defined as materials in which the individual displacement vectors are essentially random, resulting in partial compensation of dipole moments and small net spontaneous polarizations P_s. An example of a three-dimensional ferroelectric is β-$Gd_2(MoO_4)_3$.

One-dimensional ferroelectrics include the ABO_3 perovskites (where A = Na, K, Ca, Sr, Ba, and Pb and B includes many of the transition elements); the ferroelectric tungsten-bronze type of general formula $(A1)_2(A2)_4C_4(B1)_2(B2)_8O_{30}$ (KLN and $K_{6-x-y}Li_{4+x}Nb_{10+y}\cdot$

O_{30} exemplify this family); and the $LiNbO_3$ family, which includes $LiTaO_3$ and $BiFeO_3$. The latter two are also related to the perovskites. Perovskite and perovskite-like phases are presently the most important group of ferroelectrics and will be discussed in the next section.

The $BaMF_4$ family of phases belong to the two-dimensional class. Structurally, the materials are orthorhombic and made up of puckered sheets of MF_6 octahedra joined at the corners. The sheets are linked together solely by Ba^{2+} ions.[218]

The spontaneous polarizations P_s for the one-dimensional ferroelectrics are usually greater than 25×10^{-2} C/m^2. For the two-dimensional type, $10 \times 10^{-2} > P_s > 3 \times 10^{-2}$ C/m^2, and for the three-dimensional ferroelectrics $P_s < 5 \times 10^{-2}$ C/m^2.[221,222]

3.1.3. Perovskites and Perovskite-like Phases

Perovskite and perovskite-like structures are exhibited by a very large number of compounds and they offer wide crystal-chemical latitude for alteration of crystal structure and dielectric, transport, and electrical properties by suitable solid solution substitutions. For a fairly comprehensive tabulation of ternary and quaternary perovskites, some properties, and preparation of materials through about 1968, the reader is referred to Ref. 223.

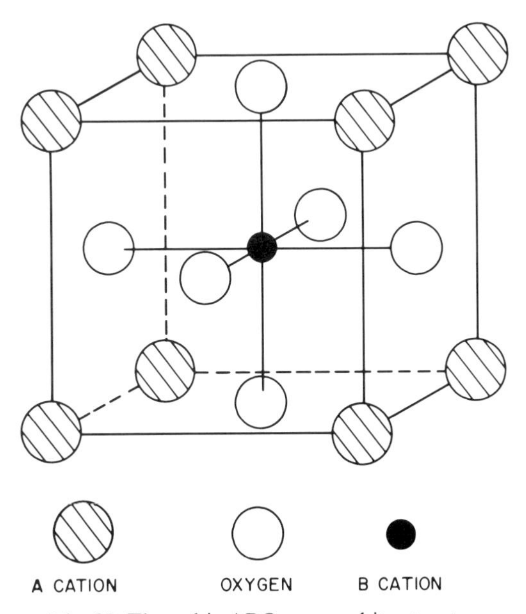

Fig. 38. The cubic ABO_3 perovskite structure.

Perovskites and perovskite-like phases are characterized by the presence of BO_6 octahedra. The manner in which the BO_6 octahedra are connected and their disposition give rise to a variety of perovskite-like phases. The cubic ABO_3 perovskite structure (prototype phase is cubic $CaTiO_3$ above 900°C) is illustrated in Figure 38. In the paraelectric region $BaTiO_3$ has this structure. The oxygen ions are octahedrally coordinated to the small B ion. The A ion is coordinated to 12 oxygens. The macroscopic crystal is built up by BO_6 octahedra sharing corners or oxygen ions. For $BaTiO_3$ below the Curie temperature the structure is tetragonally distorted, with the A and B ions displaced relative to the oxygen ions, thereby developing dipole moments. The octahedra still share corners.

$LiTaO_3$ and $LiNbO_3$ are two important isostructural ferroelectric phases that are perovskite-like (see subsequent sections). Both are rhombohedral and the structure of $LiTaO_3$ is shown in Figure 39. The TaO_6 octahedra share faces both with empty and with LiO_6 octahedra along the trigonal axis or *c* axis in the hexagonal notation. In contrast to $BaTiO_3$, adjacent stacks are connected by sharing edges. The ions are displaced along the trigonal axis, resulting in electric dipoles. The trigonal axis is also the optical axis.

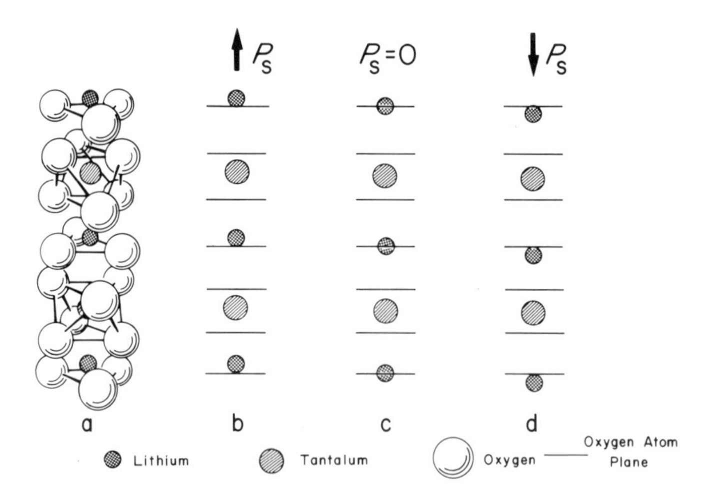

Fig. 39. (a) Stacking of TaO_6 octahedra in the ferroelectric $R3c$ phase of $LiTaO_3$ along the trigonal axis, containing in sequence Li, Ta, empty site. (b) Schematic illustration of the displacement of Li and Ta ions, resulting in a resultant spontaneous polarization. (c) Schematic illustration of ion positions in the high-temperature paraelectric $R\bar{3}c$ phase, which exhibits zero polarization. (d) As in (b), but with displacements and polarity reversed (after Abrahams[221]).

The A cation will be somewhat larger than the B cation. In the cubic perovskite phase ion contact will occur if $R_A + R_O$ equals $\sqrt{2}(R_B + R_O)$, where R_A, R_B, and R_O are the ionic radii. Goldschmidt[224] defined a tolerance factor t such that $R_A + R_O = t\sqrt{2}(R_B + R_O)$. When $0.8 < t < 0.9$ only the cubic perovskite is stable. Distorted perovskites are possible for a larger range of t. In fact, several structures are possible for a given phase, depending on the temperature. Taking $BaTiO_3$ as an example, it is rhombohedral below $-100°C$, orthorhombic up to 0°C, tetragonal from 0 to $\sim 120°C$ (Curie temperature), and cubic above 120°C Roth[225] has classified the room-temperature structures possible for $A^{2+}B^{4+}O_3$ and $A^{3+}B^{3+}O_3$ perovskites in terms of the radii of the cations. For example, the structural classification of $A^{2+}B^{4+}O_3$ perovskite phases as a function of cation radii is shown in Figure 40.

As an example of the effect of solid solution substitutions on crystal structure and properties, consider the $PbZrO_3$–$PbTiO_3$ system (PZT materials).[226,227] Several compositions within this system are important for piezoelectric transducers, and, with fairly large concentrations of La^{3+} ions (PLZT materials), are of potential for imaging and storage devices.[226] As mentioned earlier, $PbZrO_3$

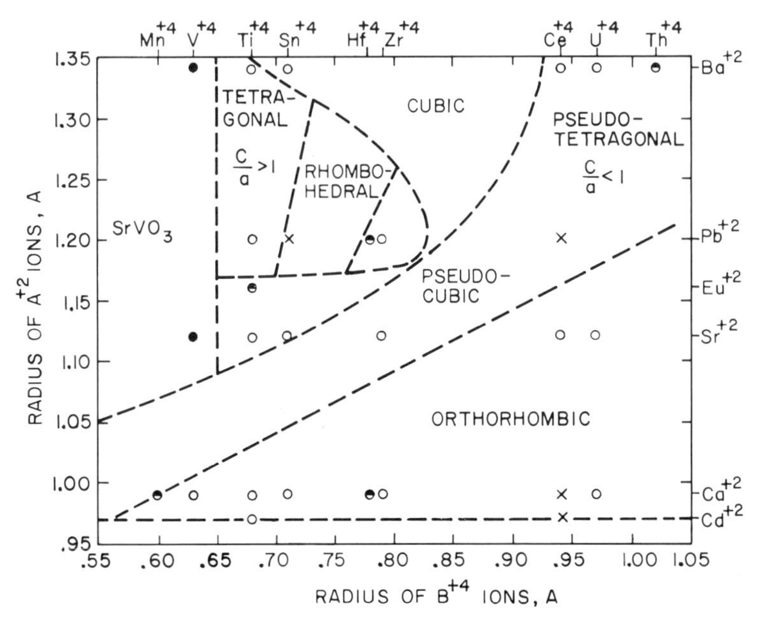

Fig. 40. Structural classification of perovskite $A^{+2}B^{+4}O_3$ phase as a function of cation radii (after Roth[225]).

(orthorhombic) is antiferroelectric (AFE) with $T_c \approx 320°C$. The addition of Ti for Zr causes the orthorhombic AFE phase to become rhombohedral and ferroelectric. At room temperature $PbZr_{0.9}Ti_{0.1}O_3$ is ferroelectric and rhombohedral. T_c is raised. Further additions of Ti increases T_c and at about $PbZr_{0.5}Ti_{0.5}O_3$ at room temperature the ferroelectric rhombohedral structure gives way to a ferroelectric tetragonal structure, which is the structure of ferroelectric $PbTiO_3$. The temperature T_c rises approximately linearly through this system to ~490°C, T_c for $PbTiO_3$. Above T_c all of the solid solutions are cubic and paraelectric.

3.1.4. Curie Temperature and Spontaneous Polarization

On the basis of an examination of precise structural and ferroelectric data for a variety of ferroelectric crystals, Abrahams *et al.*[228] derived the following empirical equation relating Curie temperature T_c and relative ionic displacement Δz of the homopolar metal ion along the polar direction (e.g., Ti in $BaTiO_3$, Sb in SbSI, and Nb in $LiNbO_3$) at $T \ll T_c$ and atmospheric pressure:

$$T_c = (2.00 \pm 0.09) \times 10^4 (\Delta z)^2 \quad °K$$

As examples, Δz for $BaTiO_3$, $Ba_6Ti_2Nb_8O_{30}$ ($T_c \approx 505°K$), and $LiNbO_3$ ($T_c \approx 1468°K$) is 0.132, 0.174, and 0.269 Å, respectively. The above equation may also be expressed as

$$T_c = (\chi/2k)(\Delta z)^2$$

where k is Boltzmann's constant and χ is a force constant equal to $5.52 \pm 0.25 \times 10^4$ dyn/cm.

The above equation allows one to estimate T_c from a knowledge of the ion displacement. Based on spontaneous polarization data for five compounds, it was shown that $P_s \approx 258(\Delta z)\ \mu C/cm^2$ and $T_c \approx 0.3P_s^2$.

The statistical theory of ferroelectricity[231] (see Volume 2, Chapter 3) relates T_c to the mass m of ions and the bonding between ions. The temperature T_c is proportional to m and inversely proportional to a force constant. Heavier ions lead to increased T_c and strong bonds to lowered T_c. This is an important result because it allows for an understanding of the effect of changes in composition on ferroelectric properties. Lowering of T_c raises the dielectric constant at room temperature. For a fuller discussion of these points the reader is referred to Volume 2, Chapter 3.

Discussions of the ionic nature of displacive ferroelectrics usually utilize formal charges, i.e., consider complete electron transfer to the anion and this suggests complete ionic bonding. For example, one can consider $LiNbO_3$ as $Li^{+1}Nb^{5+}O_3^{2-}$. Lithium-7 NMR studies and electric field gradient calculations[229,230] for $LiNbO_3$ and $LiTaO_3$ have resulted in a determination of the charge associated with the transition metal cations and oxygen. Niobium in $LiNbO_3$ possesses a net charge of $+1.59$, while each oxygen possesses a net negative charge of -0.86. For $LiTaO_3$ the net positive charge on Ta is 1.21 and each oxygen possesses a charge of -0.74. Thus the bonding within each octahedron is predominantly covalent in character, the degree of covalency being larger in $LiTaO_3$. The charge distribution has also been determined by application of the statistical theory of ferroelectricity[231] and is in agreement with the above results.

3.2. Piezoelectrics

The piezoelectric behavior of several ferroelectric perovskite-like materials is of technological importance, and this is due to the combination of high dielectric constant and spontaneous polarization exhibited by these materials. The piezoelectric effect refers to the development of an electric polarization by the application of a stress. Conversely, an electric field applied to the crystal causes it to become elastically strained. In one-dimensional notation, the polarization P is related to the mechanical stress X and electric field E through the equation

$$P = -bX + \chi E$$

where b and χ are the piezoelectric strain coefficient and dielectric susceptibility, respectively. The elastic strain e is related to b, X, and E by

$$e = -sX + bE$$

where s is the elastic compliance at constant electric field.

The piezoelectric effect is allowed or can occur in crystals which belong to 20 noncentrosymmetric point groups. Since ferroelectricity is restricted to ten noncentrosymmetric groups, a crystal may be piezoelectric without being ferroelectric. Quartz is such an example. On the other hand, $BaTiO_3$ is both ferroelectric and piezoelectric.

As is well known, piezoelectrics are used to convert mechanical energy into electrical energy and vice versa and thus have wide applications as transducer materials, even in ceramic form.* Other important applications are for wave filters and delay lines. An electromechanical coupling coefficient k is defined which is a measure of the ability to convert from mechanical to electrical energy and vice versa. It is the square root of the ratio of the mechanical energy stored to the electrical energy applied, or the square root of the electrical energy stored to the mechanical energy applied. The term k^2 does not represent the overall efficiency, which may be much lower than k^2 at low frequencies. A high coupling coefficient, of course, is desirable for most applications. Several other piezoelectric constants, such as the ratio of the open-circuit field to the applied stress, are in use which also relate to particular directions and planes or cuts for various shapes, such as plates, bars, and disks. The mechanical Q is also a quality factor which expresses the ratio (strain in-phase with stress)/(strain out-of-phase with stress). The reader is referred to Refs. 232 and 233, for example, for a fuller account of piezoelectric constants and definitions.

In addition to the magnitude of the polarization and dielectric constant, other properties of importance which determine usefulness for a particular application are ease of polling, hysteresis losses, loop squareness, temperature and frequency dependence of the above ferroelectric properties, and aging effects. Aging refers to the stability of frequency as a function of time and is particularly important for filters.[233]

Single-domain single crystals of $LiTaO_3$ and $LiNbO_3$ exhibit strong piezoelectric effects.[234] Ceramic $BaTiO_3$ is a widely used piezoelectric and solid solution additions of $PbTiO_3$ and $CaTiO_3$ improve the piezoelectric properties. Several PZT ($PbZrO_3$–$PbTiO_3$) ceramics have been shown to exhibit a combination of properties which make them technologically useful.[235] Compositions near the rhombohedral–tetragonal phase boundary (~ 42 mole % $PbTiO_3$) exhibit large enough Curie temperature (300–400°C), high coupling coefficients, and good P_s (up to 50 $\mu C/cm^2$). The addition of Nb for Ti and Zr or La for Pb decreases the conductivity, reduces the coercive field, and drastically reduces aging. Improved piezoelectric behavior is also obtained when $Pb(Ti, Zr)O_3$ is modified by the addition of

* The term ceramic is used here to refer to a dense polycrystalline aggregate. Although the polarization and hence the piezoelectric effects are less pronounced in a poled ceramic than in a single crystal, they have the advantage over single crystals from an economic viewpoint, as well as exhibiting greater mechanical strength and greater ease with which the chemical composition can be altered.

$A^{1+}B^{5+}O_3$ (A = K, Na; B = Sb, Bi) or $A^{3+}B^{3+}O_3$ (A = Bi, La; B = Fe, Al, Cr).[236] Sodium and Sb substitution yielded materials exhibiting a dielectric constant above 1500 and radial coupling coefficients (for disks) above 0.6.[236] Values of Q of several hundred are exhibited by PZT ceramics.

Single-crystal quartz is an important nonferroelectric piezoelectric for frequency control and selection and exhibits acoustic Q's in excess of 10^6 because of high purity and crystalline perfection. Commercial quartz crystals are grown hydrothermally and acoustic loss has been shown to be associated with interstitial H^+,[237] which compensates nontetravalent ions, such as Fe^{3+}, Fe^{2+}, Cu^{2+}, and Al^{3+}, which are in the lattice at Si^{4+} sites.[238] Decreasing the concentration of these ions, as well as adding Li, suppresses the incorporation of H^+ and increases Q.[239–241]

3.3. Nonlinear Optical Materials

3.3.1. Introduction

The advent of the laser has stimulated much research and development activity in such diverse fields as materials processing, instrumentation and measurement, spectroscopy, medicine, detection of air pollution, holography, optical communications, and data processing and storage. As is well known, a laser is a source of intense coherent radiation of a given frequency and lasers are available (gas, liquid, and solid, including semiconductor injection lasers) which cover a wide frequency range. The interaction of the high-intensity light waves with the electronic charge distribution of atoms or molecules comprising some crystalline materials may give rise to observable nonlinear optical effects.* Single crystals of nonlinear optical materials are used in laser systems for amplification, modulation, and conversion of laser frequencies.† Frequency conversion is of particular importance because it enables one to extend the range of available coherent light frequencies. Although there are continuing efforts directed toward discovering new optical materials for use in laser systems, a number of materials are available for performing some of the above functions. In this section we discuss several aspects of the

* For treatments of the physics of nonlinear optics see Ref. 249.

† The crystal structures of nonlinear optical materials lack a center of symmetry. For recent reviews of these materials see Refs. 250a–250d. The format here parallels to a large extent the discussions given in Refs. 250a and 250c.

nonlinear optical nature of materials relevant to relating structure and composition to nonlinear optical properties.

The interaction of the oscillating electric field of the light wave with the valence electrons or electronic charge distribution around an ion or molecule induces oscillations of the electron cloud resulting in oscillating electric dipoles. For ordinary light, i.e., light possessing small electric field strengths, the induced polarization (displacement of the electron cloud) or dipole moment per unit volume, P, is directly proportional to the magnitude of the electric field E of the light wave and is given by

$$\mathbf{P} = \chi \mathbf{E}$$

where χ is called the *linear* optical susceptibility; it is related to index of refraction n by

$$4\pi\chi = n^2 - 1$$

Three independent values of χ are required for an anisotropic crystalline solid to characterize the linear optical properties.

Laser light, possessing intense electric fields (10^5–10^6 V/cm), gives rise to a number of *nonlinear* effects. The above equation for P thus contains higher-order terms:

$$P = \chi E + dE^2 + RE^3 + \cdots$$

The second term (dE^2) is responsible for second harmonic generation (SHG). This was the first nonlinear optical effect studied (Figure 41a) and quartz was the material used for the experiment.[251] The red beam of a ruby laser ($\lambda = 694$ μm) was passed through a quartz crystal and a weak second harmonic in the ultraviolet ($\lambda = 374$ μm) was observed which moved through the crystal at a smaller velocity than the primary beam because of the higher index of refraction at the shorter wavelength (n varies with wavelength and this is referred to as normal dispersion). The conversion efficiency was very low and is due to interference effects resulting from the generation of new ultraviolet light by the primary beam at a point along the path which is exactly 180° out of phase with the ultraviolet generated earlier. In fact, it has been shown[252] that the intensity of the second harmonic drops to zero whenever the path length equals an integral number of interference distances; but it is a maximum at one-half the SHG interference distance and this latter distance is referred to as the "coherence length." For quartz this coherence length is several micrometers.

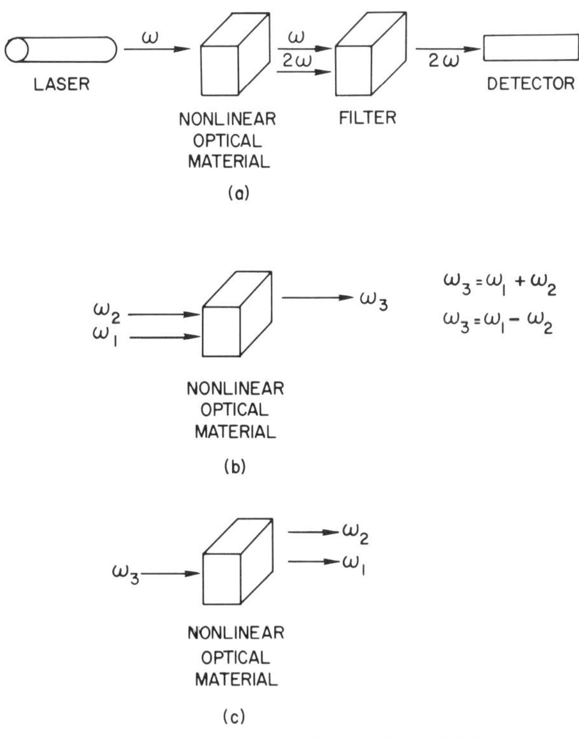

Fig. 41. Schematic illustration of (a) second harmonic generation, (b) sum and difference mixing, and (c) parametric generation (with mirrors, tunable parametric oscillator).

Birefringent crystals can be utilized to overcome the above interference problem. In a uniaxial birefringent material there is one unique axis, called the optical axis, along which the velocities of the two waves are equal.* If the birefringence of the crystal is greater than the dispersion in index between fundamental and harmonic frequencies, then by suitable choice of crystal direction, the velocity of the harmonic can be matched with the velocity of the fundamental and this leads to long coherence lengths and large increases in conversion efficiencies.(252,253) The phase matching direction is specified by the angle it makes with the optical axis and this angle is related to the indices of refraction for the ordinary and extraordinary waves of the

* Those crystals for which the velocities are the same in two different directions are referred to as biaxial.

second harmonic and the ordinary index of the fundamental. Because of the phenomenon of "walk off,"[250a–250d] the most useful materials for SHG are those uniaxial materials for which the phase matching direction is normal to the optical axis ($\theta_m = 90°$) and this situation has been termed noncritical phase matching.[254]

Sum and difference mixing for optical power amplification and parametric oscillation are illustrated in Figure 41(b, c). In the former, two frequencies are mixed to produce sum and difference frequencies. For optical parametric oscillation an input laser frequency results in two other frequencies. The laser or pump frequency is ω_3, ω_2 is referred to as the signal frequency, and ω_1 is the idler frequency (Figure 41c). In both situations all three polarization waves in the crystal must be matched for efficient conversion. Tuning over a wide frequency range can be accomplished by processes which change the refractive indices of the material. They include varying temperature and varying crystal orientation, which allows variation of the extraordinary refractive indices, pressure, and electric fields (electrooptic effect). The first parametric oscillator utilizing $LiNbO_3$ as the nonlinear optical material was reported in 1965 and it is the most widely used material for this purpose because of its transparency range and tuning versatility.[250a–250d]

3.3.2. Desirable Intrinsic Properties of Materials for Nonlinear Applications

The important nonlinear optical crystals are insulators, such as the previously described oxides, or highly resistive semiconductors at ordinary temperatures. They must be highly transparent in the optical spectrum of interest. It is apparent from the above discussion that nonlinear effects occur in noncentrosymmetric, highly polarizable crystals which possess a high index of refraction. For SHG and sum-and-difference-frequency generation to be observed, the crystals must exhibit sufficient birefringence for phase matching[252,253] (the birefringence must be equal to or exceed the normal dispersion of the refractive indices between the frequencies of interest).

The nature or source of polarization within a crystal has been used to classify nonlinear optical materials into two classes. One class comprises materials in which the polarization is a result of the relative displacement of ions, such as in ferroelectrics. The second class of materials owe their polarizability to the ease with which the valence electrons or electronic charge distribution about atoms or anionic groups can be influenced by an electric field. Examples of the latter materials are the semiconductors Se, CdSe, $ZnGeP_2$, $AgGaS_2$, and Ag_3AsS_3. Materials which exhibit large electronic polarizability usually contain atoms of large atomic number (and small electronegativity) and are therefore quite polarizable. The intrinsic energy gap of semiconductors is the upper limit, in energy, to the useful transparency region.

The requirement of acentricity and an examination of the atoms and molecular groups (and their disposition) have been useful in the search for new nonlinear optical materials.[250a–250d]* In addition, the Miller delta rule,[257] which relates the magnitude of the nonlinear optical susceptibility d to the linear optical susceptibility χ, is valuable in gaining information on the strength of the nonlinear effects to be expected from a given material. Miller's equation is

$$d = \chi^3 \Delta$$

where Δ is a constant within a factor of ± 3 for most acentric crystals examined. By use of $4\pi\chi = n^2 - 1$, it can be seen that

$$d \propto (n^2 - 1)^3$$

and large indices of refraction are required for large nonlinear optical coefficients and, thus, useful nonlinear materials.

The nonlinear optical properties of several compounds of stoichiometry $A^{I}B^{III}C_2^{VI}$ and $A^{II}B^{IV}C_2^{V}$ (Table 9) have only been determined during the last few years.[258,259] They are tetrahedrally coordinated, ordered, tetragonal phases (chalcopyrite structure type $E1_1$, space group $I\bar{4}2d$) and are the ternary analogs of the 2–6 and 3–5 diamond cubic semiconductors, respectively (Figure 42). Just as in the 2–6 and 3–5 compounds, the bonding in these phases is essentially covalent (sp^3 hybrids), although for $AgGaS_2$, $CuGaS_2$, and $CuInS_2$ noble metal d electron wave functions are involved.[260] The fact that dissimilar atoms are present leads to an ionic contribution to the bonding.

---

* The use of the existing structural information in seeking out potential nonlinear optical materials has sometimes resulted in error (in both directions) because of faulty structural work. The Giebe–Schiebe[255] test for piezoelectric coupling and the recent powder technique developed by Kurtz[256] are useful for rapidly scanning materials for the presence or absence of acentricity. The powder technique also gives limited information concerning the magnitude of d and on whether an acentric material is phase matchable.

TABLE 9
Comparison of Several Nonlinear Optical Materials[258,259]

	$d \times 10^{12}$, m/V	n	$(d^2/n^3) \times 10^{23}$, m^2/V^2
$CdGeAs_2$	236	3.58	121
$ZnGeP_2$	75	3.09	19
$AgGaSe_2$	33	2.57	6.4
CdSe	18	2.45	2.2
$AgGaS_2$	12	2.32	1.1
Ag_3AsS_3	9	2.7	0.4
$LiNbO_3$	~0.5	2.09	0.27

A figure of merit for nonlinear optical materials is d^2/n^3, where d is the largest phase-matchable nonlinear optical coefficient.[261] A comparison of the figures of merit of several materials are shown in Table 9 and Figure 43.[262,263] In the latter the useful transparency

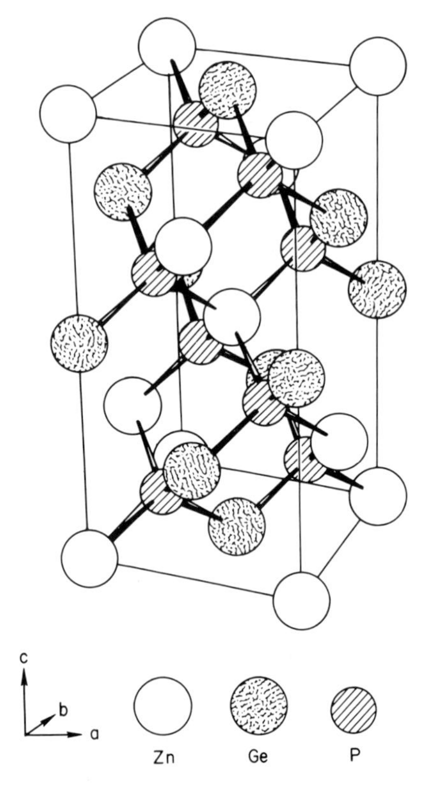

Fig. 42. The tetragonal chalcopyrite structure illustrated for $ZnGeP_2$.

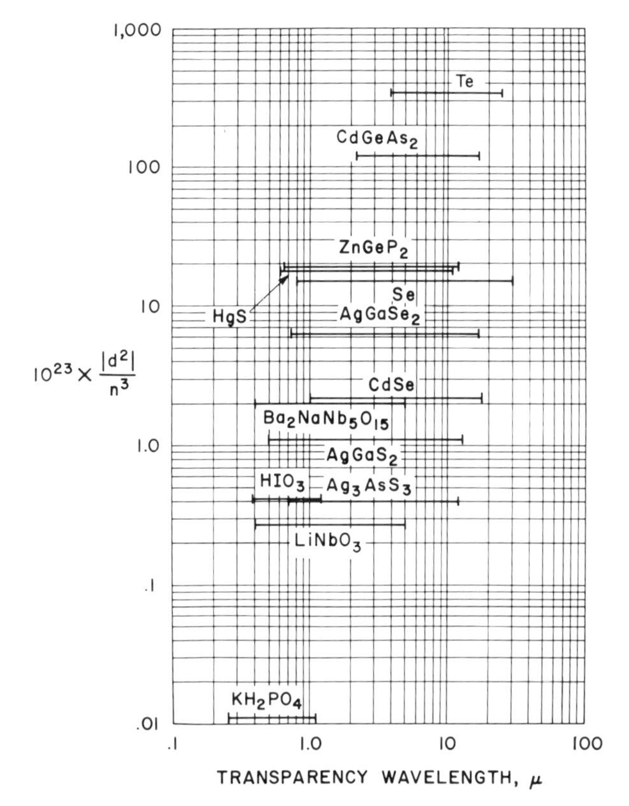

Fig. 43. Comparison of the figures of merit and useful transparency ranges of several nonlinear optical materials (after Boyd and McFee[258,262]).

ranges are also indicated. Except for $LiNbO_3$, all of the materials in Table 9 are semiconductors with large electronic polarizabilities. The materials $CdGeAs_2$, $ZnGeP_2$, and $AgGaS_2$ are particularly important for sum mixing in the infrared.[258]

3.3.3. Nonlinear Optical Susceptibilities and Ionicities

Relating the intrinsic nonlinear optical properties of materials in a straightforward manner to the microscopic features of a solid, i.e., the chemical or electronic nature of the component atoms and local crystallographic structural features, is a lofty desire because this would enable one to understand differences in behavior among isostructural phases as well as being a guide in predicting the behavior of new phases. The Miller delta rule is only partially successful in doing this and more sophisticated theories (discussed below) are required to accurately predict the nonlinearity.

As discussed in an earlier section, a characteristic feature of the perovskites and perovskite-like phases, such as $LiNbO_3$ and ferroelectric tungsten bronzes, is the presence in the structure of BO_6 octahedra which can be considered as the basic building block in these materials. This fact is the basis of an electronic band model for nonlinear behavior of displacive ferroelectrics.[264,265] This theory, in addition to relating the Miller delta to the spontaneous polarization, shows that materials exhibiting a tetragonal distortion (C_{4v}, or point group 4*mm*) are better than trigonally distorted (C_{3v}, or point group 3*mm*) materials. The magnitude of the nonlinear coefficient is predicted to be larger the smaller the principal quantum number of the B or transition metal atom.

Several other theories,[266–268] quantum mechanical in nature, have discussed the importance of the distortion of the valence electron distribution regarding d and have shown a correlation between the Miller delta and bond dipole moments.

Of particular significance with regard to relating the nonlinear susceptibilities to the microscopic nature of materials is the application of the Phillips–Van Vechten (PV) quantum dielectric theory of solids (Chapter 1, this volume)[269–273] by Levine[274–276] to the calculation of the Miller delta. He has successfully computed the Miller delta and nonlinear optical susceptibility d for a large variety of nonlinear optical materials, including those in which d electrons play a role in the bonding.

The very good agreement between calculated Δ_{ijk} for a given polarization direction and experimental values is illustrated in Figure 44 for a wide variety of nonlinear optical materials. There are two types of acentricity, i.e., two ways the atoms composing the bond can differ. The two atoms can have an electronegativity difference and they can have different atomic radii. For ternary compounds, such as $ZnGeP_2$, the nonlinearity contributed by each type of bond (e.g., Zn–P and Ge–P) is computed. Although no displacive ferroelectrics are shown in Figure 44, a similar treatment to the above has also been given by Levine for this class of nonlinear materials.[276]

Peterson[277] has performed Na and Cu NMR studies on $NaMX_2$ (α-$NaFeO_2$ structure), $CuMO_2$ ($CuFeO_2$ structure), and Cu I–III–VI chalcopyrite phases and showed that (1) the charge on the metal ion M in the $NaMX_2$ phases correlated with the average electronegativity difference between M and chalcogen, (2) the nuclear quadrupole

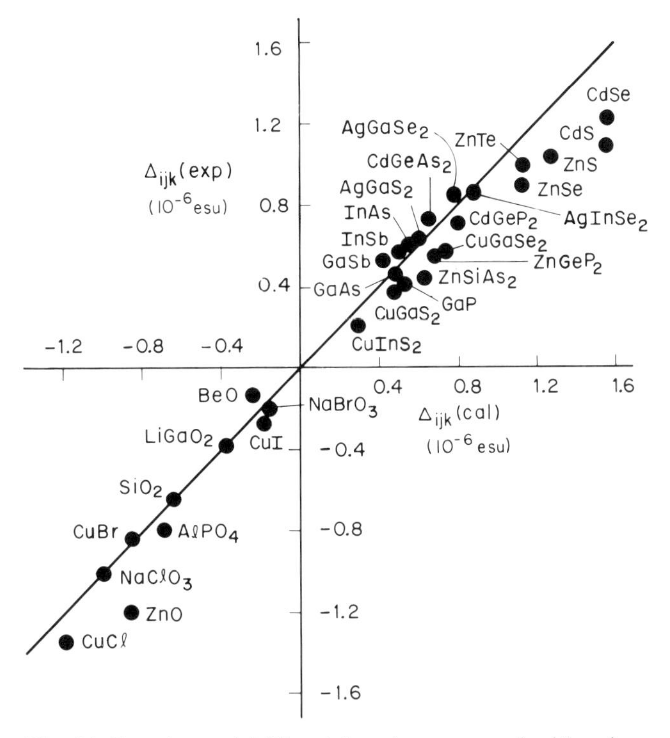

Fig. 44. Experimental Miller deltas Δ_{ijk} compared with values calculated by Levine.[276]

coupling constants correlate with the electronegativity of trivalent M atom in the $CuMO_2$ phases, and (3) the coupling constants correlate either with the electronegativity of the trivalent metal (III) or the chalcogen for the I–III–VI phases. The electronegativities used were an average of those determined by three methods, as given by Sanderson.[278]

3.4. Electrooptic and Pyroelectric Materials

Ferroelectric materials may exhibit large electrooptic and pyroelectric effects.* The electrooptic effect[243] arises when an external electric field alters the charge distribution and moves ions,

* The electrooptic effect is allowed in crystals having a center of symmetry, but the effect will be very small in moderate fields.[242]

and can be used to alter the refractive indices of a material. The effect is particularly large in ferroelectrics below the Curie temperature since the electric field changes the existing polarization and birefringence. Thus, changes in velocity (and phase) of transmitted light can be accomplished by an applied field and this effect can be utilized for modulating light over a broad frequency range, useful for optical communications. The most useful modulator materials known at the present time are single-domain ferroelectric single crystals of $LiTaO_3$ and $LiNbO_3$.[244] In particular, $LiTaO_3$ exhibits low birefringence and large electrooptic coefficients, making it very useful for electrooptic modulation. The lower birefringence of $LiTaO_3$ compared to $LiNbO_3$ has been suggested as being due to more symmetric TaO_6 octahedra compared to NbO_6 octahedra.[245]

The PLZT ceramics show large changes in birefringence by switching the polarization and these materials are of potential use in optical memories and displays.[246] However, the electrooptic properties utilized for such devices depend on the grain size.[226]

The pyroelectric effect[242,243] is the change in polarization with temperature and results in the generation of a current and may be used for detecting radiation, particularly if the Curie temperature is near room temperature. Under the influence of a strong light pulse the current is proportional to $dP/dt = (dP/dT)(dT/dt)$. The effect will be large near the Curie temperature because the temperature dependence of the spontaneous polarization dP/dT will be large. Chynoweth[247] studied this effect extensively in $BaTiO_3$ and triglycine sulfate $[(NH_2CH_2COOH)_3 \cdot H_2SO_4]$ and Glass[248] has constructed highly efficient room-temperature detectors from crystals of barium strontium niobate $[Ba_xSr_{1-x}Nb_2O_6\ (0.2 \leq x \leq 0.8)]$. The Curie temperatures for these materials are just above room temperature. Other factors which enter into the efficiency of detection have been listed by Carruthers.[250c]

3.5. Summary

The dielectric materials important for many electronic and optical applications are acentric and many possess perovskite and perovskite-like structures. These structures are characterized by BO_6 octahedra and a wide variety of phases can be obtained by virtue of the fact that the octahedra can be attached in several ways in building up a macroscopic phase. These structures afford considerable latitude for chemical substitution and thus for alteration of properties and the future for new materials based on them is indeed bright. Some ternary diamondlike semiconductors (2–4–5 and 1–3–6) have been shown to be important nonlinear optical materials for the infrared.

Nonlinear optical susceptibilities of materials can now be calculated from bond ionicity, atomic radii, and *d*-electron contribution to the bonding.

4. Mechanical Behavior

4.1. Introduction

The nature of the interatomic bonding forces, i.e., the electronic structure of the crystal, determines the cohesive strength of solids, and thus their elastic behavior, slip and twinning systems favorable for plastic deformation, and cleavage (fracture) planes. Accurate calculations of mechanical properties, such as modulus of elasticity (Young's modulus), theoretical yield, and fracture strengths, of a perfect or ideal crystal from first principles are difficult, if not impossible. Theoretical estimates have been based on assumptions concerning the nature of the binding forces between atoms, surface energies, and heats of sublimation. These estimates are important because they show which macroscopic properties are relevant for estimating the strengths of solids and for the search for new high-strength materials. In addition, correlations of melting points, atomic volumes, and other easily measurable intrinsic properties of solids which are qualitative indicators of cohesive strength are often made with a number of mechanical properties for predictive purposes. This also gives one a deeper insight into the nature of materials and a more comprehensive or unifying view of classes of materials. Several examples of these correlations are presented below.

An estimate of the theoretical breaking strength of a perfect solid based on the creation of two new surfaces shows that the maximum breaking strength is equal to $(E\gamma/a_0)^{1/2}$, where E, γ, and a_0 are respectively Young's modulus, surface energy, and equilibrium separation of atomic planes.[279] Thus substances exhibiting large moduli of elasticity, surface energy, and small atomic volume or high density of atoms per unit volume should be relatively strong. The so-called Morse function, which gives the variation of the potential energy of two atoms as a function of their separation, has also been used to estimate breaking strengths, particularly for covalently bonded solids. Electrostatic considerations have been used for ionic solids. The theoretical estimates of breaking strength for non-metallic solids based on Morse functions and electrostatic considerations are between 5 % and 15 % of E.

Frenkel[280] computed the force required to shear two planes of atoms past each other in a perfect crystal and showed that the critical yield stress (or elastic limit) is of the order $G/2\pi$, where G is the shear modulus. Experimental values of the elastic limit are 100–1000 times smaller than the above estimate. By considering the form of the interatomic forces and other configurations of mechanical stability, the theoretical shear strength could be reduced to $\sim G/30$, still well above the observed values in *ordinary* materials.[281] It is now firmly established that crystalline imperfections, such as dislocations, microscopic cracks, and surface irregularities, are primarily the reasons for the observed mechanical weakness of crystalline solids.* This aspect of the mechanical behavior of solids, including a discussion of strengthening mechanisms, is discussed in Volume 2, Chapter 7. In this section the chemical and structural aspects of mechanical behavior, i.e., bonding and crystal structure, are emphasized.

4.2. Elastic Behavior

4.2.1. Metals

The elastic modulus of pure metallic elements varies in a systematic manner as a function of position in the periodic table, as shown in Figure 45. Similarly, the compressibilities (reciprocal of the bulk moduli) also change in a periodic manner (Figure 46). Maxima in Young's modulus and minima in compressibilities occur for the transition elements. Since the above moduli reflect the strength of the bonding, these plots illustrate the importance of d electrons to the bonding; d bonds or bonds involving d electrons or orbitals are stronger than s and p bonds even though p bonds are also directional in character.

As noted in the introduction, atomic or molecular volumes indicate bonding strength. Small atomic volume (or small atom separation) indicates strong interatomic cohesion. Thus, the radii of

* It has been shown that metal "whiskers," which grow spontaneously on the surface of some metals, and some extremely thin fibers produced by other techniques exhibit mechanical strengths approaching the theoretical values, i.e., have very high elastic limits and exhibit little or no plasticity. The metal whiskers are nearly perfect single crystals. Strong man-made metallic and nonmetallic fibers are being utilized to produce mechanically strong composite materials.

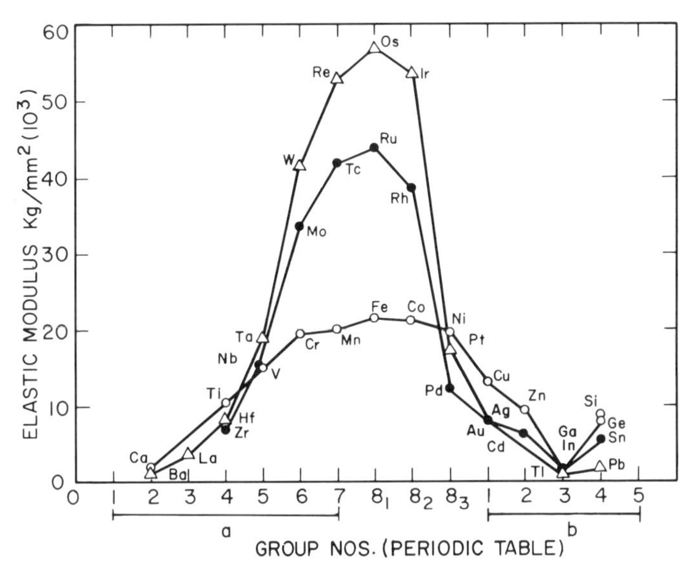

Fig. 45. Young's modulus of many of the elements as a function of position in the periodic table (after Koster[282]).

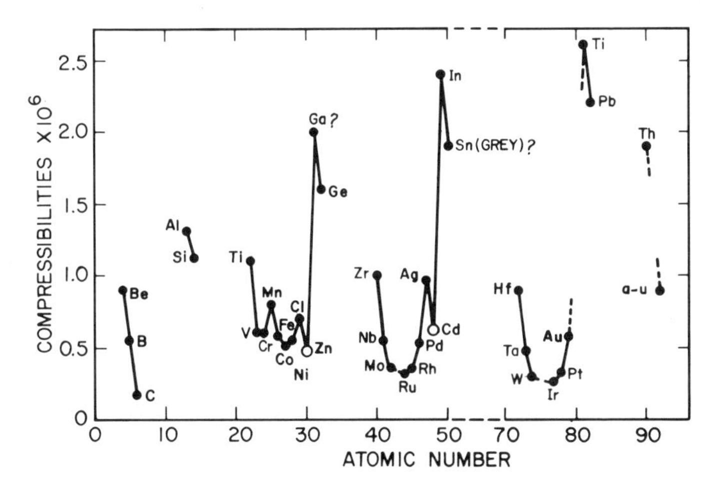

Fig. 46. Compressibilities of many of the metallic elements as a function of atomic number. Those elements (Na, K, Rb, Cs, Mg, Ca, Sr, and Ba) exhibiting compressibilities in excess of 3×10^6 cm^2/kg are not shown. For Zn and Cd the values plotted are three times the linear compressibility in the direction of the close-packed hexagonal planes (after Hume-Rothery and Coles.[283]).

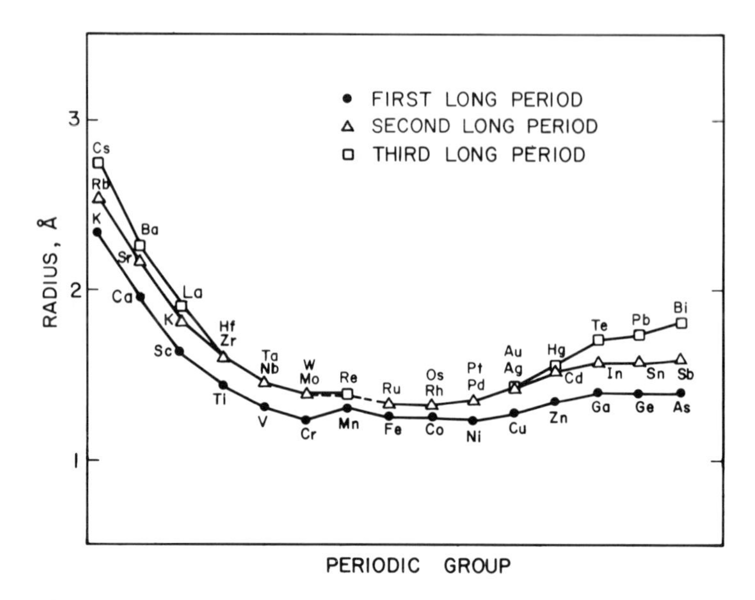

Fig. 47. Goldschmidt coordination 12 radii as a function of the periodic group. Data taken from the compilation of Dickinson.[284]

those elements exhibiting relatively stronger cohesion will be smaller, as shown in Figure 47. This plot indicates that the binding in transition elements is strong and consistent with the data shown in Figures 45 and 46. Therefore, among the elements, the transition elements exhibit the highest fracture strengths because of the high values of E and small atomic volumes.

Melting points are also a qualitative measure of cohesion. As an example, the melting points of the elements as a function of atomic number are shown in Figure 48 and show a periodic variation similar to the behavior of compressibilities (Figure 46). The melting points of the transition elements are comparatively higher and once again reflect the stronger bonding characteristic of these metals. The heats of sublimation or binding energies (reduced to the same temperature for comparison) show a similar variation as melting points with atomic number, with peaks at W, Nb, and V.[283]

The relative magnitudes of the elastic properties of solid solution alloys and intermetallic compounds are also generally related to the melting points. Table 10 contains Young's moduli for several carbides and nitrides of the transition elements, together with similar data for representatives from other classes of materials. Note that the high elastic moduli are generally consistent with the high melting points

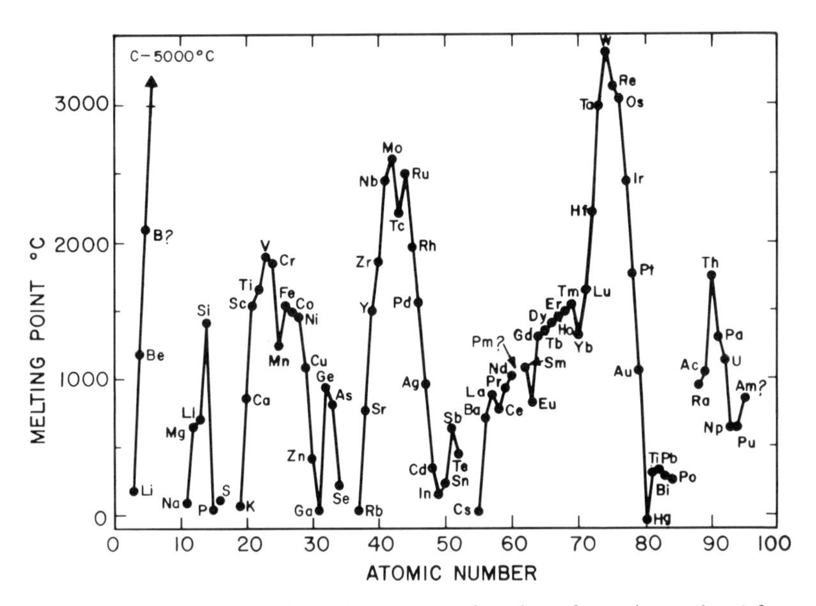

Fig. 48. Melting points of the elements as a function of atomic number (after Hume-Rothery and Coles[283]).

of these materials. However, exceptions can be found, exemplified by the modulus of ZrC, which is comparable to polycrystalline Cu (melting point 1083°C).

The above transition metal carbides and nitrides are often referred to as interstitial phases because the small C atoms occupy interstitial positions within the metallic sublattice, leading to close-packed phases. Thus, these phases exhibit small atomic or molecular volumes. In fact, phases (both metallic and nonmetallic) containing the small radius atoms or ions B, C, N, O, Be, Al, and Si generally are the strongest (Table 10).

Hardness tests, which result in plastic as well as an elastic deformation, are also qualitative indicators of the strength of materials. Nowotny[286] has suggested the following expression for the Vickers hardness H of the monocarbides of the transition elements:

$$H = K[(T_m - T)/MV^{2/3}]^{1/2}$$

where T_m, T, M, V, and K are the melting point, test temperature, molecular weight, molar volume, and a constant ($\sim$600), respectively. Thus, high T_m and small M and V favor high strength, consistent with what was said earlier.

TABLE 10
Young's Moduli and Melting Points for a Number of Materials*

	Melting point, °C	Young's modulus, $kg/mm^2 \times 10^{-3}$
NbC	3500	34.7
TaC	3880	29.0
TiC	3140	32.2
VC	2830	27.4
WC (hexagonal)	2870	72.2
ZrC	3530	14.0
B_4C	2470	46.2
TiN	2950	35
AlN	Sublimes at 2000	35
Si_3N_4	Sublimes at 1900	38.5
B	2300	44.8
Si	1410	18.2
Fe	1540	21.4
Cu	1083	13.2
Cu (70 %)–Zn (30 %)	920	10.2
Al	660	7.2
Al_2O_3	2050	~53.0
BeO	2520	35.7
SiO_2 (in air)	1700	7.4
NaCl	801	2.5
KCl	776	1.8

* Data mainly from Refs. 285 and 303.

4.2.2. Nonmetals

The elastic properties of diamondlike semiconductors Ge and Si, as well as 3–5 and 2–6 compounds, including those with the hexagonal wurtzite structure, have been correlated with interatomic distance and lattice vibrational frequencies.[287] The bonding in Ge and Si is purely covalent and while it is essentially covalent in the 3–5 and 2–6 compounds, the latter are more ionic. Dimensional analysis[288] had suggested that the elastic constants are of the order of q^2/r^4, where q is the electronic charge and r some characteristic length of the lattice in question. Keyes defined an elastic constant C_0 equal to q^2/b^4, where b is the distance between nearest-neighbor atoms, and showed that the bulk modulus for the above materials varies as q^2/b^4. Keyes also defined a reduced bulk modulus as being the ordinary bulk modulus divided by C_0. The fundamental lattice vibration frequency correlates well with the reduced bulk modulus for all except the 2–6 compounds and this is attributed to the increased ionicity of the 2–6 compounds.

Mitra and Marshall[289] showed that the bulk moduli of 16 alkali halides with the NaCl structure are also equal to a constant/a^n, where a is the lattice constant and n is closer to three than to four. Whereas Keyes's constant is given by q^2, no such simple constant could be given for the alkali halides, which are essentially ionically bonded phases. Kittel derived the following expression for the bulk modulus K, applicable to the above alkali halides:

$$K = (\alpha q^2/18R_0{}^4)[(R_0/\rho) - 2]$$

where R_0, α, and ρ are the equilibrium nearest-neighbor ion separation, Madelung constant, and repulsive range parameter, respectively. Cohesive energies calculated on the basis of this equation are in very good agreement with observed values.

An empirical expression for computing compressibilities of nonmetallic crystalline solids has been derived[290] which has been shown to give excellent agreement with measured compressibilities for a wide variety of materials, including ternaries, and representing nine different crystal structures. The compressibility κ is given by

$$\kappa = [(V/U)/q/m)]Z$$

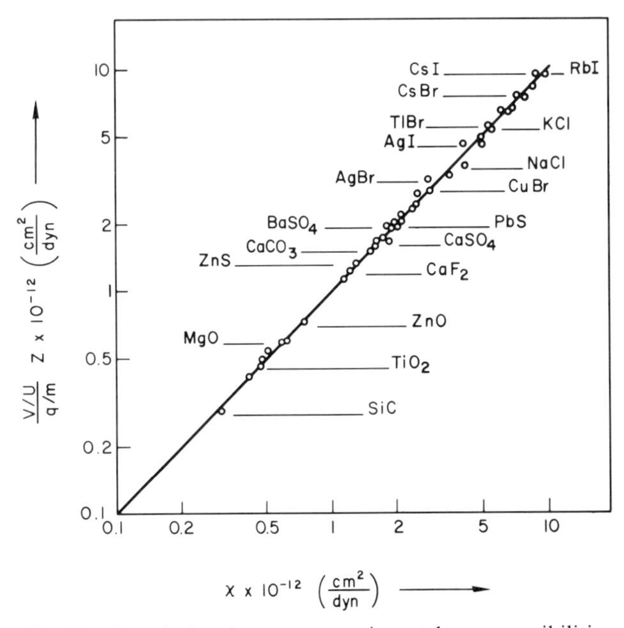

Fig. 49. Correlation between experimental compressibilities κ for 45 solids (19 identified) with those calculated from $\kappa = [(v/U)/(q/m)]Z$ (after Plendl *et al.*[290]).

where U is the cohesive energy, V is the molecular volume, m is the number of component atoms, q is the number of atoms per molecule, and Z is the maximum valence or number of bonds per molecule. The ratio Z/U is the reciprocal of the bond strength or cohesive energy per bond. Thus, large cohesive energies lead to small ratios of Z/U and hence to low compressibilities. The computed compressibilities for 45 compounds, compared with experimentally determined values, are shown in Figure 49. Only 19 of the compounds have been identified

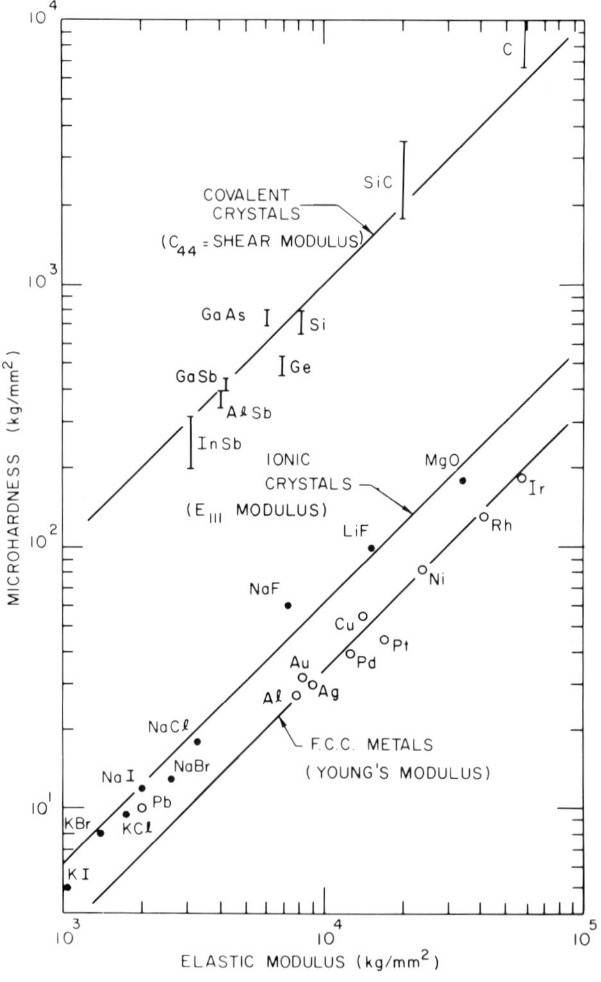

Fig. 50. Hardness versus elastic modulus for several fcc metals, essentially covalent crystals, and ionic materials (after Gilman[293] and Chin[294]).

on this figure. The above expression for compressibility has also been discussed in relation to hardness of nonmetallic solids.[290]

In general the elastic moduli of the alkali halides (purely ionic solids) are smaller than those for the metallic elements, alloys, and covalently bonded materials (Table 10 and Figure 50).

4.3. Plastic Behavior

The plastic deformation of crystalline solids proceeds by a process of slip and/or twinning on certain crystal planes and in certain crystal directions. In metals the slip planes (denoted by $\{\ \}$) are usually those having the highest atomic density and they are the most widely spaced. The slip directions (denoted by $\langle\ \rangle$) in the plane are those having the highest linear atomic density. A particular combination of slip plane and slip direction is referred to as a slip system.

In order to illustrate the role of bonding, slip and twinning systems and cleavage planes for several groups of materials are tabulated in Table 11. For example, the preferred slip systems for the ionic phases, such as NaCl, MgO and LiF, are $\{110\}\langle 110\rangle$, and they differ from those preferred by the isostructural metals, such as Cu and Ag [$\{111\}\langle 110\rangle$], although the latter mode becomes operative in the ionic materials at elevated temperatures, perhaps because the bonding becomes more diffuse.[291,292] The $\{110\}\langle 110\rangle$ slip is preferred in crystals of low ionic polarizability (NaCl, MgO, LiF, etc.), while $\{100\}\langle 110\rangle$ slip is preferred in isostructural materials of high ionic polarizability (PbS, PbSe, etc.).[291,292] In the ionic materials the preferred $\{110\}$ planes are not the most densely packed planes. The densest are the cube planes, but they are not preferred because of the strong repulsive interactions between like ions which would occur during shear on these planes. On the other hand, $\{100\}$ slip can occur in the less ionic materials (PbS, PbSe, etc.). Similarly, the formation of stacking faults and twins is not energetically favorable in highly ionic fcc crystals because of the repulsion between neighboring planes of similar atoms.

The basal plane in the hcp metals Zn and Cd is the most densely packed plane and slip generally occurs in this plane and in the $\langle 11\bar{2}0\rangle$ direction. Prismatic and pyramidal systems, in addition to the basal system, have been shown to operate at room and elevated temperatures, depending on the mode of deformation. The c/a ratio plays an important role in determining the ease with which other slip systems

TABLE 11

Slip Systems for a Select Group of Materials Illustrating the Effect of Bonding Type(291,292,300–302)

Material examples	Crystal structure	Nature of bonding	Slip mode (number of systems)	Twin mode (number of systems)	Remarks
Cu, Ag, Pb	FCC	Metallic	{111}⟨110⟩ (12)	{111}⟨112⟩ (12)	—
Fe, Nb, Mo	BCC	Metallic	{110}⟨111⟩ (12) {112}⟨111⟩ (12) {123}⟨111⟩ (24)	{112}⟨111⟩ (12)	—
NbC, TaC, VC	FCC(NaCl)	Some ionic, but classically metallic in behavior	{110}⟨110⟩ (6) {111}⟨110⟩ (12) {110}⟨001⟩ (6)	—	Macroscopically brittle at room temperature; ionic contribution to bonding (charge transfer) deduced from photoelectron spectroscopy and interpretation of heats of formation[298,299]; the stability of these phases (high melting points, for example) attributed to the ionic and *d*-orbital contributions; {111}⟨110⟩ and {110}⟨001⟩ slip systems operative at elevated temperatures
NaCl	FCC	Ionic	{110}⟨110⟩ (6) {110}⟨110⟩ (6) {111}⟨110⟩ (12)	—	{100}⟨110⟩ and {111}⟨110⟩ systems operative at elevated temperatures
MgO, LiF	FCC(NaCl)	Mainly ionic	{110}⟨110⟩ (6) {100}⟨110⟩ (6) {111}⟨110⟩ (12)	—	{110}⟨110⟩ slip system preferred in crystals of *low ionic polarizability*; {100}⟨110⟩ and {111}⟨110⟩ operative at elevated temperatures
PbS, PbSe, PbTe	FCC(NaCl)	Covalent + ionic	{100}⟨110⟩ (6)	—	{100}⟨110⟩ slip systems preferred in crystals of *high ionic polarizability*; cleavage plane at room temperature is {100}

Si, Ge	Diamond cubic	Covalent	{111}⟨110⟩ (12)	Cleavage plane {111}
III–V compounds (GaAs, InSb)	Diamond cubic	Largely covalent, some ionic contribution	{111}⟨110⟩ (12)	Cleavage plane {110}
CsCl, CsBr	BCC	Ionic	{110}⟨001⟩ (6) {110}⟨111⟩ (12)	{110}⟨001⟩ slip systems preferred in crystals tending toward *ionic* bonding (includes LiTl, MgTl, AuZn, AuCd); {110}⟨111⟩ slip systems preferred in crystals tending toward *metallic* bonding (example: CuZn; see above for Fe, etc.)
CaF_2, BaF_2	Cubic	Essentially ionic	{100}⟨110⟩ {100}⟨110⟩ + {110}⟨110⟩	{100}⟨110⟩ preferred at low temperatures; {100}⟨110⟩ + {110}⟨110⟩ preferred at elevated temperatures; cleavage plane {111}

can operate. In the ideal hcp structure, i.e., ideal packing of equal-size rigid spheres, the c/a ratio is 1.632. Zinc and Cd have c/a ratios of 1.856 and 1.886, while the ratios are 1.624, 1.587, 1.568, 1.59, 1.586, 1.617, and 1.624 for Mg, Ti, Be, Zr, Hg, Re, and Co, respectively. As the c/a ratio decreases, the packing densities on other planes become relatively larger and therefore they are more likely to operate, the slip direction still being $\langle 11\bar{2}0 \rangle$.

Gilman[293] has suggested that the major factor controlling the yield strength of crystals is the intrinsic "plastic resistance" to dislocation motion. This plastic resistance is expected to be related to the nature of bonding and the crystal structure. Since hardness is related to the yield strength* and the elastic constants are related to bonding behavior, Gilman obtained a plot of hardness versus modulus for face-centered cubic metals and for crystals with the zincblende structure exhibiting strong covalent bonding, as shown in Figure 50. The figure indicates that as a group, the covalent-bonded crystals are much harder than the *pure elements.* Recently Chin[294] obtained a similar plot for ionic crystals; this is included in Figure 50. Somewhat surprisingly, the behavior of the ionic crystals is much closer to that of metals than that of the covalently bonded crystals.

In a polycrystalline material the deformation of a particular grain is influenced by the deformation of neighboring grains. Deformation-induced lattice rotation of the individual grains gives rise to changes in orientation and development of texture. Several independent sets of deformation modes (slip and twinning systems) must operate within each grain to maintain continuity at the grain boundaries. Five independent deformation modes must operate simultaneously[295] and it is that combination of five which requires a minimum of work of deformation.[296] The macroscopic brittleness of some polycrystalline materials is due to the nature of the bonding because it results in an insufficient number of independent slip and/or twinning systems. The $\{110\}\langle 110 \rangle$ slip mode, predominant in NaCl, MgO, and LiF, for example, possesses only two independent slip

---

* Marsh[304] has shown empirically that the yield pressure P under a Vickers hardness indentor is related to the yield stress Y by the following equation:

$$P/Y = C + K[3/3 - \lambda)] \ln[3/(\lambda + 3\mu - \lambda\mu)]$$

where $\mu = (1 + \nu)Y/E$, $\lambda = 6(1 - 2\nu)Y/E$, ν is Poisson's ratio, E is Young's modulus, and C and K are constants. This equation is useful for estimating Y of glasses and brittle metals, but not for crystalline ceramics.

TABLE 12
Ratio of Bulk Modulus K to Shear Modulus G and Elongation Prior to Fracture for Some Polycrystalline FCC Metals (after Pugh[297])

	K/G	Poisson's ratio	Elongation, %
Calcium	2.33	0.31	60
Strontium	1.97	0.28	—
Aluminum	2.74	0.34	50
Thorium	1.74	0.26	—
Nickel	2.52	0.32	30
Rhodium	1.77	0.26	Small
Iridium	1.74	0.26	Small
Palladium	4.27	0.39	40
Platinum	5.25	0.44	40
Copper	3.00	0.35	60
Silver	3.44	0.37	60
Gold	6.14	0.42	50
Lead	7.37	0.44	64

systems, even though there are a total of six slip systems. The criterion for independent slip systems is given by Groves and Kelly.[305]

The ratio of bulk modulus K to shear modulus G (modulus of rigidity) has been shown to indicate to some extent intrinsic ductility or malleability.[297] This is illustrated for fcc metals in Table 12. Those metals with low K/G are expected to be intrinsically brittle, as exemplified by Ir and Rh. Bulk modulus K is related to the cleavage stress, while G is related to the shear stress for slip. A similar correlation exists for hcp metals, even though they are inherently less ductile because of the smaller number of slip systems.[297] On the other hand, elemental metals with complex structures (α-Mn for example) and many intermetallic compounds are intrinsically brittle and this may be due to a lack of easy slip modes and may bear no simple relation to the elastic properties, as signified by K/G. The bcc metals have approximately the same value of K/G and Poisson's ratio and exhibit equal ductility.[297] However, it must be borne in mind that the above has not been normalized to the melting points of the metals.

4.4. Summary

The preceding discussion focused on the chemical and structural aspects of the mechanical behavior of crystalline solids. Several general conclusions emerge which serve as a guide for the qualitative comparison of groups of materials and for the preparation of new structural materials. High-strength materials are to be found among those exhibiting high melting points, high hardness, and high surface free energies, and containing atoms whose radii are small. Small radii leads to dense atomic packing and small interatomic distances and thus small atomic volumes. Materials possessing the above properties are alloys and compounds which contain transition metal atoms and small-radius atoms or ions, such as B, C, N, Si, and O. The bonding in the high-strength materials is more or less directional in character and the compounds are generally brittle. Dislocations are difficult to move in directionally bonded materials, i.e., shear is difficult. The bonding in purely ionic crystals is nondirectional in character and is therefore comparatively weaker. It is well known that the nature of the stress system imposed on a solid, the strain rate, and temperature affect mechanical behavior and the above should be viewed in this light, as well as in the light of the role of imperfections.

Acknowledgments

It is a pleasure to acknowledge my colleagues G. D. Boyd, J. R. Carruthers, G. Y. Chin, T. H. Geballe, F. J. Di Salvo, A. G. Gossard, L. M. Holmes, B. F. Levine, S. Mahajan, D. McWhan, G. E. Peterson, M. Robbins, L. R. Testardi, and R. H. Willens for reading and commenting on several sections of this review. I am particularly indebted to B. F. Levine for use of Figure 44 prior to publication, and G. Y. Chin for use of his data for ionic crystals shown in Figure 50.

References

1. C. Herring, in *Magnetism* (G. T. Rado and H. Suhl, eds.), Vol. IV, Academic, New York (1966).
2. N. S. Hush and M. H. L. Pryce, *J. Chem. Phys.* **28**, 244 (1958).
3. C. Kittel, *Introduction to Solid State Physics*, 3rd ed. Wiley, New York (1966).
4. R. J. Weiss, *Solid State Physics for Metallurgists*, Pergamon, Oxford (1963).
5. B. T. Matthias, A. M. Clogston, H. J. Williams, E. Corenzwit, and R. C. Sherwood, *Phys. Rev. Letters* **7**, 7 (1961).
6. B. T. Matthias and R. M. Bozorth, *Phys. Rev.* **109**, 604 (1958).
7. A. J. Bradley and J. Thewlis, *Proc. Roy. Soc.* A**115**, 456 (1927).
8. J. A. Oberteuffer, J. A. Marcus, L. H. Schwartz and G. P. Felcher, *Phys. Rev.* **2**(3), 670 (1970).
9. S. H. Charap, in *Magnetism and Metallurgy* (A. Berkowitz and E. Knuller, eds.), Vol. I, Academic, New York (1969).

10. K. N. R. Taylor, *Contemp. Phys.* **10**(5), 423 (1970).
11. L. Corliss, J. Hastings, and R. Weiss, *Phys. Rev. Letters* **3**, 211 (1959).
12. V. N. Bykov, V. S. Golovkin, N. V. Ageev, V. A. Levdik, and S. I. Vinogradov, *Soviet Phys.—Doklady* **4**, 1970 (1959).
13. G. E. Bacon, *Acta Cryst.* **14**, 823 (1961).
14. G. Shirane and W. J. Takei, *J. Phys. Soc. Japan*, **17**, Suppl. B-III, 35 (1962).
15. P. J. Brown, C. Wilkinson, J. B. Forsythe, and R. Nathans, *Proc. Phys. Soc. (London)* **85**, 1185 (1965).
16. M. K. Wilkinson, E. O. Wollan, W. C. Koehler, and J. W. Cable, *Phys. Rev.* **127**, 2080 (1962).
17. W. C. Koehler, R. M. Moon, A. L. Trego, and A. R. Mackintosh, *Phys. Rev.* **151**, 405 (1966).
18. M. A. Ruderman and C. Kittel, *Phys. Rev.* **96**, 99 (1954).
19. T. Kasuya, *Progr. Theoret. Phys. (Kyoto)* **16**, 45 (1956).
20. K. Yosida, *Phys. Rev.* **106**, 893 (1957); **107**, 369 (1957).
21. V. Jaccarino and L. R. Walker, *Phys. Rev. Letters* **15**, 528 (1965); see also K. C. Brog and W. H. Jones, Jr., *Phys. Rev. Letters* **24**, 58 (1970).
22. A. M. Clogston, B. T. Matthias, M. Peter, H. J. Williams, E. Corenzwit, and R. C. Sherwood, *Phys. Rev.* **125**, 541 (1962).
23. G. K. Wertheim, V. Jaccarino, J. H. Wernick, and D. N. E. Buchanan, *Phys. Rev. Letters* **12**, 24 (1964).
24. G. K. Wertheim and J. H. Wernick, *Acta Met.* **15**, 297 (1967).
25. R. Nathans, M. T. Pigott, and C. G. Shull, *J. Phys. Chem. Solids* **6**, 38 (1958); S. J. Pickart and R. Nathans, *Phys. Rev.* **123**, 1163 (1961).
26. P. A. Beck, *Met. Trans.* **2**, 2015 (1971).
27. H. Okamoto and P. A. Beck, *Monatsh. Chem.* **3**, 907 (1972).
28. K. Ono, Y. Ishikawa, and A. Ito, *J. Phys. Soc. Japan* **17**, 1747 (1962).
29. C. G. Shull and M. K. Wilkinson, *Phys. Rev.* **97**, 304 (1955).
30. M. J. Marcinkowski and R. M. Poliak, *Phil. Mag.* **8**, 1023 (1963).
31. M. K. Wilkinson, N. S. Gingrich, and C. G. Shull, *J. Phys. Chem. Solids* **2**, 289 (1957).
32. J. H. Wernick, S. E. Haszko, and W. J. Romanov, *J. Appl. Phys.* **32**, 2495 (1961).
33. N. S. Satya Murthy, R. J. Begum, C. S. Somanathan, and M. R. L. N. Murthy, *J. Appl. Phys.* **40**, 1870 (1969).
34. R. C. Sherwood, E. A. Nesbitt, J. H. Wernick, D. D. Bacon, A. J. Kurtzig, and R. Wolfe, *J. Appl. Phys.* **42**, 1704 (1971).
35. L. Creveling, H. L. Luo, and G. S. Knapp, *Phys. Rev. Letters* **18**, 851 (1967).
36. L. Creveling and H. L. Luo, *Phys. Rev.* **176**, 614 (1968).
37. G. Y. Chin, R. C. Sherwood, J. H. Wernick, D. R. Mendorf, and G. S. Knapp, *Phys. Letters* **27A**, 302 (1968).
38. G. Y. Chin, J. H. Wernick, R. C. Sherwood, and D. R. Mendorf, *Solid State Commun.* **6**, 153 (1968).
39. E. P. Wohlfarth, *J. Appl. Phys.* **39**, 1061 (1968).
40. W. B. Pearson, *Z. Krist.* **126**, 362 (1968).
41. H. Claus, A. K. Sinha, and P. A. Beck, *Phys. Rev. Letters* **26A**, 38 (1968).
42. A. Narath and A. C. Gossard, *Phys. Rev.* **183**, 391 (1969); A. Narath, A. C. Gossard, and J. H. Wernick, *Phys. Rev. Letters* **20**, 795 (1968).
43. D. D. Dunlap, J. B. Darby, Jr., and C. W. Kimball, *Phys. Letters* **25A**, 431 (1967).

44. R. L. Cohen, R. C. Sherwood, and J. H. Wernick, *Phys. Letters* **26A**, 462 (1968).
45. R. L. Cohen, J. H. Wernick, K. W. West, R. C. Sherwood, and G. Y. Chin, *Phys. Rev.* **188**, 684 (1969).
46. G. Y. Chin, in *Advances in Materials Research*, Vol. 5 (H. Herman, ed.), Wiley, New York (1971).
47. G. Y. Chin, J. H. Wernick, R. C. Sherwood, and D. R. Mendorf, unpublished research.
48. B. G. LeFevre and E. A. Stacke, Jr., in *Advances in X-Ray Analysis*, Vol. 12 (C. S. Barrett, G. R. Mallett, and J. B. Newkirk, eds.), Plenum, New York (1969), p. 113.
49. T. Taoka, K. Yosukochi, and R. Honda, in *Mechanical Properties of Intermetallic Compounds* (J. H. Westbrook, ed.), Chapter 8, Wiley, New York (1960).
50. J. Crangle, *Phil. Mag.* **5**, 355 (1965); J. Crangle and W. R. Scott, *J. Appl. Phys.* **36**, 921 (1965); R. M. Bozorth, P. A. Wolff, D. D. Davis, V. B. Compton, and J. H. Wernick, *Phys. Rev.* **122**, 1157 (1961); M. J. Zuckermann, *Solid State Commun.* **9**, 1861 (1971).
51. G. E. Bacon and J. Crangle, *Proc. Roy. Soc.* A**273**, 387 (1963).
52. R. S. Tebble and D. J. Craik, *Magnetic Materials*, Wiley–Interscience, New York (1969).
53. O. Dahl, *Z. Metallk.* **128**, 133 (1936).
54. L. E. Orgel, *An Introduction to Transition-Metal Chemistry*, Methuen, London (1960); C. S. G. Phillips and R. J. P. Williams, *Inorganic Chemistry*, Part 2, Oxford Univ. Press (1966).
55. G. Blasse, *Philips Res. Rep.* Suppl. 3 (1964); J. Smit and H. P. J. Wijn, *Ferrites*, Wiley, New York (1959); E. J. W. Verwey and E. L. Heilmann, Jr., *Chem. Phys.* **15**, 174 (1947); F. de Boer, J. H. van Santen, and E. J. W. Verwey, *Chem. Phys.* **18**, 1032 (1950).
55a. S. Geller, *Z. Krist.* **125**, 1 (1967).
55b. L. Suschow, M. Kokta, and V. J. Flynn, *J. Solid State Chem.* **2**, 137 (1970); **5**, 85 (1972).
56. M. Robbins and L. Darcy, *J. Phys. Chem. Solids* **27**, 741 (1966).
57. M. Robbins, G. K. Wertheim, R. C. Sherwood, and D. N. E. Buchanan, *J. Phys. Chem. Solids* **32**, 717 (1971).
58. N. Menyuk, K. Dwight, R. J. Arrott, and A. Wold, *J. Appl. Phys.* **37**, 1387 (1966).
59. A. H. Cooke, C. J. Ellis, K. A. Gehring, M. J. M. Leask, D. M. Martin, B. M. Wanklyn, M. R. Wells, and R. L. White, *Solid State Commun.* **8**, 689 (1970).
60. B. W. Mangum, J. N. Lee, and H. W. Moos, *Phys. Rev. Letters* **27**, 1517 (1971).
61. E. Cohen, L. A. Riseberg, W. A. Nordland, R. D. Burbank, R. C. Sherwood, and L. G. Van Uitert, *Phys. Rev.* **186**, 476 (1969).
62. L. Neel, *J. Phys. Radium* **15**, 225 (1954).
63. S. Taniguchi and M. Yamamoto, *Sci. Rep. Res. Inst., Tohuku Univ.*, Ser. A, **6**, 330 (1954).
64. H. J. Bunge and H. G. Muller, *Z. Metallk.* **48**, 26 (1957).
65. S. Chikazumi, K. Suzuki, and H. Iwata, *J. Phys. Soc. Japan* **12**, 1259 (1957); S. Chikazumi, *J. Appl. Phys.* **29**, 346 (1958).
66. S. Chikazumi, K. Suzuki, and H. Iwata, *J. Phys. Soc. Japan* **15**, 250 (1960).
67. N. Tamagawa, Y. Nakagawa, and S. Chikazumi, *J. Phys. Soc. Japan* **17**, 1256 (1962).

68. G. Y. Chin, *J. Appl. Phys.* **36**, 2915 (1965).
69. G. Y. Chin, *Mat. Sci. Eng.* **1**, 77 (1966).
70. G. Y. Chin and E. A. Nesbitt, *J. Appl. Phys.* **37**, 1214 (1966).
71. G. Y. Chin, E. A. Nesbitt, J. H. Wernick, and L. L. Vanskike, *J. Appl. Phys.* **38**, 2623 (1967).
72. A. T. English, G. Y. Chin, and A. R. Von Neida, *J. Appl. Phys.* **38**, 997 (1967).
73. A. R. Von Neida, G. Y. Chin, and A. T. English, *J. Appl. Phys.* **39**, 610 (1968).
74. I. G. Austin and N. F. Mott, *Science*, **168**, 71 (1970).
75. I. G. Austin and D. Elwell, *Comtemp. Phys.* **11**(5), 455 (1970).
76. J. Appel, in *Solid State Physics* (F. Seitz, D. Turnbull, and H. Ehrenreich, eds.), Vol. 21, p. 193, Academic, New York (1968).
77. D. Adler, in *Solid State Physics* (F. Seitz, D. Turnbull, and H. Ehrenreich, eds.), Vol. 21, p. 1, Academic, New York (1968); *IBM J. Res. Dev.* **1970** (May), 261.
78. D. B. McWhan and J. P. Remeika, *Phys. Rev.* B **2**(9), 3734 (1970), and references therein.
79. A. Jayaraman, D. B. McWhan, J. P. Remeika, and P. D. Dernier, *Phys. Rev.* B **2**(9), 3751 (1970).
80. A. Menth and J. P. Remeika, *Phys. Rev.* B **2**(9), 3756 (1970).
81. A. C. Gossard, D. B. McWhan, and J. P. Remeika, *Phys. Rev.* B **2**(9), 3762 (1970).
82. P. D. Dernier and M. Marezio, *Phys. Rev.* B **2**(9), 3771 (1970).
83. N. F. Mott, *Proc. Phys. Soc. (London)* **62**, 416 (1949).
84. N. F. Mott, *Can. J. Phys.* **34**, 1356 (1956); *Phil. Mag.* **6**, 287 (1961); *Adv. Phys.* **16**, 49 (1967); *Rev. Mod. Phys.* **40**, 677 (1968).
85. H. G. Drickamer, *Solid State Phys.* **17**, 1 (1965).
86. A. Jayaraman, V. Narayanamurti, E. Bucher, and R. G. Maines, *Phys. Rev. Letters* **25**(6), 368 (1970); **25**, 1430 (1970).
87. R. E. Newham and Y. M. de Haan, Laboratory for Insulation Research, MIT, Progress Report #26, 1960 (unpublished).
88. D. B. McWhan, J. P. Remeika, T. M. Rice, W. F. Brinkman, J. P. Maita, and A. Menth, *Phys. Rev. Letters* **27**(14), 941 (1971).
89. P. K. Baltzer, P. K. Wojtowicz, M. Robbins, and E. Lopatin, *Phys. Rev.* **151**(2), 367 (1966).
90. P. W. Anderson, *Phys. Rev.* **102**, 1008 (1956).
91. M. Robbins, P. K. Baltzer, and E. Lopatin, *J. Appl. Phys.* **39**(2), 662 (1968).
92. N. Menyuk, K. Dwight, R. J. Arnott, and A. Wold, *J. Appl. Phys.* **37**, 1387 (1966).
93. H. W. Lehmann and M. Robbins, *J. Appl. Phys.* **37**, 1389 (1966); *J. Appl. Phys.* **38**(3), 946 (1967).
94. P. F. Bongers, C. Haas, A. M. J. G. van Run, and G. Zanmarchi, *J. Appl. Phys.* **40**(3), 958 (1969).
95. I. W. Shepherd, *Solid State Commun.* **8**, 1835 (1970).
96. G. Busch and P. Wachter, *Phys. Kondens. Materie*, **5**, 232 (1966).
97. T. R. McGuire and M. W. Shafer, *J. Appl. Phys.* **35**, 984 (1964).
98. B. T. Matthias, R. M. Bozorth, J. H. Van Vleck, *Phys. Rev. Letters* **7**, 160 (1961).
99. S. Van Houten, *Phys. Letters* **2**, 215 (1962).
100. F. H. Holtzberg, T. R. McGuire, S. Methfessel, and J. C. Suits, *Phys. Rev. Letters* **13**, 18 (1964); *J. Appl. Phys.* **35**, 1033 (1964).

101. G. Busch, J. Junod, R. G. Morris, J. Muheim, and W. Stutius, *Phys. Letters* **11**, 9 (1964).
102. J. J. Ryne and T. R. McGuire, *IEEE Trans. Magnetics* **MAG-8**(1), 105 (1972).
103. V. L. Moruzzi and D. T. Teaney, *Solid State Commun.* **1**, 127 (1963).
104. P. Schwob and O. Vogt, *Physics Letters* **24A**, 242 (1967).
105. G. Busch, P. Schwob, and O. Vogt, *Phys. Letters* **20**, 602 (1966); D. B. McWhan, P. C. Souers, and G. Jura, *Phys. Rev.* **143**, 385 (1966).
106. T. Kasuya, *IBM J. Res. Dev.* **14**, 214 (1970); T. Kasuya and A. Yanase, *Rev. Mod. Phys.* **40**, 684 (1968); *J. Phys. Soc. Japan* **25**, 1025 (1968).
107. G. Will, S. J. Pickart, H. A. Alperin, and R. Nathans, *J. Phys. Chem. Solids* **24**, 1969 (1963).
108. U. Enz, J. F. Fast, S. van Houten, and J. Smit, *Philips Res. Repts.* **17**, 451 (1962).
109. F. Holzberg, T. R. McGuire, S. Methfessel, and J. C. Suits, *Phys. Rev. Letters* **13**, 18 (1964); *J. Appl. Phys.* **35**, 1033 (1964); **37**, 976 (1966); S. Methfessel and D. C. Mattis, in *Handbuch der Physik*, Vol. 28, #1 (S. Flugge, ed.), Springer-Verlag, Berlin (1968); S. von Molnar and S. Methfessel, *J. Appl. Phys.* **38**, 959 (1967).
110. M. Föex, *Compt. Rend.* **223**, 1126 (1946); *J. Rech. Centre Natl. Rech. Sci. Lab. Bellevue (Paris)* **4**, 238 (1951).
111. D. B. McWhan, A. Menth, J. P. Remeika, W. F. Brinkman, and T. M. Rice, to be published in *Phys. Rev.*
112. D. C. Mattis, in *The Theory of Magnetism*, Harper and Row, New York (1965).
113. T. Oguchi, *Phys. Rev.* **133**, A1098 (1964).
114. N. D. Mermin and H. Wagner, *Phys. Rev. Letters* 17, 1133 (1966).
115. H. E. Stanley and T. A. Kaplan, *Phys. Rev. Letters* 17, 913 (1966).
116. M. E. Lines, *J. Appl. Phys.* **40**(3), 1352 (1969).
117. A. Narath, *Phys. Rev.* **131**, 1929 (1963); A. Narath and H. L. Davis, *Phys. Rev.* **137**, A163 (1965).
118. E. Legrand and R. Plumier, *Phys. Stat. Sol.* **2**, 317 (1962); R. Plumier, *J. Appl. Phys.* **35**, 950 (1964).
119. R. J. Birgeneau, J. Skalyo, Jr., and G. Shirane, *J. Appl. Phys.* **41**(3), 1303 (1970).
120. R. J. Birgeneau, H. J. Guggenheim, and G. S. Shirane, *Phys. Rev. Letters* **22**, 720 (1969).
121. J. Skalyo, Jr., G. Shirane, R. J. Birgeneau, and H. J. Guggenheim, *Phys. Rev. Letters* **23**, 1394 (1969).
122. G. K. Wertheim, H. J. Guggenheim, H. J. Levinstein, D. N. E. Buchanan, and R. C. Sherwood, *Phys. Rev.* **173**, 614 (1968).
123. L. J. de Jongh, A. C. Botterman, F. R. de Boer, and A. R. Miedema, *J. Appl. Phys.* **40**(3), 1363 (1969).
124. M. F. Mostafa and R. D. Willett, *Phys. Rev.* **B4**(7), 2213 (1971).
125. A. I. Gubanov, *Fiz. Tverd. Tela* **2**, 502 (1960).
126. K. Handrick, *Phys. Stat. Sol.* **32**, K55 (1969).
127. S. Mader and A. S. Nowick, *Appl. Phys. Letters* **7**, 57 (1965).
128. B. C. Giessen and R. H. Willens, *Phase Diagrams*, Vol. 3, Academic, New York (1970); T. R. Anantharaman and C. Suryanarayana, *J. Mat. Sci.* **6**, 1111 (1971).
129. P. Duwez, R. H. Willens, and R. C. Crewdson, *J. Appl. Phys.* **36**, 2267 (1965).
130. M. Weiner, Magnetic moments of amorphous Pd–Co–Si alloys, Ph.D thesis, Calif. Inst. Tech., Pasadena, California, 1968.
131. T. E. Sharon and C. C. Tsuei, *Solid State Commun.* **9**, 1923 (1971).

132. P. Duwez and S. C. H. Lin, *J. Appl. Phys.* **38**, 4096 (1967).
133. C. C. Tsuei, G. Longworth, and S. C. H. Lin, *Phys. Rev.* **170**, 603 (1968).
134. A. K. Sinha, *J. Appl. Phys.* **42**, 338 (1971).
135. R. Hasegawa, *Phys. Rev.* B **3**(5), 1631 (1971).
136. A. K. Sinha and P. Duwez, *J. Appl. Phys.* **43**(2), 431 (1972).
137. J. Bardeen, L. N. Cooper, and J. H. Schrieffer, *Phys. Rev.* **108**, 1175 (1957).
138. B. T. Matthias, *Progress in Low Temperature Physics*, Vol. II, Interscience, New York (1957).
139. B. T. Matthias, T. H. Geballe, and V. B. Compton, *Rev. Mod. Phys.* **35**, 1 (1963).
140. B. W. Roberts, in *Intermetallic Compounds* (J. H. Westbrook, ed.), Wiley, New York (1967).
141. B. T. Matthias, *Phys. Rev.* **97**, 74 (1955).
142. F. J. Morin and J. P. Maita, *Phys. Rev.* **129**, 1115 (1963).
143. D. Pines, *Phys. Rev.* **109**, 280 (1958).
144. A. E. Dwight, in *Columbium Metallurgy* (D. L. Douglas and F. W. Kunz, eds.), Met. Soc. AIME—Interscience (1961).
145. S. Geller, *Acta Cryst.* **9**, 885 (1956).
146. M. Weger, *Rev. Mod. Phys.* **36**, 173 (1964).
147. J. Labbe and J. Freidel, *J. Phys. (Paris)* **27**, 153 (1966).
148. W. E. Blumberg, J. Eisinger, V. Jaccarino, and B. T. Matthias, *Phys. Rev. Letters* **5**, 149 (1960).
149. A. M. Clogston and V. Jaccarino, *Phys. Rev.* **121**, 1357 (1961).
150. A. M. Clogston, A. C. Gossard, V. Jaccarino, and Y. Yafet, *Phys. Rev. Letters* **9**, 262 (1962).
151. B. G. Silbernagel, M. Weger, W. G. Clark, and J. H. Wernick, *Phys. Rev.* **153**, 535 (1967).
152. R. G. Shulman, B. J. Wyluda, and B. T. Matthias, *Phys. Rev. Letters* **1**, 278 (1958).
153. R. H. Willens, T. H. Geballe, A. C. Gossard, J. P. Maita, A. Menth, G. W. Hull, Jr., and R. R. Soden, *Solid State Commun.* **7**, 837 (1969).
154. M. Bernasson, P. Descouts, R. Flükiger, and A. Treyband, *Solid State Commun.* **8**, 837 (1970).
155. E. Ehrenfreund, A. C. Gossard, and J. H. Wernick, *Phys. Rev.* B **4**(9), 2906 (1971).
156. W. L. McMillan, *Phys. Rev.* **167**, 331 (1968).
157. B. T. Matthias, T. H. Geballe, L. D. Longinotti, E. Corenzwit, G. W. Hull, R. H. Willens, and J. P. Maita, *Science* **156**, 645 (1967).
158. N. V. Ageev, N. E. Alekseevskii, N. N. Mikhailov, and V. F. Shamrai, *JETP Letters* **6**, 329 (1967).
159. S. Foner, E. J. McNiff, Jr., B. T. Matthias, and E. Corenzwit, in *Proc. of LT 11*, St. Andrews University, 11b, 1025 (1968).
160. G. Arrhenius, E. Corenzwit, R. Fitzgerald, G. W. Hull, H. L. Luo, B. T. Matthias, and W. H. Zachriasen, *Proc. Nat. Acad. Sci.* **61**, 621 (1968).
161. L. R. Testardi, J. E. Kunzler, H. J. Levinstein, J. P. Maita, and J. H. Wernick, *Phys. Rev.* B (1), 107 (1971).
162. B. W. Batterman and C. S. Barrett, *Phys. Rev. Letters* **13**, 390 (1964).
163. R. Mailfert, B. W. Batterman, and J. J. Hanak, *Phys. Stat. Solid* **32**, K67 (1969); L. J. Vieland, R. W. Cohen and W. Rehwald, *Phys. Rev. Letters* **26**, 373 (1971).

164. B. N. Kodess, V. B. Kuritzin, and B. N. Tretjako, *Phys. Letters* **37A**(5), 415 (1971).
165. J. Perel, B. W. Batterman, and E. I. Blount, *Phys. Rev.* **166**, 616 (1968); G. Shirane and J. D. Axe, *Phys. Rev.* **B4**, 2957 (1971).
166. L. R. Testardi, R. R. Soden, E. S. Greiner, J. H. Wernick, and V. G. Chirba, *Phys. Rev.* **154**, 399 (1967).
167. L. R. Testardi, *Phys. Rev.* B **3**(1), 95 (1971).
168. L. R. Testardi, J. J. Hauser, and M. H. Read, *Solid State Commun.* **9**, 1829 (1971).
169. C. W. Chu, E. Bucher, A. S. Cooper, and J. P. Maita, *Phys. Rev.* **B4**, 320 (1971); L. R. Testardi, *Phys. Letters* **35A**(2), 117 (1971).
170. A. C. Lawson and W. H. Zachariasen, *Phys. Letters* **38A**(1), 1 (1972).
171. A. G. Knapton, *J. Inst. Metals* **87**, 62 (1958).
172. E. Nembach, K. Tachikawa, and S. Takano, *Phil. Mag.* **21**, 869 (1970).
173. G. Aschermann, E. Friederich, E. Justi, and J. Kramer, *Z. Phys.* **42**, 349 (1941).
174. B. T. Matthias, *Phys. Rev.* **92**, 874 (1953).
175. N. Pessall and J. K. Hulm, *Physica* **2**, 311 (1966).
176. N. Pessall, R. E. Gold, and H. A. Johansen, *J. Phys. Chem. Solids* **29**, 19 (1968).
177. J. K. Hulm, M. S. Walker, and N. Pessall, *Physica* **55**, 60 (1971).
178. T. H. Geballe, B. T. Matthias, J. P. Remeika, A. M. Clogston, V. B. Compton, J. P. Matia, and H. J. Williams, *Physics* **2**, 293 (1966).
179. J. J. Hopfield, *Physica* **55**, 41 (1971).
180. T. B. Reed, H. C. Gatos, W. J. LaFleur, and J. T. Roddy, in *Metallurgy of Advanced Electronic Materials* (G. E. Brock, ed.), Interscience, New York, 1963.
181. R. Hagner and E. Saur, *Naturwiss.* **49**, 444 (1962).
182. R. H. Willens, E. Buehler, and B. T. Matthias, *Phys. Rev.* **159**(2), 327 (1967).
183. G. W. Webb, L. J. Vieland, R. E. Miller, and Wicklund, *Solid State Commun.* **9**, 1769 (1971).
184. L. D. Hartsough and R. H. Hammond, *Solid State Commun.* **9**, 885 (1971).
185. L. L. Oden and R. E. Siemens, *J. Less Common Metals* **14**, 33 (1968).
186. S. Geller, *Acta Cryst.* **9**, 885 (1956).
187. E. A. Wood, V. B. Compton, B. T. Matthias, and E. Corenzwit, *Acta Cryst.* **11**, 604 (1958).
188. R. H. Willens and E. Buehler, *Appl. Phys. Letters* **7**, 25 (1965).
189. C. C. Tsuei, H. Yen, and P. Duwez, *Phys. Letters* **34A**, 80 (1971); H. L. Luo, M. F. Merriam, and D. C. Hamilton, *Science* **145**, 581 (1964); C. C. Tsuei and L. R. Newkirk, *Phys. Rev.* **183**, 619 (1969).
190. W. Y. K. Chen and C. C. Tsuei, *Phys. Rev.* B **5**(3), 901 (1972).
191. R. H. Willens and E. Buehler, *J. Appl. Phys.* **38**(1), 405 (1967).
192. M. B. Maple, *Solid State Commun.* **8**, 9915 (1970).
193. B. T. Matthias, M. Peter, H. J. Williams, A. M. Clogston, E. Corenzwit, and R. C. Sherwood, *Phys. Rev. Letters* **5**, 542 (1960).
194. T. H. Geballe, B. T. Matthias, J. P. Remeika, A. M. Clogston, V. B. Compton, J. P. Maita, and H. J. Williams, *Physics* **2**, 293 (1966).
195. A. A. Abrikosov and L. P. Gorkov, *Zh. Eksperim. i Teor. Fiz.* **39**, 1781 (1960); *Soviet Phys.—JETP* **12**, 1243 (1961).
196. R. A. Ferrel, *Phys. Rev. Letters* **3**, 262 (1959).
197. P. W. Anderson, *Phys. Rev. Letters* **3**, 325 (1959).
198. G. Boato, G. Gallinaro, and C. Rizzuto, *Phys. Letters* **5**, 20 (1963); *Rev. Mod. Phys.* **3C**, 162 (1964); *Phys. Rev.* **148**, 535 (1966).

199. J. G. Huber and M. B. Maple, *Solid State Commun.* **1970**, 1987.
200. A. B. Kaiser, *J. Phys.* C **3**, 409 (1970).
201. T. Sugawara, *Physics* **55**, 143 (1971).
202. P. G. Degennes, *J. Phys. Radium* **23**, 510 (1962).
203. H. E. Barz, A. S. Cooper, E. Corenzwit, M. Marezio, B. T. Matthias, and P. H. Schmidt, *Science* **175**, 884 (1972).
204. B. T. Matthias, M. Marezio, E. Corenzwit, A. S. Cooper, and H. E. Barz, *Science* **175**, 1465 (1972).
205. R. Cheverel, M. Sergent, and J. Prigent, *J. Solid State Chem.* **3**, 515 (1971).
206. J. C. Phillips, *Phys. Rev. Letters* **28**, 1196 (1972).
207. R. A. Hein, J. W. Gibson, R. Mozelsky, R. C. Miller, and J. K. Hulm, *Phys. Rev. Letters* **12**, 230 (1964).
208. C. S. Koonce, M. L. Cohen, J. F. Schooley, W. R. Hosler, and E. R. Pfeiffer, *Phys. Rev.* **163**, 380 (1967).
209. J. K. Hulm, C. K. Jones, D. W. Deis, H. A. Fairbank, and P. A. Lawless, *Phys. Rev.* **169**, 388 (1968).
210. N. E. Phillips, B. B. Triplett, R. D. Clear, H. E. Simon, J. K. Hulm, C. K. Jones, and R. Mazelsky, *Physica* **55**, 571 (1971).
211. J. Appel, *Physica* **55**, 577 (1971).
212. E. Ehrenfreund and T. H. Geballe, to be published.
213. J. J. Hauser, *Phys. Rev.* B **3**(5), 1611 (1971).
214. W. Cochran, *Adv. Phys.* **9**, 387 (1960); **18**, 157 (1969).
215. J. D. Axe, *Solid State Commun.* **5**, 413 (1967).
216. G. Shirane, J. D. Axe, J. Harada, and J. P. Remeika, *Phys. Rev.* B **2**, 155 (1970).
217. C. Kittel, *Phys. Rev.* **82**, 729 (1951); M. H. Cohen, *Phys. Rev.* **84**, 369 (1951).
218. M. Eibschutz, H. J. Guggenheim, S. H. Wemple, I. Camlibel, and M. DiDomenico, Jr., *Phys. Letters* **29A**, 409 (1969); M. DiDomenico, Jr., M. Eibschutz, H. J. Guggenheim, and I. Camlibel, *Solid State Commun.* **7**, 1119 (1969).
219. T. Yamada and H. Iwasaki, *Appl. Phys. Letters* **21**, 89 (1972).
220. H. Thomas, *IEEE Trans. on Magnetics* **MAG-5**, 874 (1969).
221. S. C. Abrahams and E. T. Keve, *Ferroelectrics* **2**, 129 (1971).
222. S. C. Abrahams, in *MTP International Review of Science; Physical Chemistry*, Series 1, Volume II, *Chemical Crystallography* (J. M. Robertson, ed.), Butterworths, University Park Press, Md. (1973).
223. F. S. Gallusso, *Structure, Properties, and Preparation of Perovskite-Type Compounds*, Pergamon, Oxford (1969).
224. V. M. Goldschmidt, *Skriften Norske Videnskapsakad. Oslo, I. Mat.–Naturv. Kl.* **1926**, No. 8.
225. R. S. Roth, *J. Res. NBS* **58**, RP2736 (1957).
226. G. H. Haertling, *Am. Ceram. Soc. Bull.* **43**, 875 (1964); C. C. Land and P. D. Thacher, *Proc. IEEE* **57**, 751 (1969).
227. E. Sawaguchi, *J. Phys. Soc. Japan* **8**, 615 (1953).
228. S. C. Abrahams, S. K. Kurtz, and P. B. Jamieson, *Phys. Rev.* **172**, 551 (1968).
229. G. E. Peterson, P. M. Bridenbaugh, and P. Green, *J. Chem. Phys.* **46**, 4009 (1967).
230. G. E. Peterson and P. M. Bridenbaugh, *J. Chem. Phys.* **48**, 3402 (1968).
231. M. E. Lines, *Phys. Rev.* **2**, 698 (1970).
232. C. S. Brown, R. C. Kell, R. Taylor, and L. A. Thomas, *Proc. Inst. Electrical Engineers*, Paper #3798, 99 (1962).

233. H. Jaffe, *IEEE Trans. on Electron Devices* **ED-16**, 557 (1969); B. Jaffe, W. R. Cook, Jr., and H. Jaffe, *Piezoelectric Ceramics*, Academic, New York (1971).
234. A. A. Ballman, *J. Am. Chem. Soc.* **98**, 112 (1965).
235. B. Jaffe, R. S. Roth, and S. Morzullo, *J. Appl. Phys.* **25**, 809 (1954).
236. T. Ikeda and T. Okano, *J. Appl. Phys. Japan* **3**, 63 (1964).
237. D. M. Dodd and D. B. Fraser, *J. Phys. Chem. Solids* **26**, 673 (1965).
238. E. D. Kolb, D. A. Pinnow, T. C. Rich, N. C. Lias, E. E. Crudenski, and R. A. Laudise, *Mat. Res. Bull.* **7**(5), 397 (1972).
239. J. C. King, A. A. Ballman, and R. A. Laudise, *J. Phys. Chem. Solids* **23**, 1019 (1962).
240. A. A. Ballman, R. A. Laudise, and D. W. Rudd, *Appl. Phys. Letters* **8**, 53 (1966).
241. A. A. Ballman, D. M. Dodd, N. A. Keubler, R. A. Laudise, D. L. Wood, and D. W. Rudd, *Appl. Optics* **7**, 1387 (1968).
242. J. F. Nye, *Physical Properties of Crystals*, Oxford (1957).
243. E. Fatuzzo and W. J. Merz, *Ferroelectricity*, Interscience, New York, 1967.
244. R. T. Denton, F. S. Chen, and T. S. Kensel, *J. Appl. Phys.* **38**, 1611 (1967).
245. G. E. Peterson, J. R. Carruthers, and A. Carnevale, *J. Chem. Phys.* **53**, 2436 (1970).
246. L. K. Anderson, *Ferroelectrics* **3**, 69 (1972).
247. A. G. Chynoweth, *J. Appl. Phys.* **27**, 78 (1956); *Phys. Rev.*, **102**, 705 (1956); *Phys. Rev.* **117**, 1235 (1960).
248. A. M. Glass, *J. Appl. Phys.* **40**, 4699 (1969).
249. P. A. Franken and J. F. Ward, *Rev. Mod. Phys.* **35**, 23 (1963); N. Bloembergen, *Nonlinear Optics*, Benjamin, New York (1965); D. A. Kleinnan, *Phys. Rev.* **128**, 1761 (1962); F. N. H. Robinson, *Bell System Tech. J.* 46, 913 (1967); S. A. Alhmanor and R. V. Khokhlov, *Soviet Phys.—Usp.* **9**, 210 (1966); A. Yariv, *Quantum Electronics*, Wiley, New York (1967); G. C. Baldwin, *An Introduction to Nonlinear Optics*, Plenum, New York (1969).
250a. J. C. Bergman and S. K. Kurtz, *Mat. Sci. Eng.* **5**, 235 (1970).
250b. R. A. Laudise, in *Proc. of the Chania Conf. on Electronic Materials, Chania, Greece, 1970*, to be published.
250c. J. R. Carruthers, *Encyclopedia of Chemical Technology, Supplement Volume*, Wiley, (1971).
250d. J. A. Giordmaine, *Physics Today* **21**(11), 38 (1969).
251. P. A. Franken, A. E. Hill, C. W. Peters, and G. Weinreich, *Phys. Rev. Letters* **7**, 118 (1961).
252. P. D. Maker, R. W. Terhune, M. Nisenhoff, and C. M. Savage, *Phys. Rev. Letters* **8**, 21 (1962).
253. J. A. Giordmaine, *Phys. Rev. Letters* **8**, 19 (1962).
254. M. V. Hobden, *J. Appl. Phys.* **38**, 4365 (1967).
255. E. Giebe and A. Scheibe, *Z. Physik* **33**, 760 (1925).
256. S. K. Kurtz and T. T. Perry, *J. Appl. Phys.* **39**, 3798 (1968).
257. R. C. Miller, *Appl. Phys. Letters* **5**, 17 (1964).
258. G. D. Boyd, E. Buehler, and F. G. Storz, *Appl. Phys. Letters* **18**(7), 301 (1971); G. D. Boyd, W. B. Gandrud, and E. Buehler, *Appl. Phys. Letters* **18**(10), 446 (1971); G. D. Boyd, H. Kasper, and J. H. McFee, *IEEE J. Quantum Electronics* **QE-7**(12), 563 (1971); G. D. Boyd, E. Buehler, F. G. Storz, and J. H. Wernick, *IEEE J. Quantum Electronics* **QE-8**(4), 419 (1972).
259. R. L. Byer, H. Kildal, and R. S. Feigelson, *Appl. Phys. Letters* **19**, 237 (1971).

260. J. L. Shay, B. Tell, H. M. Kasper, and L. M. Schiavone, *Phys. Rev.* **B 5**, 5003 (1972).
261. S. E. Harris, *Proc. IEEE* **57**, 2096 (1969).
262. G. D. Boyd and J. H. McFee, to be published.
263. B. Tell, J. L. Shay, and H. M. Kasper, *Phys. Rev.* B, **6**, 3008 (1972).
264. M. DiDomenico, Jr. and S. H. Wemple, *J. Appl. Phys.* **40**, 720 (1969).
265. S. H. Wemple and M. DiDomenico, Jr., *J. Appl. Phys.* **40**, 735 (1969).
266. F. N. H. Robinson, *Bell System Tech. J.* **46**, 193 (1967).
267. S. S. Jha and N. Bloembergen, *IEEE J. Quantum Electronics* **QE-4**, 670 (1968); *Phys. Rev.* **171**, 891 (1968).
268. M. C. Flytzanis and J. Ducuing, *Phys. Letters* **25A**, 315 (1968).
269. J. C. Phillips, *Phys. Rev. Letters* **20**, 550 (1968); *Phys. Rev.* **168**, 905 (1968); **168**, 917 (1968).
270. J. C. Phillips, *Phys. Rev.* **166**, 832 (1968).
271. J. A. Van Vechten, *Phys. Rev.* **182**(3), 891 (1969).
272. J. A. Van Vechten, *Phys. Rev.* **187**, 1007 (1969).
273. J. C. Phillips, *Covalent Bonding in Crystals, Molecules and Polymers*, Univ. of Chicago Press, Chicago, Ill. (1969).
274. B. F. Levine, *Phys. Rev. Letters* **22**, 787 (1969).
275. B. F. Levine, *Phys. Rev. Letters* **25**, 440 (1970).
276. B. F. Levine, to be published.
277. G. E. Peterson, in *Magnetic Resonance*, Plenum, New York (1970).
278. R. T. Sanderson, *Inorganic Chemistry*, Reinhold, New York (1967).
279. E. Orowan, *Rep. Prog. Phys.* **12**, 185 (1949).
280. J. Frenkel, *Z. Physik* **37**, 572 (1926).
281. J. K. Mackenzie, Thesis, Bristol, 1949.
282. W. Koster, *Z. Metallk.* **39**, 1 (1948).
283. W. Hume-Rothery and B. R. Coles, *Atomic Theory for Students of Metallurgy*, The Institute of Metals, London (1969).
284. S. K. Dickinson, Jr., Ionic, covalent, and metallic radii of the chemical elements, AFCRL-70-0727, Dec. 1, 1970.
285. C. J. Smithells, *Metals Reference Book*, 4th ed. Vol. 3, Plenum, New York (1967).
286. H. O'Neill, *Hardness Measurement of Metals and Alloys*, 2nd ed., Chapman and Hall, London (1967).
287. R. W. Keyes, *J. Appl. Phys.* **33**, 3371 (1962).
288. H. B. Huntington, *Solid State Phys.* **7**, 213 (1958).
289. S. S. Mitra and R. Marshall, *J. Chem. Phys.* **41**, 3158 (1964).
290. J. N. Plendl, S. S. Mitra, and P. J. Gielisse, *Phys. Stat. Sol.* **12**, 367 (1965).
291. M. J. Buerger, *Am. Min.* **15**, 21, 35 (1930).
292. J. J. Gilman, *Acta Met.* **7**, 608 (1959).
293. J. J. Gilman, *Austr. J. Phys.* **13**, 327 (1969).
294. G. Y. Chin, to be published.
295. R. von Mises, *Z. Angew. Math. Mech.* **8**, 161 (1928).
296. G. I. Taylor, *J. Inst. Metals* **62**, 307 (1938).
297. S. F. Pugh, *Phil. Mag.* **45**, 823 (1954).
298. L. Ramqvist, *J. Appl. Phys.* **42**(5), 2113 (1971).
299. J. C. Phillips, *J. Appl. Phys.* **43**(8), 3560 (1972).
300. G. E. Hollox and D. J. Roewecliffi, *J. Mat. Sci.* **6**, 1261 (1971).
301. F. R. N. Nabarro, Z. S. Basinski, and D. B. Holt, *Adv. Phys.* **13**(50), 193 (1964).

302. W. A. Rochinger and A. H. Cottrell, *Acta Met.* **4**, 109 (1956).
303. A. Kelly, *Strong Solids*, Clarendon Press, Oxford (1966), and references therein.
304. D. M. Marsh, *Proc. Roy. Soc.* A **279**, 420 (1964).
305. G. W. Groves and A. Kelly, *Phil. Mag.* **8**, 877 (1963).
306. J. Orchotsky and K. Schröder, *J. Appl. Phys.* **43**, 2413 (1972).

5

Introduction to Chemical and Structural Defects in Crystalline Solids

Morris E. Fine
Department of Materials Science
Northwestern University
Evanston, Illinois

1. Introduction

The concept of the crystalline solid defines the perfect crystal to be a solid in which the atoms are arranged in a three-dimensional periodic array of perfect order. A basic group of atoms or molecules, themselves in a definite arrangement, is repeated in space according to a regular pattern to give the crystal structure. That atoms are arranged in crystals in regular patterns was deduced during the latter part of the 18th century by early crystallographers, such as R. J. Haüy, from the occurrence of macroscopic planar faced crystals in nature. These ideas were confirmed when diffraction of crystals by X-rays was discovered by Von Laue in 1912 and by the subsequent determination of structure of crystals beginning with the Braggs.

The understanding that masses of most solid substances are assemblages of a large number of crystals or grains is quite old. It was clearly stated by Savart in 1829.[1] This led to the first crystal defect to be recognized; namely the grain boundary or interface between two differently oriented grains of the same phase in a piece of solid material. We will return to this subject later. Each type of defect will be mentioned in this introduction and then be discussed in detail later in this chapter.

Soon after Von Laue's discovery of X-ray diffraction Darwin[2] in 1914 realized that crystals themselves were not perfect, because the measured intensities of the diffracted beams from rocksalt crystals were substantially greater than the theoretical values. Darwin[2–4] postulated that real crystals are really made up of a "mosaic composed of small crystalline fragments which are approximately parallel to each other, but whose orientations are distributed through an angular range of many minutes of arc on either side of a mean direction. All atoms within each fragment are regarded as being in perfect arrangement." In modern terminology one speaks of subgrains rather than a mosaic structure.

The idea of point defects in crystals goes back to Frenkel,[5] who in 1926 proposed the existence of point defects to explain the observed values of ionic conductivity in crystalline solids. In a crystal of composition MX such as a monovalent metal halide or a divalent metal oxide or sulfide, volume ionic conductivity occurs by motion of positive or negative ions in the lattice under the influence of an electric field. If the crystal were perfect, imperfections, such as vacant lattice sites or interstitial atoms, would need to be created for ionic conductivity to occur. A great deal of energy is required to dislodge an ion from its normal lattice position and thus the current in perfect crystals would be very, very small under normal voltages. To get around this difficulty, Frenkel proposed that point defects existed in the lattice prior to the application of the electric field. This, of course, has been substantiated by subsequent work and the concept of point defects in all classes of solids, metals, ionic crystals, covalent crystals, semiconductors, etc., is an important part of the physics and chemistry of crystalline solids, not only with respect to ionic conductivity but also with respect to diffusion, radiation damage, creep, and many other properties.

Point defects, that is, interstitial atoms or ions and vacant lattice sites, lead to the idea of nonstoichiometric compounds since, for example, in a compound of nominal composition MX there is the possibility of $M > X$ or $X > M$. The solid, of course, must be electrically neutral and if $X > M$ because, for example, there are cation vacancies, charge compensation from some other defect must be present. Nonstoichiometric compounds are in fact very common.

One of the first nonstoichiometric compounds where the defects associated with the nonstoichiometry were identified is wüstite. The nominal composition of wüstite is FeO and the structure is that of rocksalt. Wüstite exists in equilibrium only above 570°C. It exists at 1350°C over the composition range $Fe_{0.95}O$ to $Fe_{0.84}O$.[6] The stoichiometric composition FeO, which would contain only divalent cations, Fe^{2+}, is not stable at any temperature at atmospheric pressure. Quenching from above 570° retains wüstite at low temperatures as a metastable phase. Since there is an excess oxygen, the deviation from stoichiometry could occur from either extra oxygen ions in interstitial positions or from vacant iron ion sites, giving a deficiency in cations. In 1933 Jette and Foote[7] compared the lattice parameter and the density of a number of different compositions of wüstite. Their results could be explained only by cation vacancies. Subsequent work of many others further verified this conclusion. The charge balance is maintained by the presence of two trivalent iron ions, Fe^{3+}, for each divalent cation vacancy. The trivalent ions are another form of point defect. The presence of Fe^{2+} and Fe^{3+} in wüstite gives rise to itinerant electrons and wüstite is a good conductor.

Actually, the defect structure in some wüstite compositions is even more complicated. In $Fe_{0.9}O$ the vacancies and Fe^{3+} ions cluster together in a specific arrangement, to be described later, giving a volumetric defect.[8] The clusters themselves at low temperatures are spaced in a periodic array to form a supercell. In addition, within the vacancy clusters in highly nonstoichiometric wüstite some cations occupy tetrahedral sites rather than octahedrally coordinated sites as in the perfect rocksalt structure. It is probably mostly Fe^{3+} ions which occupy tetrahedral sites; the Fe^{3+} ions are smaller than the Fe^{2+} ions and the tetrahedral sites are smaller than the octahedral sites in wüstite.

While the structure of wüstite will be fully described and discussed later in this chapter, it has been briefly mentioned in the introduction to indicate the variety of imperfections which may be expected in a highly nonstoichiometric compound.

In the perfectly ordered alloy or the perfectly structured compound each atom species occupies one sublattice of the structure.[9] For example, the Cu_3Au structure consists of four interpenetrating simple cubic lattices. The Cu atoms are on three of the sublattices and Au atoms are on the fourth sublattice. Occurrence of atoms on the wrong sublattice is another kind of defect. Such configurational defects might arise from thermal disorder, deviations from stoichiometry like excess Au or Cu atoms, or growth errors which occurred during the ordering. One kind of growth error is the antiphase domain, a planar defect formed when two domains of order which are out of step meet. For example, in one domain the Cu atoms may be on the α, β, and γ sublattices while the Au atoms are on the δ sublattice and in the second domain the Cu atoms may be on the α, γ, and δ sublattices while the Au atoms are on the β sublattice. Configurational defects may also be arranged to form a regular pattern. A number of examples will be discussed.

Impurity atoms are also point defects. They may occupy interstitial lattice sites in the host lattice, such as C atoms in the Fe lattice to form an interstitial solid solution, or they may replace atoms of the host lattice, such as Br^- replacing Cl^- when NaBr is added to NaCl. This is a substitutional solid solution. Such impurity atoms are generally not randomly arranged. Ordering or clustering in small regions often occurs, leading again to linear, planar, and volumetric defects.

The dislocation or line imperfections in crystals evolved over the period 1928–1934 through consideration of mechanical properties by a number of investigators independently: Orowan, Taylor, Dehlinger, Polanyi, and Yamaguchi.[10] The theoretical shear strength of perfect crystals is many orders of magnitude greater than the actual shear strength of very pure metal crystals, as well as that of rocksalt crystals, and this led a number of people to propose the existence of imperfections in real crystals to account for this low strength. For example, the theoretical strength of gold is about 100,000 psi while the actual strength is less than 100 psi. The imperfection which evolved from these considerations was the dislocation or a line through the crystal where the atoms or ions are not in their normal positions in the crystal. The motion of this line on its slip plane causes permanent deformation by shear and since the atoms along the dislocation are not in their low-energy positions, it moves under application of only a small shear stress. Equal numbers of atoms enter and leave the dislocation as it moves and thus energy to create an imperfection is not required in order for the dislocations to move. At elevated temperatures dislocations may move out of their slip plane by the climb process, which involves vacancy motion. This is important in creep of solids. A line imperfection in a continuum solid having the properties of a dislocation had been mathematically treated by Volterra many years earlier.

The dislocation theory of plastic deformation was confirmed by direct observation of dislocations during the 1950's, first by decorating them with impurity atoms and later directly by transmission electron microscopy, and for the first time a basic understanding of the mechanical properties of materials emerged.

This brief review has indicated many of the different kinds of defects which exist in otherwise perfect crystalline solids. Table 1 gives a classification of crystal defects. The purpose of this chapter is to briefly introduce each category of defect to give a broad picture of the complete field so that the individual chapters in this treatise may be placed in proper perspective. The classification is convenient for

TABLE 1
Classification of Imperfections in Crystals

1. Point defects: vacant lattice sites, interstitial atoms, impurity atoms, atoms on wrong sublattices, ions with different valences
2. Line defects: dislocations, impurity atoms segregated to dislocations
3. Planar defects: surfaces, grain boundaries, small-angle grain boundaries, cell walls, planar stacking faults, antiphase domain walls, shear structures, segregation of impurity atoms to interfaces
4. Volumetric defects: clusters of atoms of different chemical composition, ordered regions, clusters of point defects, divacancies, trivacancies, impurity-atom vacancy or interstitial complexes, small regions of disorder, extended long-range configurational defects, initial stages of spinodal decomposition.

separating the various elements of defects, but as has been pointed out, interactions among defects to form more complex defects is quite common.

2. Point Defects

The first question to be examined is why points defects should occur. Because of entropy considerations, there will always be a certain equilibrium concentration of each kind of point defect. While the internal energy is lowest for the perfect crystal, there is an added configuration entropy term ($-TS_v$) due to the presence of vacancies and the many possible ways of arranging them. Thus the equilibrium concentration increases with temperature. The equilibrium concentration of vacancies or Schottky defects, C_v, in metal or elemental crystals is given by the well-known expression

$$C_v = n_v/N \cong e^{-\varepsilon_v/kT} \tag{1}$$

where N is the total number of atoms, n_v is the number of vacant lattice sites, and ε_v is the energy of formation of a vacancy. Thus the concentration of vacancies increases exponentially with temperature. The occurrence of a vacancy in metals or elemental crystals does not require a different defect to balance the charge since each atom in the lattice carries no net charge. An essentially identical expression exists for interstitial atoms with ε_i, n_i, and N_i for the interstitial replacing the values for the vacancy. In metals, which have a close-packed structure, the energy to form an interstitial is much higher than the energy to form a vacancy (about 4 eV compared to 1 eV in copper) and thus the equilibrium concentration of vacancies, about 10^{-4} at 1300°K, is many orders of magnitude greater than the equilibrium concentration of interstitials, which is about 10^{-15} at 1300°K.[11] The ratio of concentration of interstitials to that of vacancies is expected to be higher in non-close-packed crystals such as Si than in metals.

Textbooks often mention the Frenkel defect along with the Schottky defect as possible defects in metals or elemental crystals. The Frenkel defect consists of an interstitial atom and a vacant lattice site. The interstitial atom is imagined to originate from an atom in a lattice site, thereby creating a vacancy. The individual vacancy and individual interstitial atom mentioned in the previous paragraph are imagined to originate from the surface. Therefore their formation energies are those for single defects, while the formation energy of the Frenkel defect is the energy required to create a vacant lattice site plus an interstitial atom. The formation energy of the Frenkel defect ε_F is thus considerably larger than that of the single vacant lattice site or interstitial atom. The number of Frenkel defects,

$$C_F = n_F/(NN_i)^{1/2} \cong e^{-\varepsilon_F/2kT} \tag{2}$$

Since $\varepsilon_F/2$ is much greater than ε_v in metals, pairs of vacancies and interstitials are not expected. The concentration of vacancies in annealed metals is much larger than that of interstitials. In addition, however, there is the possibility that interstitial atoms, and also vacancies, are produced by plastic deformation of metals. Such defects are nonequilibrium and would disappear on annealing, that is, heating to a sufficiently elevated temperature.

The presense of vacancies in metals was verified first by Nenno and Kauffman[12] and later by Simmons and Balluffi[13] by comparing the thermal expansion of the specimen as a whole with the thermal expansion of the lattice parameter. The former includes the creation of vacancies on heating, while the latter is the expansion of the perfect lattice and thus the deviation between them increases exponentially with temperature. The difference in aluminum gives a vacancy concentration of 1×10^{-3} at the melting point of aluminum. The energy of formation of a vacancy in aluminum is 0.6 eV. Vacant lattice sites have been directly observed using the field ion microscope. An example is shown in Figure 1.

Excess vacancies may be obtained at lower temperatures by quenching the solid after heating it to a high temperature. Such excess vacancies affect many different properties, such as diffusion, where they allow solid-state reactions to take place at low temperatures. The important age hardening reaction in aluminum-base alloys requires quenched-in vacancies. An Al–2 at. % Cu alloy becomes much stronger

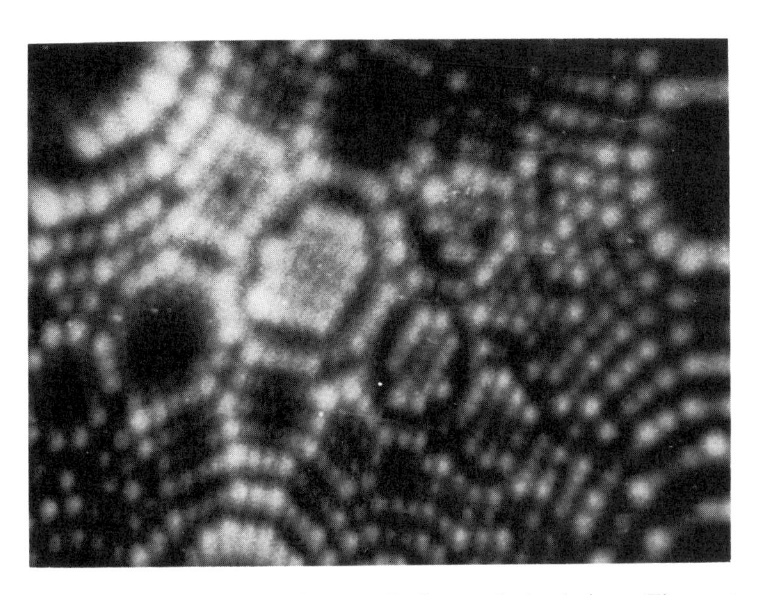

Fig. 1. Field ion microscope photograph of vacancies in platinum. The most obvious vacancy is in the upper left quadrant of the photograph. From Prof. Erwin W. Mueller, The Pennsylvania State University, State College, Pa.

when quenched from 520°C and aged at room temperature, due to formation of clusters of copper atoms called Guinier–Preston zones. If the equilibrium concentration of vacancies were present, it would require about 2000 yr for the reaction to reach half-completion. The actual time is 3 hr.[14] Vacancies increase the resistivity of metals and resistivity changes are a particularly convenient way of studying the annealing out of excess vacancies. Nonequilibrium vacant lattice sites, as well as interstitials, may be introduced by irradiation and by plastic deformation.

Consider next a stoichiometric compound of composition MX. The following pairs of defects are possible:[15] (1) equal numbers of vacancies on M and X sites (Schottky–Wagner defect), (2) equal numbers of vacancies and interstitials on M sites (M-site Frenkel defect), (3) equal numbers of vacancies and interstitials on X sites (X-site Frenkel defect), and (4) equal numbers of M and X interstitials.

A fifth type of imperfection in MX compounds is for an M atom to exchange positions with an X atom, that is, the configurational defect mentioned in the introduction.

The equilibrium concentration of Schottky–Wagner (SW) defects is

$$C_{\mathrm{SW}} \cong e^{-\varepsilon_{\mathrm{SW}}/2kT} \tag{3}$$

where ε_{SW} is the energy required to create an M and X vacancy pair. A similar expression holds for the M and X Frenkel-type defects with ε_{FM} or ε_{FX} replacing ε_{SW} in Eq. (3). Again the defect with the lowest formation energy will predominate.

The cation M is usually much smaller than the anion X. The formation energy to form an X interstitial is generally large and therefore the third and fourth defects are rather unlikely. The Schottky–Wagner-type defects predominate in the alkali halides and PbS.[15] The M Frenkel-type defect predominates in AgBr and AgCl below 600°K. The equilibrium concentration of SW point defects in NaCl is in the range 10^{-3}–10^{-4} near the melting point and the formation energy is about 2 eV.

The interchange of an M and an X atom in the structure is only likely if the M atoms and the X atoms are nearly the same size and if the ionic contribution to the binding is small, i.e., the atoms have nearly the same electronegativity. It occurs in "NiAl",[16] which exists over a wide range of composition, when there is excess Ni. When there is excess Al there is some substitution of Al on the Ni lattice sites but the dominant defect is Ni vacancies.

Nonstoichiometric compounds have already been mentioned. Many examples exist besides Fe_xO where there are cation vacancies.[15] In zinc oxide excess Zn may be dissolved into ZnO as interstitial atoms. These give electrons to the conduction band, thereby increasing the electronic conductivity, and the Zn interstitial becomes a positively charged "impurity" center. These positively charged impurity centers and the electron concentration in the conduction band must, of course, be equal to maintain electrical neutrality.

When impurity or alloying elements or ions are added to crystals they may substitute for one of the host atoms in the host crystal, giving a substitutional solid solution, or they may take up unoccupied interstitial positions in the lattice, giving an interstitial solid solution. Which occurs depends mainly upon the relative sizes of the host atoms or ions and the alloying atoms or ions. Alloying atoms or ions which are relatively small are expected to form interstitial solid solutions. The individual substitutional or interstitial atom or ion may be regarded as a point defect. Such impurity atoms affect a large number of properties of the host crystal. For example, in metals they scatter conduction electrons and decrease the conductivity and they interact with dislocations to increase the strength. Interstitial atoms are more potent strengtheners than substitutional atoms. In semiconductors the whole field of doping concerns the effect of impurity atoms on the electrical properties. The impurity atom, from a technical point of view, is the most important point defect.

In polar crystals impurity ions of different valence require a compensating defect. For example, if Li^+ is substituted for Ni^{2+} in NiO, then compensating Ni^{3+} ions are created.[15] The presence of Ni ions of two valences increases the electronic conductivity. In alkali halides where the cations exist in only one state of valence, the solubility of cations of different valence is very small because a high-energy compensating defect must be created.

The conduction process in substances such as NiO doped with Li or Fe_xO where cations of different valence of the same ion are present has been considered on the basis of the electron hopping from, for example, an Fe^{2+} to an Fe^{3+} with the ions exchanging their identities, or, alternately, on the basis of a hole in the valence electron band. Electronic conduction in ZnO with excess Zn has already been mentioned as occurring by electrons in the conduction band. In the case of doping of covalent semiconductors such as Ge or Si with trivalent or quintavalent atoms the hopping model, of course, has no validity. The defects created at ordinary temperatures are holes in the valence band or electrons in the conduction band.

Association of defects to form clusters and ordering of defects, a subject which has been mentioned in the introduction, reduce the formation energy per defect. Clustering and ordering of point defects will be discussed in later sections.

3. Dislocations

The theoretical shear strength of crystals is $G/3$ to $G/30$, where G is the shear modulus;[17–21] the actual shear strength of pure metal crystals is much smaller. For example, the shear modulus of iron is 12×10^6 psi (8.3×10^{10} N/m^2). Taking $G/10$ predicts a shear strength of 12×10^5 psi (8.3×10^9 N/m^2). Plastic deformation of metals takes place primarily by slip on definite slip planes and in definite slip directions. The critical resolved shear stress of iron, that is, the shear stress resolved in $\langle 111 \rangle$ on the (110) plane at room temperature, is only about 2×10^3 psi (1.4×10^7 N/m^2). For pure metals, iron, like other body-centered cubic transition metals, is rather strong. Face-centered cubic metals like copper are much weaker; $G/10$ for copper is 6.7×10^5 psi, while the critical resolved shear stress in $\langle 110 \rangle$ (111) is only 3×10^2 psi. Most nonmetallic crystals are brittle at low temperatures but are quite ductile at elevated temperatures, i.e., $T > 0.05 T_m$.

The theoretical shear strength of ice is about 14×10^4 psi, while glaciers deform at stresses as low as 2 psi. However, a number of nonmetallic crystals with high or rather high melting points are ductile at room temperature. Examples are NaCl, AgCl, and MgO. The critical resolved shear stress of NaCl is about only 100 psi, while the theoretical strength is about 10×10^4 psi.

As mentioned in the introduction, it is this discrepancy between the theoretical strength of bulk crystals and their actual strength which led to the concept that real crystals contain imperfections making them weak. The particular imperfection invented because it had the desired properties to account for the discrepancy was the dislocation or linear crystalline defect. It has been possible to grow dislocation-free, small-diameter whisker crystals and these indeed have the theoretical strength.[22]

The simplest kind of line imperfection to consider is the edge dislocation. Referring to Figure 2, imagine that the top of a perfect crystal is cut part way through, a layer of atoms is added to the depth of the cut, and then the crystal is welded together on the top. This results in an extra layer of atoms in the top of the crystal, as shown in Figure 2, for a simple cubic lattice. The line marking the edge of the

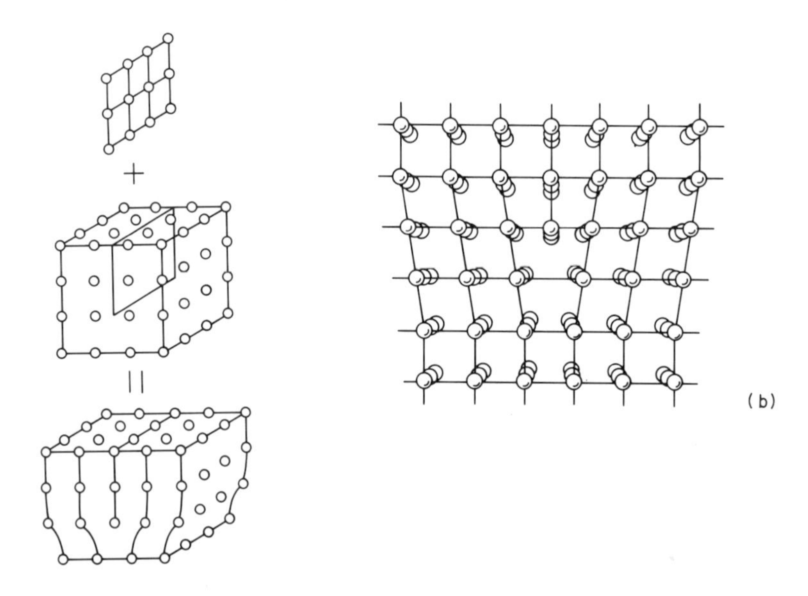

Fig. 2. Schematic drawings of edge dislocations. (a) Creation of an edge dislocation by insertion of an extra layer; from Weertman and Weertman.[19] (b) Atom rows in vicinity of an edge dislocation; from Hirth and Lothe.[21]

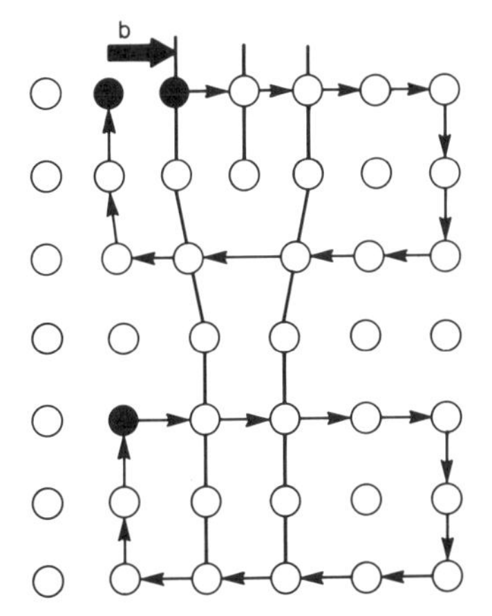

Fig. 3. Burgers circuit around an edge dislocation (upper part of figure) and circuit around a region of perfect material (lower part of figure). There is a closure error **b** in the circuit around the edge dislocation, from Weertman and Weertman.(19)

extra layer of atoms is called an edge dislocation. It was the first kind of dislocation proposed. One of the properties of a dislocation is its Burgers vector **b**. The Burgers vector may be determined by trying to follow a path around the individual dislocation of a certain number of lattice distances in each direction. The closure error (there will be none if the path does not enclose a dislocation) is **b**, as shown in Figure 3 for an edge dislocation. In an edge dislocation **b** is normal to the dislocation line. In general **b** may assume any orientation with respect to the dislocation. When **b** is parallel to the dislocation the dislocation is called a screw dislocation. Figure 4 depicts a screw dislocation and an associated Burgers circuit. The screw dislocation is along the row of atoms marked a, normal to the atom plane shown. The dislocations depicted in the previous figures are for simple cubic crytals. This is for convenience. The atom structure about the dislocations will be somewhat more complicated in more complex crystal structures, but the principles remain the same.

Photographs of actual dislocations taken by transmission electron microscopy of thin-foil metal samples are shown in Figure 5. In Figure 5(a) the dislocations are mainly in edge orientation; in Figure 5(b) they are mainly in screw orientation. Dislocations of intermediate orientation are also seen.

The relation between dislocations and plastic deformation of crystalline solids has been noted. Production of slip steps by motion of a pair of edge dislocations is depicted in Figure 6 with the two edge

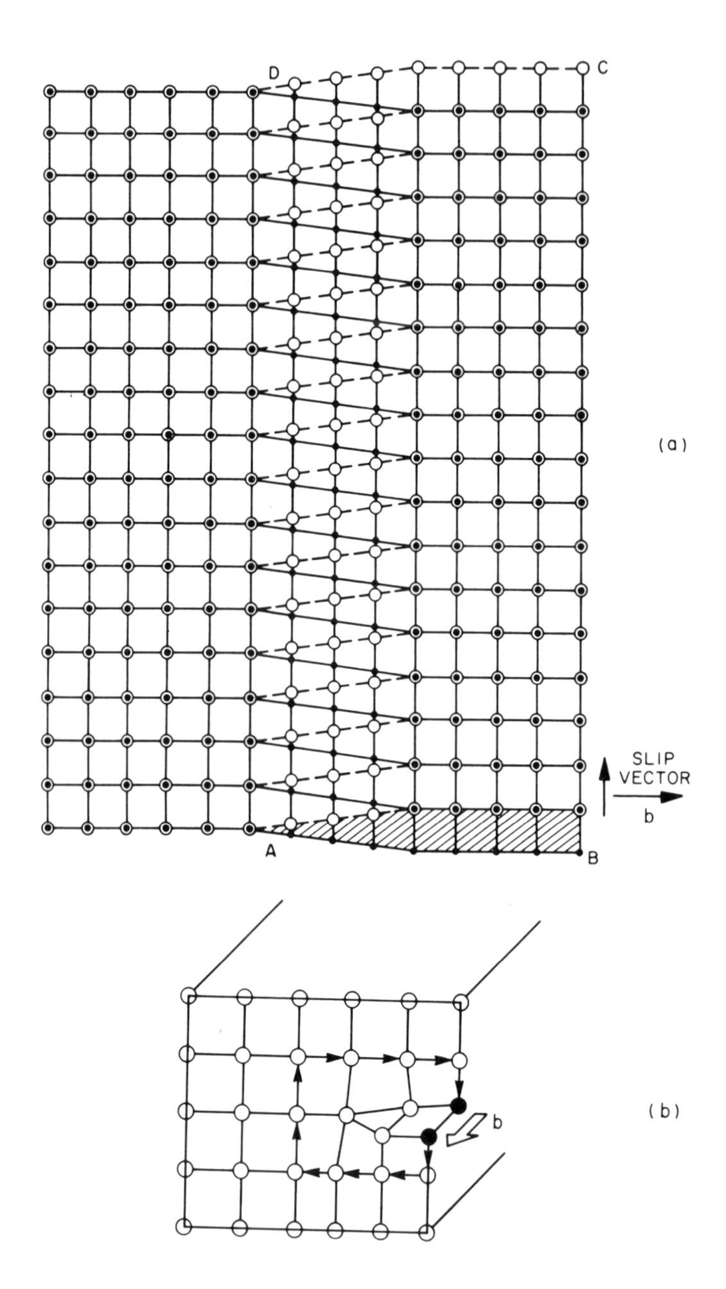

Fig. 4. Schematic drawing of a screw dislocation and a Burgers circuit around a screw dislocation. (a) Arrangement of atoms around a screw dislocation. The plane of the figure is the slip plane and the screw dislocation lies along AD; from Read.[18] (b) Burgers circuit around a screw dislocation; from Weertman and Weertman.[19]

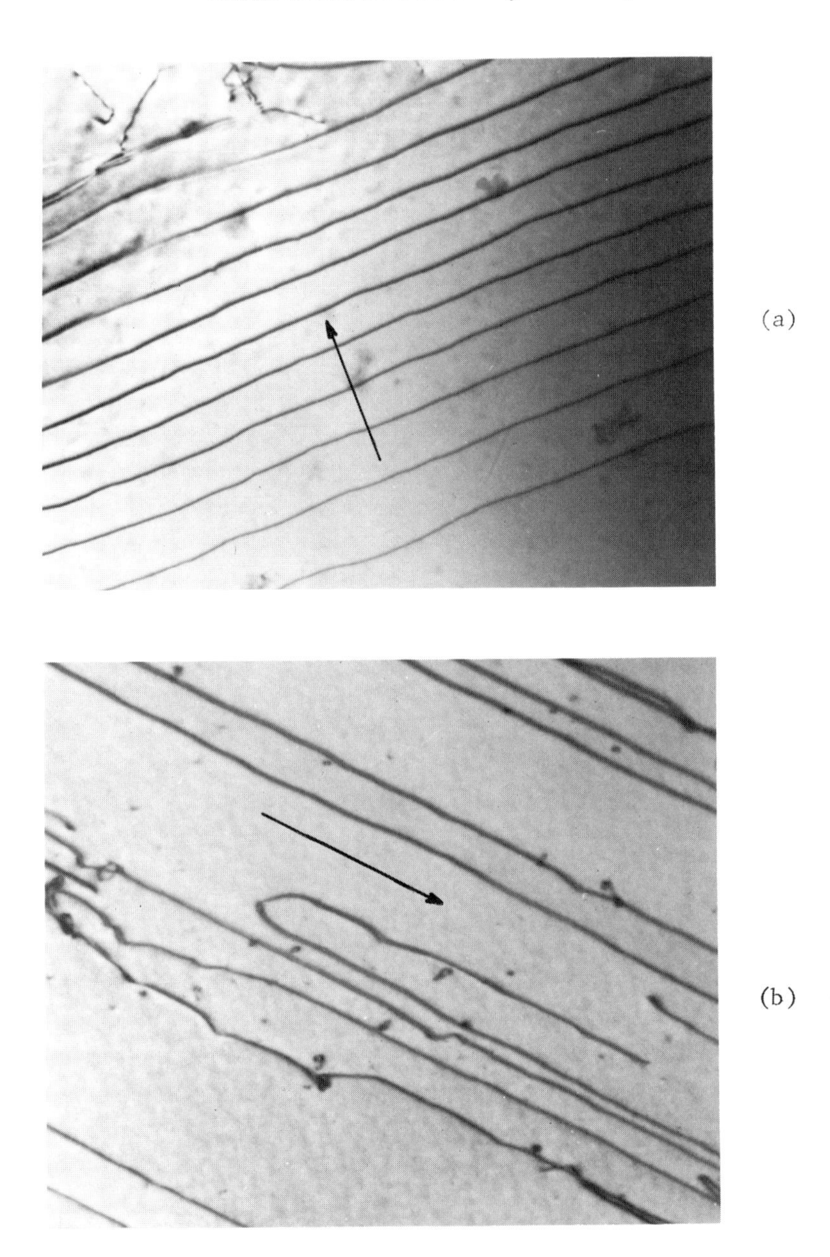

Fig. 5. Transmission electron microscope photographs of dislocations in Ti-base Al alloys. The Burgers vectors ($\mathbf{b} = [\bar{1}\bar{1}20]$) are indicated by the arrows. In (a) the dislocations are mainly in edge orientation (high interstitials, 0.87 % Al); 20,000 ×. In (b) they are mainly in screw orientation (5.2 % Al). From Dr. Takeo Sakai, Northwestern University, Evanston, Illinois.

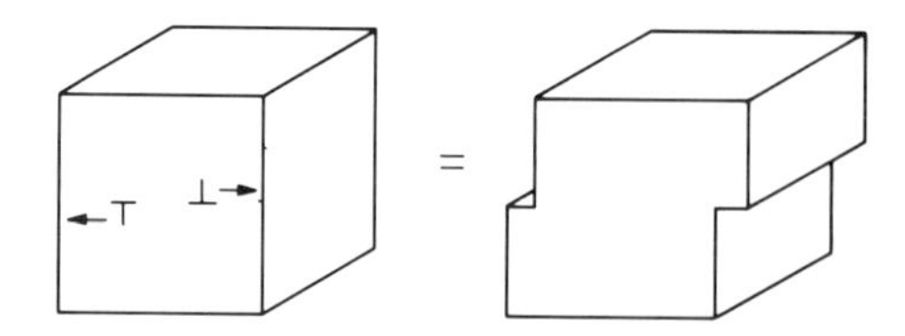

Fig. 6. Slip steps produced by motion of two edge dislocations of opposite sign; from Weertman and Weertman.[19]

dislocations denoted by the conventional symbol $\perp$. The vertical line represents the extra half-plane of atoms, while the horizontal line represents the slip plane. Both dislocations lie on the same slip plane. These dislocations are of opposite sign and move in opposite directions from an applied stress. When they move out of the crystal they leave steps which are $\mathbf{b}$ in magnitude and direction; $\mathbf{b}$ is the slip vector of the dislocation.

Dislocations move under low applied stress because the atoms along the dislocation are not in the low-energy positions corresponding to the crystal structure. As they move, an equal number of atoms enter the dislocation and leave the dislocation. The amount of good material and bad material is constant. This process gives rise to a lattice friction stress called the Peierls stress. The Peierls stress is very small in face-centered cubic metals. It is much smaller than the flow stress for motion of dislocations, i.e., the critical resolved shear stress. The flow stress is determined by interaction among dislocations and interactions of dislocations with impurity and alloying atoms. The Peierls stress in nonmetallic crystals is large and commonly controls the flow stress.

Dislocations are nonequilibrium defects. Since the atoms or ions near dislocations are not in their lowest-energy positions, the region near the dislocation is in a higher energy state than the perfect crystal. A length of dislocation b long, which is of atomic dimensions, has an energy of roughly 3–4 eV in metals[17–21] and thus a dislocation line in a real macroscopic crystal which may be 10^8 atoms long represents a very large amount of energy which may be gained if the dislocation moves out of the crystal. In view of this, one needs to consider why real crystals which have not been deformed previously contain dislocations. They arose from growth accidents while the crystal was being formed. Although single crystals which are free of dislocations have been prepared of many metals and nonmetals, ordinarily prepared and annealed metal crystals contain approximately 10^6 dislocation lines/cm^2. The number is generally less in ionic and covalent crystals.

The energy of a dislocation line is usually divided into two parts, the core energy and the energy from the elastic stress field about the dislocation. Atoms near the core or center of the dislocation are rather far from their equilibrium positions and thus give rise to the core energy. Atoms further away from the dislocation core are elastically displaced from their equilibrium positions and elasticity theory may be used to determine the elastic stress field about the dislocation and the dislocation energy due to the elastic stress field. Referring to Figure 2, in the edge dislocation the atoms above the dislocation are in compression while those below are in tension, that is, the atoms above the dislocation are closer together than the equilibrium spacing and those below are farther apart. At the dislocation core the displacement of the atoms from their equilibrium positions is so great that linear elasticity cannot be used; however, the strains diminish rapidly with increase in distance from the dislocation and beyond a few atom distances linear elasticity may be used.

In the expressions for the stress field about dislocations it is convenient to use ordinary coordinates x, y, and z and also cylindrical coordinates r, θ, and z. The z-coordinate axis is taken along the dislocation, $r^2 = x^2 + y^2$, and $\tan\theta = y/x$. For an edge dislocation x is chosen parallel to **b** which lies in the slip plane and y is normal to the slip plane. The origin and sign are selected so that the extra half-plane lies in the $x = 0$ plane below the $y = 0$ plane for positive **b** and above the $y = 0$ plane for negative b. The stress field about an edge dislocation[17–21] is given by the following equations, where ν is Poisson's ratio:

$$\sigma_{xx} = \frac{Gb}{2\pi(1-\nu)} \frac{y(3x^2 + y^2)}{(x^2 + y^2)^2} \tag{4a}$$

$$\sigma_{yy} = \frac{Gb}{2\pi(1-\nu)} \frac{y(x^2 - y^2)}{(x^2 + y^2)^2} \tag{4b}$$

$$\sigma_{xy} = \frac{Gb}{2\pi(1-\nu)} \frac{x(x^2 - y^2)}{(x^2 + y^2)^2} \tag{4c}$$

$$\sigma_{rr} = \frac{Gb}{2\pi(1-\nu)} \frac{\sin\theta}{r} \tag{4d}$$

$$\sigma_{r\theta} = \frac{Gb}{2\pi(1-\nu)} \frac{\cos\theta}{r} \tag{4e}$$

Here σ_{xx}, σ_{yy}, and σ_{rr} are normal stresses and σ_{xy} and $\sigma_{r\theta}$ are shear stresses. The normal stress σ_{xx}, which is larger than σ_{yy}, has a different sign above and below the slip plane, where $y = 0$, as previously discussed and shown in Figure 2. The shear stress σ_{xy} is greatest on the slip plane

As can be seen from Figure 4, there are only shear stresses about a screw dislocation and thus $\sigma_{xx} = \sigma_{yy} = \sigma_{rr} = 0$. The shear stress is expressed most simply in cylindrical coordinates:[17–21]

$$\sigma_{z\theta} = Gb/2\pi r \tag{5}$$

Note that the shear stress is not dependent on θ for a screw dislocation.

The elastic stress field about the dislocation falls off as r, being infinite at the center of the dislocation, but, of course, Eq. (4a)–(4e) and (5) do not hold in the dislocation core.

Dislocations of orientation intermediate between edge and screw (mixed dislocation) may be resolved into screw and edge components; therefore their stress fields will be intermediate between those of edge and screw dislocations.

The strain energy associated with edge and screw dislocations may be determined from Eqs. (4a)–(4e) and (5). The results for the strain energy per unit length of dislocation are[17–21]

$$\begin{aligned} E_s(\text{edge}) &= [Gb^2/4\pi(1 - \nu)] \ln(R/r_0) \\ E_s(\text{screw}) &= (Gb^2/4\pi) \ln(R/r_0) \end{aligned} \tag{6}$$

Poisson's ratio is about 0.3, so that the energy of edge dislocations is about a third larger than the energy of screw dislocations. The energy of a mixed dislocation may be resolved into screw and edge components and thus is intermediate between E_s(edge) and E_s(screw).

In (6) r_0 is the core radius and is about $5|\mathbf{b}|$.[17–21] The strain energies from a pair of dislocations of opposite sign cancel at large distances; therefore R in (6) is usually taken to be the subgrain radius. Taking $R = 10^5 b$ gives E_s of about 0.3 eV for a dislocation in aluminum one atom diameter long.[17–21]

The long-range stress field about a dislocation loop also cancels and thus the strain energy of a curved dislocation is approximately $(Gb^2/4\pi) \ln(\rho/r_0)$, where ρ is the radius of curvature.[17–21]

Equation (6) gives only the strain energy of the dislocation. The total energy of the dislocation must include the energy of the core: $E = E_s + E(\text{core})$. The energy E(core) is approximately $Gb^2/10$, which is about 10–20% of E_s.[17–21]

The dislocation energy is proportional to $\mathbf{b}^2$. This is a very important property of dislocations because it means that two dislocations each of Burgers vector $\mathbf{b}_1$ have a total energy lower than that of a single dislocation of Burgers vector $2\mathbf{b}_1$, and therefore a dislocation of Burgers vector $n\mathbf{b}_1$ is expected to split into n dislocations of Burgers vector $\mathbf{b}_1$.

When a sufficiently large external force is applied to a crystalline material containing dislocations the dislocations move, accomplishing plastic deformation, as previously discussed (see Figure 6). If an external stress is applied to the crystal, then forces are applied to the dislocations which it contains. The force on a dislocation per unit length of dislocation is[17–21]

$$F = \sigma_r b \tag{7}$$

where σ_r is the resolved stress on the slip plane in the slip direction. This force is always normal to the dislocation line.

Due to the stress field around dislocations, there is an interaction between dislocations. Dislocations of like sign attract and dislocations of opposite sign repel. The force between two parallel screw dislocations and the force between two parallel edge dislocations are given, respectively, by[17–21]

$$F_r = G\mathbf{b}_1 \cdot \mathbf{b}_2/2\pi r \tag{8a}$$

and

$$F_r = G\mathbf{b}_1 \cdot \mathbf{b}_2/2\pi(1 - \nu)r \tag{8b}$$

where r is the distance between the dislocations.

A dislocation is defined as being perfect when $\mathbf{b}$ is a lattice translation vector and imperfect when it is not a lattice translation vector. A lattice translation vector is any vector which joins two points of the crystal lattice and when the line is extended these points are repeated at regular intervals along the line. Since plastic deformation by slip does not change the crystal structure, the Burgers vectors, that is, the slip vectors for the dislocations which accomplish slip, must be lattice translation vectors and the dislocations must be perfect dislocations. Furthermore, dislocations with the smallest possible lattice translation vectors, the primitive lattice translation vectors, are expected because these have the least energy. The Burgers vector for face-centered cubic metals and NaCl structured ionic crystals is thus $\frac{1}{2}a\langle 110\rangle$ and it is $\frac{1}{2}a\langle 111\rangle$ for body-centered cubic metals, where a is the lattice parameter of the unit cell.

Motion of imperfect dislocations causes local change in structure, such as twinning or generation of a different structure as in martensitic transformations.

Perfect dislocations may sometimes split into partial dislocations to form an extended dislocation such as $\mathbf{b} = \mathbf{b}_1 + \mathbf{b}_2$, where $\mathbf{b}_1$ and $\mathbf{b}_2$ are Burgers vectors of imperfect dislocations. This occurs, for example, in face-centered cubic metals, as shown in Figure 7, where $\mathbf{b} = \frac{1}{2}a[\bar{1}01]$, $\mathbf{b}_1 = \frac{1}{6}a[\bar{2}11]$, and $\mathbf{b}_2 = \frac{1}{6}a[\bar{1}\bar{1}2]$. In order for this dissociation to occur, $|\mathbf{b}|^2$ must be greater than $(|\mathbf{b}_1|^2 + |\mathbf{b}_2|^2)$. The magnitude of the $\frac{1}{2}a[\bar{1}01]$ vector is $a/\sqrt{2}$ and $|\mathbf{b}|^2$ is $a^2/2$, while the magnitude of the

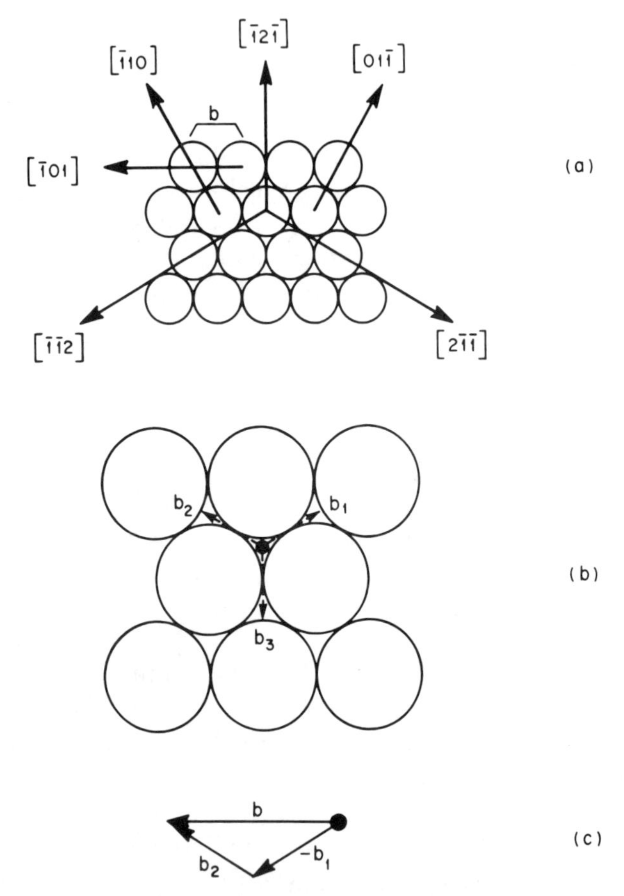

Fig. 7. Splitting of perfect dislocation into partial dislocations in fcc metal; from Weertman and Weertman.[19] (a) The slip plane (111) and important directions in the slip plane. ⟨111⟩ are the directions of the Burgers vectors of perfect dislocations and ⟨211⟩ are the directions of the partial dislocations. (b) The three possible partial Burgers vectors in the (111) plane. (c) How the two Burgers vectors of the partial dislocations vectorially add to give the Burgers vector of the perfect dislocation.

$\frac{1}{6}a[211]$ and $\frac{1}{6}a[\bar{1}\bar{1}2]$ vectors are $a/\sqrt{6}$, and $(|\mathbf{b}_1|^2 + |\mathbf{b}_2|^2)$ is $a^2/3$, in keeping with the observed dissociation. The region separating b_1 and b_2 contains a stacking fault in the crystal. Stacking faults will be defined and discussed in the next section, but there is a stacking fault energy associated with the stacking fault region between the two partials. The stacking fault energies γ in some face-centered cubic metals are approximately 20 ergs/cm^2 in austenitic (face-centered cubic) stainless steel alloys and approximately 225 ergs/cm^2 in aluminum.(17–21) Since the energy is lowered by the dissociation, the two partials repel each other elastically and tend to move apart; however, the wider the separation w, the greater the energy of the stacking fault region. There is thus an equilibrium separation of the partials. In general, the equilibrium spacing between the partials in an extended dislocation is(17–21)

$$w = (\mathbf{b}_1 \cdot \mathbf{b}_2)/2\pi\gamma \tag{9}$$

For the actual partials in face-centered cubic metals being discussed, $w = Ga^2/24\pi\gamma$. The spacing for aluminum, which has a high stacking fault energy, computes to about one atom diameter. Extended dislocations are thus not expected in aluminum and they are not observed. The width for austenitic stainless steel computes to about 1×10^{-6} cm. Extended dislocations are commonly observed in austenitic stainless steel by transmission electron microscopy.

Since the $\frac{1}{6}a\langle 211\rangle$ partial dislocations in face-centered cubic metals are not perfect dislocations, the motion of the lead partial of an extended dislocation creates a stacking fault; the trailing dislocation destroys it since $\mathbf{b}$ of the total dislocation $(\mathbf{b}_1 + \mathbf{b}_2 = \mathbf{b})$ is the slip vector of a perfect dislocation. During slip the two partials (and the total dislocation) move at the same speed in a homogeneous material with w remaining constant.

Many other examples of extended dislocations exist. Their further discussion will be deferred to the section on planar defects except for the case of superlattice dislocations. The ordered Cu_3Au structure and antiphase domains have already been described in the introduction. While $(a/2)\langle 110\rangle$ is a lattice translation vector in the face-centered cubic structure, it is not a lattice translation vector in ordered Cu_3Au. Motion of a dislocation with $\mathbf{b} = (a/2)\langle 110\rangle$ in the ordered Cu_3Au creates an antiphase domain boundary. Motion of a second dislocation through the structure with the same $\mathbf{b}$ restores the perfect order since $a\langle 110\rangle$ is a lattice translation vector; the structure of Cu_3Au is simple cubic. Pairs of imperfect dislocations with, for example, $\mathbf{b} = (a/2)[110]$ are thus expected. Such pairs, called superlattice dislocations or superdislocations, have lower energies than single perfect dislocations with $\mathbf{b} = a\langle 110 \rangle$ in view of the dislocation energy being proportional to $|\mathbf{b}|^2$. The two imperfect dislocations are parallel, have the same sign, and thus repel each other. As with the previous extended dislocation discussed, there is a stacking fault between the pair, an antiphase domain in the present case, and the total antiphase domain energy per unit length of extended dislocation will again be γw. The separation w will be given by Eq. (9). For Cu_3Au a separation of 130 Å has been observed.[23] The $(a/2)\langle 110 \rangle$ dislocations may, for example, themselves be split into partials with $\mathbf{b} = (a/6)\langle 211 \rangle$ to give a group of four dislocations. Figure 8 is a transmission electron micrograph of a superdislocation.

Edge dislocations can glide only on the slip plane containing the Burgers vector of the dislocation. Such motion is called conservative motion. The extra partial plane of atoms (Figure 2) is normal to the

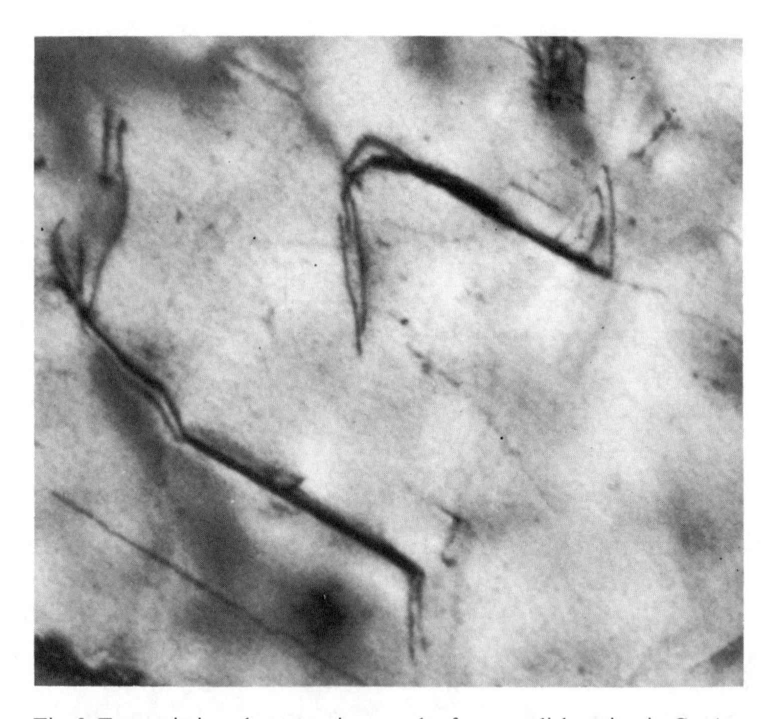

Fig. 8. Transmission electron micrograph of a superdislocation in Cu_3Au. As explained in the text, the splitting is a function of the stacking fault energy and this varies with dislocation orientation. The splitting is observed to change abruptly when the dislocation changes slip plane. Magnification 50,000 ×. From Dr. D. E. Mikkola, Northwestern University, Evanston, Illinois.

slip plane. Edge dislocation motion normal to the slip plane requires extension or shrinkage of the extra plane. This nonconservative motion, called dislocation climb, requires motion of vacancies to the dislocation or away from the dislocation. Thus dislocation climb is closely related to atomic diffusion by atom interchange with a vacancy and, like diffusion, is thermally activated. Dislocation climb is only an important deformation process at elevated temperatures where creep occurs and the diffusion rates are sufficiently high for the process to take place in a reasonable amount of time. The activation energies for self-diffusion are in the range 1–4 eV per atom. In creep, solids slowly deform under application of a stress too small to cause rapid plastic deformation. Extension of the extra plane by one lattice spacing to the adjacent parallel slip plane requires diffusion of a whole row of atoms to and a whole row of vacancies away from the edge of the extra plane, i.e., the edge dislocation. Conversely, shrinkage of the extra plane requires diffusion in the opposite sense. Atoms diffuse singly to and away from the dislocations. If an atom is added to or subtracted from an edge dislocation, a dislocation jog is created at that point. Such jogs spread sideways as more point defects are added.

Since the Burgers vector of an edge dislocation is normal to the dislocation line, conservative motion of an edge dislocation can only occur on the plane which contains both the dislocation and the Burgers vector. The Burgers vectors of screw dislocations are parallel to the dislocation line and thus slip can occur on any plane containing the dislocation line. There is no extra partial plane of atoms with a screw dislocation. A real dislocation generally is at least partly edge in character and therefore such a mixed dislocation has a definite slip plane.

Slip of a dislocation on a different slip plane from the original slip plane is called cross-slip. Cross-slip may then occur for screw dislocations or screw segments of curved mixed dislocations. In order for cross-slip to occur, a dislocation must not be extended on the original slip plane or a constriction must form in an extended dislocation. The latter is a process which occurs with some difficulty and thus extensive cross-slip is only expected in crystals which have narrow dislocations such as aluminium and most body-centered cubic metals. Body-centered cubic metals such as iron have wavy slip lines because cross-slip on a number of different slip planes is possible: $\{110\}$, $\{112\}$, and $\{123\}$. The slip direction is always $\langle 111 \rangle$.

One of the properties of a dislocation concerns the manner of termination. A dislocation can terminate only at a surface (such as an external surface, a grain boundary, or an interface between two phases contained in the material) or by coming together with other dislocations. Dislocations may come together at a point known as a node. The Burgers vector at a node is conserved, i.e., $\sum_i \mathbf{b}_i = 0$, where the sum is over all of the dislocations which join at the node. A group of dislocation nodes are shown in Figure 9.

A usual metal crystal contains about 10^6 dislocation lines/cm^3. Moving all of these out of the crystal under the influence of an applied stress does not account for the observed shear strain. Suppose our crystal is a cube 1 cm on a side and it is oriented in such a way that a shear stress is applied on the slip plane in a possible slip direction. Suppose one-third of the 10^6 dislocation lines have this slip direction and are oriented favorably for slip (this is clearly an overestimate), then taking $|\mathbf{b}|$ equal to 3×10^{-8} cm gives a total shear strain of 1%. Shear strains much greater than this are observed, sometimes exceeding 100% in single crystals. Clearly, dislocation multiplication processes

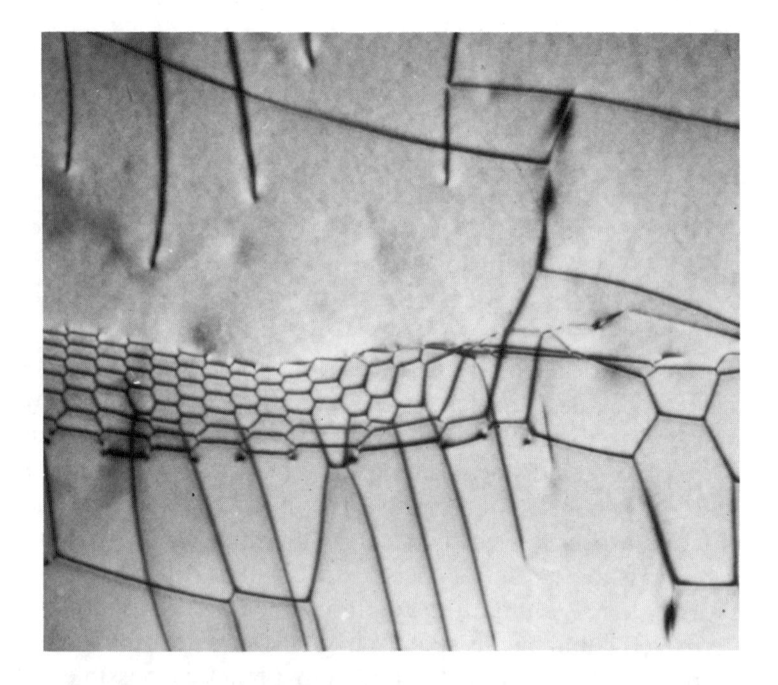

Fig. 9. Low-energy dislocation array, subgrain boundary, in Ti–5.2 at.% Al annealed for 10 hr at 850°C. Many dislocation nodes are observed in the subgrain boundary. Magnification 27,000×. From Dr. Takeo Sakai, Northwestern University, Evanston, Illinois.

must be operative. There is other evidence for the operation of dislocation multiplication processes. In the absence of a dislocation multiplication process, a slip height of only b would occur for each active slip plane, or at most several b if several dislocations happened to be on the same slip plane. Actual slip heights may be $1000b$. Finally, the dislocation density may be determined directly using transmission electron microscopy. Crystals which have undergone extensive plastic deformation contain 10^{11} dislocation lines/cm^3. Figure 10 shows the dislocation structure of a heavily cold-worked metal.

The simplest dislocation multiplication process is the Frank–Read source, which is illustrated in Figure 11. Imagine that a segment of a dislocation line BC lies on the slip plane, leaving the slip plane at B and C, so that B and C act as anchoring points, or alternately B and C may be impurity particles or dislocation nodes. Under the influence of the applied stress the dislocation segment bows out, expands, and finally breaks loose from the anchoring points to form a dislocation

Fig. 10. Transmission electron micrograph of heavily cold-rolled copper. The reduction was 70% at room temperature. Dislocation density is in the range 10^{10}–10^{11} dislocation lines/cm^2. The dislocation density is inhomogeneous. The dark regions contain a high dislocation density where individual dislocations cannot be resolved. The light regions have a much lower dislocation density. Magnification 35,000×. From Prof. M. Meshii, Northwestern University, Evanston, Illinois.

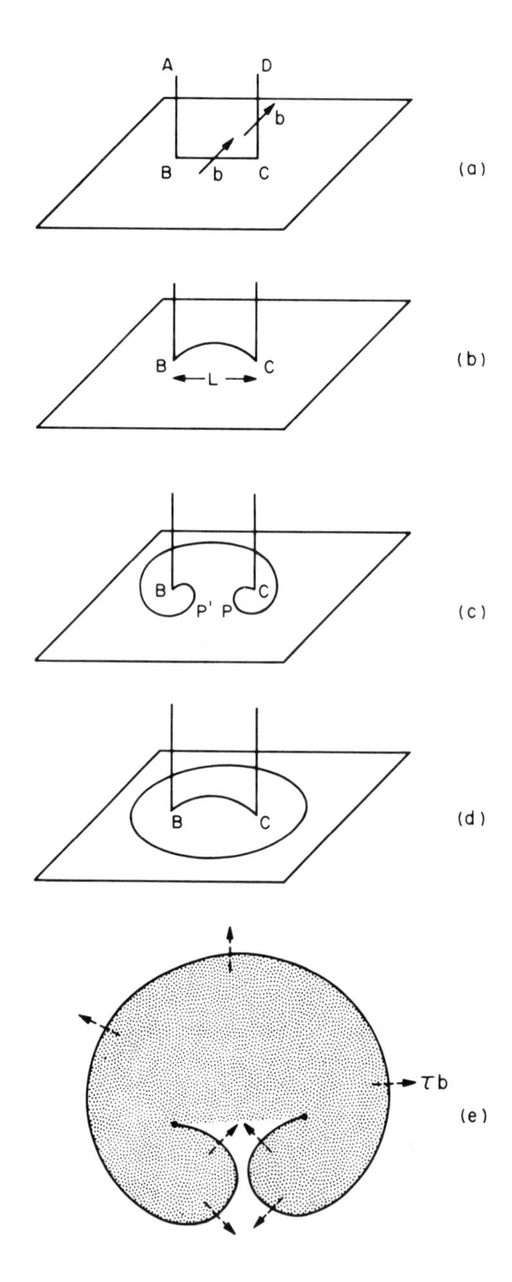

Fig. 11. Operation of a Frank–Read Source. In (a) the dislocation leaves the slip plane at *B* and *C*, which act as anchoring points. Under an applied stress the dislocation segment *BC* expands, finally forming a loop around *BC*, leaving a dislocation segment between *B* and *C*. (e) The dislocation bowed out just before the loop breaks loose from the pinning points; from Weertman and Weertman[19] and Read.[18]

ring, leaving a dislocation segment between B and C identical to the original segment. Thus a single source can create many dislocations. The stress to operate such a source is[17–21]

$$\sigma_s \cong 2Gb/l \qquad (10)$$

where l is the distance between B and C. Taking $l = 10^{-3}$–10^{-4} cm corresponds to the observed yield stress in pure metals.

The cross-slip of screw dislocations is important in the dislocation multiplication process, for a dislocation configuration similar to that shown in Figure 11 can be formed by cross-slip of a dislocation segment from one slip plane to a neighboring slip plane. By this process, an initial dislocation can give rise to many Frank–Read-type sources. Grain boundaries are also sources of dislocations. As discussed in the next section, grain boundaries can be considered as arrays of dislocations.

Dislocation sources under the action of an applied stress may generate a large number of dislocations. In a single crystal, in the absence of obstacles, these dislocations will readily move out of the crystal forming slip steps on the surface. But obstacles are present. In a single crystal, impurity particles and other dislocations are obstacles. In polycrystalline materials the grain boundaries act as obstacles because the slip planes are not continuous and change orientation across the grain boundary. Dislocations pile up at an obstacle. The lead dislocation feels not only the applied stress σ but also a stress from the other dislocations in the pileup so that the actual stress on the lead dislocation is $n\sigma$, where n is the number of dislocations in the pileup.[17–21] Such pileups can cause generation of new dislocations from sources across grain boundaries which, without pileups, would require a much higher applied σ for operation. The strength of solids is generally observed to decrease with grain size, in keeping with this model.

If a cold-worked material is heated, then more thermal energy is available to aid dislocation motion over obstacles and to aid dislocation climb. Annealing takes place and the dislocation density is reduced by two processes, recovery and recrystallization. In recovery the internal energy and dislocation density are reduced by straightening of curved dislocation lines and annihilation of dislocations of opposite sign. The remaining dislocations form stable arrays such as cell structures and low-angle boundaries such as that shown in Figure 9. In recrystallization, grains which have low dislocation density grow at the expense of the cold-worked grains. Recrystallization is a type of phase transformation and the activation energy for recrystallization is close to that for self-diffusion. Recovery takes place at lower temperatures than recrystallization. Actually, dynamic recovery occurs during plastic deformation of many metals at room temperature and even lower.

Dislocations of opposite sign attract each other and thus if the dislocation mobility is sufficiently high, this force is sufficiently high to bring them together for dislocation annihilation. The annihilation process is easily visualized with the aid of Figure 2. Imagine a second edge dislocation with its extra plane in a lower corner of the figure. The two extra planes can join together, annihilating both dislocations. If the two dislocations lie on the same slip plane, then only slip is required. If they do not, then climb is also required for annihilation. Screw dislocations of opposite sign also annihilate and since a screw dislocation can slip on any plane containing the dislocation, recovery is more rapid for screw dislocations. Mixed dislocations, of course, require climb to leave their primary slip plane.

During recovery, networks and low-energy arrays of dislocations form (Figure 9). The arrangement of the dislocations in a crystal is called the substructure and crystals which have been given a recovery anneal often have dislocation cells which divide the crystal into subgrains. This subject will be brought up again in the section of this chapter on surfaces. For now it suffices to discuss why such dislocation arrays form, that is, why forming certain arrangements of dislocations leads to a reduction in the energy due to the dislocations in the crystal. During plastic deformation the various parts of the crystal become rotated with respect to one another and the dislocations are distributed so that atomic planes are bent and twisted. During recovery small regions of perfect crystal form while the dislocations congregate into groups which separate the perfect crystallites. The elastic stress field about an edge dislocation is not symmetric, as previously discussed [see Figure 2 and Eqs. (4a)–(4e)]; there is tension below the end of the extra plane and compression above this region. A particularly low-energy array of edge dislocations consists of a row of equally spaced same-signed dislocations normal to the slip plane. Then the compressive stress field of one dislocation acts to partially cancel the tensile stress field of its nearest neighbor.

This section on dislocations will be concluded with a brief discussion of the interaction of point defects and dislocations. Interaction of vacancies and edge dislocations has already been discussed.

The vacancies are attracted to the compression side of the edge dislocation, where they are annihilated, leaving a dislocation jog. Nucleation of jogs is difficult because the dislocation energy must be increased, but once a jog has been formed then other vacancies can move into the dislocation with spreading apart of the sides of the jog. The dislocation energy is not increased by the spreading. Vacancies are not attracted to straight screw dislocations, but if the screw dislocation is bent so that it has a localized edge component, vacancies are attracted to this region, which grows. The end result is helical dislocations. Figure 12 shows helical "screw" dislocations in a Mg–base Li alloy.

Solute atoms also interact with dislocations. Interstitial atoms are attracted to edge dislocations; the interstitial sites located at the end of the extra plane can accommodate an interstitial atom more readily than a normal interstitial site in the lattice. Therefore there is a lowering of internal energy when an interstitial atom such as C in Fe

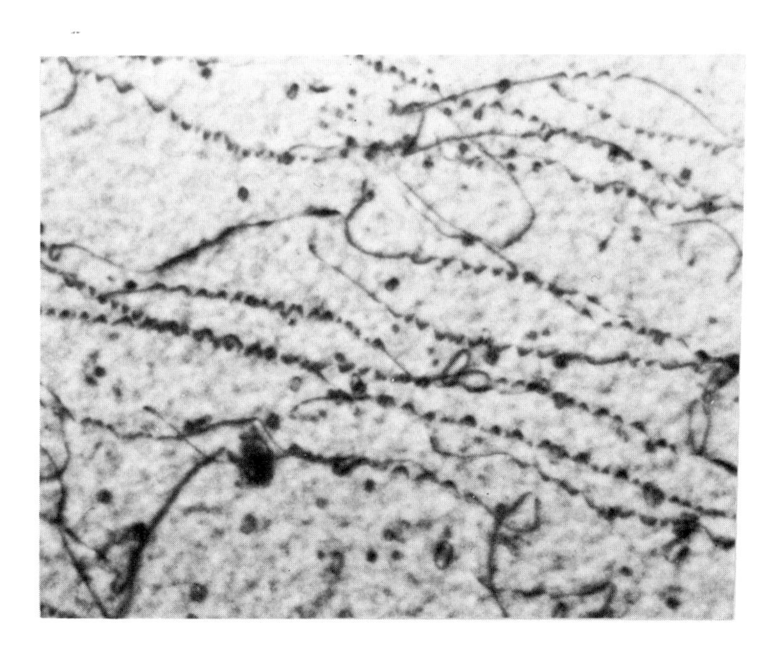

Fig. 12. Transmission electron micrograph of helical dislocations in a Mg–11 at.% Li alloy. The dislocations were initially mainly in the screw orientation. They became helical by condensation of point defects produced by deformation on screw dislocations. Magnification 20,000 ×. From Dr. Akio Urakami, Northwestern University, Evanston, Illinois.

diffuses from a normal interstitial site to an edge dislocation. Substitutional atoms also interact with edge dislocations and are attracted to them. Atoms larger than the host atoms find lower-energy positions in the region around edge dislocations where the stress field is tensile and, conversely, smaller atoms find low-energy positions where the stress field is compressive. Interaction of solute atoms with dislocations also occurs through a modulus effect. If the impurity atom changes the local shear modulus G, then the energy of any dislocation nearby will be affected because G appears in the equation for the energy. Thus impurities that reduce G will be attracted to dislocations and those that increase G will be repelled. This effect is important for both edge and screw dislocations. There may also be electrostatic interaction between solute atoms and dislocations.

The equilibrium concentration of solute C_d near a dislocation is given by[17–21]

$$C_d/C \cong e^{-\varepsilon_i/kT} \tag{11}$$

where C is the concentration of solute in the crystal far from a dislocation and ε_i is the interaction energy between a solute atom and the dislocation. If ε_i is negative, $C_d/C > 1$, while the converse is true if ε_i is positive. This equation holds only as long as the dislocations are not close to being saturated with impurity atoms.

When solute atoms have segregated to dislocations such dislocations are said to have solute atmospheres. Dislocation atmosphere gives rise to a number of important effects, such as yield drops in stress–strain curves and localized chemical attack.

4. Planar Defects

While the atoms in the interiors of grains are substantially in a perfect crystalline arrangement, atoms at or near surfaces are not; because insufficient atoms are present to form the perfect crystalline pattern and complete the chemical bonding, there are "dangling" bonds. Surface atoms have higher free energies, giving rise to a surface energy and since the material tends to reduce its surface to lower its free energy, the surface has surface tension. The true surface energy of a phase is the energy per unit area of the clean surface between the phase and its vapor. For crystalline solids these are generally in the range 500–2000 ergs/cm^2.[24,25] The free surface energies of liquids and glasses are, of course, less than those of crystalline solids of the same composition.[25]

If more than one phase is present in a crystalline material, then there must be interfaces between the various phases which are present and each kind of interface has a surface or interfacial energy which depends upon the orientations of the two crystals which face each other across the boundary. Such interfaces may be regarded as incoherent, semicoherent, or coherent. In a coherent interface there is perfect matching in the arrangement and spacing of the atoms in the planes of the two phases which face each other across the interface. In an incoherent interface there is no such good fit and in a semi-coherent interface there are regions of good fit separated by restricted regions of poor fit. The interface energy is, of course, least for the coherent boundary and may be as low as 15 ergs/cm^2. While most surfaces between solid phases are incoherent, many important examples of coherent interfaces exist. A few examples will be mentioned. The nickel-base "superalloys" used extensively for their high-temperature strength for such applications as jet engine turbine blades have a precipitate of nominal composition Ni_3Al (other alloying elements replace Ni and Al in the structure) in a matrix of a nickel-base solid solution. The "Ni_3Al" precipitates have the Cu_3Au structure, which has already been described, while the Ni-base matrix is face-centered cubic. The Cu_3Au and fcc structures are closely related; the Cu_3Au structure, although it is simple cubic, derives from the fcc structure by ordering. If the lattice parameters of the "Ni_3Al" precipitate and the Ni solid solution matrix are nearly the same, then the lattices of the precipitate and matrix are continuous (the precipitate and matrix are coplanar) and there is merely a change in atom composition across the interfaces, which are $\{100\}$ planes in both phases. The GP zones which are important in precipitation hardening of Al-base alloys, the α' precipitates (nomial composition NiAl, CsCl structure) used to harden stainless steels, and $MgFe_2O_4$ precipitates in MgO all form coherent interfaces with their matrix phases, and the lattices are coplanar in each case. The $MgFe_2O_4$ precipitates in MgO, shown in Figure 13, have the spinel structure, while the matrix has the NaCl structure. Both structures have nearly cubic close-packing of the O^{2-} anions with the cations in interstitial positions and the O^{2-} sublattice is continuous between the precipitate and matrix. In this case[26–28] the precipitates are octahedral shaped and the interfaces are $\{111\}$ planes. There is a discontinuity in cation site occupancy and cation composition across the interface.

For the coherent interfaces just described the lattices of the two phases are coplanar. The simplest example of a coherent interface

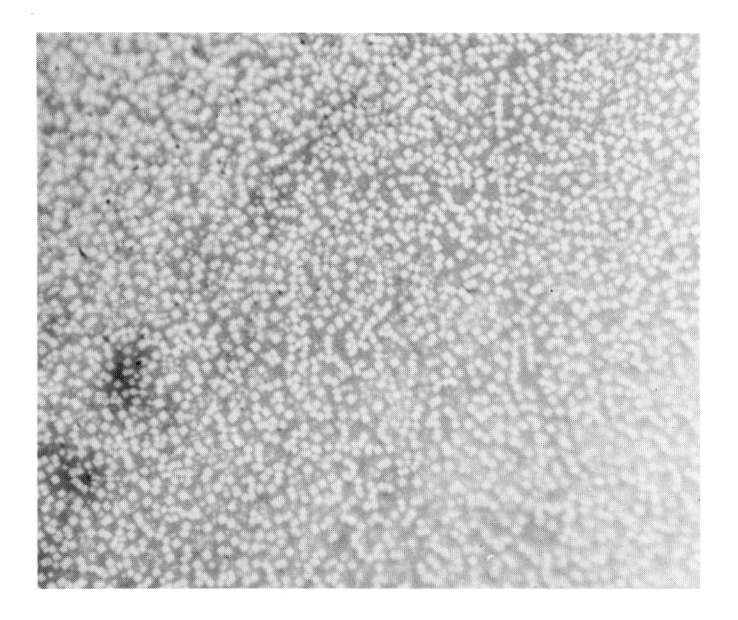

Fig. 13. Magnesioferrite precipitates in MgO containing 4.25 cat.% Fe solution treated in air at 1400°C and aged 75 hr at 1000°C. Magnification 750×. From G. W. Groves.[26]

where the two phases are not coplanar is a {111} plane of an fcc structure facing an (0001) plane of an hcp structure. The arrangements of atoms on these planes, shown in Figure 7, are identical and if the repeat distances are nearly the same, then a coherent interface may form with $\langle 110 \rangle_{fcc} \parallel \langle 12\bar{1}0 \rangle_{hex}$. The fcc and hcp phases coexist in cobalt and the interface is the one just described.

Another example is TiB_2 precipitates in TiC.[29,30] Here TiC and TiB_2 have the fcc (NaCl) and hexagonal structures, respectively, with the C and B atoms in interstitial positions; {111} and (0001) planes containing only Ti atoms occur with atom spacings differing by only 1%. The precipitates form as very thin platelets 10 Å thick with the orientation just given. The precipitates nucleate at dislocation nodes and as the precipitate plates grow, the dislocations split into partials with the TiB_2 taking the place of a stacking fault between the partials. A transmission electron microscope photograph of the structure is shown in Figure 14. A few tenths of a per cent of B added to TiC or VC, where a similar precipitate forms, causes a large increase in the high-temperature strength of the carbides because the precipitates anchor the dislocations.

In order for coherent interfaces to occur, the atom spacings in the planes which face each other across the interface must be nearly the same. The mismatch between the two structures must be taken up by lattice strain for a coherent interface to occur, and if the mismatch is too great, it is energetically cheaper to form a semicoherent interface. The epitaxial growth of one phase on the surface of another was treated theoretically[31] and for a layer 100 atom planes thick a semicoherent interface is expected if the mismatch is greater than 0.5 %. Thus in many cases of reported epitaxial growth the interface is, no doubt, semicoherent rather than coherent. The structure of semicoherent interfaces will be discussed again in the context of low-angle grain boundaries.

As previously noted, most masses of crystalline phases are made up of many crystals and these are separated by grain boundaries. Grain boundaries emerging on the surface of a polished sample of polycrystalline cadmium are shown in Figure 15(a). The fracture in polycrystalline α-Al_2O_3 is largely intergranular. Figure 15(b) is a scanning electron microscope photograph of the fracture surface of α-Al_2O_3 containing 10 vol. % Ni and the crystal faces are clearly seen.

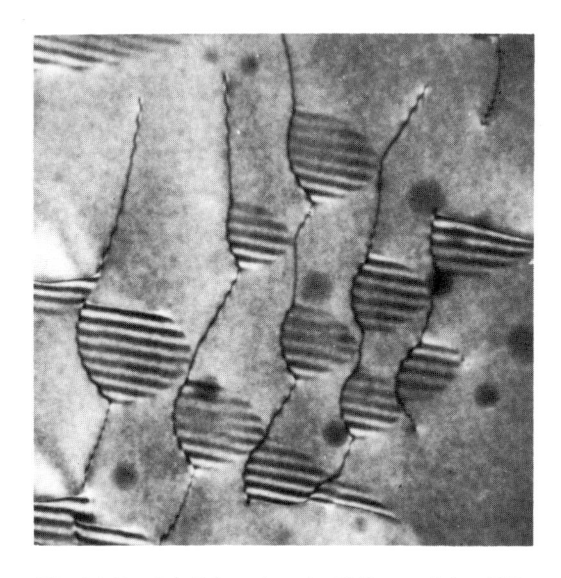

Fig. 14. Partial dislocations in TiC containing TiB_2 platelets in an array parallel to the (110) foil plane. The platelets are about 2 μm in diameter. From Venables.[30]

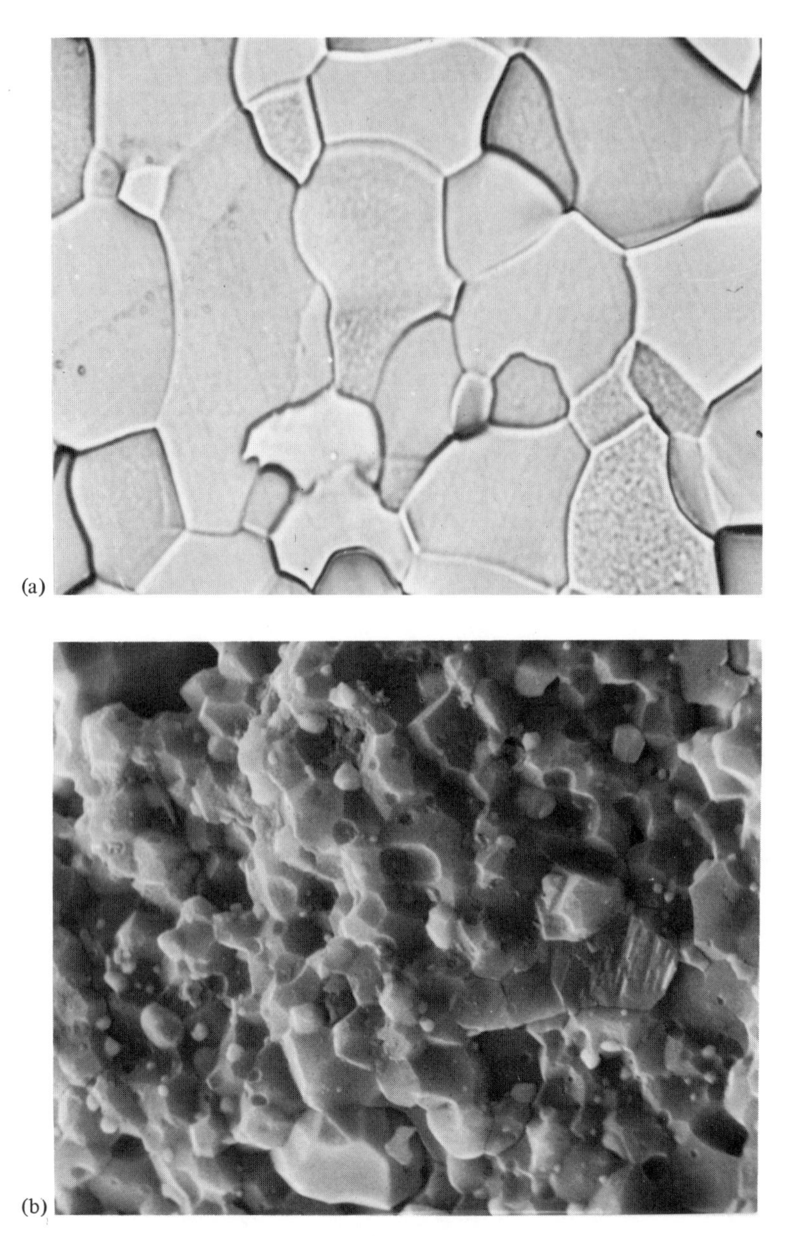

Fig. 15. (a) Grain boundaries in cadmium. The grains are on different levels because the rate of etching of the polished surface (in 18% HNO_3 in ethyl alcohol) varies with the grain orientation. Magnification 350×. From Dr. E. A. Grey, University of Liverpool, Liverpool, England. (b) Scanning electron micrograph of fracture surface of polycrystalline α-Al_2O_3 containing 10 vol.% Ni dispersed as small particles. The fracture is largely through the grain boundaries and facets of the α-Al_2O_3 and nickel grain surfaces are clearly seen (2500×). From J. Slater, Northwestern University, Evanston, Illinois.

The atoms near and at the boundaries, like a surface with another phase, are not in the perfect crystalline arrangement of the crystal interior, giving rise to a surface energy which depends on the relative orientation of the two grains which face each other across the interface, generally increasing with the angle of misorientation. Figure 16(a) shows that a small-angle tilt boundary consists of a wall of parallel edge dislocations which are periodically spaced. The spacing between dislocations, if θ, the tilt angle, is small, is[17–21]

$$d = b/[2\sin(\theta/2)] \cong b/\theta \tag{12}$$

Similarly, a low-angle twist boundary consists of a grid of screw dislocations, as shown in Figure 16(b). Such low-angle boundaries are semicoherent boundaries consisting of regions of good fit and regions of poor fit, i.e., the grid of dislocations. The long-range stress fields about such dislocations cancel and the stress field vanishes roughly a distance d from an interface dislocation. Thus formations of low-angle grain boundaries or cell walls, as previously noted in the discussion of dislocations, are low-energy arrays of dislocation and are expected to occur in crystals particularly after annealing (Figure 9). For a simple low-angle tilt boundary the grain boundary energy is[17–21]

$$\gamma \cong \theta(A - B\log\theta) \tag{13}$$

where $B = Gb/4\pi(1 - \nu)$ and $A = (E_c/b) - B\ln 5$, where E_c is the core energy.

Equation (13) breaks down when the dislocations in the boundary are so closely spaced that their cores overlap. For an angle of 15°, d computed from Eq. (12) is $4b$ and this is the maximum possible limit of applicability of Eq. (13). Equation (13) holds for Ge to about 15°,[17–21] but holds for Cu only to 5°[32] because the dislocation core is narrower in Ge in keeping with the difference in the nature of the atom binding.[21] The grain boundary energies for (100) tilt boundaries of 5° and 15° in Cu are about 250 and 400 egs/cm^2.

There are special orientations of the two grains where matching across the interface is good so that low values of grain boundary energies occur for certain angles of tilt or twist. For grain boundaries at and near these orientations the boundary consists of regions of good fit separated by grain boundary misfit dislocations. The Burgers vectors of the misfit dislocations must take up the mismatch between the two planes which face each other across the boundary. Such low-energy cusps in the energy versus angle curves have been observed and the grid of misfit dislocation may be seen by transmission electron microscopy.[33]

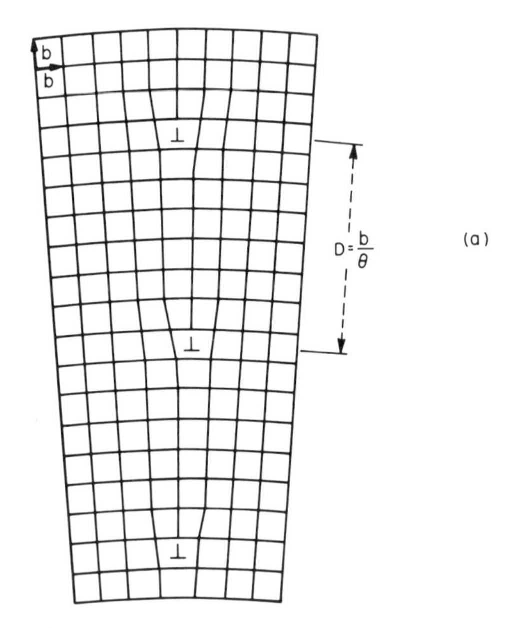

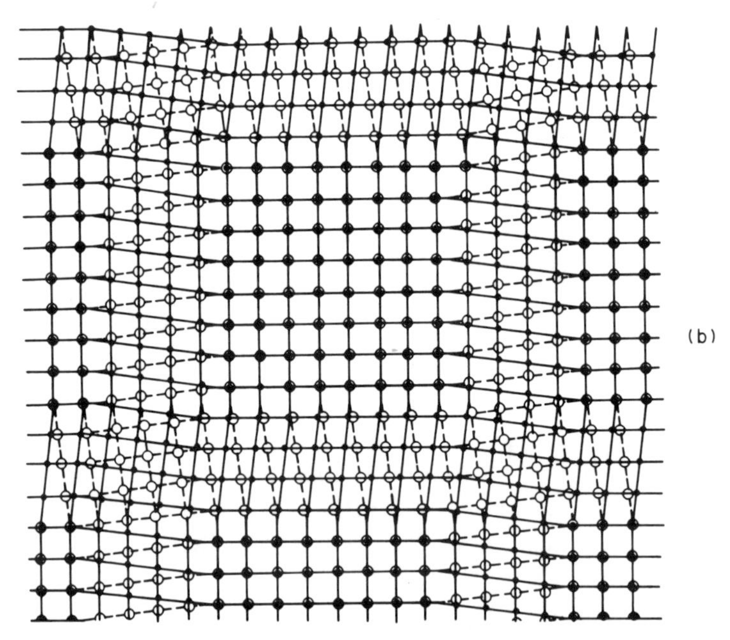

Fig. 16. Dislocation models of small-angle grain boundaries; (a) simple tilt boundary, (b) simple twist boundary. The former is an array of edge dislocations, the latter is an array of screw dislocations. From Read.[18]

The semicoherent interface between different phases also consists of regions of good fit separated by a dislocation grid and the interfacial energy may be computed on this basis. Such dislocations are called van der Merwe dislocations.[31]

Stacking faults and antiphase domains, which have already been discussed in conjunction with partial dislocations, are other types of planar defect. The various crystals can be considered as composed of a stacking of planar arrangements of atoms according to a regular sequence. The simplest case may be visualized as stacking of identical close-packed planes of atoms. Such planes are shown in Figure 7 and the stacking is depicted in Figure 17. Stacking of such planes according to the sequence *ABC ABC* results in the cubic close-packed or fcc structure and *AB AB AB* results in the hexagonal close-packed structure. The close-packed planes are {111} in fcc and [0001] in hcp crystals. The rocksalt and spinel structures have cubic close-packing of the large anion lattice with the small cations in interstititial positions; the corundum structure has hexagonal close-packing of the anion lattice, again with cations in interstitial positions. A stacking fault is a mistake in the packing sequence such as *ABC ABC AB*/*ABC ABC* in

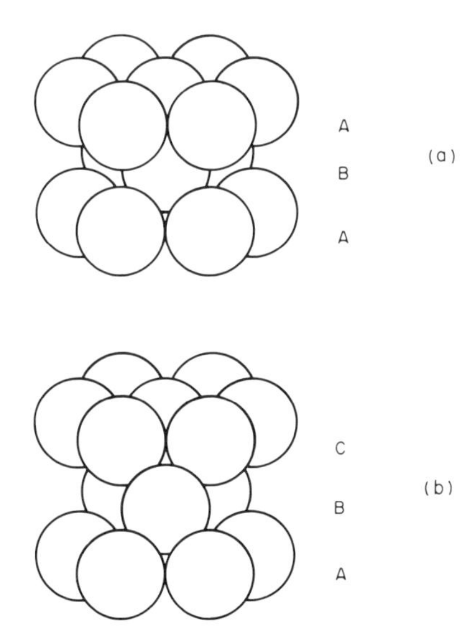

Fig. 17. Models showing how close-packed planes of atoms are stacked to give (a) the hexagonal close-packed structure and (b) the face-centered cubic or cubic close-packed structure. From Weertman and Weertman.[19]

an fcc crystal. The missing layer, *C* in the example given, results in four layers of hcp stacking. Such a stacking fault may be created by $(1/6)a\langle 211\rangle$ slip of one $\{111\}$ layer over another and this is thus the stacking sequence observed on traversing a direction normal to the region between the two partial dislocations in an extended dislocation in an fcc crystal. Slip on $\{111\}$ of $(1/6)a\langle 211\rangle$ gives an intrinsic fault and the partial dislocation (Shockley partial) is glissile. An intrinsic fault may also be formed by removing a partial plane of atoms, thus forming an edge dislocation $\frac{1}{3}a\langle 111\rangle$ at the end of the extra planes. Such partial dislocations are Frank partials and are sessile. Insertion of part of an extra plane of atoms gives an extrinsic stacking fault, *ABC ABC/B/ABC ABC* with two "missing" planes in the sequence and the partial dislocation at the edge of the extra plane is also a sessile Frank partial with a Burgers vector $\frac{1}{3}a\langle 111\rangle$.

A twin boundary may also be formed by changing the stacking sequence such as *ABCABC|BACBA*. The lattices on each side of the twin boundary are mirror related. Extrinsic and intrinsic stacking faults and twinning may arise from growth accidents when the crystal was formed and the stacking fault planes may extend completely through the crystal with no associated dislocations. Stacking faults are common in crystals which have low stacking fault energy. Stacking fault energies in metals vary from about 5 $ergs/cm^2$ for some copper-base alloys to perhaps 225 $ergs/cm^2$ in Al. Annealing twins are common in fcc metals with relatively low stacking fault energies such as Cu and Ag. They are not seen in Al. The twin boundary appears similar to a grain boundary but it is a planar surface inside a grain.

Twinning has long been known in naturally occurring minerals and nonmetallic crystals and many examples in each crystal system exist.[34] Repeated or polysynthetic twinning occurs in a number of minerals such as calcite, the triclinic feldspars, and magnetite, giving the minerals a banded appearance. Extended dislocations and the attendant stacking fault regions also occur in many nonmetallic crystals. As previously mentioned, the structure of corundum, α-Al_2O_3, is based on hexagonal close-packing of the large O^{2-} ions; two-thirds of the octahedral interstices are filled with Al^{3+} ions. At high temperatures, above 900°C, slip occurs by dislocation motion on (0001) planes in $\langle 11\bar{2}0\rangle$-type directions. The perfect dislocations in corundum have Burgers vectors $a\langle 11\bar{2}0\rangle$. These split into four partials of Burgers vectors $\frac{1}{3}a\langle 11\bar{2}0\rangle$,[25,35] as shown in Figure 18. This splitting is justified both by the reduction in dislocation energy and

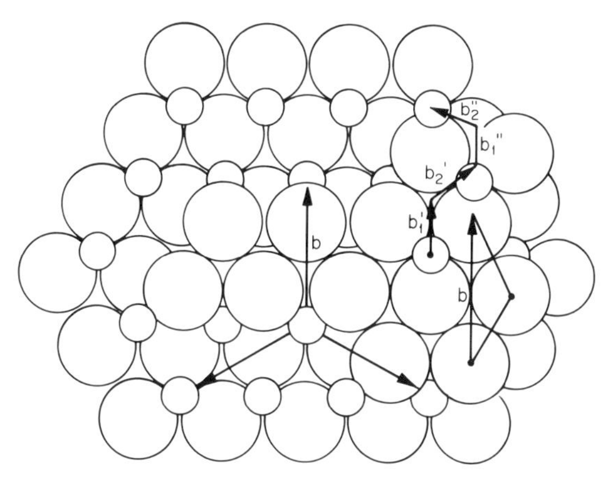

Fig. 18. Ball model of Al_2O_3 showing various slip vectors. The large balls represent O^{2-} and the small balls Al^{3+}. The slip vector (Burgers vector) of a perfect dislocation is represented by **b**. As explained in the text, the perfect dislocation splits into four partials with Burgers vectors $\mathbf{b}_1'$, $\mathbf{b}_2'$, $\mathbf{b}_1''$, and $\mathbf{b}_2''$. From Kingery.[25]

because $a\langle 11\bar{2}0\rangle$ motion moves one O^{2-} ion directly over another, which would give a higher Peierls stress. The dislocation motion is through the troughs in the structure. The result is three different stacking fault regions bounded by the four partials: $\mathbf{b}_1'$–SF_1–$\mathbf{b}_2'$–SF_2–$\mathbf{b}_1''$–SF_3–$\mathbf{b}_2''$.

Antiphase domains in ordered Cu_3Au have already been discussed. In the Cu_3Au structure there are four interpenetrating simple cubic lattices. In CuAu ordered below 385° there are only two sublattices, α and β, and the Au and Cu occupy alternate (001) planes, making the structure tetragonal. As in Cu_3Au, antiphase domains result when the ordered phase is formed from the disordered phase on cooling or aging isothermally below the ordering temperature. Ordering starts simultaneously in different parts of the disordered phase and antiphase boundaries result when two domains of different configuration meet, i.e., (1) Au on α, Cu on β and (2) Au on β, Cu on α. Such an antiphase domain in CuAu is depicted in Figure 19. A boundary in ordered Cu_3Au is shown in Figure 20. An antiphase boundary can be formed in a perfect CuAu-type crystal by a shift of half a face diagonal, $\frac{1}{2}a\langle 110\rangle$ in the $\{111\}$ plane, i.e., by motion of a dislocation of Burgers vector $\frac{1}{2}a\langle 110\rangle$ on the $\{111\}$ plane. Thus superdislocations are expected in CuAu and other ordered structures[36] such as Cu_3Au. Antiphase domains often occur on specific

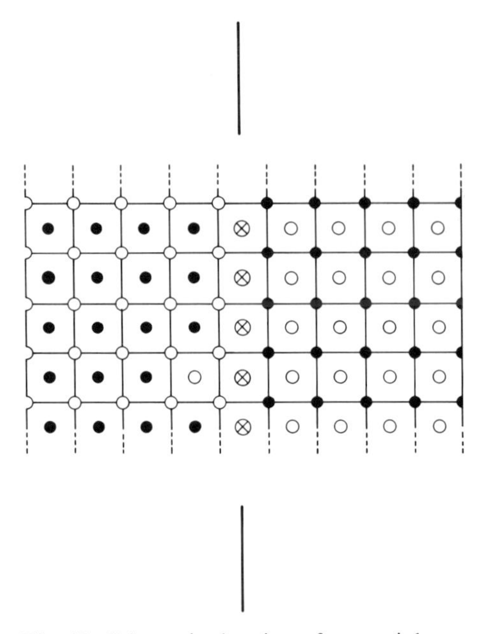

Fig. 19. Schematic drawing of an antiphase boundary in the CuAu structure. The Au atoms are represented by open circles and the Cu atoms by closed circles. At the antiphase boundary where the atoms exchange positions, the atoms may be either Au or Cu as represented by the symbol ⊗. The upward vertical direction is [001], while [100] is normal to the plane of the paper. The antiphase boundary shown is on an (001) plane.

crystallographic planes in order to reduce the wrong bonds across the interface; however, this is not always the case.[37]

In CuAu and NiMn, which have the same structure under certain circumstances, a periodic arrangement of antiphase boundaries occurs.[38–40] In CuAu between 385 and 410°C the antiphase boundaries are periodically spaced 20 Å apart, giving an orthorhombic structure. The structure may be directly seen by transmission electron microscopy.[38–41] Such periodic antiphase domain structures or long-period superlattices are really crystal structures with large unit cells and the question arises whether the "defects" in such structures should really be classified as defects.[41] Periodic spacing of defects to give large unit-cell structures, such as the Wadsley and Magneli shear phases, will be taken up again in the section on volumetric defects.

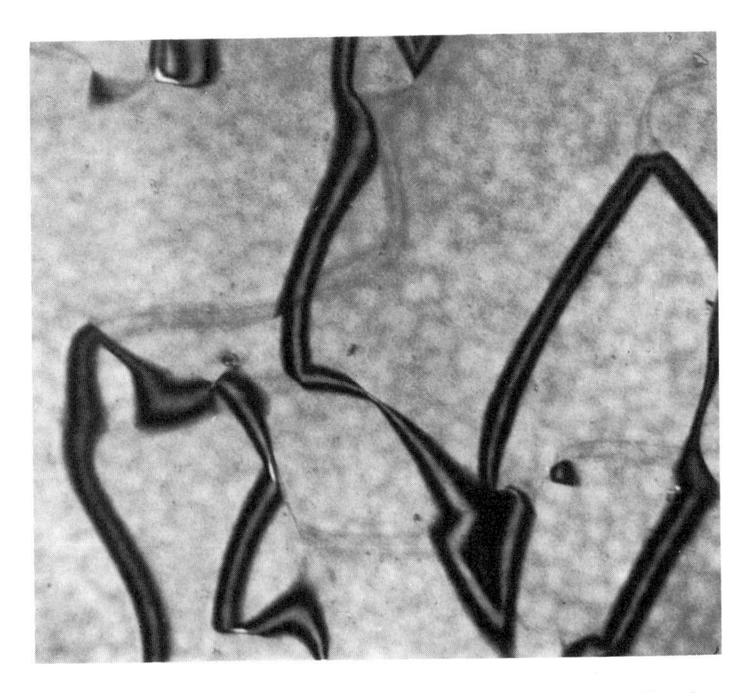

Fig. 20. Antiphase domain boundaries in ordered Cu_3Au. Magnification 70,000×. From D. E. Mikkola, Northwestern University, Evanston, Illinois.

A domain boundary in a ferromagnetic, ferrimagnetic, or antiferromagnetic material is similar to the antiphase domain boundary except the ordering is among magnetic moment directions and these change from one low-energy direction to an equivalent low-energy direction across the boundary. Magnetic domain wall boundary energies in metals are typically a few ergs/cm^2 while antiphase domain wall energies are perhaps 25–250 ergs/cm^2.

Point defects segregate to planar defects just as they do to linear defects. Adsorption on surfaces and the related catalysis of reactions on surfaces are well-known phenomena. The amount of segregation is related to the reduction in surface tension γ for the boundary by the Gibbs adsorption equation, which for a multicomponent system is

$$\sum_i \Gamma_i \, d(\ln \alpha_i x_i) = -d\gamma/kT \tag{14}$$

where Γ_i is the number of adsorbed atoms of component i per unit area of surface, α_i and x_i are the activity coefficient and atom fraction of the ith component in the bulk phase, and k is Boltzmann's constant.

Assuming that (a) component 1, the solvent, obeys Raoult's law, (b) component 2 is variable in composition and obeys Henry's law, and (c) all other components are constant and not adsorbed, then Eq. (14) reduces to, when $dx_1 = -dx_2$,

$$\frac{\Gamma_2}{x_2} - \frac{\Gamma_1}{x_1} = -\frac{1}{kT}\frac{\partial \gamma}{\partial x_2}\bigg|_{x_3, x_4, \ldots} \tag{15}$$

If $\partial\gamma/\partial x_2$ is negative, then adsorption of component 2 occurs at the boundary, but if $\partial\gamma/\partial x_2$ is positive, the boundary has less of component 2 than material away from the boundary. Reduction in γ through adsorption occurs because solute or impurities bond to surface atoms having unsatisfied or "dangling" bonds. There may also be a reduction in strain energy because atoms at the boundary are not in a close-packed arrangement. A solute or impurity atom thus may fit better at the boundary than in the interior of the crystal, similar to segregation about an edge dislocation. For example, Ca^{2+} segregates preferentially to grain boundaries in polycrystalline sapphire, while Mg^{2+} does not.[42] The ionic diameter of Ca^{2+} is 0.9 Å, while that of Mg^{2+} is 0.65 Å. Segregation of impurities to grain boundaries is a source of brittle fracture in metals. In slowly cooled steels containing 0.03 at. % Sb and 0.034 at. % P, using Auger spectroscopy, surface compositions of approximately 10 and 6 at. % Sb and P, respectively, were determined.[43,44] These correspond to $\partial\gamma/\partial x_2$ of approximately -1000 ergs/cm^2-at. % solute in bulk phase.[45]

Segregation also occurs on stacking faults. As already mentioned, the intrinsic stacking fault between two partial dislocations in an fcc metal consists of a thin ribbon of an hcp metal. Many metals when added to Cu, Ag, or Au reduce the stacking fault energy[46] because the solute atoms prefer to be in this stacking fault ribbon, thereby reducing the stacking fault energy. This gives rise to an increase in strength through what is called Suzuki locking. In order to move an extended dislocation with such segregation, the dislocation must be torn away from the segregation. For example, when 9 at. % Al is added to copper from the lowering in γ a concentration of 18 at. % Al is computed from Eq. (15).[47]

5. Volumetric Defects

A point defect is small in all dimensions; a dislocation is small in two dimensions and is extended in one dimension. A planar defect is small in one dimension and is extended in one or two dimensions; a volumetric defect is not small in any dimension and may be extended in one, two, or three dimensions.

The simplest such defect is perhaps the vacancy cluster.[48,49] Two vacancies attract to form a divacancy; the binding energy in metals is typically 0.05–0.2 eV. A third vacancy may be added to form a trivacancy and so forth. While divacancies and trivacancies might be classified as point defects, they have been included under volumetric defects in this chapter. The clusters may grow to a size where they collapse and form a dislocation loop. Additional vacancies may condense on the loops so that they grow to a size where they are visible in the transmission electron microscope. Examples are given in Figure 1, which shows (a) stacking fault disks in Al and (b) stacking fault tetrahedra in Au.

Vacancies and impurity atoms may also attract to form complexes. One expects large substitutional atoms and interstitial impurity atoms to be attracted to vacancies because of reduction in total strain energy similar to the attraction of impurity atoms to dislocations to form "atmospheres." Segregation about dislocations may reach the stage where a distinct second phase forms decorating the dislocation. Dislocations in MgO decorated by $MgFe_2O_4$ precipitates are shown in Figure 22.

Dislocations and interfaces catalyze reactions in solids,[14] dislocation nodes and grain corners being particularly effective catalytic sites. When a particle forms on a dislocation or on a surface a portion of dislocation or surface is destroyed. The energy gain from this reduces the energy required to form the nucleus. For example, suppose a β particle forms in α on a preexisting α–γ surface (α, β, and γ represent three different phases). Then the surface energy which must be supplied to form the particle is

$$E_s = A_{\alpha\beta}\gamma_{\alpha\beta} + A_{\beta\gamma}\gamma_{\beta\gamma} - A_{\beta\gamma}\gamma_{\alpha\gamma} \tag{16}$$

where $A_{\alpha\beta}$ and $A_{\beta\gamma}$ are the areas of the $\alpha\beta$ and $\beta\gamma$ surfaces which are formed. The area $A_{\alpha\gamma}$ of the surface which is destroyed, of course, equals $A_{\beta\gamma}$. The γ's represent the surface tensions of the designated interfaces. Clearly if $\gamma_{\alpha\gamma} > \gamma_{\beta\gamma}$, catalysis of nucleation occurs.

In solid solution phases which contain a substantial amount of an alloying element where there is a deviation from ideality, the lowest free-energy state is a nonrandom one having local clustering or local ordering of the atoms or ions. Positive deviations are associated with like-atom clustering, while negative deviations are associated with ordering. Deviations from randomness may be determined by measure-

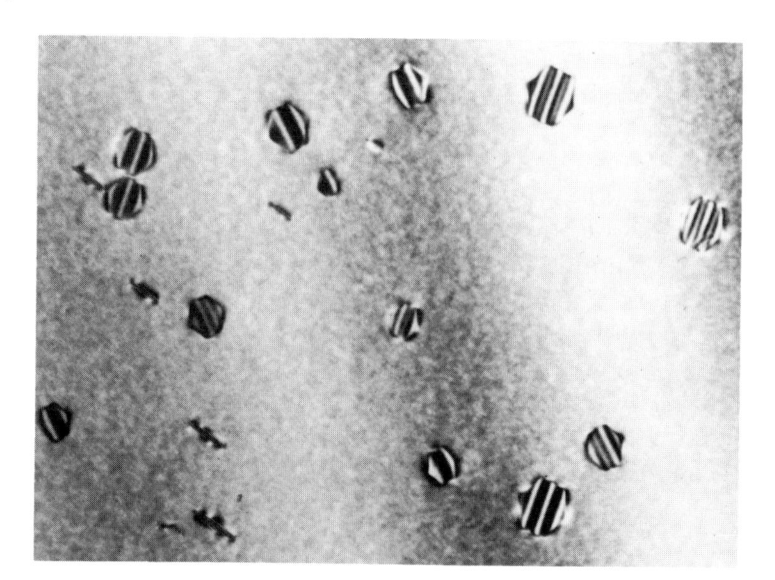

(a)

(b)

Fig. 21. Defects formed by condensation of vacancies onto dislocation loops formed by collapse of small vacancy clusters. (a) Dislocation loops in 99.999% Al, quenched from 610°C into liquid N_2, then aged 1 hr at −80°C and 2 hr at 60°C. The defect is a Frank sessile dislocation enclosing an intrinsic stacking fault. Magnification 27,000×. From Dr. K. Y. Chen, Northwestern University, Evanston, Illinois. (b) Stacking fault tetrahedra in 99.999% Au quenched from 1038° into −35°C brine and aged at 25°C. The tetrahedra (some are truncated) are bound by intrinsic stacking faults and dislocations at the edges called stair rod dislocations. Magnification 70,000×. From Dr. J. A. McComb, Northwestern University, Evanston, Illinois.

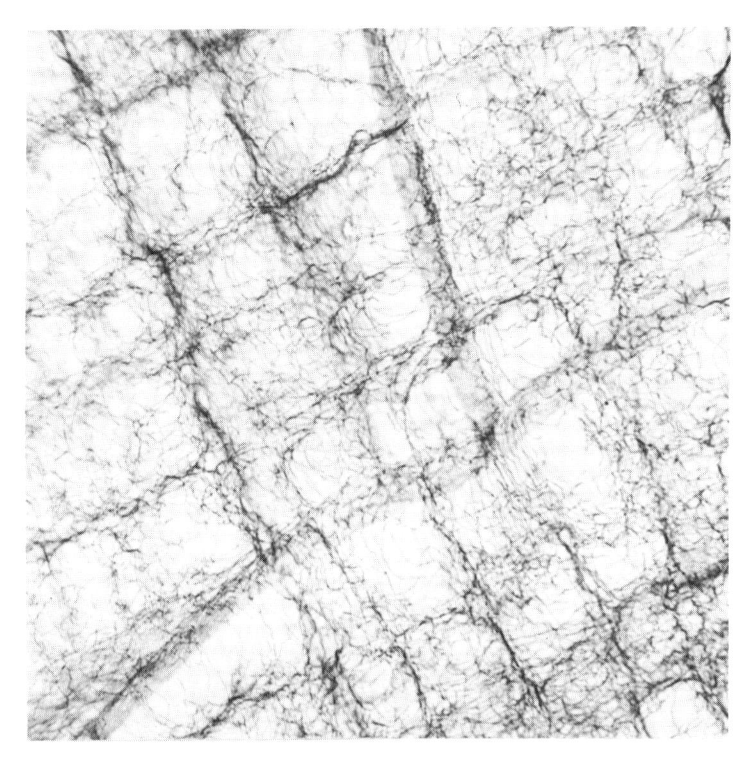

Fig. 22. Magnesioferrite precipitates decorating dislocations in MgO–1.35 cat.% Fe; single crystal was aged 64 hr at 900°C (50 ×). From Dr. G. W. Groves, Northwestern University, Evanston, Illinois.

ment of diffuse X-ray scattering which gives pair probability values for first, second, third, etc. neighboring atoms. Using a computer, the pair probabilities may be converted to the actual local atomic arrangement.[50–52] In Cu-base Ni alloys the analysis shows the presence of plate-shaped Ni-rich clusters. These give rise to superparamagnetism.

Wüstite, $Fe_{1-x}O$ has a positive deviation from ideality in the partial molar free energy of Fe below 1100°C.[53] As previously mentioned, clustering of cation vacancies is observed and the clustered vacancy defects are in turn ordered.[8] The X-ray diffraction pattern of $Fe_{1-x}O$ at room temperature shows spots due to a superstructure in addition to the spots from the NaCl structure. Quantitative analysis of the intensities of the superstructure and the main peaks in a single crystal of $x = 0.098$ quenched to room temperature led to the structure shown in Figure 23. The cluster of 13 octahedral vacancies and four associated tetrahedral cations shown gave substantial agreement with the observed intensities. In order to maintain local charge balance,

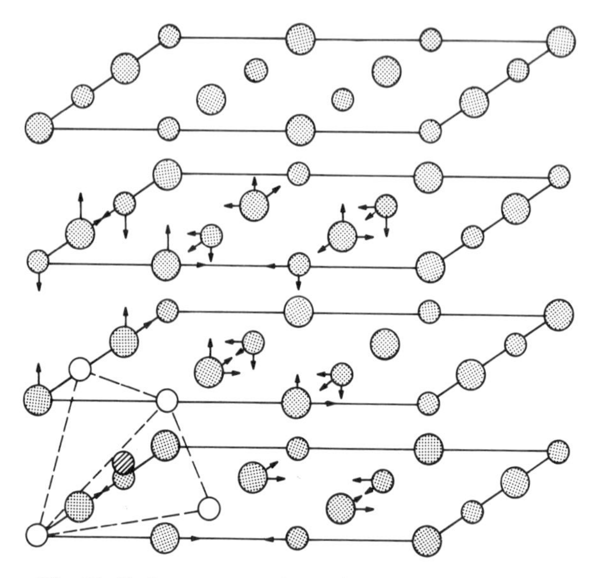

Fig. 23. Cation vacancy cluster in $Fe_{0.902}O$. Cations in tetrahedral interstitial sites are associated with the cluster. One octant of the defect is shown. The defect consists of 13 octahedral vacancies (open circles) and four associated tetrahedral cations (shaded circles). The large and small filled circles represent, respectively, oxygen and iron ions in octahedral sites. The ions surrounding the defect are displaced as indicated by the arrows. Fe^{3+} ions in the vicinity of the defect and in the tetrahedral sites neutralize the excess negative charge. From Koch and Cohen.[8]

many cations in and near the defects must be Fe^{3+}. The tetrahedral cations are thought to be Fe^{3+} because the tetrahedral site is smaller than the octahedral site. These defect clusters, which exist even at high temperatures[8] where wüstite is stable, are arranged periodically at low temperatures to give the superlattice lines. For compositions near $x = 0.06$ planar regions of the ordered structure alternate with regions of near-stoichiometric FeO to give a modulated structure.

Confirmation of defect clustering in $Fe_{1-x}O$ is given by the magnetic properties. Random cation vacancies would be expected to weaken the antiferromagnetic interaction and lower the antiferromagnetic Néel temperature; however, T_N is almost unaffected by variation in x.[54,55] Clustering of the cation vacancies leaves the remaining part of the cation sublattice intact so that the antiferromagnetic interaction is not weakened. Clustering of cation vacancies also occurs in $V_{1-x}O$.[56]

An interesting ordered structure occurs in TiO.[57] At the stoichiometric composition about 15 % of the Ti^{2+} and O^{2-} sites are vacant and these are most likely random above 1000°C, the structure being that of NaCl. The ordered structure is monoclinic with every third (110) plane having half the Ti^{2+} and half the O^{2-} ions alternately missing. Transmission electron micrographs of ordered TiO show a substructure; the TiO is divided into parallel lamallae 300–1000 Å wide. Each platelet is a different crystallographic variant of the cubic to monoclinic transformation and the arrangement probably occurs to minimize strain energy.

When excess Al^{3+} is added to $MgAl_2O_4$ (Mg–Al spinel), vacancies occur on octahedral interstitial sites.[58,59] Because of local charge compensation requirements, the region of the defect clusters must have a different composition from the perfect lattice. $MgAl_2O_4$, a normal spinel, will tolerate a considerable excess of Al^{3+} at elevated temperatures but not at low temperatures. This gives rise to precipitation on aging. However, crystals grown by the Verneuil oxyhydrogen flame fusion technique having MgO to Al_2O_3 ratio less than 1:4 are single phase as prepared but diffuse X-ray scattering occurs. Study of this diffuse scattering in crystals having MgO to Al_2O_3 ratios in the range 1:1.5 to 1:3.5[58,59] suggested isolated octahedral vacancies and the associated lattice distortion in 1:1.5 crystals and chains or strings of defects along the six $\langle 110 \rangle$ directions in 1:3.5 crystals. Since excess Al^{3+} requires reduction in the number of Mg^{2+} ions, the octahedral vacancies are created by Al^{3+} ions occupying the tetrahedral sites vacated by the Mg^{2+} ions.

In a number of oxide phases where wide ranges of solid solubility had been reported discrete series of stoichiometric phases are thought to occur.[60,61] Based on the observed diffraction patterns, homologous series of phases have been proposed in the oxide systems of Mo, W, V, Ti, and Nb. An example is the series W_nO_{3n-2}; $W_{40}O_{118}$ and $W_{24}O_{70}$ belong to this series. In many cases the structures may be thought to arise from a parent structure by operation of a defect mechanism, crystallographic shear on periodically spaced planes. This shear, which changes the coordination about some of the cations, may occur by nonconservative motion of an imperfect dislocation array.[41] Motions of dislocations in the array remove or extend layers of atoms in the structure. For example, in one model[62,63] for producing W_nO_{3n-2} from WO_3, random oxygen vacancies are imagined to form and then aggregate into disks. These collapse and shear occurs, producing crystallographic shear plane regions surrounded by dislocation loops as shown in Figure 24. In the perfect structure the WO_6 octahedra are stacked as indicated. The anion vacancies are shown on the left-hand side of the figure and the right side shows the structure after collapse and shear. The collapsed and sheared regions are then imagined to expand by nonconservative dislocation motion by vacancies diffusing to the dislocation loops.[64] The desired composition is obtained by proper spacing and arrangement of the shear planes. Periodically spaced dislocations have been observed in non-stoichiometric WO_3,[41] in keeping with the model.

Various defects occur in the sheared structures.[60] Dislocation loops have already been mentioned. Twinning has been observed where the directions of shear planes change. There is also evidence for deviations from strict periodicity in the shear planes. Thus some shear structures have ranges in composition and it is easy to imagine a continuous structural change through such a phase region

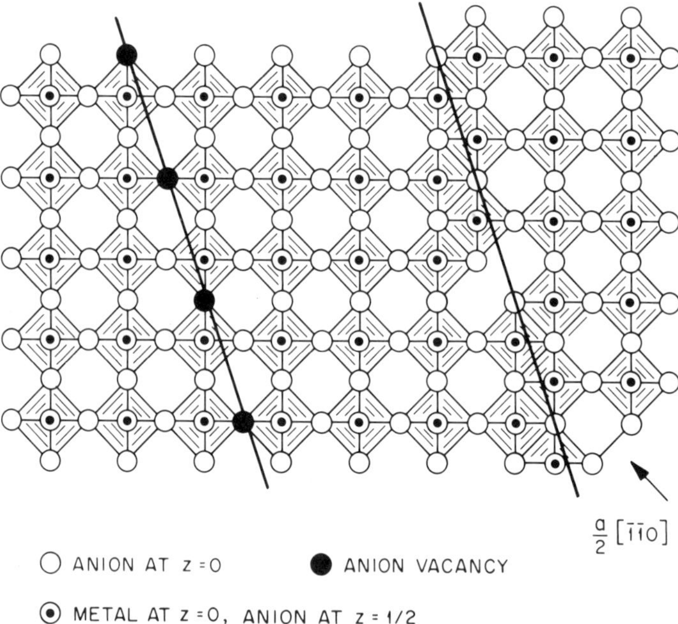

Fig. 24. Anderson and Hyde model for forming a crystallographic shear plane in WO_3. The perfect structure consists of WO_6 octahedra stacked as shown. In the Anderson and Hyde model anion vacancies are imagined to collect as shown on the left-hand side of the figure and then collapse to give the structure shown on the right. Along the fault the stacking arrangement of the WO_6 tetrahedra is not the same as in good material and also the structure is sheared along the fault plane by the vector $\frac{1}{2}a[\bar{1}\bar{1}0]$. From Hyde and Bursill.[63]

as in phase regions for ordered metallic alloys where long-period superlattices occur.[65] Study of the shear structures is a very active field of research and a much clearer understanding of the structural changes occurring in these phase regions will no doubt emerge.

Precipitation or ex-solution processes occur extensively in solids because many solid phases have a decreasing composition range of stability as the temperature is lowered. Since the precipitates and impurity particles are second-phase particles and not defects, except for the interfaces formed, they are beyond the scope of this chapter; however, certain aspects of precipitation need to be considered.

Heterogeneous nucleation was previously mentioned. When the precipitates have structures which are very closely related or identical to the matrix phase and the atom spacings are almost the same, then low-energy interface boundaries may result and the precipitates nucleate homogeneously; catalysis is not required for easy nucleation. The initial stages of such precipitation gives precipitates which are only a few atoms in size. This is often called a preprecipitation stage. The Guinier–Preston "zones" which form in Al-base Ag alloys is an example of this. These are octahedron-shaped enriched solute regions having the same crystal structure as the matrix.[66] The magnesioferrite precipitates which form in MgO are a similar case. The precipitate and matrix have continuous anion lattices; they differ only in the cation lattices. Again the initial tiny "particles" may be considered as a volumetric defect in the matrix lattice since they are probably smaller than a unit cell of magnesioferrite.[26–28]

A metastable solid solution may decompose without nucleation through fluctuations which are large in extent but small in degree, as first proposed by Gibbs. When the second derivative of the free energy with respect to composition $\partial^2 f/\partial c^2$ is negative, then the theory of Cahn for spinodal decomposition predicts that the decomposition begins with periodic fluctuations in composition in the parent phase.[67] This is a form of clustering. Spinodal decomposition occurs in a number of systems like Al–Zn,[68] SnO_2–TiO_2,[69,70] and $CoFe_2O_4$–Co_3O_4.[71] It is responsible for the high coercive force in the Al–Ni–Co magnets[67] as well as in $CoFe_2O_4$–Co_3O_4 magnets.

Acknowledgments

The author is deeply indebted to Drs. M. Meshii and J. B. Cohen for many helpful suggestions on the content of this review chapter and for providing many of the photomicrographs.

References

1. C. S. Smith, in *A History of Metallography*, pp. 199–200, Univ. of Chicago Press (1960).
2. C. G. Darwin, The theory of X-ray reflexion I, II, *Phil. Mag.* **27**, 315–333, 675–690 (1914).
3. C. G. Darwin, the reflexion of X-rays from imperfect crystals, *Phil. Mag.* **43**(Series 6), 800–829 (1922).
4. W. L. Bragg, C. G. Darwin, and R. W. James, The intensity of reflexion of X-rays by crystals, *Phil. Mag.* **1**, 897–922 (1926).
5. J. Frenkel, Über die Wärmebewegung in festen und flüssigen Körpern, *Z. Physik* **35**, 652–669 (1926).
6. L. S. Darken and R. W. Gurry, The system iron–oxygen. I. The wüstite field and related equilibria, *J. Am. Chem. Soc.* **67**, 1398–1412 (1945).
7. E. R. Jette and F. Foote, An X-ray study of the wüstite (FeO) solid solutions, *J. Chem. Phys.* **1**, 29–36 (1933).
8. F. B. Koch and J. B. Cohen, The defect structure of $Fe_{1-x}O$, *Acta Cryst.* **B25**, 275–287 (1969).
9. J. B. Cohen, The order–disorder transformation, in *Phase Transformations*, pp. 561–620, American Society for Metals, Cleveland (1970).
10. E. Orowan, Dislocations in plasticity, in *The Sorby Centennial Symposium on the History of Metallurgy* (C. S. Smith, ed.), pp. 359–376, Metallurgical Society Conference Vol. 27, Gordon and Breach, New York (1965).
11. T. Broom and R. K. Ham, The effect of lattice defects on some physical properties of metals, in *Vacancies and Other Point Defects in Metals and Alloys*, Monograph No. 23, The Institute of Metals, London (1958).
12. S. Nenno and J. W. Kauffman, Detection of equilibrium vacancy concentrations in aluminum, *Phil. Mag.* **4**(48), 1382–1384 (1959).
13. R. O. Simmons and R. W. Balluffi, Measurements of equilibrium vacancy concentrations in aluminum, *Phys. Rev.* **117**, 52–61 (1960).
14. M. E. Fine, *Introduction to Phase Transformations in Condensed Systems*, pp. 109–111, Macmillan, New York (1964).
15. F. A. Kröger and H. J. Vink, Relations between the concentrations of imperfections in crystalline solids, in *Solid State Physics* (F. Seitz and D. Turnbull, eds.), Vol. 3, p. 310, Academic, New York (1956).
16. A. J. Bradley and A. Taylor, An X-ray analysis of the nickel–aluminum system, *Proc. Roy. Soc.* (*London*) **159**, 56–72 (1937).
17. A. H. Cottrell, *Dislocations and Plastic Flow in Crystals*, Oxford Univ. Press, London (1953).
18. W. T. Read, *Dislocations in Crystals*, McGraw-Hill, New York (1953).
19. J. Weertman and J. R. Weertman, *Elementary Dislocation Theory*, Macmillan, New York (1964).
20. J. Friedel, *Dislocations*, Addison-Wesley, Reading, Mass. (1964).
21. J. P. Hirth and J. Lothe, *Theory of Dislocations*, McGraw-Hill, New York (1968).
22. S. S. Brenner, Properties of whiskers, in *Growth and Perfection of Crystals* (R. H. Doremus *et al.*, eds.), pp. 157–190, Wiley, New York (1958).
23. M. J. Marcinkowski, in *Electron Microscopy and Strength of Crystals* (G. Thomas and J. Washburn, eds.), p. 333, Interscience, New York (1963).

24. A. H. Cottrell, *An Introduction to Metallurgy*, p. 339, Edward Arnold, London (1967).
25. W. D. Kingery, *Introduction to Ceramics*, pp. 194, 568, Wiley, New York (1960).
26. G. W. Groves and M. E. Fine, Solid solution and precipitation hardening in Mg–Fe–O alloys, *J. Appl. Phys.* **35**, 3587–3593 (1964).
27. G. P. Wirtz and M. E. Fine, Precipitation and coarsening of magnesioferrite dilute solutions of iron in MgO, *J. Am. Ceram. Soc.* **51**, 402–406 (1968).
28. E. W. Kruse III and M. E. Fine, Precipitation strengthening of MgO by $MgFe_2O_4$, *J. Am. Ceram. Soc.* **55**, 32–37 (1972).
29. W. S. Williams, Dispersion hardening of titanium carbide by boron doping, *Trans. TMS–AIME* **236**, 211 (1966).
30. J. D. Venables, The nature of precipitates in boron-doped TiC, *Phil. Mag.* **16**, 873 (1967).
31. J. H. van der Merwe, Crystal interfaces. Part I. Semi-infinite crystals, *J. Appl. Phys.* **34**, 117 (1963); Crystal interfaces. Part II. Finite overgrowths, *J. Appl. Phys.* **34**, 123 (1963).
32. N. A. Gjostein and F. N. Rhines, Absolute interfacial energies of [001] tilt and twist grain boundaries in copper, *Acta Met.* **7**, 319 (1959).
33. T. Schober and R. W. Balluffi, Quantitative observation of misfit dislocation arrays in low-angle and high-angle twist grain boundaries, *Phil. Mag.* **21**, 109 (1970).
34. E. W. Dana and W. E. Ford, *A Textbook of Minerology*, 4th ed., pp. 179–194, Wiley, New York (1932).
35. M. L. Kronberg, Plastic deformation of single crystals of sapphire: Basal slip and Twinning, *Acta Met.* **5**, 507 (1957).
36. C. S. Barrett and T. B. Massalski, *The Structure of Metals*, 3rd ed., pp. 270–284, McGraw-Hill, New York (1966).
37. H. Berg and J. B. Cohen, Long-range order and ordering kinetics in $CoPt_3$, *Met. Trans.* **3**, 1797–1805 (1972).
38. D. W. Pashley and A. E. B. Presland, The observation of anti-phase boundaries during the transition from CuAu I to CuAu II, *J. Inst. Met.* **87** (1958–59).
39. S. Ogawa, D. Watanabe, H. Watanabe, and T. Komoda, The direct observation of the long period of the ordered alloy CuAu(II) by means of electron microscope, *Acta Cryst.* **11**, 872 (1958).
40. V. Krasevec, P. Delavignette, and S. Amelinckx, Superstructure due to periodic twinning in quenched NiMn alloy, *Mat. Res. Bull.* **2**, 1029 (1967).
41. S. Amelinckx and J. Van Landuyt, The use of electron microscopy in the study of extended defects related to nonstoichiometry, in *The Chemistry of Extended Defects in Non-Metallic Solids* (L. Eyring and M. O'Keeffe, eds.), pp. 295–320, North-Holland, Amsterdam (1970).
42. H. L. Marcus and M. E. Fine, Grain boundary segregation in MgO-doped Al_2O_3, *J. Am. Ceram. Soc.* **55**, 568 (1972).
43. H. L. Marcus and P. W. Palmberg, Auger fracture surface analysis of a temper embrittled 3340 steel, *Trans. TMS–AIME* **245**, 1664 (1969).
44. P. W. Palmberg and H. L. Marcus, An auger spectroscopic analysis of the extent of grain boundary segregation, *Trans. ASM* **62**, 1016 (1969).
45. M. E. Fine and H. L. Marcus, Segregation to an interface and brittle fracture of metals, *Met. Trans.* **2**, 1474 (1971).
46. A. Howie and P. R. Swann, Direct measurements of stacking-fault energies from observations of dislocation nodes, *Phil. Mag.* **6**, 1215 (1961).

47. R. W. Guard and M. E. Fine, Surface thermodynamic treatment of absorption on a dislocation-Suzuki locking, *Trans. TMS–AIME* **233**, 1383–1388 (1965).
48. H. G. Van Bueren, *Imperfections in Crystals*, 2nd ed., pp. 34–36, 275–280, 306–310, 562, 581, 610–611, North-Holland, Amsterdam (1961).
49. R. M. J. Cotterill, in *Vacancy Clusters in Pure and Impure FCC Metals* (R. M. J. Cotterill, M. Doyama, J. J. Jackson, and M. Meshii, eds.), pp. 97–162, Academic, New York (1965).
50. P. C. Gehlen and J. B. Cohen, Computer simulation of structure associated with local order in alloys, *Phys. Rev.* **139**, A844–A855 (1965).
51. J. E. Gragg, Jr., P. Bardhan, and J. B. Cohen, The "Gestalt" of local order, in *Critical Phenomena in Alloys, Magnets and Superconductors*, (R. E. Mills, E. Ascher, R. I. Jaffee, eds.), Chapter 6, Part 3, pp. 309–337, McGraw-Hill, New York (1971).
52. J. B. Cohen, A brief review of the properties of ordered alloys, *J. Mat. Sci.* **4**, 1012–1022 (1969).
53. R. J. Ackermann and R. W. Sandford, Argonne National Laboratory Rept. ANL-7250.
54. F. B. Koch and M. E. Fine, Magnetic properties of Fe_xO as related to the defect structure, *J. Appl. Phys.* **38**, 1470–1471 (1967).
55. M. E. Fine and F. B. Koch, Néel transformation in near-stoichiometric Fe_xO, *J. Appl. Phys.* **39**, 2478–2479 (1968).
56. P. S. Bell and M. H. Lewis, Nonstoichiometric vacancy order in vanadium monoxide, *Phys. Stat. Sol.* **7**, 431 (1971).
57. D. Watanabe, O. Terasaki, A. Jostsons, and J. R. Castles, Electron microscope study on the structure of low-temperature modification of titanium monoxide phase, in *The Chemistry of Extended Defects in Non-Metallic Solids* (L. Eyring and M. O'Keeffe, eds.), pp. 238–256, North-Holland, Amsterdam (1970).
58. H. Jagodzinski and H. Saalfeld, Kationenverteilung und Strukturbeziehungen in Mg–Al–Spinellen, *Z. Krist.* **110**, 197 (1958).
59. H. Jagodzinski and K. Haefner, On order–disorder in ionic nonstoichiometric crystals, *Z. Krist.* **125**, 188 (1967).
60. A. Magneli, Structural order and disorder in oxides of transition metals of the titanium, vanadium, and chromium groups, in *The Chemistry of Extended Defects in Non-Metallic Solids* (L. Eyring and M. O'Keeffe, eds.), pp. 148–162, North-Holland, Amsterdam (1970).
61. A. Wadsley, in *Non-Stoichiometric Compounds* (L. Mandelcorn, ed.), Chapter 3, Academic, New York (1963).
62. J. S. Anderson and B. G. Hyde, On the possible role of dislocations in generating ordered and disordered shear structures, *J. Phys. Chem. Solids* **29**, 1393 (1967).
63. B. G. Hyde and L. A. Bursill, Point, line and planar defects in some nonstoichiometric compounds, in *The Chemistry of Extended Defects in Non-Metallic Solids* (L. Eyring and M. O'Keeffe, eds.), pp. 347–378, North-Holland, Amsterdam (1970).
64. P. Delavignette and S. Amelinckx, Large dislocation loops in antimony telluride, *Phil. Mag.* **6**, 601 (1961).
65. H. Sato and R. S. Toth, in *Alloying Behavior Effects in Concentrated Solid Solutions*, AIME Series 29 (T. B. Massalski, ed.), Gordon & Breach, New York (1965).
66. J. E. Gragg, Jr. and J. B. Cohen, The structure of Guinier–Preston zones in Al–5 at.% Ag, *Acta Met.* **19**, 507–519 (1971).

67. J. W. Cahn, Spinodal decomposition, *Trans. TMS–AIME* **242**, 166–180 (1968).
68. K. B. Rundman and J. E. Hilliard, Early stages of spinodal decomposition in an Al–Zn alloy, *Acta Met.* **15**, 1025 (1967).
69. V. S. Stubican and A. H. Schultz, Phase separation by spinodal decomposition in the tetragonal system, *J. Am. Ceram. Soc.* **53**, 211–214 (1970).
70. A. H. Schultz and V. S. Stubican, Separation of phases by spinodal decomposition in the systems Al_2O_3–Cr_2O_3 and Al_2O_3–Cr_2O_3–Fe_2O_3, *J. Am. Ceram. Soc.* **53**, 613–616 (1970).
71. M. Takahashi, J. R. C. Guimarães, and M. E. Fine, Spinodal decomposition in the system $CoFe_2O_4$–Co_3O_4, *J. Am. Ceram. Soc.* **54**, 281–295 (1971).

6

Defect Equilibria in Solids

George G. Libowitz
Materials Research Center
Allied Chemical Corporation
Morristown, New Jersey

1. Introduction

1.1. Native Defects

An ideal crystal consists of a perfectly ordered arrangement of atoms, ions, or molecules. However, in any real crystal, at temperatures above absolute zero, there are always imperfections or defects in the crystal lattice, as discussed in Chapter 5. This chapter will deal with defects whose distribution and concentration in the lattice are governed by the laws of thermodynamics.† In pure crystals such defects are called native defects. The existence of native defects in a lattice arises from a tendency of a crystal to increase its entropy or degree of disorder. As defects are introduced into a crystal, the entropy ΔS will increase. The number of defects will be limited, however, by the enthalpy necessary to form the defects, ΔH. The actual number of defects present at any temperature is that which gives a minimum in the free energy G of the crystal according to the relation

$$G = G^* + \Delta H - T\,\Delta S = G^* + N_D\,\Delta H_D - TN_D\,\Delta S_v - T\,\Delta S_c(N_D) \quad (1)$$

where G^* is the free energy of the theoretically perfect crystal, N_D is the number of defects, ΔH_D is the enthalpy change per defect, ΔS_v is the change in vibrational entropy per defect, and $\Delta S_c(N_D)$ is the change in configurational entropy, which is a function of the number of defects.

† Several chapters in Volume 2 will deal with defects in relation to properties; this chapter will serve to introduce a number of topics covered in more detail elsewhere in the Treatise. In particular, the reader is referred to the chapter by Brebrick on semiconductors (Volume 2, Chapter 5).

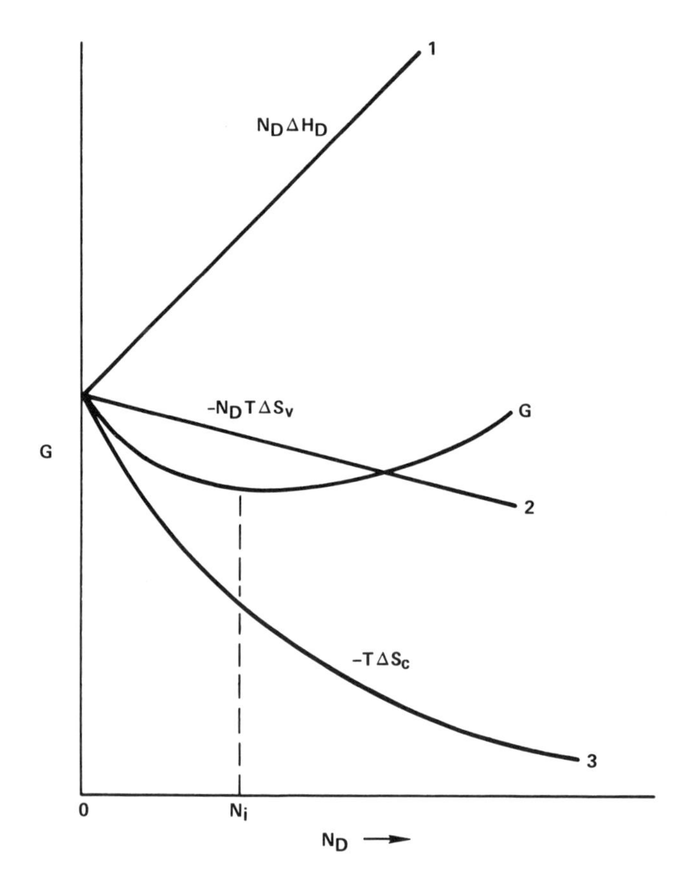

Fig. 1. Variation of free energy of a crystal with defect concentration.

The manner in which each of the terms in Eq. (1) changes with defect concentration is shown in Figure 1. It can be seen that the effect of increasing entropy on the free energy is counterbalanced by the increasing enthalpy such that a minimum in the free energy is obtained when the terms represented by curves 1, 2, and 3 are summed. The concentration of defects corresponding to the minimum in the free energy curve, N_i, is sometimes referred to as the intrinsic defect concentration or the intrinsic disorder. N_i is dependent upon the temperature and the particular values of ΔH_D and ΔS_v, which are different for each type of defect in a crystal.

Although the change in vibrational entropy is assumed to be positive in Fig. 1, its sign depends upon the type of defect under consideration. However, even in cases where it is negative, the change in configurational entropy is usually sufficient to give rise to a minimum in G.

1.2. Law of Mass Action and Point Defects

Consider a chemical reaction

$$bB + cC \rightleftharpoons dD + eE$$

where B, C, D, and E are chemical species and b, c, d, and e are the corresponding numbers of moles of each species. Then, according to the law of mass action,

$$(a_D{}^d)(a_E{}^e)/(a_B{}^b)(a_C{}^c) = K \tag{2}$$

where a_B is the chemical activity of the B species, and so on, and K is an equilibrium constant. Equation (2) can be written more generally for any number of species as

$$\prod_i a_i^{m_i} = K \tag{3}$$

where m_i, the number of moles of the ith species, is negative for chemical species appearing on the left-hand side of the chemical equation.

The law of mass action may be applied to reactions involving the formation and interaction of lattice defects which meet the criteria discussed in Section 1.1. Equation (3) is applicable when a crystal is in equilibrium with its ambient. At low concentrations of defects, concentrations may be used in place of activities in Eq. (3). In such cases, brackets usually are used to denote concentration, so that Eq. (2) may be written

$$[D]^d[E]^e/[B]^b[C]^c = K$$

The simplest types of native atomic defects are point defects. These may be vacancies (unoccupied normal sites in the lattice), interstitial defects (atoms situated on normally unoccupied interstices), and substitutional defects (one type of atom substituting for another on a normal site). All these defects may be treated as chemical entities in the application of the law of mass action. For example, the formation of Frenkel defects (vacancy plus interstitial) in a lattice of A atoms may be written as follows:

$$A_A \rightleftharpoons A_I + V_A \tag{4}$$

where A_A represents an A atom on a normal site, A_I is an A atom on an interstitial site, and V_A is an A-atom vacancy. The application of Eq. (3) to this reaction can be written

$$(a_I)(a_V)/(a_A) \simeq (a_I)(a_V) = K_F \tag{5}$$

where a_I, a_V, and a_A are the activities of interstitial A atoms, A vacancies, and A atoms on normal sites, respectively, and K_F is the equilibrium constant for the formation of Frenkel defects. For small concentrations of defects, the activity of A atoms on normal sites may be taken as unity.

At higher defect concentrations, a_A may differ from unity to a significant degree. In addition, it is necessary to treat lattice sites as chemical entities. In writing the symbol A_A, the A atom is implicitly assumed to be associated with a normal site. Therefore, formation of a vacancy by removal of an A atom from a crystal by the reaction

$$A_A \rightleftharpoons A(\text{gas}) + V_A$$

may be viewed as dissociation of an A atom and a normal site. Similarly, interstices also must be treated as chemical entities, and strictly speaking, Eqs. (4) and (5) should be written

$$A_A + V_I \rightleftharpoons A_I + V_A, \qquad \text{and} \qquad (a_I)(a_V)/(a_A)(a_{VI}) = K_F \tag{6}$$

where V_I represents an unoccupied interstice. As in the case of a_A, a_{VI} is taken as unity at low defect concentrations.

1.3. Electronic Defects

Electronic defects, such as electrons and holes, in a crystal also may be treated as chemical entities, and thus they also obey the law of mass action. Just as there are native atomic defects in a crystal, thermal energy will cause electrons to dissociate from a chemical bond leaving behind positive holes. In terms of the energy band model, thermal excitation will give rise to electrons in the conduction band and holes in the valence band. The reaction may be written as follows:

$$0 \rightleftharpoons e^- + h^+$$

and from Eq. (3),

$$np = K_i \tag{7}$$

where n is the electron concentration, p is the hole concentration, and K_i is the intrinsic ionization constant.

Point defects may ionize to create electronic defects. In general, metal interstitials, nonmetal vacancies, and substitutional atoms having more electrons than the normal atom on that site will act as donors and contribute electrons to the crystal, while nonmetal interstitials, metal vacancies, and substitutional atoms having less electrons than the normal atom will act as acceptors and supply positive holes to the crystal. The detailed reactions for these ionization processes are discussed in later sections.

1.4. Energetics of Defect Formation

The relationship between standard free energy and equilibrium constant for a chemical reaction is as follows:

$$\Delta G^\circ = -RT \ln K = \Delta H^\circ - T\Delta S^\circ$$

Consequently, the equilibrium constant can be written

$$K = [\exp(\Delta S^\circ/R)] \exp(-\Delta H^\circ/RT) \tag{8}$$

where ΔH° and ΔS° are the standard enthalpy and entropy of reaction, respectively. For a defect reaction such as Eq. (4) or Eq. (6), the equilibrium constant can be written

$$K_F = [\exp(\Delta S_F/R)] \exp(-\Delta H_F/RT)$$

where ΔH_F is the enthalpy of formation of Frenkel defects, and ΔS_F is the change in vibrational entropy of the crystal caused by removal of A atoms from normal sites and their placement into interstices.

For the case of Eq. (7), we may write for the equilibrium constant

$$K_i = [\exp(\Delta S_i^\circ/R)] \exp(-\Delta E_i^\circ/RT) \tag{9}$$

Since the PV term is negligible in the intrinsic ionization process, E_i° may be substituted for the enthalpy. E_i° is the energy necessary to remove an electron from the valence band and place it into the conduction band. Consequently, it is the width of the band gap.

2. Native Defects

2.1. Defect Equilibria in Elemental Crystals

In elemental crystals the native defects are vacancies and interstitials. Vacancies may be formed by removing atoms from normal sites and placing them on the surface according to the reaction

$$A_A \rightleftharpoons A(\text{surface}) + V_A \tag{10}$$

However, the addition of atoms to the surface does not increase the total number of surface atoms, since atoms previously on the surface are now interior atoms. Therefore the formation of vacancies in this manner increases the total number of sites. Equation (10) can be written

$$0 \rightleftharpoons V_{\mathrm{A}} \tag{10a}$$

Interstitial defects can be formed by placing surface atoms into interstices:

$$\mathrm{A(surface)} + V_I \rightleftharpoons \mathrm{A}_I \tag{11}$$

The removal of an atom from the surface essentially leaves the total number of surface atoms constant because interior atoms then become surface atoms. However, the total number of atoms on normal sites decreases and Eq. (11) should be written

$$\mathrm{A_A} + V_I \rightleftharpoons \mathrm{A}_I \tag{12}$$

The total number of sites also decreases with formation of interstitial defects in this manner, and at high concentrations of defects, one must consider the decrease in the number of interstices due to the decrease in the total number of sites. If α is the number of interstices per normal lattice site,† then Eq. (12) should be written

$$\mathrm{A_A} + (1 + \alpha)V_I \rightleftharpoons \mathrm{A}_I \tag{12a}$$

Accordingly, Eq. (10a) also should be rewritten:

$$0 \rightleftharpoons V_{\mathrm{A}} + \alpha V_I \tag{10b}$$

Vacancies and interstitials also may be formed without changing the total number of sites by forming Frenkel defects [see Eq. (4)].

If the mass action expressions are written for each defect formation reaction (substituting concentrations for activities),

$$[V_{\mathrm{A}}][V_I]^{\alpha} = K_V \tag{13}$$

$$[A_I]/[\mathrm{A_A}][V_I]^{1+\alpha} = K_I \tag{14}$$

$$[\mathrm{A}_I][V_{\mathrm{A}}]/[V_I][\mathrm{A_A}] = K_{\mathrm{F}} \tag{15}$$

it can be seen that

$$K_{\mathrm{F}} = K_V K_I \tag{16}$$

† The value of α depends upon the crystal structure. For example, for a face-centered cubic lattice, α is one for octahedral interstices and two for tetrahedral interstices.

The values of the equilibrium constant may be computed from Eq. (8). Burton[1] has shown that vacancy formation entropies in cubic metals range from 3.6 to 5.2 e.u. For the case of copper, a value of 3.8 e.u. was computed. The energy of vacancy formation has been estimated[2] to be about 23 kcal/mole. Thus at 1000°K, K_V for copper is found to be 6×10^{-5} from Eq. (8). Assuming $[V_I] \simeq 1$, this value would also correspond to the vacancy concentration according to Eq. (13).

Because of close packing in fcc metals, the energy of interstitial formation in copper should be much higher than that of the vacancy; Johnson and Wilson[3] calculated a value of 63 kcal/mole. Using this value and 1.6 e.u.[2] for the entropy of interstitial formation, K_I is found to be 4×10^{-14} at 1000°K. For $[A_A] \simeq [V_I] \simeq 1$, this value also corresponds to the interstitial concentration. Therefore, at equilibrium at 1000°K, the concentration of interstitials in copper is 10^{-9} that of vacancies. The equilibrium constant for Frenkel defect formation K_F is found to be 2×10^{-19} at 1000°K.

It is interesting to compare the results for a metal with those obtained for a covalent element such as silicon. The energy of vacancy formation,[4] 49 kcal/mole, is higher than for copper, as would be expected for covalent bonding. Using 4.4 e.u.[5] for the entropy of vacancy formation, a value of 4×10^{-11} is obtained for K_V at 1000°K. Thus, the equilibrium concentration of vacancies in a covalent material such as silicon is much lower than in a metal such as copper. Because the size of the interstices in the diamondlike structure of silicon is the same as that of normal lattice sites, it would be expected that in this case the energy of interstitial formation is comparable to that of vacancy formation. The value calculated by Bennemann,[4] 25 kcal/mole, is actually lower than the vacancy formation energy. Therefore, for silicon the equilibrium concentration of interstitials is greater than that of vacancies.

2.2. Defect Equilibria in Binary Compounds

In binary compounds the sublattice of both components may contain the point defects discussed in Section 1.2. In addition, interstitials of either component may be present. Any one of these point defects will give rise to an excess or deficiency of one of the components. Therefore, in order to maintain the stoichiometric composition, there must be at least two types of defects, having opposite effects on the stoichiometry, present in equivalent concentrations.

Because of differences in defect formation energies, the concentrations of all other defects usually are negligible compared to the two predominant types.

Let us consider a generalized binary compound MX_s, where M represents the more electropositive component, X the more electronegative component, and s the stoichiometric ratio of X to M. The nine possible ways that two types of defects may exist in the MX_s lattice are listed in Table 1. These conjugate pairs of defects are called

TABLE 1

Intrinsic Defects in Binary Compounds

Type	Name	Example
$V_M + V_X$	Schottky	TiO
$V_M + M_I$	Frenkel	ZnO
$X_I + V_X$	Anti-Frenkel	LaH_2
$X_M + M_X$	Antistructure	AuZn
$V_M + M_X$	*V–S*	NiAl
$X_M + V_X$	*S–V*	—
$X_I + M_I$	Interstitial	—
$X_I + M_X$	*I–S*	—
$X_M + M_I$	*S–I*	—

intrinsic defects in binary compounds. Because of strong Coulomb repulsions, substitutional defects (M_X or X_M) do not occur to any measurable extent in ionic crystals. However, they may occur in metallic or covalent compounds. Only the first five defects listed in the table have been observed experimentally and an example of a compound containing each type of defect is given.

2.2.1. Defect Formation Equations

Equations for the formation of each of these intrinsic defects may be formulated. For Schottky defects,

$$0 \rightleftharpoons V_M + sV_X + \alpha V_I; \qquad K_S = [V_M][V_X]^s[V_I]^\alpha \tag{17}$$

The α interstices (in this case α is the number of interstices per M site) are generated because the total number of sites is increased on formation of Schottky defects. However, if no interstitial defects are present, the concentration of V_I with respect to metal atom sites is merely α, and Eq. (17) can be written

$$K_S' = [V_M][V_X]^s \tag{17a}$$

For Frenkel and anti-Frenkel defects we can write, respectively,

$$M_M + V_I \rightleftharpoons V_M + M_I; \qquad K_F = [V_M][M_I]/[M_M][V_I] \tag{18}$$

$$X_X + V_I \rightleftharpoons X_I + V_X; \qquad K_{AF} = [X_I][V_X]/[X_X][V_I] \tag{19}$$

In these cases, the total number of sites remains constant, as it also does in the formation of antistructure defects:

$$M_M + X_X \rightleftharpoons X_M + M_X; \qquad K_{as} = [X_M][M_X]/[M_M][X_X] \tag{20}$$

The situation for the V–S defect is somewhat more complicated:

$$\begin{aligned} sM_M &\rightleftharpoons sM_X + (1 + s)V_M + \alpha V_I \\ K_{VS} &= [M_X]^s[V_M]^{(1+s)}[V_I]^\alpha/[M_M]^s \end{aligned} \tag{21}$$

The total number of sites increases, but if no interstitial defects are present, $V_I = \alpha$, and

$$K'_{VS} = [M_X]^s[V_M]^{(1+s)}/[M_M]^s \tag{21a}$$

For every s substitutional defects introduced into the lattice, $(s + 1)$ M vacancies must be formed to maintain the stoichiometric composition. The reason for this may become clear by considering, for example, the intermetallic compound NiAl, for which $s = 1$. According to Eq. (21), two Ni vacancies should be produced for each Ni substitutional (Ni_{Al}) formed. When a Ni substitutional defect is present, the lattice contains two extra Ni atoms: One is the Ni on the Al site, and the other is due to the fact that the Al atom normally occupying that site is gone. Consequently, it takes two Ni vacancies to bring the lattice back to the stoichiometric composition.

Although the last four intrinsic defects listed in Table 1 never have been observed, for the sake of completeness it is desirable to show their formation reactions.

For S–V defects

$$X_X \rightleftharpoons X_M + (s + 1)V_X + \alpha V_I$$

With no interstitials

$$K_{SV} = [X_M][V_X]^{(s+1)}/[X_X]$$

For interstitial defects the total number of sites would decrease:

$$M_M + sX_X + (\alpha + s + 1)V_I \rightleftharpoons M_I + sX_I,$$

$$K_I = [M_I][X_I]^s/[M_M][X_X]^s[V_I]^{\alpha+s+1}$$

For I–S defects

$$(s+1)X_X + M_M + (s+\alpha+1)V_I \rightleftharpoons M_X + (s+1)X_I$$
$$K_{IS} = [M_X][X_I]^{(s+1)}/[X_X]^{s+1}[M_M][V_I]^{(s+\alpha+1)}$$

For S–I defects

$$(s+1)M_M + sX_X + (s+\alpha+1)V_I \rightleftharpoons (s+1)M_I + sX_M$$
$$K_{SI} = [M_I]^{(s+1)}[X_M]^s/[M_M]^{(s+1)}[X_X]^s[V_I]^{(s+\alpha+1)}$$

2.2.2. Intrinsic Defect Concentrations

At the stoichiometric composition, the defects in each conjugate pair are present in equivalent concentrations. These concentrations may be referred to as the intrinsic defect concentrations N_i and they can be calculated from the appropriate equilibrium constants. For the case of Schottky defects, if there are N_i M vacancies in the lattice, then there are sN_i X vacancies at the stoichiometric composition. Expressing concentration as the ratio of defects to M-atom sites, at the stoichiometric composition Eq. (17a) may be written

$$K_S' = N_i(sN_i)^s \tag{22}$$

If the constant s is incorporated into the equilibrium constant, we have

$$K_S'' = N_i^{(s+1)}, \qquad \text{where} \quad K_S'' = K_S'/s^s \tag{23}$$

Similar relationships are obtained for the other intrinsic defects. For Frenkel defects

$$K_F = N_i^2/(1-N_i)(\alpha - N_i) \tag{24}$$

For anti-Frenkel defects

$$K_{aF} = N_i^2/(s-N_i)(\alpha - N_i) \tag{25}$$

For antistructure defects

$$K_{as} = N_i^2/(1-N_i)(s-N_i) \tag{26}$$

For V–S defects

$$K_{VS}'' = N_i^{2s+1}/(s - N_i - sN_i)^s \tag{27}$$

where $K_{VS}'' = K_{VS}'/s^s(s+1)^{s+1}$. For S–V defects

$$K_{SV}' = N_i^{s+2}/(s - N_i - sN_i)$$

where $K'_{SV} = K_{VS}/(s + 1)^{s+1}$. For interstitial defects

$$K_I' = N_i^{s+1}/(\alpha - N_i - sN_i)^{\alpha+s+1}$$

where $K_I' = K_I/s^s$. For I–S defects

$$K'_{IS} = N_i^{s+2}/(s - N_i)^{s+1}(\alpha - sN_i - N_i)^{s+\alpha+1}$$

where $K'_{IS} = K_{IS}/(s + 1)^{s+1}$. For S–I defects

$$K'_{SI} = N_i^{2s+1}/(1 - N_i)^{s+1}(s\alpha - sN_i - N_i)^{s+\alpha+1}$$

where $K'_{SI} = K_{SI}/(s + 1)^{s+1}s^{\alpha}$.

From Eq. (8) it can be seen that the intrinsic defect concentrations will increase with increasing temperature and they will be low for high enthalpies of defect formation. The application of these equations to some specific systems would be illustrative. From thermodynamic measurements[6] on cerium hydride CeH_2, it was deduced[7] that the intrinsic defects were anti-Frenkel defects ($V_H + V_I$) and a value of 3.0×10^{-4} was computed for K_{aF} at 600°C. This compound has the fluorite structure which contains one octahedral interstice per Ce atom. Therefore $\alpha = 1$. Since the compound is a dihydride, $s = 2$. Equation (25) then can be written

$$K_{aF} = N_i^2/(2 - N_i)(1 - N_i) = 3.0 \times 10^{-4}$$

giving a value of 2.4×10^{-2} for N_i at 600°C. This means that one out of every 42 interstices is occupied by a hydrogen atom, and one out of every 84 hydrogen atom positions is vacant.

Density and X-ray measurements[8] on TiO revealed that the intrinsic defects were Schottky defects ($V_{Ti} + V_O$) and the intrinsic defect concentration in a sample annealed at 1300°C was 0.140. TiO has the NaCl-type structure. Using Eq. (23) with $s = 1$, K''_S is calculated to be 0.020. Since there are two tetrahedral interstices per metal atom, $\alpha = 2$, and $K_S' = 0.080$.

2.3. Nonstoichiometry—Equilibria with External Phases

If the concentrations of each type of point defect in a conjugate pair are not equivalent, the crystal becomes nonstoichiometric. If there is an excess of M vacancies, X interstitials, or X substitutionals, the compound will have a surplus of X (or deficiency of M) over the ideal stoichiometric composition. This condition will be referred to as a *positive* deviation from stoichiometry. Conversely, for a *negative* deviation from stoichiometry, X vacancies, M interstitials, or M substitutionals must be in excess. According to Eqs. (17)–(21), an increase in the concentration of one of the point defects in a conjugate pair will lead to a corresponding decrease in the other.

The formula for a nonstoichiometric compound may be written $MX_{s+\delta}$, where δ, the degree of deviation from stoichiometry, may be positive or negative. In terms of defect concentrations, δ may be expressed as follows:

$$\delta = \frac{[X_I] - [V_X] + s([V_M] - [M_I]) + (s + 1)([X_M] - [M_X])}{1 + [M_I] + [M_X] - [V_M] - [X_M]} \tag{28}$$

The concentrations of point defects, and therefore the direction and degree of nonstoichiometry, are determined by the thermodynamic activities of external phases in equilibrium with the nonstoichiometric compound. For example, the concentration of X vacancies may be controlled by maintaining the crystal in equilibrium with another phase containing X at a definite activity. The external phase may be X_2 gas (e.g., O_2). The equation for formation of X vacancies then may be written

$$X_X \rightleftharpoons \tfrac{1}{2}X_2(\text{gas}) + V_X \tag{29}$$

The pressure of X_2 gas will then determine the activity and concentration of vacancies.

The concentration of X vacancies may also be controlled by the activity of the M component in an external phase. For example,

$$M(\text{external phase}) \rightleftharpoons M_M + sV_X + \alpha V_I \tag{29a}$$

In this case the total number of sites in the crystal increases.

In most experimental work on nonstoichiometric compounds the extent of nonstoichiometry usually is determined by the pressure of the nonmetal gas in equilibrium with the crystal. Therefore, the remaining discussion in this section will be in terms of X_2 gas and the activity of the X component. However, it should be kept in mind that equivalent relationships can be obtained for the M component, and that the external phase need not be X_2 gas but may be any compound containing the X component.

The equilibrium constant for Eq. (29) is

$$K_{VX} = P_{X_2}^{1/2}[V_X]/[X_X] = P_{X_2}^{1/2}[V_X]/(s - [V_X]) \tag{30}$$

At high defect concentrations and, therefore, large negative deviations from stoichiometry the concentration of the conjugate defect (either

V_M, X_I, or X_M) will be negligible. Thus, according to Eq. (28), we have

$$\delta = -[V_X]$$

since the concentrations of all other defects are assumed to be insignificant. Consequently, Eq. (30) can be written

$$P_{X_2}^{1/2} = (s + \delta)K_{VX}/(-\delta) \tag{31}$$

The equation for the formation of M vacancies by adding X to the crystal is

$$(s/2)X_2(\text{gas}) \rightleftharpoons sX_X + V_M + \alpha V_I, \qquad K_{VM} = [X_X]^s[V_M][V_I]^\alpha/P_{X_2}^{s/2} \tag{32}$$

At large deviations from stoichiometry, it can be assumed that M vacancies are the only defects present in significant quantity. Then $[X_X] \simeq 1$, $[V_I] = \alpha$, and, according to Eq. (28),

$$\delta = s[V_M]/(1 - [V_M])$$

Equation (32) then may be rewritten as

$$P_{X_2}^{1/2} = [\delta/K'_{VM}(s + \delta)]^{1/s} \tag{33}$$

Relationships between δ and P_{X_2} may be obtained in a similar manner for the other point defects. For X interstitials

$$\tfrac{1}{2}X_2 + V_I \rightleftharpoons X_I \tag{34}$$

$$K_{XI} = [X_I]/(\alpha - [X_I])P_{X_2}^{1/2} \tag{35}$$

$$P_{X_2}^{1/2} = (1/K_{XI})[\delta/(\alpha - \delta)] \tag{36}$$

For M interstitials

$$M_M + sX_X + (1 + \alpha)V_I \rightleftharpoons M_I + (s/2)X_2 \tag{37}$$

$$K_{MI} = [M_I]P_{X_2}^{s/2}/(\alpha - [M_I])^{1+\alpha} \tag{38}$$

$$P_{X_2}^{1/2} = [K_{MI}(\alpha s + \alpha\delta + \delta)^{1+\alpha}/(-\delta)(s + \delta)^\alpha]^{1/s} \tag{39}$$

For X substitutionals

$$\tfrac{1}{2}(s + 1)X_2(\text{gas}) \rightleftharpoons sX_X + X_M + \alpha V_I \tag{40}$$

$$K_{XM} = [X_M]\alpha^\alpha/P_{X_2}^{(s+1)/2} \tag{41}$$

$$P_{X_2}^{1/2} = (1/K'_{XM})[\delta/(1 + s + \delta)]^{1/(s+1)} \tag{42}$$

For M substitutionals

$$(s + 1)X_X + M_M + \alpha V_I \rightleftharpoons \tfrac{1}{2}(s + 1)X_2(\text{gas}) + M_X \tag{43}$$

$$K_{MX} = P_{X_2}^{(s+1)/2}[M_X]/(1 - [M_X])^{s+1}\alpha^\alpha \tag{44}$$

$$P_{X_2}^{1/2} = \frac{(K'_{MX})^{1/(s+1)}(s+1)(s+\delta)}{[(-\delta)(s+1+\delta)^s]^{1/(s+1)}} \tag{45}$$

2.3.1. Effect of Intrinsic Defect Concentration on Nonstoichiometry

Equations (17)–(21) can be obtained by appropriate combinations of Eqs. (29)–(45). For example, consider the case of anti-Frenkel defects. Comparison of Eqs. (19), (25), (30), and (35) shows that

$$K_{aF} = K_{VX}K_{XI} = N_i^2/(s - N_i)(\alpha - N_i) \tag{46}$$

When both X vacancies and X interstitials coexist in the lattice, the deviation from stoichiometry δ becomes [from Eq. (28)]

$$\delta = [X_I] - [V_X]$$

From Eqs. (30) and (35), we can write

$$\delta = \frac{\alpha P_{X_2}^{1/2} K_{XI}}{P_{X_2}K_{XI} + 1} - \frac{sK_{VX}}{K_{VX} + P_{X_2}^{1/2}} \tag{47}$$

Let us consider a one-to-one compound with one interstice per metal atom so that $s = \alpha = 1$. Rearranging terms in Eq. (47) yields

$$K_{XI}(1 - \delta)P_{X_2} - \delta(K_{XI}K_{VX} + 1)P_{X_2}^{1/2} - K_{VX}(1 + \delta) = 0$$

Using Eq. (46) and the fact that $N_i \ll 1$, we obtain

$$N_i^2(1 - \delta)P_{X_2} - \delta K_{VX}P_{X_2}^{1/2} - K_{VX}^2(1 + \delta) = 0$$

Solving for P_{X_2} yields

$$P_{X_2} = K_{VX}^2\{\delta^2 + 2N_i^2(1 - \delta^2) \pm \delta[\delta^2 + 4N_i^2(1 - \delta^2)]^{1/2}\}/2N_i^4(1 - \delta)^2$$

At the stoichiometric composition, $\delta = 0$ and

$$P_{X_2}^\circ = K_{VX}^2/N_i^2$$

where $P_{X_2}^\circ$ is the pressure of X_2 gas in equilibrium with the M crystal at the stoichiometric composition. Therefore

$$P_{X_2}/P_{X_2}^\circ = \{\delta^2 + 2N_i^2(1 - \delta^2) \pm \delta[\delta^2 + 4N_i^2(1 - \delta^2)]^{1/2}\}/2N_i^2(1 - \delta)^2 \tag{48}$$

Similar relations may be derived for the other conjugate pairs of intrinsic defects.

The ratio of the equilibrium X_2 pressure to that in equilibrium with the stoichiometric composition as computed from Eq. (48) is

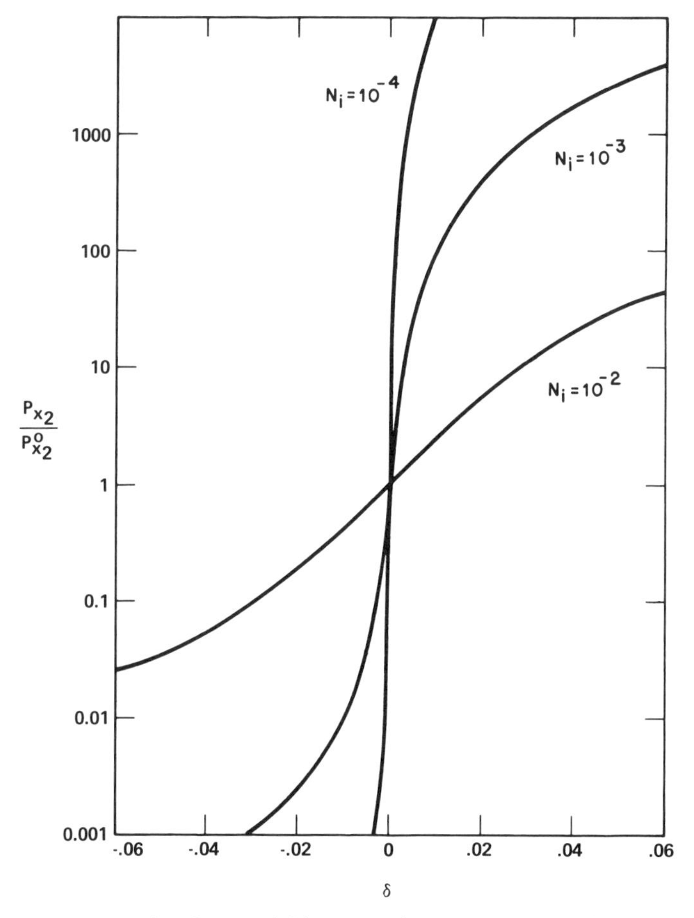

Fig. 2. Deviation from stoichiometry of a compound as a function of pressure of one of the components and intrinsic defect concentration.

shown as a function of δ for various values of N_i in Figure 2. It can be seen that the tendency of a compound to deviate from stoichiometry is greater for those compounds which have high intrinsic defect concentrations. For example, for a hypothetical MX compound containing anti-Frenkel defects, in order to obtain a 0.7% deviation from the stoichiometric composition ($\delta = 0.007$), it is necessary to increase or decrease) the X_2 pressure by a factor of 5000 when the intrinsic defect concentration is 10^{-4}. On the other hand, only a twofold increase (or decrease) is required to cause the same deviation from stoichiometry if $N_i = 0.01$.

2.3.2. Characterization of Defects from Pressure (or Activity) Measurements

If a binary compound MX_s exhibits negative deviations from the stoichiometric composition, the predominant defects may be either X vacancies, M interstitials, or M substitutionals. Comparison of Eqs. (31), (39), and (45) shows that for each defect, the equilibrium pressure is equal to a constant times some function of δ. However, in each case the function of δ is different. Therefore, the nature of the defect responsible for the negative deviation from stoichiometry may be deduced by comparing experimental data of pressure (or activity) as a function of δ with each of Eqs. (31), (39) and (45) at large deviations from stoichiometry. This may be done by calculating the equilibrium constant for each experimental point assuming each type of defect. The equilibrium constant should not vary with composition (or δ) for the correct defect.

As an illustration, these equations will be applied to uranium hydride UH_3, which is normally deficient in hydrogen. In this case $s = 3$ and $\alpha = 1$. Table 2 lists hydrogen pressures[9] in equilibrium

TABLE 2

Equilibrium Constants for Formation of Point Defects in Uranium Hydride

$-\delta$	$P_{H_2}^{1/2}$, atm$^{1/2}$	K_{VH}	K_{UI}	K'_{UH}
0.045	5.59	0.26	2.7	0.61
0.046	5.39	0.26	2.5	0.59
0.056	4.24	0.25	1.5	0.49
0.057	4.18	0.25	1.5	0.49
0.062	3.82	0.25	1.2	0.45
0.064	3.73	0.26	1.2	0.45
0.069	3.39	0.25	1.0	0.41
0.073	3.24	0.26	0.9	0.40
0.081	3.03	0.27	0.8	0.39

with $UH_{3+\delta}$ for negative values of δ, and the equilibrium constants calculated therefrom using Eqs. (31), (39), and (45). It can be seen that K_{VH} remains quite constant with composition, while K_{UI} and K'_{UH} both decrease monotonically with increasing δ. Clearly then, the predominant defects in hydrogen-deficient uranium hydride are hydrogen vacancies.

Compounds exhibiting large positive deviations from stoichiometry may be treated in the same manner. In this case, the possible defects are M vacancies, X interstitials, and X substitutionals. Here again, a different functional relationship between activity and δ is obtained for each type of defect as can be seen from Eqs. (33), (36), and (42). An example of the use of these equations to deduce the nature of the defects is given for the intermetallic compound AuZn. From Zn activity measurements[10] as a function of composition, equilibrium constants were calculated[11] for each point defect. AuZn has the CsCl-type structure where $\alpha = 6$ and $s = 1$. The results are shown in Table 3. It can be seen that $K'_{V\mathrm{Au}}$ and $K_{\mathrm{Zn}I}$ change with composition

TABLE 3

Equilibrium Constants for Formation of Point Defects in AuZn

δ	a_{Zn}	$K'_{V\mathrm{Au}}$	$K_{\mathrm{Zn}I}$	$K'_{\mathrm{Zn(Au)}}$
0.0188	0.0197	0.94	0.16	24
0.0454	0.0294	1.5	0.26	26
0.0636	0.0358	1.7	0.30	24
0.0842	0.0409	1.9	0.35	24
0.1057	0.0458	2.1	0.39	24
0.1810	0.0641	2.4	0.49	20

by factors of two and three, respectively, while $K'_{\mathrm{Zn(Au)}}$ remains constant to within 8 %. Apparently, then, the predominant defects are Zn substitutionals. This has been corroborated by density measurements.[10]

2.4. Ionization of Defects

As mentioned in Section 1.3, point defects in a crystal may ionize. For example, in the hypothetical compound MX, when X atoms are removed to form X vacancies, the electrons which were associated with the X ion in the crystal (or with the bond in covalent compounds) remain behind. At low temperatures these electrons are associated with the vacancy, but at high temperatures they become dissociated from the vacancy (ionize) and are free to move through the crystal. The equation for ionization of an X vacancy is written

$$V_{\mathrm{X}} \rightleftharpoons V_{\mathrm{X}}^{+} + e^{-} \tag{49}$$

In a similar manner, metal vacancies will ionize to form holes leaving behind negatively charged vacancies:

$$V_M \rightleftharpoons V_M^- + h^+ \tag{50}$$

If the charge on the ion which normally occupies the vacant site is greater than unity, the vacancy may become ionized to a correspondingly higher charge. For example, in cobalt monoxide singly charged Co vacancies may form:[12]

$$\tfrac{1}{2}O_2 \rightleftharpoons O_O + V_{Co}^- + h^+ + \alpha V_I \tag{51}$$

At oxygen pressures below 10^{-6} atm, however, the cobalt vacancies become doubly ionized:

$$V_{Co}^- \rightleftharpoons V_{Co}^{2-} + h^+ \tag{52}$$

Equilibrium constants for the ionization of vacancies may be generally written as follows:

$$K_{49} = [V_X^{z+}]n^z/[V_X] \qquad \text{and} \qquad K_{50} = [V_M^{z-}]p^z/[V_M]$$

where n and p are the electron and hole concentrations, respectively, and z represents the degree of ionization or charge on the vacancy.

Similar equations may be written for the ionization of the other types of point defects:

$$M_I \rightleftharpoons M_I^{z+} + ze^-; \qquad K_{53} = [M_I^{z+}]n^z/[M_I] \tag{53}$$

$$X_I \rightleftharpoons X_I^{z-} + zh^+; \qquad K_{54} = [X_I^{z-}]p^z/[X_I] \tag{54}$$

For vacancies and interstitials, defects leading to positive deviations from stoichiometry (excess X) ionize to form holes, while defects corresponding to negative deviations from stoichiometry ionize to form electrons. The opposite is usually true for substitutional defects. Because of strong Coulomb repulsions, substitutional defects do not occur to any significant extent in ionic compounds. In covalent compounds, when the more electronegative element X replaces the more electropositive component, the excess electrons are easily ionized:

$$X_M \rightleftharpoons X_M^{z+} + ze^-; \qquad K_{55} = [X_M^{z+}]n^z/[X_M] \tag{55}$$

Conversely, a more electropositive element M replacing a more electronegative component X will have a deficiency in electrons and will form holes:

$$M_X \rightleftharpoons M_X^{z-} + zh^+; \qquad K_{56} = [M_X^{z-}]p^z/[M_X] \qquad (56)$$

An example of such substitutional defects is seen in bismuth telluride[13] Bi_2Te_3. Excess tellurium is due to Te substitutional defects in the lattice which ionize to form electrons:

$$5Te(\text{Te-rich liq.}) \rightleftharpoons 3Te_{Te} + 2Te_{Bi}^{+} + 2e^- + 2\alpha V_I$$

while excess Bi causes the formation of holes:

$$5Bi(\text{Bi-rich liq.}) \rightleftharpoons 2Bi_{Bi} + 3Bi_{Te}^{-} + 3h^+ + 2\alpha V_I$$

2.4.1. Pressure Dependence of Defect Concentration

The degree of ionization of defects frequently can be determined by measuring the defect concentration as a function of pressure of the more volatile component and using the mass action law. For example, for the case of cobalt vacancies in CoO, the equilibrium constant for Eq. (21) is

$$K_{51} = [O_O][V_{Co}^-][V_I]^\alpha p/P_{O_2}^{1/2} \qquad (57)$$

Since $[h^+] = [V_{Co}^-]$, $V_I = \alpha$, a constant, and at low defect concentrations, $[O_O] \simeq 1$, Eq. (57) can be rewritten:

$$[V_{Co}^-] = (K'_{51})^{1/2} P_{O_2}^{1/4} \qquad (58)$$

On the other hand, if doubly ionized vacancies are predominant, the vacancy formation equation is

$$\tfrac{1}{2}O_2 \rightleftharpoons O_O + V_{Co}^{2-} + 2h^+ + \alpha V_I \qquad (59)$$

$$K_{59} = [O_O][V_{Co}^{2-}][V_I]^\alpha p^2/P_{O_2}^{1/2} \qquad (60)$$

$p = 2[V_{Co}^{2-}]$, and therefore

$$[V_{Co}^{2-}] = (K'_{59}/4)^{1/3} P_{O_2}^{1/6}$$

Thus, if the vacancy concentration is measured as a function of pressure by either thermogravimetric or solid-state coulometric titration techniques and $\ln[V_{Co}^{2-}]$ is plotted against $\ln P_{O_2}$, a straight line with a slope of $\frac{1}{4}$ should be obtained if the vacancies are singly ionized, while doubly ionized vacancies should yield a slope of $\frac{1}{6}$.

2.4.2. Electrical Conductivity Studies

The degree of ionization also may be determined from the variation of electrical conductivity with pressure at low defect concentrations. If we substitute p for $[V_{Co}^-]$ in Eq. (57) and $\frac{1}{2}p$ for $[V_{Co}^{2-}]$ in Eq. (60), we obtain

$$p = (K'_{51})^{1/2} P_{O_2}^{1/4} \qquad \text{for singly ionized vacancies}$$

$$p = (2K'_{59})^{1/3} P_{O_2}^{1/6} \qquad \text{for doubly ionized vacancies}$$

If the mobility of holes remains constant with oxygen pressure (as it usually does at low defect concentrations), then the electrical conductivity is proportional to the hole concentration. Consequently, conductivity will be proportional to $P_{O_2}^{1/4}$ for singly ionized vacancies and to $P_{O_2}^{1/6}$ for doubly ionized vacancies. Figure 3 shows the data of Fisher and Tannhauser[12] at three temperatures. It can be seen that at the higher oxygen pressures ($P_{O_2} > 10^{-3}$ atm) the slopes of the curves are $\frac{1}{4}$, indicating singly charged vacancies, while at low pressures ($P_{O_2} < 10^{-5}$ atm) the slopes become $\frac{1}{6}$, indicative of doubly charged vacancies.

At low defect concentrations electrical conductivity studies also may be used to indicate the nature of the defects present. For example, in studying monoclinic zirconia Vest *et al.*[14] found that at higher oxygen pressures the conductivity was proportional to $P_{O_2}^{1/5}$. The equation for the formation of fully ionized zirconium vacancies in ZrO_2 is

$$\begin{gathered} O_2 \rightleftharpoons 2O_O + V_{Zr}^{4-} + 4h^+ + \alpha V_I, \\ K_{61} = [V_{Zr}^{4-}][O_O]^2[V_I]^\alpha p / P_{O_2} \end{gathered} \tag{61}$$

Since $[V_{Zr}^{4-}] = \frac{1}{4}p$ and $[V_I] = \alpha$, a constant, and assuming $[O_O] \simeq 1$, we can write

$$\sigma \propto p = (4K_{61})^{1/5} P_{O_2}^{1/5}$$

If the vacancies were less than fully ionized, the conductivity would be proportional to $P_{O_2}^{1/4}$ or greater. On the other hand, if the defects were oxygen interstitials rather than Zr vacancies, the defect formation equation would be

$$\tfrac{1}{2}O_2 + V_I \rightleftharpoons O_I^{z-} + zh^+, \qquad K_{62} = [O_I^{z-}]p^z / P_{O_2}^{1/2}[V_I]$$

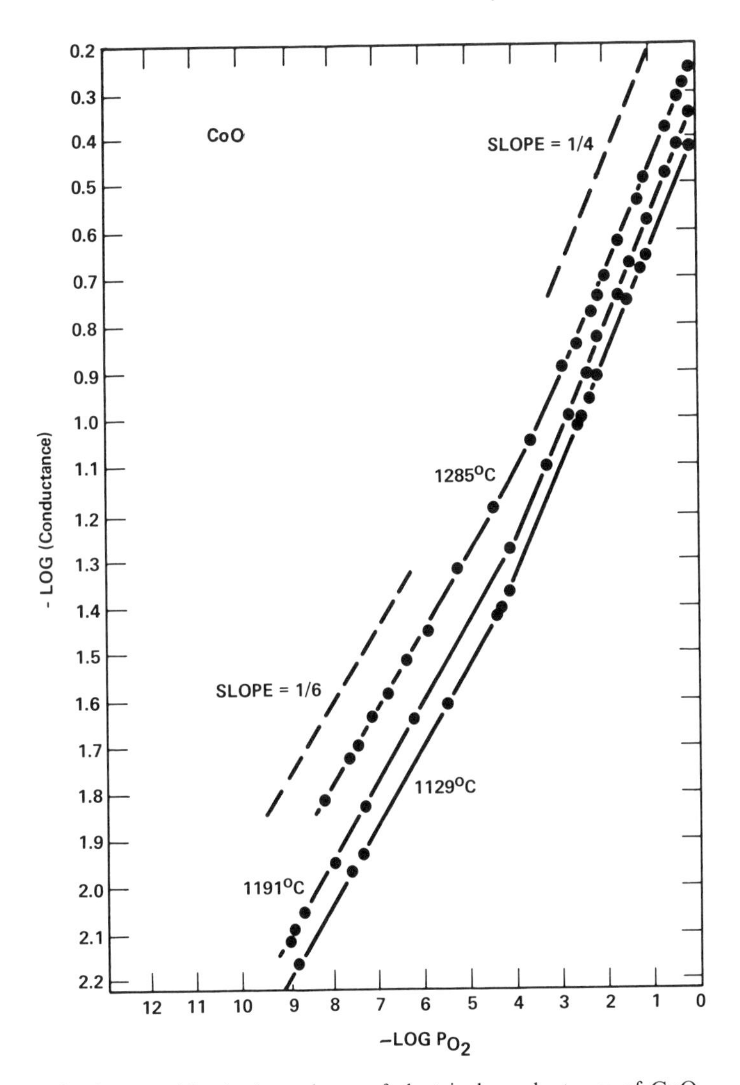

Fig. 3. Logarithmic dependence of electrical conductance of CoO on oxygen pressure (data of Fisher and Tannhauser[12]).

For low defect concentrations, $[V_I] \simeq \alpha$ and

$$\sigma \propto p \propto P_{O_2}^{1/2z}$$

Since z can be either one or two, oxygen interstitials can give only $P_{O_2}^{1/2}$ or $P_{O_2}^{1/4}$ dependence. Therefore, since $P_{O_2}^{1/5}$ dependence was observed experimentally for the conductivity, the authors concluded that the predominant defects must be fully ionized Zr vacancies.

2.4.3. Brouwer's Approximation Method

The defect equilibria equations in Sections 2.1–2.3 all were written assuming no ionization of defects. Such equations are applicable to metallic and possibly some covalent crystals. However, for semiconductors and ionic crystals, equations including ionization and electronic charges must be taken into account. This was done in Sections 2.4.1 and 2.4.2, where some information regarding the compounds under consideration was deduced. However, in order to have a complete description of a system, it is necessary to write equations for all possible processes including defect formation, ionization, and charge compensation. For example, if we consider a crystal containing Schottky defects, the expressions for the formation of intrinsic defects are given by Eqs. (17) and (17a), and for ionization of vacancies by Eqs. (49) and (50). When the crystal is in equilibrium with an external phase (e.g., X_2 gas) Eqs. (29) and (30) are applicable. For a 1:1 compound (i.e., $s = 1$) the equilibrium constants for the reactions can be written

$$\ln K_S = \ln[V_M] + \ln[V_X] \tag{62}$$

$$\ln K_{49} = \ln n + \ln[V_X{}^+] - \ln[V_X] \tag{63}$$

$$\ln K_{50} = \ln p + \ln[V_M{}^-] - \ln[V_M] \tag{64}$$

$$\ln K_{29} = \ln K_{VX} = \tfrac{1}{2}\ln P_{X_2} + \ln[V_X] \tag{65}$$

$$\ln K'_{VM} = \ln[V_M] - \tfrac{1}{2}\ln P_{X_2} \tag{66}$$

In addition, we may write the intrinsic ionization constant from Eq. (7):

$$\ln K_i = \ln n + \ln p \tag{67}$$

The electroneutrality condition (i.e., in any crystal the total negative charge must equal the total positive charge in order to maintain electroneutrality) is

$$n + [V_M{}^-] = p + [V_X{}^+] \tag{68}$$

Equations (62)–(68) completely define the defect concentrations in a crystal containing Schottky defects. When the equilibrium constants are known, there are seven equations and seven unknowns, $[V_M]$, $[V_X]$, $[V_M{}^-]$, $[V_X{}^+]$, n, p, and P_{X_2}. Equations (62)–(67) are linear relations between logarithms of concentrations and logarithms of the equilibrium constants, while Eq. (68) is not. Because of this, the solution of the seven simultaneous equations becomes algebraically complicated and tedious. For cases where there is further ionization of the defects (e.g., $V_M^- \rightleftharpoons V_M^{2-} + h^+$), or where there are more than two types of defects present, additional equations are introduced and the solution becomes even more complicated. Brouwer[15] has proposed a graphical approximation method for solving such relations, which has been further developed by Kröger and Vink.[16] The method entails dividing the range of concentrations into regions such that charge compensation involves only two types of defects and the neutrality condition (68) is thus simplified. For large negative deviations from stoichiometry, the concentrations of V_M^- and holes become negligible with respect to $[V_X^+]$ and n, and Eq. (68) may be approximated by

$$\ln n = \ln[V_X^+] \tag{69}$$

Using this relation, the concentrations of all defects may be expressed in terms of X_2 pressure and the equilibrium constants. Graphical plots of logarithm of concentration against logarithm of X_2 pressure, or more conveniently, against the variable R (where $R = K'_{VM}P_{X_2}^{1/2}$) yields straight lines. The region of large negative deviation from stoichiometry is designated as region I in Figure 4. From Eq. (66) we see that $[V_M] = R$, and from Eq. (62), $[V_X] = K_S/R$. Equations (63) and (69) yield $n = [V_X^+] = \{K_{49}[V_X]\}^{1/2} = (K_{49}K_S/R)^{1/2}$. From Eq. (67), $p = K_i/n = K_i(R/K_{49}K_S)^{1/2}$ and from Eq. (64), $[V_M^-] = K_{50}V_M/p = (RK_{49}K_S)^{1/2}K_{50}/K_i$. Thus, we see that the concentrations of defects $[D]$ can be expressed in powers of R, and the $\ln[D]$ versus $\ln R$ plots in Figure 4 have slopes of $\pm\frac{1}{2}$ or ± 1.

Since the equilibrium constant for the formation of ionized Schottky defects $K_{S(i)}$ is frequently known, rather than K_S', it can be used in the expressions above by substituting $K_{S(i)}K_i/K_{49}K_{50}$ for K_S'. The concentrations of defects as a function of R in region I are summarized in the first column of Table 4.

TABLE 4

Concentrations of Defects in Various Composition Regions of a Crystal Containing Schottky Defects

Defect	Region I	Region II $K_{S(i)} \gg K_i$	Region II′ $K_i \gg K_{S(i)}$	Region III
$[V_M]$	R	R	R	R
$[V_M^-]$	$(RK_{S(i)}K_{50}/K_i)^{1/2}$	$K_{S(i)}^{1/2}$	$K_{50}R/K_i^{1/2}$	$(K_{50}R)^{1/2}$
$[V_X]$	$K_{S(i)}K_i/K_{49}K_{50}R$	$K_{S(i)}K_i/K_{49}K_{50}R$	$K_{S(i)}K_i/K_{49}K_{50}R$	$K_{S(i)}K_i/K_{49}K_{50}R$
$[V_X^+]$	$(K_{S(i)}K_i/K_{50}R)^{1/2}$	$K_{S(i)}^{1/2}$	$K_i^{1/2}K_{S(i)}/K_{50}R$	$K_{S(i)}/(K_{50}R)^{1/2}$
n	$(K_{S(i)}K_i/K_{50}R)^{1/2}$	$K_{S(i)}^{1/2}K_i/K_{50}R$	$K_i^{1/2}$	$K_i/(K_{50}R)^{1/2}$
p	$(K_iK_{50}R/K_{S(i)})^{1/2}$	$K_{50}R/K_{S(i)}^{1/2}$	$K_i^{1/2}$	$(K_{50}R)^{1/2}$

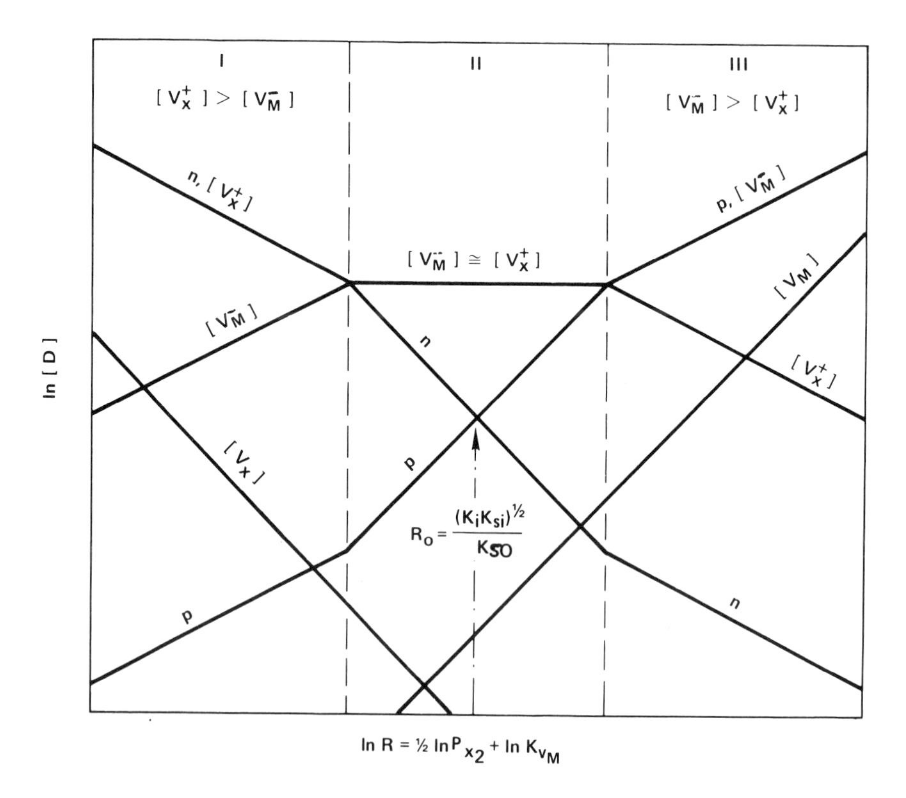

Fig. 4. Defect concentrations as a function of X_2 pressure in the compound MX for $K_{S(i)} > K_i$.

At large positive deviations from the stoichiometric composition, $[V_M^-] \gg n$ and $p \gg [V_X^+]$, and therefore Eq. (68) becomes

$$\ln[V_M^-] = \ln p \tag{70}$$

The expressions for the concentrations of defects derived from Eqs. (62)–(67) and Eq. (70) in this region (region III in Figure 4) are summarized in the last column of Table 4.

The concentrations of defects in the vicinity of the stoichiometric composition (designated region II here) depends upon the value of the intrinsic defect equilibrium constant K_S' (or $K_{S(i)}$) and upon the intrinsic ionization constant K_i. Two possible cases may be considered. In large-band-gap materials such as ionic conductors, $K_{S(i)} > K_i$. Therefore, the concentration of Schottky defects is greater than that of electronic defects and Eq. (68) is approximated by

$$\ln[V_M^-] = \ln[V_X^+] \tag{71}$$

Consequently, as can be seen from the first column of Table 4, the boundary between region I and region II occurs when

$$(RK_{S(i)}K_{50}/K_i)^{1/2} = (K_{S(i)}K_i/K_{50}R)^{1/2}$$

and therefore

$$R_{I,II} = K_i/K_{50} \tag{72}$$

From the last column in Table 4, we obtain

$$(K_{50}R)^{1/2} = K_{S(i)}/(K_{50}R)^{1/2}$$

and therefore

$$R_{II,III} = K_{S(i)}/K_{50} \tag{73}$$

Therefore, the width of region II in Figure 4 is

$$\ln R_{II,III} - \ln R_{I,II} = \ln(K_{S(i)}/K_i) \tag{74}$$

From Eqs. (62)–(67) and Eq. (71), the concentrations given under region II in Table 4 are obtained. The stoichiometric composition is rigorously defined by the condition

$$[V_M] + [V_M^-] = [V_X] + [V_X^+] \tag{75}$$

However, since in region II, $[V_M^-] \simeq [V_X^+] \gg [V_M], [V_X]$, for $K_{S(i)} \gg K_i$, the deviation from stoichiometry is relatively insensitive to R or to large changes in X_2 pressures. In this case, then, the stoichiometric composition is more nearly defined by the point where $n = p$, which occurs at $R_0 = (K_iK_{S(i)})^{1/2}/K_{50}$.

An example of a compound having $K_{S(i)} \gg K_i$ to which this treatment has been applied[17] is KBr. At 600°C, Kröger[17] gives values of 8×10^{-14} for $K_{S(i)}$ and 3×10^{-35} for K_i. From Eq. (74), it can be seen that the bromine pressure (since $R \propto P_{Br_2}^{1/2}$) changes by a factor of 10^{43} over region II in Figure 4. Combination of Eqs. (72), (73), and (75) and the expressions in Table 4 shows that the stoichiometry δ varies by only 10^{-3} over this region. Although the vacancy concentration remains approximately constant, the concentration of holes (or electrons) changes by a factor of about 10^{21} over the same region.

In electronic semiconductors the condition $K_i \gg K_{S(i)}$ usually holds, and the relative concentrations of defects in the region near the stoichiometric composition become different than those shown in Figure 4. When $K_i \gg K_{S(i)}$, then $n, p \gg [V_M^-], [V_X^+]$, and the neutrality condition (68) becomes $n = p$. Using this relation and Eqs. (62)–(67), the expressions under the columns labeled region II′ were obtained.

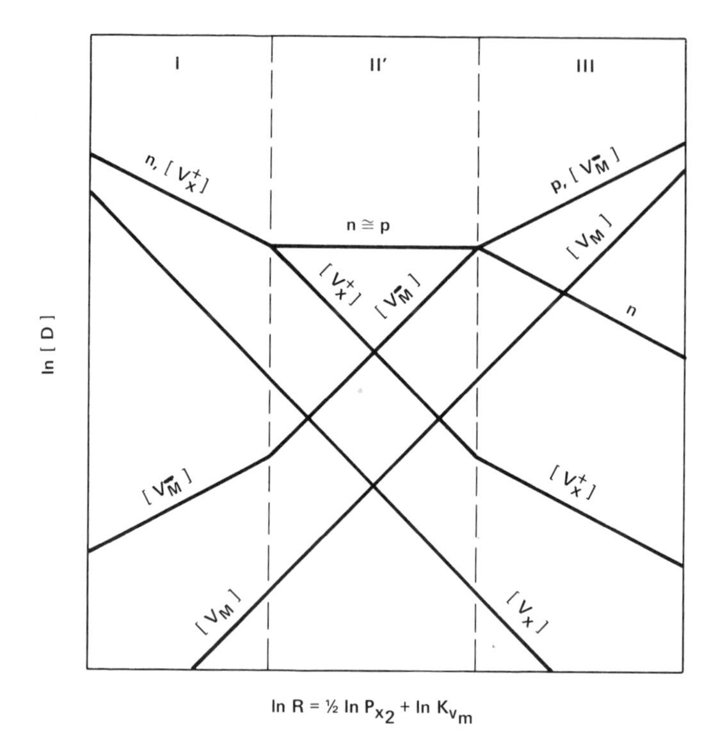

Fig. 5. Defect concentrations as a function of X_2 pressure in the compound MX for $K_i > K_{S(i)}$.

The corresponding graphical representation of this case is given in Figure 5. Region II′ ranges from $R_{I,II'} = K_{S(i)}/K_{50}$ to $R_{II',III} = K_i/K_{50}$. In this case the electronic properties within region II′ are insensitive to composition, while the vacancy concentrations vary with composition. The stoichiometric composition corresponds to $[V_M^{-}] = [V_X^{+}]$.

Brouwer's method may also be used for compounds containing other intrinsic defects, in which case Eqs. (62)–(68) would be different. It is also an extremely useful method of predicting behavior of materials containing impurities or foreign atoms. The application of the method to this latter case is discussed in Section 3.1.2.

2.4.4. Partial Equilibrium

In the discussion above, complete equilibrium was assumed at all times. However, the electronic properties of semiconductors prepared at elevated temperatures are frequently measured at room temperature or below. If a sample is cooled rapidly, the point defects will be frozen in, but equilibria involving electronic defects usually will be maintained. Therefore the total number of each type of point defect will remain constant, but the electrons and holes will redistribute themselves among the point defects. For example, for the case of Schottky defects discussed in the previous section,

$$[V_M^-] + [V_M] = [V_M]_{\text{total}}; \qquad [V_X^+] + [V_X] = [V_X]_{\text{total}} \tag{76}$$

where $[V_M]_{\text{total}}$ and $[V_X]_{\text{total}}$ have the same values as they did at the particular temperature and pressure from which the sample was quenched. However, the ratios of $[V_M^-]$ to $[V_M]$ and $[V_X^+]$ to $[V_X]$ are different at the lower temperatures.

Equations (76) introduce sum relations in addition to the electroneutrality condition. Thus, in order to use Brouwer's approximation method, assumptions must be made for each sum relation, thereby making the treatment of partial equilibrium more complex than that of complete equilibrium. The situation becomes even more complicated if incomplete quenching occurs. More detailed discussions of incomplete equilibrium are given by Kröger[17] and Van Gool.[18]

2.5. Relationship between Mass Action Law and Statistical Thermodynamics

2.5.1. Point Defects

The total free energy of a crystal MX_s containing point defects may be written as follows:

$$\begin{aligned} G = {} & N\mu_{MX_s} + N_{VM}g_{VM} + N_{VX}g_{VX} + N_{IM}g_{IM} + N_{IX}g_{IX} \\ & + N_{M(X)}g_{M(X)} + N_{X(M)}g_{X(M)} \\ & - kT\ln\left[\frac{N!}{N_{VM}!N_{X(M)}!(N - N_{VM} - N_{X(M)})!}\right. \\ & \times \frac{(sN)!}{N_{VX}!N_{M(X)}!(sN - N_{VX} - N_{M(X)})!} \\ & \left.\times \frac{(\alpha N)!}{N_{IM}!N_{IX}!(\alpha N - N_{IM} - N_{IX})!}\right] \end{aligned} \tag{77}$$

where N_{VM}, N_{VX}, N_{IM}, N_{IX}, $N_{M(X)}$, and $N_{X(M)}$ are the numbers of M vacancies, X vacancies, M interstitials, X interstitials, and M substitutional and X substitutional defects, respectively, in the crystal, and g_{VM}, g_{VX}, g_{IM}, g_{IX}, $g_{M(X)}$, and $g_{X(M)}$ are the corresponding free energies of formation (other than the contribution from configurational entropy) of each defect. $N\mu_{MX_s}$ is the free energy of the ideal defect-free crystal. The ln term represents the configurational entropy of the crystal due to the presence of defects. It can be seen that M vacancies and X substitutionals are distributed over the N metal-atom sites, X vacancies and M substitutionals over the (sN) X-atom sites, and interstitials over the αN interstices in the lattice.

Let us assume that the concentrations of all defects other than Schottky defects are negligible. Then Eq. (77) can be written (using Stirling's approximation)

$$G = N\mu_{MX_s} + N_{VM}g_{VM} + N_{VX}g_{VX} - kT[N \ln N - N_{VM} \ln N_{VM} - (N - N_{VM}) \ln(N - N_{VM}) + sN \ln(sN) - N_{VX} \ln N_{VX} - (sN - N_{VX}) \ln(sN - N_{VX})] \quad (78)$$

$$(\partial G/\partial N_{VX})_{N_{VM}} = g_{VX} + kT[\ln N_{VX} - \ln(sN - N_{VX})] \quad (79)$$

Since $sN = N_X + N_{VX}$, where N_X is the number of X atoms in the lattice, we have

$$\partial N_{VX}/\partial N_X = -1 \quad (80)$$

The chemical potential of the X component in the crystal, μ_X, is

$$\mu_X \equiv \partial G/\partial N_X = (\partial G/\partial N_{VX})(\partial N_{VX}/\partial N_X)$$

and from Eqs. (79) and (80), we obtain

$$\mu_X = -g_{VX} + kT \ln[(sN - N_{VX})/N_{VX}] \quad (81)$$

If we consider M vacancies to be formed by addition of X atoms to the crystal, the total number of sites increases, and therefore N is not constant. Consequently, in obtaining $(\partial G/\partial N_{VM})_{N_{VX}}$ from Eq. (78), we must utilize the fact that $\partial N/\partial N_{VM} = 1$, since

$$N = N_{VM} + N_M \quad (82)$$

where N_M is the number of metal atoms in the crystal. Therefore from Eq. (78) we obtain

$$(\partial G/\partial N_{VM})_{N_{VX}} = \mu_{MX_s} + g_{VM} - kT[\ln N - \ln N_{VM} + s \ln(sN) - s \ln(sN - N_{VX})] \quad (83)$$

$$\mu_X = \partial G/\partial N_X = (\partial G/\partial N_{VM})(\partial N_{VM}/\partial N_X) \quad (84)$$

$$sN = N_X + N_{VX}$$

Therefore, from Eq. (82), we can write

$$N_{VM} = N_M - [(N_{VX} - N_X)/s], \qquad \partial N_{VM}/\partial N_X = 1/s$$

Substituting this and Eq. (83) into Eq. (84) yields

$$\mu_X = [(\mu_{MX_s} + g_{VM})/s] - kT\ln[(N/N_{VM})^{1/s}(sN)/(sN - N_{VX})] \tag{85}$$

and equating Eqs. (81) and (85) gives

$$\ln[(N_{VM}/N)(N_{VX}/sN)^s] = -(sg_{VX} + \mu_{MX_s} + g_{VM})/kT$$

and in terms of molar concentrations this becomes

$$[V_M][V_X]^s = \exp[-(g_{VM} + sg_{VX} + \mu_{MX_s})/RT] \tag{86}$$

where $(g_{VM} + sg_{VX} + \mu_{MX_s})$ is the free energy of formation of a mole of M vacancies plus s moles of X vacancies by adding a mole of MX_s to the crystal, and therefore it is the free energy of formation of Schottky defects. Thus, Eq. (86) is the mass action relation for Schottky defects [identical to Eq. (17a)] with the right-hand side representing the equilibrium constant K_S'.

The mass action law equations for the other intrinsic defects can be derived in a similar manner from Eq. (77).

Equations (31), (33), (36), (39), (42), and (45) relating activity to deviation from stoichiometry δ at large values of δ also can be derived from Eq. (77). For example, if a crystal of $MX_{s\pm\delta}$ is in equilibrium with X_2 gas, then

$$\mu_X = \tfrac{1}{2}\mu_{X_2}(g) = \tfrac{1}{2}\mu_{X_2}^0(g) + RT\ln P_{X_2}^{1/2}$$

Substituting this into Eq. (81) and taking the standard state of X as X_2 gas at 1 atm, we can write

$$\tfrac{1}{2}\ln P_{X_2} = -g_{VX}/RT + \ln[(sN - N_{VX})/N_{VX}]$$

Utilizing Eq. (28), this becomes (when only X vacancies are present)

$$P_{X_2}^{1/2} = [(s + \delta)/(-\delta)]\exp(-g_{VX}/RT)$$

which is identical with Eq. (31) if $K_{VX} = \exp(-g_{VX}/RT)$. Similarly, Eq. (33) may be obtained from Eq. (85) with $K_{VM} = \exp[-(\mu_{MX_s} + g_{VM})/RT]$. The other relationships also may be derived from Eq. (77) in an analogous manner.

2.5.1.1. Energetic Considerations

The free energies of formation g_{VM}, g_{VX}, g_{IM}, g_{IX}, $g_{M(X)}$, and $g_{X(M)}$ all consist of an entropy term (change in vibrational entropy of the crystal due to the presence of that defect) and an enthalpy term as mentioned in Section 1.4. It should be pointed out that the value of each of the free energies is dependent upon a particular standard state of one of the components. For example, if an X vacancy is formed by removal of an X atom from the crystal, it is with respect to X at some standard state (e.g., X_2 gas). Conversely, if the same defect is formed by Eq. (29a), the standard state of the external metal phase must be specified.

On the other hand, the free energies of formation of intrinsic defects are absolute values, independent of standard state. For instance, in the case of anti-Frenkel defects, X vacancies are formed by the removal of X atoms from the crystal [Eq. (29)] to some standard state of X (e.g., X_2), and X interstitials may be formed by the addition of X atoms from the same standard state of X to interstices in the crystal [Eq. (34)]. Thus, the energy to form an anti-Frenkel defect is just the energy necessary to remove an X atom from a normal site and place it into an interstice, and the particular standard state chosen is irrelevant. That is to say, Eq. (19) is the sum of Eqs. (29) and (34) and $g_{aF} = g_{VX} + g_{IX}$.

As a second illustration of the interrelationship of defect energies, let us consider the case of Schottky defects. As mentioned above, g_{VX} represents the free energy of formation of X vacancies by Eq. (29). However, g_{VM} is the free energy of formation of M vacancies by the following reaction:

$$\mathrm{M_M} \rightleftharpoons \mathrm{M(std\ state)} + V_\mathrm{M} \tag{87}$$

In order to obtain the equation for the formation of Schottky defects [Eq. (17)], it is necessary to add the reaction for formation of MX_s,

$$\mathrm{M(std\ state)} + (s/2)\mathrm{X_2(std\ state; gas)} \rightleftharpoons \mathrm{MX}_s = \mathrm{M_M} + s\mathrm{X_X} + \alpha V_I \tag{88}$$

to Eq. (87) plus s times Eq. (29). This is the reason for the term μ_{MX_s} in Eq. (86). It is seen that in summing Eqs. (87) and (88) and s times Eq. (29), the species M(std state) and X_2(gas) cancel, and therefore the formation of Schottky defects is also independent of the standard state of the components. Equation (17) also can be formed by Eq. (32) plus s times Eq. (29). However, Eq. (32) is the sum of Eqs. (87) and (88).

For substitutional defects let us consider a case where $s = 1$. The term $g_{M(X)}$ is the free energy of formation of an M substitutional by removal of an X atom from its normal site to some standard state of X and the addition of an M atom to that site from M in its standard state:

$$g_{\mathrm{M(X)}} + \mathrm{X_X} + \mathrm{M(std\ state)} \rightleftharpoons \tfrac{1}{2}\mathrm{X_2(std\ state)} + \mathrm{M_X}$$

and conversely,

$$g_{X(M)} + M_M + \tfrac{1}{2}X_2(\text{std state}) \rightleftharpoons M(\text{std state}) + X_M$$

Consequently, the free energy of formation of antistructure defects [Eq. (20)] is $g_{M(X)} + g_{X(M)}$, independent of standard state. Equation (20) also may be obtained by Eqs. (40) and (43); i.e., $K_{as} = K_{XM}K_{MX}$.

2.5.2. Intrinsic Ionization

If one considers the distribution of electrons and holes over the energy states of a crystal, it may be shown by Fermi statistics that the number of electrons at any energy level j is given by

$$n_j = g_j/\{1 + \exp[(\varepsilon_j - \varepsilon_F)/kT]\} \tag{89}$$

where g_j represents the degeneracy of the jth level, ε_j is the energy of that level, and ε_F is the Fermi level. When $\varepsilon_j = \varepsilon_F$, $n_j = g_j/2$. Therefore ε_F is the energy at which the probability of occupation of states is $\frac{1}{2}$.† In terms of thermodynamic quantities, it is also the electrochemical potential or the partial molar free energy of electrons. This can be seen by writing the configurational entropy for the distribution of electrons among available energy states

$$S = k \ln\left[\prod_j g_j!/n_j!(g_j - n_j)!\right]$$

and using Stirling's approximation:

$$S = k\left[\sum_j g_j \ln\left(\frac{g_j}{g_j - n_j}\right) + \sum_j n_j \ln\left(\frac{g_j - n_j}{n_j}\right)\right]$$

Utilizing Eq. (89) and letting $N = \sum_j n_j$ be the total number of electrons and $E = \sum_j n_j\varepsilon_j$ the total energy of the system, this equation can be rewritten

$$S = k\sum_j g_j \ln\left[1 + \exp\left(\frac{\varepsilon_F - \varepsilon_j}{kT}\right)\right] + \frac{E}{T} - \frac{N\varepsilon_F}{T} \tag{89a}$$

By definition, the electrochemical potential of electrons μ_e is

† The meaning of ε_F is easily seen for conductors. However, in semiconductors the Fermi level may be situated in the band gap. More precisely, then, ε_F is the energy below which the probability of states being occupied is $>\frac{1}{2}$ and above which the probability is $<\frac{1}{2}$.

$(\partial G/\partial N)_{P,T}$ and since $G = E + PV - TS$,

$$\mu_e = -T(\partial S/\partial N)_{E,V}$$

By differentiating Eq. (89a), we see that $\mu_e = \varepsilon_F$. Thermodynamically, two systems are in equilibrium when their electrochemical potentials are equal and in semiconductors we see that this corresponds to the Fermi levels being equal.

When $(\varepsilon_j - \varepsilon_F)/kT \gg 1$, so that the jth level is only slightly populated, Eq. (89) becomes

$$n_j = g_j \exp[-(\varepsilon_j - \varepsilon_F)/kT] \tag{90}$$

which is the classical distribution law. If Eq. (90) is applied to the conduction band of an intrinsic semiconductor, it becomes

$$n = \mathcal{N}_c \exp[-(\varepsilon_c - \varepsilon_F)/kT]$$

where $\mathcal{N}_c$ is the density of states in the conduction band and ε_c is the energy at the bottom of the conduction band.

For the case of holes, p_j is merely $g_j - n_j$, and therefore

$$p_j = g_j/\{1 + \exp[(\varepsilon_F - \varepsilon_j)/kT]\} \tag{91}$$

As in the case of electrons, it can be shown that

$$p = \mathcal{N}_v \exp[-(\varepsilon_F - \varepsilon_v)/kT]$$

where ε_v is the energy of the highest level in the valence band and $\mathcal{N}_v$ is the density of states. Therefore

$$np = \mathcal{N}_c \mathcal{N}_v \exp[-(\varepsilon_c - \varepsilon_v)/kT] \tag{92}$$

The right-hand side of this equation represents the equilibrium constant for intrinsic ionization [see Eq. (7)] and $\varepsilon_c - \varepsilon_v$ is merely the band gap in the semiconductor.

If electrons and holes are localized on atoms so that conduction occurs by a hopping mechanism, the distribution of electrons and holes over atoms in the lattice would be taken into account, using an analysis similar to that in Section 2.5.1. An equation similar to Eq. (92) is then obtained, except that $\mathcal{N}_c$ and $\mathcal{N}_v$ would represent the number of atoms over which electrons and holes, respectively, could be distributed, and the energy would still be that necessary to remove an electron from a chemical bond.

Equations (89)–(91) also may be applied to donor or acceptor levels in semiconductors so that the expressions for the equilibrium constants of ionization reactions such as Eqs. (49), (50), and (53)–(56)

may be determined. For those defects which act as electron donors (V_X, M_I, X_M) the equilibrium constant is

$$K_D = \mathcal{N}_c \exp[-(\varepsilon_c - \varepsilon_D)/kT]$$

while for defects which act as electron acceptors (V_M, X_I, M_X)

$$K_A = N_v \exp[-(\varepsilon_A - \varepsilon_v)/kT]$$

where ε_D and ε_A are energies of the donor and acceptor levels, respectively.

At large electron concentrations (i.e., $\varepsilon_j \sim \varepsilon_F$), the classical distribution approximation is no longer valid and Eqs. (90) and (91) are not applicable. However, the mass action law can still be utilized if activities instead of concentrations are used, but an explicit expression for the activity coefficient must be obtained.[19]

2.6. Defect Interactions

2.6.1. Activity Coefficients

In previous sections it has been assumed that the enthalpies and entropies of defect formation are independent of composition. However, if there is interaction among the defects, this assumption is no longer valid. If the degree of interaction among defects is not negligible, additional terms for interaction enthalpy and entropy between defects must be introduced into Eq. (77). In terms of the mass action law, this means that activities and activity coefficients must be used instead of concentrations. For example, let us consider the formation of X vacancies [Eq. (29)]:

$$K'_{VX} = P_{X_2}^{1/2} a_{VX}/a_{XX} = P_{X_2}^{1/2} \gamma_{VX}[V_X]/\gamma_{XX}[X_X]$$

where γ_{VX} and γ_{XX} are the activity coefficients of X vacancies and X atoms on normal sites, respectively. From Eq. (30) it can be seen that

$$K_{VX} = P_{X2}^{1/2}[V_X]/[X_X] = (\gamma_{XX}/\gamma_{VX})K'_{VX} \tag{93}$$

Since the activity coefficients are functions of concentration, K_{VX} is no longer constant (with composition) if concentrations rather than activities are used and if interactions are significant. The explicit dependence of activity coefficient on concentration depends upon the particular model considered for the interaction of defects. The simplest case is one in which only one type of defect is present in the lattice and the distribution of defects is independent of interaction energy (Bragg–Williams approximation). The system then may be considered as a regular solution of X vacancies and X atoms on normal sites and the activity coefficients are[20]

$$\gamma_{VX} = \exp\{-(s - [V_X])^2 Z_X \xi_{VX}/2skT\},$$
$$\gamma_{XX} = \exp\{-[V_X]^2 Z_X \xi_{VX}/2skT\}$$

where ξ_{VX} is the interaction energy between vacancies and Z_X is the coordination number of X sites. Using these activity coefficients and appropriate standard states,[21] it can be shown that

$$a_X = P_{X_2}^{1/2} = K_{VX}[(s + \delta)/(-\delta)] + Z_X \xi_{VX} \delta / sRT \tag{94}$$

By comparing this expression with Eq. (31), it is seen that an additional term has been added which is a function of interaction energy and δ.

In a similar manner relationships for the other point defects can be derived. For interstitial defects regular solutions of interstitials and vacant interstices are considered, and for substitutional defects regular solutions of substitutional defects and normal lattice atoms are considered. Expressions equivalent to Eq. (94) have been derived[22] for each point defect and the extra terms due to interaction are tabulated in Table 5. Despite the fact that the defect distribution is

TABLE 5

Interaction Energy Terms in Expressions for X Activities in $MX_{s+\delta}$ For Bragg–Williams Approximation

Type of defect	Interaction energy term
X Vacancies	$(Z_X \xi_{VX}/sRT)\delta$
M Vacancies	$(Z_M \xi_{VM}/2sRT)[\delta(2s + \delta)/(s + \delta)^2]$
X Interstitials	$(Z_I \xi_{IX}/\alpha RT)\delta$
M Interstitials	$(Z_I \xi_{IM}/2\alpha sRT)[\delta(2s + \delta)/(s + \delta)^2]$
X Substitutionals	$[Z_M \xi_{X(M)}/2(s + 1)RT][\delta(2s + \delta + 2)/(s + 1 + \delta)^2]$
M Substitutionals	$[Z_X \xi_{M(X)}/2s(s + 1)RT][\delta(2s + \delta + 2)/(s + 1 + \delta)^2]$

assumed to be random and there is no additional configurational entropy term due to interaction, application of this treatment to the Pt–S system[23] $FeS_{1+\delta}$[24] TiO[25] and various metal hydrides[21,22,26] and carbides[27] has given good agreement with data. There have been many extensions of the simple Bragg–Williams approximation including, among others, a quasichemical treatment[28] in which the number of defect pairs is assumed to be proportional to the exponential of the interaction energy, a consideration of the variation of interaction energy with concentration of defects,[29] and the effect of site blocking by existing defects.[30]

In crystals containing more than one type of defect, the situation will be even more complicated because interactions among unlike defects also must be considered. For cases in which charged defects are present, there obviously will be Coulomb interactions among defects. Therefore, positively charged defects will tend to be surrounded by negatively charged defects in the crystal and vice versa. Let us consider the case of ionized Schottky defects (V_M^- and V_X^+) in MX_s with $s = 1$:

$$K_{S(i)} = [V_M^-][V_X^+] = [\exp(\Delta S_{S(i)}/R)] \exp(-\Delta H_{S(i)}/RT) \qquad (95)$$

Because of interactions, concentrations cannot be used and we must write

$$K_{S(i)} = a_{V_M^-} a_{V_X^+} = \gamma_{V_M^-} \gamma_{V_X^+} [V_M^-][V_X^+] \qquad (96)$$

Lidiard[31] has shown that the Debye–Hückel theory for liquid electrolytes can be applied to the case of charged defects in ionic solids. The activity coefficient for a defect i can be written:

$$\gamma_i = \exp(-q^2 z_i^2 / 2\kappa RT l_D)$$

where $l_D = (\kappa RT / 4\pi q^2 v \sum x_i z_i^2)$ and $z_i q$ is the charge on the defect, κ is the dielectric constant of the crystal, v is the molar volume, and x_i is the concentration of the defect i. Here l_D is the Debye length or the radius of the sphere of oppositely charged defects surrounding each charged defect. For the case of ionized Schottky defects (with $s = 1$)

$$\gamma_{V_M^-} = \gamma_{V_X^+} = \gamma_V = \exp(-q^2/2\kappa RT l_D); \qquad l_D = (\kappa RT / 4\pi q^2 v [V])$$

Substituting this into Eq. (96) yields

$$[V_M^-][V_X^+] = K_{S(i)}/\gamma_V^2 = [\exp(\Delta S_{S(i)}/R)] \exp\{-[\Delta H_{S(i)} - (q^2/\kappa RT l_D)\}$$

Comparison of this expression with Eq. (95) shows that the enthalpy of Schottky defect formation is decreased by interactions, and therefore the concentration of defects should increase. However, the effect is usually quite small. For example, in NaCl at high temperatures, the change in $\Delta H_{S(i)}$ is only 0.8 kcal/mole[31] from a total value of about 48 kcal/mole.

2.6.2. Association of Defects

Another approach to treating defect interactions is to consider the association of simple point defects into new defect species. As an illustration, consider the case of charged Schottky defects interacting to form a defect pair. The reaction is

$$V_M^- + V_X^+ \rightleftharpoons (V_M^- V_X^+)$$
$$K_{97} = [(V_M^- V_X^+)]/[V_M^-][V_X^+] = [\exp(\Delta S/R)] \exp(-\Delta H/RT) \tag{97}$$

where the concentration of defect pairs $(V_M^- V_X^+)$ is explicitly taken into consideration. The entropy of pair formation consists of a configurational part, which depends upon the number of ways the associate can be formed in the lattice, and a vibrational part, which is a function of the change in vibrational entropy of the crystal due to the presence of pairs rather than isolated point defects.

Since the association of defects is usually an exothermic reaction (rather than endothermic as in point defect formation), ΔH is negative. In alkali halides the enthalpy of Schottky defect pairing is about -20 kcal/mole.

Association of neutral defects also may occur, due either to reduction of strain energy or to covalent interactions. An example of the former is the formation of divacancies in metals. The presence of a vacancy causes strain in the lattice. By placing two vacancies in adjacent lattice sites, some of the strain is relieved and the enthalpy gained by this process is that of divacancy formation. Values of this quantity in metals range from about -2 to -6 kcal/mole.[32]

An example of covalent interaction of defects is the pairing of vacancies in alkali halides (see also Volume 2, Chapter 2). Neutral F centers (anion vacancies) may form divacancy pairs with an enthalpy of association of about -4 kcal/mole. Neutral cation vacancies also may attract each other with an even greater association enthalpy, ranging from about -40 kcal/mole for iodides to about -75 kcal/mole for chlorides.[17] This is even larger than the Coulomb attraction between oppositely charged vacancies.

Associate species which are larger than pairs also may form. For example, in a compound MX_2 which contains singly charged X vacancies and doubly charged metal vacancies, the following reaction may occur to form a "trimer":

$$V_M^{2-} + 2V_X^+ = (V_X^+ V_M^{2-} V_X^+); \qquad K = [(V_X^+ V_M^{2-} V_X^+)]/[V_M^{2-}][V_X^+]^2$$

Such a trimer is called an M center in alkali halides. Like defects also may form clusters:

$$nV_X \rightleftharpoons (nV_X); \qquad K = [(nV_X)]/[V_X]^n$$

where n is the number of defects in the cluster. Clusters may grow by the following reaction:

$$(nV_X) + V_X \rightleftharpoons ((n+1)V_X)$$

which may eventually lead to precipitation of a new phase. The photographic process is believed to be related to association of Ag interstitials in the silver salt.

The general effect of association is to increase the overall concentration of defects. As pairs are formed by Eq. (97), Eq. (17) is driven to the right. Therefore, the total (single plus associated) concentration of each type of vacancy will be greater.

When using Brouwer's method, the association reactions discussed here also must be considered if they occur to any significant extent. For example, if reaction (97) is important, the additional equation

$$ln[V_M^-] + ln[V_X^+] + ln\, K_{97} = ln(V_M^- V_X^+)$$

would be included along with the other seven in the analysis carried out in Section 2.4.3, and the additional unknown to be solved would be the concentration of the associate pair $(V_M^- V_X^+)$.

Finally, it should be mentioned that defects need not necessarily occupy adjacent sites to be considered part of an associate. Defects may be two or more atomic distances apart. We may write the reaction

$$\left(\sum_i x_i\right) V_M + \left(\sum_i x_i\right) V_X \rightleftharpoons x_1(V_M V_X)_1 + x_2(V_M V_X)_2 + \cdots + x_n(V_M V_X)_n$$

where $(V_M V_X)_1$ is a vacancy pair one atomic distance apart, $(V_M V_X)_2$ is a vacancy pair two atomic distances apart, and so on. Here $x_1, x_2, \ldots$ are the corresponding numbers of each type of pair. The conditions and equations relating to this type of association are discussed more fully by Kröger.(17)

3. Multicomponent Systems

3.1. Equilibria Involving Foreign Atoms

Foreign atoms or impurities may enter a crystal lattice either interstitially or substitutionally. Consequently, they also may be considered as lattice defects such that their interactions with native defects and normal lattice atoms can be treated by the mass action law.

Impurities will generally affect the electrical properties of crystals by introducing ionizable centers. Interactions between foreign atoms and native defects also will affect solubility relationships and defect concentrations.

3.1.1. Elemental Crystals

Interstitial defects in covalent elemental crystals will usually act as electron donors due to simple ionization: for example, copper in germanium:

$$\mathrm{Cu}_I \rightleftharpoons \mathrm{Cu}_I{}^+ + e^-; \qquad K = n[\mathrm{Cu}_I{}^+]/[\mathrm{Cu}_I]$$

The copper interstitial also may become doubly ionized:

$$\mathrm{Cu}_I{}^+ \rightleftharpoons \mathrm{Cu}_I^{2+} + e^-; \qquad K = n[\mathrm{Cu}_I^{2+}]/[\mathrm{Cu}_I{}^+]$$

Substitutional impurities may act as donors or acceptors depending upon the number of valence electrons on the impurity atom as compared to the host atoms. For example, phosphorus in silicon will act as a donor:

$$P_{\mathrm{Si}} \rightleftharpoons P_{\mathrm{Si}}^+ + e^-; \qquad K = n[P_{\mathrm{Si}}^+]/[P_{\mathrm{Si}}]$$

while Al will act as an acceptor:

$$\mathrm{Al}_{\mathrm{Si}} \rightleftharpoons \mathrm{Al}_{\mathrm{Si}}^- + h^+; \qquad K = p[\mathrm{Al}_{\mathrm{Si}}{}^-]/[\mathrm{Al}_{\mathrm{Si}}]$$

Reactions also may be written for the incorporation of foreign atoms F into a crystal A:

$$F(\text{gas}) + V_I \rightleftharpoons F_I; \qquad K = [F_I][V_I]/P_F$$

or (98)

$$F(\text{gas}) \rightleftharpoons F_A; \qquad K_{98} = [F_A]/P_F$$

The presence of impurities also will affect the concentrations of charged native defects. Consider an elemental crystal A in which vacancies act as acceptors:

$$V_A \rightleftharpoons V_A{}^- + h^+; \qquad K_{99} = p[V_A{}^-]/[V_A] \tag{99}$$

If the crystal contains an impurity such as phosphorus in silicon which may act as a donor, the general equation may be written:

$$F_A = F_A{}^+ + e^-; \qquad K_{100} = n[F_A{}^+]/[F_A] \tag{100}$$

The overall neutrality condition for such a crystal would be

$$n + [V_A{}^-] = p[F_A{}^+]$$

but at relatively high concentrations of impurities, the following approximation can be made:

$$n = [F_A{}^+] \tag{101}$$

From Eqs. (100), (98), and (101), we obtain

$$n^2 = K_{100}K_{98}P_F \tag{102}$$

and from Eq. (99) and the intrinsic ionization relation, Eq. (7), we may write

$$n = K_i[V_A{}^-]/K_{99}$$

Substituting this into Eq. (102) yields

$$[V_A{}^-] = \{(K_{98}K_{100})^{1/2}K_{99}/K_i\}P_F^{1/2}$$

Thus, we see that the concentration of ionized vacancies increases with the square root of F gas pressure (and therefore with concentration of F). In general, the effect of foreign atoms will be to increase the concentration of native defects of opposite charge and to decrease the concentration of native defects of like charge.

If two types of foreign atoms coexist in a semiconductor, a similar situation may occur. This was demonstrated in a classic experiment by Reiss and Fuller[33] in which a single crystal of silicon containing boron was immersed in molten tin containing lithium. Since boron substitutes for silicon atoms in the lattice, it acts as an acceptor, while lithium dissolves interstitially and therefore becomes a donor according to the following relations:

$$\begin{array}{l} B_{\mathrm{Si}} \rightleftharpoons B_{\mathrm{Si}}^- + h^+ \\ \mathrm{Li(Si)} \rightleftharpoons \mathrm{Li}_I{}^+ + e^- \\ \qquad\qquad\qquad\quad \downarrow\uparrow \\ \qquad\qquad\qquad (h^+e^-) \end{array} \tag{103}$$

The holes and electrons will interact as determined by the intrinsic ionization constant K_i. Because intrinsic ionization is small at low temperatures (<400°C), holes will remove electrons by the vertical reaction shown, thus driving Eq. (103) to the right and correspondingly increasing the solubility of lithium. It was found that for boron concentrations in excess of the normal lithium solubility in silicon, the solubility of lithium increased with the concentration of boron. An analytical expression for the concentration of $\mathrm{Li}_I{}^+$ as a function $[\mathrm{B}_{\mathrm{Si}}^-]$ can be obtained by simultaneously solving the following two mass action relations:

$$K_{\mathrm{BSi}} = [\mathrm{B}_{\mathrm{Si}}^{-}]p \qquad \text{and} \qquad K_{103} = n[\mathrm{Li}_I^{\ +}]/\mathrm{Li}(\text{external})$$

along with the neutrality condition

$$[\mathrm{B}_{\mathrm{Si}}^{-}] + n = [\mathrm{Li}_I^{\ +}] + p$$

and Eq. (7).

In some cases a foreign atom may enter the lattice both as an interstitial and substitutional defect. For example, Cu, which forms interstitials as mentioned above, also may substitute for Ge atoms in Ge and act as an electron acceptor:

$$\mathrm{Cu}_{\mathrm{Ge}} \rightleftharpoons \mathrm{Cu}_{\mathrm{Ge}}^{+} + p$$

This may give rise to the following distribution reaction:

$$\mathrm{Cu}_{\mathrm{Ge}} + V_I \rightleftharpoons \mathrm{Cu}_I + V_{\mathrm{Ge}}$$

3.1.2. Compounds

When considering impurities in compounds rather than in elemental crystals, two additional factors must be taken into account. First, there is more than one type of site available for substitutional impurities. In binary compounds MX_s foreign atoms may enter either the M or X sublattice. Second, the nature and concentration of native defects may be changed by the presence of impurities.

In predominantly covalent compounds the behavior of impurities is similar to that in covalent elemental crystals. For example, in gallium arsenide, group I or II elements on Ga sites or group III or IV elements on As sites will act as acceptors, while group IV or V elements on Ga sites and group VI or VII elements on As sites will act as donors. Thus, a group IV element such as Si may act as either a donor or acceptor depending upon which sublattice it enters:

$$\mathrm{Si}_{\mathrm{Ga}} \rightleftharpoons \mathrm{Si}_{\mathrm{Ga}}^{+} + e^{-}; \qquad \mathrm{Si}_{\mathrm{As}} \rightleftharpoons \mathrm{Si}_{\mathrm{As}}^{-} + h^{+} \tag{104}$$

In compounds with more ionic character the incorporation of impurities will sometimes generate point defects in order to maintain electroneutrality. A case in point is divalent calcium substituting for tetravalent zirconium in ZrO_2. In order to compensate for the negatively charged calcium substitutional (since the Ca^{2+} ion occupies a site which is normally occupied by an ion of charge 4+), positively charged oxygen vacancies are produced:

$$\mathrm{CaO} \rightleftharpoons \mathrm{Ca}_{\mathrm{Zr}}^{2-} + \mathrm{V}_{\mathrm{O}}^{2+} + \mathrm{O}_{\mathrm{O}} \tag{105}$$

For compounds in which one of the components has more than one stable valence state (e.g., transition metals), charge compensation of foreign atoms may occur by a change in oxidation state of that component. Such compounds are called controlled-valency semiconductors. A well-known illustration of such a material is nickel oxide containing lithium which substitutes for nickel in the lattice. In order to maintain electroneutrality, for each Li^+ ion present a normal divalent Ni^{2+} ion becomes converted to a Ni^{3+} ion. The Ni^{3+} ion in the NiO lattice behaves as a hole:

$$Li_2O + \tfrac{1}{2}O_2(\text{gas}) \rightleftharpoons 2Li_{Ni}^- + 2h^+ + 2O_O \qquad (106)$$

An alternative way of writing this reaction which shows the actual, rather than relative, charge on the ions is as follows:

$$Li_2O + \tfrac{1}{2}O_2 + 2Ni_{Ni}^{2+} \rightleftharpoons 2Li_{Ni}^+ + 2Ni_{Ni}^{3+} + 2O_O^{2-}$$

Impurity reactions such as Eqs. (104), (105), or (106) may be combined with the previously given reactions for the formation and ionization of native defects to compute concentrations of all defects present under various conditions. Brouwer's approximation method is used when necessary. A discussion of all the possibilities which may occur is far beyond the scope of this chapter. However, one simple case will be considered for illustrative purposes.

Assume that a compound MX consists of divalent M and X atoms (e.g., ZnS) and the lattice contains a fixed concentration of a singly ionized impurity F_M (e.g., Ag).[18] If the pressure of M vapor is sufficiently high to maintain an excess of X vacancies, we have

$$M(\text{gas}) \rightleftharpoons M_M + V_X; \qquad K'_{29a} \simeq [V_X]/P_M \qquad (107)$$

$$V_X \rightleftharpoons V_X^+ + e^-; \qquad K_{49} = n[V_X^+]/[V_X]$$

$$V_X^+ \rightleftharpoons V_X^{2+} + e^-; \qquad K_{108} = n[V_X^{2+}]/[V_X^+] \qquad (108)$$

$$[F_M^-] = [F_M]_{\text{total}} \qquad (109)$$

We may write the following neutrality condition:

$$n + [F_M^-] = 2[V_X^{2+}] + [V_X^+]$$

Therefore, four approximations may be made as shown at the top of each column in Table 6 and the ranges in Figure 6. The concentration dependence of each defect on the pressure of M vapor is given in Table 6. Since V_X is directly proportional to P_M in all ranges [as given by Eq. (107)], it is not included in the table. The assumption is made

TABLE 6

Concentrations of Defects in $M^{2+}X^{2-}$ Containing a Singly Charged Impurity F_M^- †

	$[F_M^-] = 2[V_X^{2+}]$	$[F_M^-] = [V_X^+]$	$n = [V_X^+]$	$n = 2[V_X^{2+}]$
n	$(2QK_{107}/[F_M^-])^{1/2}P_M^{1/2}$	$(Q/[F_M^-])P_M$	$(QP_M)^{1/2}$	$(2QK_{107})^{1/3}P_M^{1/3}$
$[V_X^+]$	$(Q[F_M^-]/2K_{107})^{1/2}P_M^{1/2}$	$[F_M^-]$	$(Q)^{1/2}P_M^{1/2}$	$(Q^2/2K_{107})^{1/3}P_M^{1/3}$
$[V_X^{2+}]$	$[F_M^-]/2$	$(K_{107}[F_M^-]^2/Q)P_M^{-1}$	K_{107}	$(QK_{107}/4)^{1/3}P_M^{1/3}$

† Here $Q = K'_{29a}K_{49}$.

in Figure 6 that $K_{108} > [F_M]_{total}$ and therefore, at low M pressures, $n > [V_X^+]$. It is seen from Figure 6 that the region $n = 2[V_X^{2+}]$ is never reached. However, if we begin with the assumption that $K_{108} < [F_M]_{tot}$ and $[V_X^+] > n$ at low M pressures, then we would pass from the region where the compensation mechanism is $[F_M^-] = 2[V_X^{2+}]$ directly to the region where it is $n = 2[V_X^{2+}]$.

The situation becomes much more complex if the values of K_i and $K_{S(i)}$ are high enough to make the concentrations of holes and M vacancies significant, and if, let us say, F_M had two ionization states. Then the electroneutrality condition would be

$$[F_M^-] + 2[F_M^{2-}] + n + [V_M^-] + 2[V_M^{2-}] = [V_X^+] + 2[V_X^{2+}] + p$$

There would then be 15 compensation regions which would require consideration.

A more detailed review of the various possibilities which have been observed with foreign atoms is given by Kröger,[17] and a further discussion of foreign atoms in semiconductors will be found in the article by Brebrick (Volume 2, Chapter 5).

3.1.3. Association

Just as association of native defects occurs as described in Section 2.6.2, foreign atoms also may associate with point defects or other foreign atoms. All the situations discussed in Section 2.6.2 also hold for defect–impurity and impurity–impurity associations.

In metals the associations between impurity atoms and vacancies have important effects on the rates of age-hardening in alloys by affecting the diffusion of vacancies. The association enthalpy of a vacancy–impurity pair in metals may range from about 2 to 10 kcal/

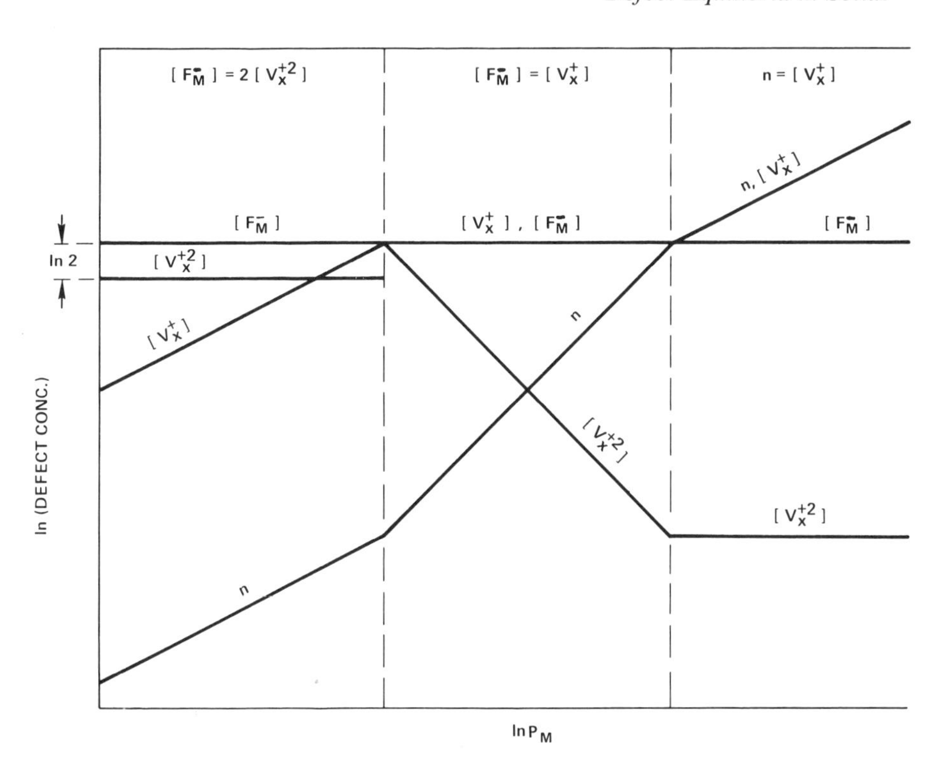

Fig. 6. Defect concentrations as a function of M pressure in the compound MX containing the impurity atom *F*.

mole. Pairing of one type of foreign atom by another also will have pronounced effects on the diffusion rates of impurities.

The formation of associates will affect the solubility of impurities in the same manner that the hole–electron equilibrium does, as was illustrated for the case of Li and B in Si. For example, Ga in Ge will increase the solubility of Li far more than expected from the known value of the intrinsic ionization constant K_i. This can be explained by the formation of a $(\mathrm{Li}_I{}^+\mathrm{Ga}_{\mathrm{Ge}}^-)$ associate as shown:

$$\begin{array}{l} \mathrm{Li}_{\mathrm{ext}} \rightleftharpoons \mathrm{Li}_I{}^+ + e^- \\ \mathrm{Ga}_{\mathrm{Ge}} \rightleftharpoons \mathrm{Ga}_{\mathrm{Ge}}^- \quad + \quad h^+ \\ \qquad\qquad \downarrow\uparrow \qquad\qquad \downarrow\uparrow \\ \qquad (\mathrm{Li}_I{}^+\mathrm{Ga}_{\mathrm{Ge}}^-) \quad (h^+e^-) \end{array} \tag{110}$$

Both vertical reactions will drive Eq. (110) to the right, thus increasing the solubility of Li, but since $K_a = [(\mathrm{Li}_I{}^+\mathrm{Ga}_{\mathrm{Ge}}^-)]/[\mathrm{Li}_I{}^+][\mathrm{Ga}_{\mathrm{Ge}}^-]$ is much larger than K_i, the effect of association will be much greater.

The formation of associates may have pronounced and sometimes unusual effects on the electrical properties of semiconductors. For example, Ga substituting for Zn in ZnS would be expected to form Ga_{Zn}^{-} and thus to increase the electron concentration and thereby decrease the hole concentration. However, Van Gool[18] has pointed out that if the following associate is formed:

$$Ga_{Zn}^{+} + V_{Zn}^{2-} \rightleftharpoons (Ga_{Zn}^{+} V_{Zn}^{2-})^{-}$$

the negatively charged associate can provoke the formation of holes according to the electroneutrality equation:

$$[(Ga_{Zn}^{+} V_{Zn}^{2-})] = p$$

thus increasing the p-type conductivity.

As in the case of native defects, reactions for association of impurity atoms must be taken into account when computing the concentrations of all defects in defining equilibrium conditions in a defect solid.

3.2. Multicomponent Compounds

Multicomponent compounds can accommodate many more types of defects than simple binary compounds. As we have seen, there are six types of point defects possible in binary compounds. In a ternary compound of general formula ML_rX_s, where L may represent a second type of electropositive component and r is the stoichiometric ratio of L to M (e.g., in $MgAl_2O_4$, M = Mg, L = Al, X = O, $r = 2$, and $s = 4$) the following point defects may occur: V_M, V_L, V_X, M_I, L_I, X_I, X_M, L_M, X_L, M_L, M_X, and L_X, for a total of 12. Furthermore, in many ternary compounds (e.g., spinels) there are two types of interstitial sites, which adds three more possible defects.

Theoretically, there may be two types of intrinsic defect combinations, conjugate pairs and triplets. An example of the latter is

$$0 \rightleftharpoons V_M + rV_L + sV_X + \alpha V_I + \alpha' V_I'$$

$$K_t = [V_M][V_L]^r[V_X]^s[V_I]^\alpha[V_I']^{\alpha'}$$

where V_I' represents a second type of interstice. However, the occurrence of triplets is not too likely in most compounds and it is usually only necessary to consider conjugate pairs.

When considering nonstoichiometry in ternary compounds it is possible to have two types of defects in excess. Two examples are nickel ferrite, where both O vacancies and Ni atoms on Fe sites occur (Ni_{Fe}) to give a compound whose overall composition is

$Ni_{1.03}Fe_{1.97}O_{3.98}$,[34] and nonstoichiometric chalcopyrite, where density measurements[35] indicate that the apparent deficiency in sulfur in $CuFeS_{2-\delta}$ is probably due to Cu and Fe interstitials in the lattice.

By measuring oxygen activity as a function of δ in lithium ferrite, $LiFe_5O_{8-\delta}$, and using a mass action law treatment, Tretyakov and Rapp[36] deduced that the defects responsible for the apparent oxygen deficiency were lithium interstitials. The reaction may be written as follows:

$$8O_O + 5Fe_{Fe} + 6Li_{Li} + 6(\alpha + 1)V_I \rightleftharpoons 4O_2 + 5Fe_{Li}^{2+} + 6Li_I^{+} + 16e^-$$

Since this is a controlled-valency semiconductor, the 16 electrons are actually Fe^{2+} ions on Fe^{3+} sites.

In order to completely define equilibria with external phases in ternary compounds, it is necessary to specify the activities of two of the components of the compound. Schmalzried[37] treats ternary compounds as pseudobinaries in which each component is a binary compound. According to this assumption, chalcopyrite, $CuFeS_2$, may be viewed as $CuS \cdot FeS$, and spinel as $MgO \cdot Al_2O_3$. We can use the general formula ML_2O_4 for a spinel in order to illustrate Schmalzried's approach. The formula may be rewritten: $(1 + \alpha)MX \cdot (1 + \beta)L_2X_3$, where $(1 + \beta)/(1 + \alpha) = 1 + y$, and y is an indication of the degree of nonstoichiometry. For $\alpha = \beta = 0$, the compound is stoichiometric. If $\beta > \alpha$, the deviation from stoichiometry is positive and vice versa. Since most compounds with the spinel structure have some ionicity, substitutional defects of M or L on the X sublattice or X on the M or L sublattices do not occur. Table 7 summarizes all possible conjugate defect pairs in such compounds and their effects on the stoichiometry.

TABLE 7

Possible Types of Defect Pairs in Ternary Ionic Compounds

	M_I	L_I	V_M	V_L	M_L	L_M	V_X	X_I
M_I	—	—	$y = 0$	$y < 0$	$y < 0$	—	—	$y < 0$
L_I	—	—	$y > 0$	$y = 0$	$y < 0$	—	—	$y < 0$
V_M	—	—	—	—	—	$y > 0$	$y > 0$	—
V_L	—	—	—	—	—	$y > 0$	$y < 0$	—
M_L	—	—	—	—	—	$y = 0$	$y < 0$	-
L_M	—	—	—	—	—	—	—	$y > 0$
V_X	—	—	—	—	—	—	—	$y = 0$
X_I	—	—	—	—	—	—	—	—

Schmalzried[37] shows that Brouwer's graphical method may be used for ternary compounds, by simultaneously solving the following mass action equations:

$$K_F = [M_I][V_M]/[M_M][V_I], \qquad K_a = [L_I][V_L]/[L_L][V_I]$$

$$K_b = [M_I][V_L]/[M_L][V_I], \qquad K_c = [L_I][V_M]/[L_M][V_I]$$

$$K_d = ([V_M]/[M_M])([V_L]/[L_L])^2([V_X]/[X_X])^4$$

$$[L_M] + [V_M] - [M_L] - [M_I] - \tfrac{1}{4}([V_X] - [X_I]) = \tfrac{3}{4}y$$

$$[L_M] + [L_I] - [M_L] - [V_L] + \tfrac{1}{2}([V_X] - [X_I]) = \tfrac{1}{2}y$$

It was found convenient to compute the concentrations as a function of the activity of the MX phase, a_{MX}, since it could be more easily measured than y. This was done by using the following relationships:

$$MX + \tfrac{2}{3}L_I \rightleftharpoons \tfrac{1}{3}L_2X_3 + M_I; \qquad K_e = (a_{M_2X_3}^{1/3}[M_I]/a_{MX}[L_I]^{2/3})$$

$$a_{MX}a_{M_2X_3} = K_f, \qquad M_I/a_{MX}^{4/3}[L_I]^{2/3} = K_e/K_f^{1/3} = K_g$$

By introducing the relations for ionization of defects and for intrinsic ionization [Eq. (7)] and using the following reaction:

$$MX \rightleftharpoons M_I^{2+} + 2e^- + \tfrac{1}{2}X_2; \qquad K_h = n^2[M_I^{2+}]P_{X_2}^{1/2}/a_{MX}$$

the concentrations of electronic defects also may be computed, either as a function of a_{MX} (with P_{X_2} constant) or as a function of P_{X_2} (with a_{MX} constant).

Using Schmalzried's method and measurements of electrical conductivity as a function of P_{O_2}, George and Grace[38] showed that the majority defects in $CaTiO_3$ were doubly ionized oxygen vacancies V_O^{2-} and electrons.

4. Extended Defects

In recent years it has been shown[39] that in many defect compounds, particularly oxides, the predominant defects are not random point defects, but rather the defect structure may consist of complexes, microdomains, block structures, or crystallographic shear planes. These extended defects will be described in more detail by Eyring and Tai in Chapter 4 of Volume 3.

A complex defect may be considered a combination of point defects. For example, in $UO_{2+\delta}$ there is evidence[40] that the excess oxygen is due to defects which consist of a combination of three oxygen interstitials, two oxygen vacancies, and two holes (U_U^{5+} ions). A complex defect differs from an associate (as discussed in Section

2.6.2) in that it is always assumed to exist in the combined state in the lattice. Equilibria between the complex and isolated defects are not considered.

The mass action law also may be applied to complex defects. Assuming that the defects in $FeO_{1+\delta}$ were complex defects consisting of two Fe vacancies plus an Fe interstitial,[41] Kofstad and Hed[42] wrote an equation similar to the following for the formation of complexes:

$$2Fe_{Fe} + \tfrac{1}{2}O_2 + 5V_I \rightleftharpoons (V_{Fe} - Fe_I - V_{Fe})^- + O_O + Fe_{Fe}^+ \qquad (111)$$

Five interstices are used because it is assumed that each complex defect will block an additional seven interstices for occupation by other complexes, but the addition of oxygen to the lattice will create two additional interstices ($\alpha = 2$). Also it is necessary to write a hole as Fe_{Fe}^+ since the normal Fe_{Fe} ions used to form holes must be explicitly taken into consideration at high concentrations of defects. Using Eq. (111), Kofstad and Hed obtained good agreement with experimental data of P_{O_2} as a function of deviation from stoichiometry. A more rigorous treatment of this system was carried out[43] using a modification of Eq. (77). Recent X-ray work[44] has shown that the defects in $FeO_{1+\delta}$ are more complex than considered here. However, the mass action law may be applied to other complex defects in the same manner.

An interesting application of the law of mass action, in which a type of complex defect is considered, is the treatment of vitreous oxides, such as silica, by Huggins and Huggins.[45] Although vitreous silica does not have a definite crystal structure, there is a considerable degree of short-range order. The authors have pointed out that silica consists of local structural groupings, designated "structons," which are listed in Table 8. The normal structons correspond to atoms on normal sites and the defect structons to the point defects in our previous discussions. The prime signs on the oxygen atoms in the formulas of the structons represent the number of bridge bonds to neighboring silicon atoms. Equations for the formation of defect structons may be written as follows:

$$\begin{aligned} 2\langle O''(4Si)\rangle &\rightleftharpoons \langle O'(Si)\rangle^- + \langle O'''(3Si)\rangle^+ \\ K_{112} &\simeq [\langle O'(Si)\rangle^-][\langle O'''(3Si)\rangle^+] \end{aligned} \qquad (112)$$

$$\begin{aligned} \tfrac{1}{2}O_2 + \langle O''(2Si)\rangle &\rightleftharpoons 2\langle O'(Si)\rangle^- + 2h^+ \\ K_{113} &\simeq p^2[\langle O'(Si)\rangle^-]^2/P_{O_2}^{1/2} \end{aligned} \qquad (113)$$

TABLE 8

Structural Units in Vitreous Silica

Structural species	Formula	Structural diagram	Charge
Normal structons:			
Si-centered structon	⟨Si(4 · O″)⟩	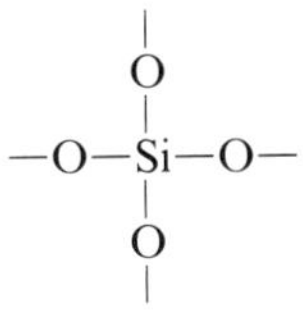 —O—Si(—O—)(—O—)—O—	0
Oxygen-centered structon	⟨O″(2Si)⟩	Si—O—Si	0
Defect structons:			
Si-centered structon with one oxygen neighbor nonbridging	⟨Si(O′, 3 · O″)⟩	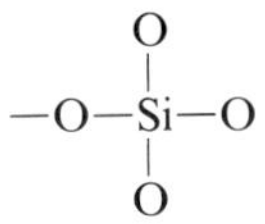 —O—Si(O)(O)—O	0
Si-centered structon with one oxygen neighbor forming a three-way bridge	⟨Si(O‴, 3 · O″)⟩	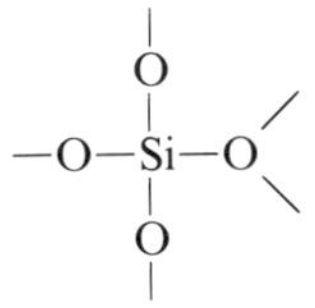 —O—Si(—O—)(—O—)—O(/)(\\)	0
Nonbridging oxygen structon	$\langle O'(Si)\rangle^-$	Si—O	−1
Oxygen-centered structon with three-way bridging	$\langle O'''(3Si)\rangle^+$	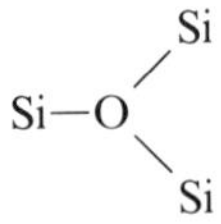 Si—O(Si)(Si)	+1

The electroneutrality condition is

$$[\langle O'(Si)\rangle^-] + n = [\langle O'''(3Si)\rangle^+] + p \tag{114}$$

From Eqs. (112)–(114) and Eq. (7), the authors were able to compute the concentration of each defect structon, as well as n and p, as a function of oxygen pressure using Brouwer's approximation method.

They also were able to obtain the effects of foreign alkali metal atoms (which are normally present as impurities in silica) on the concentration of each defect structon as was done for point defects in Section 3.1.2.

Microdomains are regions of one type of structure dispersed in another. Anderson[46] has indicated how a statistical thermodynamic approach might lead to expressions for defect equilibria in such systems. Indications are that randomly dispersed microdomains would be quite small (not much larger than point defects) in most cases. As yet, no attempt has been made to treat equilibria of crystallographic shear planes.

Acknowledgment

The author would like to thank the Kennecott Copper Corporation for providing the facilities at Ledgemont Laboratory during the writing of this chapter.

References

1. J. J. Burton, Vacancy-formation entropy in cubic metals, *Phys. Rev.* **B5**, 2948–2957 (1972).
2. A. C. Damask and G. J. Dienes, *Point Defects in Metals*, Gordon and Breach, New York (1963).
3. R. A. Johnson and W. D. Wilson, in *Interatomic Potentials and Simulation of Lattice Defects* (P. C. Gehlen, J. R. Beeler, and R. I. Jaffee, eds.), pp. 301–319, Plenum Press, New York (1972).
4. K. H. Bennemann, New methods for treating lattice defects in covalent crystals, *Phys. Rev.* **137A**, 1497–1514 (1965).
5. R. A. Swalin, Theoretical calculation of the enthalpies and entropies of diffusion and vacancy formation in semiconductors, *J. Phys. Chem. Solids* **18**, 290–296 (1961).
6. W. L. Korst and J. C. Warf, Rare earth hydrogen systems. I. Structural and thermodynamic properties, *Inorg. Chem.* **5**, 1719–1726 (1966).
7. G. G. Libowitz and J. B. Lightstone, in *Proc. 6th Rare Earth Research Conf., Gatlinburg, Tenn.*, Air Force Report AFOSR 67-1214, pp. 132–144 (1967).
8. M. D. Banus and T. B. Reed, in *The Chemistry of Extended Defects in Non-Metallic Solids* (L. Eyring and M. O'Keeffe, eds.), pp. 488–522, North-Holland, Amsterdam (1970).
9. G. G. Libowitz and T. R. P. Gibb, High pressure dissociation studies of the uranium hydrogen system, *J. Phys. Chem.* **61**, 793–795 (1957).
10. J. P. Pemsler and E. J. Rapperport, Thermodynamic properties of solid Au–Zn alloys by atomic absorption spectroscopy, *Met. Trans.* **2**, 79–84 (1971).
11. G. G. Libowitz, Point defects and thermodynamic properties in CsCl-type intermetallic compounds, *Met. Trans.* **2**, 85–93 (1971).
12. B. Fisher and D. S. Tannhauser, Electrical properties of cobalt monoxide, *J. Chem. Phys.* **44**, 1663–1672 (1966).
13. T. C. Harman, B. Paris, S. E. Miller, and H. L. Goering, Preparation and some physical properties of Bi_2Te_3, Sb_2Te_3, and As_2Te_3, *J. Phys. Chem. Solids* **2**, 181–190 (1957); G. R. Miller and C. Li, Evidence for the existence of antistructure defects in bismuth telluride by density measurements, *J. Phys. Chem. Solids* **26**, 173–177 (1965); R. F. Brebrick, Homogeneity ranges and Te_2-pressure along the three-phase curves for Bi_2Te_3(c) and a 55–58 at % Te Peritectic Phase, *J. Phys. Chem. Solids* **30**, 719–731 (1969).
14. R. W. Vest, N. M. Tallan, and W. C. Tripp, Electrical properties and defect structure of zirconia: I. Monoclinic phase, *J. Am. Ceram. Soc.* **47**, 635–640 (1964).

15. G. Brouwer, A general asymptotic solution of reaction equations common in solid state chemistry, *Philips Res. Rep.* **9**, 366–376 (1954).
16. F. A. Kröger and H. J. Vink, in *Solid State Physics, Advances in Research and Applications* (F. Seitz and D. Turnbull, eds.), pp. 307–435, Academic Press, New York (1956).
17. F. A. Kröger, *The Chemistry of Imperfect Crystals*, North-Holland, Amsterdam (1964).
18. W. Van Gool, *Principles of Defect Chemistry of Crystalline Solids*, Academic Press, New York (1966).
19. F. W. G. Rose, On the mass action laws in degenerate semiconductors, *Proc. Phys. Soc. London* **71**, 699–701 (1958); A. J. Rosenberg, Activity coefficients of electrons and holes at high concentrations, *J. Chem. Phys.* **33**, 665–667 (1960).
20. E. A. Guggenheim, *Mixtures*, Oxford Univ. Press, London (1952).
21. C. E. Messer and G. W. Hung, Dissociation pressures in the system LaH_2–LaH_3, 250–350°C, *J. Phys. Chem.* **72**, 3958–3962 (1968).
22. J. B. Lightstone and G. G. Libowitz, Interaction between point defects in nonstoichiometric compounds, *J. Phys. Chem. Solids* **30**, 1025–1036 (1969).
23. J. S. Anderson, The conditions of equilibrium of nonstoichiometric chemical compounds, *Proc. Roy. Soc.* (*London*) **A185**, 69–89 (1946).
24. J. C. Ward, Interaction between cation vacancies in pyrrhotite, *Solid State Commun.* **9**, 357–359 (1971).
25. M. Hoch, Order–disorder reactions in α-Ti(O) and TiO, *J. Phys. Chem. Solids* **24**, 157–159 (1963).
26. G. G. Libowitz, Nonstoichiometry and lattice defects in transition metal hydrides, *J. Appl. Phys.* **33**, 399–405 (1962); G. G. Libowitz and J. G. Pack, The gadolinium–hydrogen system at elevated temperatures. Vacancy interactions in gadolinium dihydride, *J. Phys. Chem.* **73**, 2352–2356 (1969).
27. M. Hoch, in *Phase Stability in Metals and Alloys* (P. Rudman, J. Stringer, and R. I. Jaffee, eds.), pp. 419–429, McGraw-Hill, New York (1967).
28. G. G. Libowitz, Nonstoichiometry in metal hydrides, *Advances in Chem. Series* No. 39, pp. 74–86 (1963).
29. A. L. G. Rees, Statistical mechanics of two-component interstitial solid solutions, *Trans. Faraday Soc.* **50**, 335–342 (1954).
30. W. A. Oates, J. A. Lambert, and P. T. Gallagher, Monte Carlo calculations of configurational entropies in interstitial solid solutions, *Trans. Met. Soc. AIME* **245**, 47–54 (1969).
31. A. B. Lidiard, Vacancy pairs in ionic crystals, *Phys. Rev.* **112**, 54–55 (1958).
32. K. P. Chik, D. Schumacher, and A. Seeger, in *Phase Stabilities in Metals and Alloys* (P. Rudman, J. Stringer, and R. I. Jaffee, eds.), pp. 449–467, McGraw-Hill, New York (1967).
33. H. Reiss and C. S. Fuller, Influence of holes and electrons on the solubility of lithium in boron-doped silicon, *Trans. Met. Soc. AIME* **206**, 276–282 (1956).
34. R. Parker and M. S. Smith, The solubility of nickel in nickel ferrite, *J. Phys. Chem. Solids* **21**, 76–80 (1961).
35. R. Adams, P. Russo, R. Arnott and A. Wold, Preparation and properties of the systems $CuFeS_{2.00-x}$ and $Cu_{1.00-x}Fe_{1.00+x}S_{2.00-y}$, *Mat. Res. Bull.* **7**, 93–100 (1972).
36. Y. D. Tretyakov and R. A. Rapp, Nonstoichiometric and defect structures in pure nickel oxide and lithium ferrite, *Trans. Met. Soc. AIME* **245**, 1235–1241 (1969).

37. H. Schmalzried, Point defects in ternary ionic crystals, in *Progress in Solid State Chemistry*, Vol. 2, pp. 265–303, Pergamon Press (1965).
38. W. L. George and R. E. Grace, Formation of point defects in calcium titanate, *J. Phys. Chem. Solids* **30**, 881–887 (1969).
39. L. Eyring and M. O'Keeffe (eds.), *The Chemistry of Extended Defects in Non-Metallic Solids*, North-Holland, Amsterdam (1970).
40. B. T. M. Willis, Positions of the oxygen atoms in $UO_{2.13}$, *Nature* **197**, 755–756 (1963).
41. W. L. Roth, Defects in the crystal and magnetic structure of ferrous oxide, *Acta Cryst.* **13**, 140–149 (1960).
42. P. Kofstad and A. Z. Hed, Defect structure model for wustite, *J. Electrochem. Soc.* **115**, 102–105 (1968).
43. G. G. Libowitz, in *Mass Transport in Oxides* (J. B. Wachtman and A. D. Franklin, eds.), pp. 109–118, Natl. Bur. Std. Special Publ. 296 (1968).
44. F. Koch and J. B. Cohen, The defect structure of $Fe_{1-x}O$, *Acta Cryst.* **B25**, 275–287 (1969).
45. R. A. Huggins and M. L. Huggins, Structural defect equilibria in vitreous silica and dilute silicates, *J. Solid State Chem.* **2**, 385–395 (1970).
46. J. S. Anderson, in *Problems of Nonstoichiometry* (A. Rabenau, ed.), pp. 1–76, North-Holland, Amsterdam (1970).

7

*Characterization of Solids—Chemical Composition**

W. Wayne Meinke
Analytical Chemistry Division
Institute for Materials Research
National Bureau of Standards
Washington, D.C.

1. Introduction

In any series of studies, experimental or theoretical, in the chemistry, the physical chemistry, or the chemical physics of solids it is very important that there be reliable, descriptive, analytical information available about the materials used in the studies.[1,2] Such information is obtained through the process called "characterization," which has been given the following definition by the National Academy of Sciences—National Research Council Committee on Characterization of the Materials Advisory Board:[3]

> "Characterization describes those features of the composition and structure (including defects) of a material that are significant for a particular preparation, study of properties, or use, and suffice for reproduction of the material."

Only when optimal information is given on the identity and the location of the atoms in a particular material can one be confident that the material can be reproduced, and therefore that the measurements and theories involving this material will have lasting significance.

This principle is easy to accept as a general statement of philosophy. However, there are many examples of elaborate scientific studies in which it was belatedly discovered that the effects measured were a function of unknown impurities in the material, of surface phenomena, etc.

Selective characterization of materials used in studies of properties is essential because composition and structure effectively determine the properties of materials. However, chemical analysis, i.e., the determination of chemical composition, can present a bewildering maze of information to the materials investigator interested in obtaining basic characterization data. He needs perspective as to the possible utility of different types of analytical techniques so that he can make intelligent decisions as to the analytical approaches that will be most significant for his problems.

It is this perspective that this chapter is designed to furnish. The Report of the Materials Advisory Board Committee[3] presented an integrated summary of pertinent information on many analytical techniques that can be used in determining the composition of solids.* Unfortunately, the report has had only limited distribution and its information is now somewhat dated.

This chapter is modeled after the MAB Report, updating and expanding the coverage to correspond to the state of the art in January 1973. A short description of various analytical techniques used to obtain composition information on materials is followed by a tabular summary of sensitivities and precisions of these techniques. The next section describes the application of these techniques to specific characterization problems such as stoichiometry, homogeneity, and oxidation state, as well as survey and quantitative methods of measuring impurities. Finally, examples are given from the literature of detailed studies on a number of high-purity materials to illustrate the present state of the art of characterization of practical samples.

2. Current Capability for Determination of Chemical Composition

2.1. Introduction

Any meaningful discussion of current capabilities in the measurement of composition must focus on the real world of practical materials.

---

* The original members of the Panel on Composition of the MAB Committee that prepared this report were: M. S. Sadler, DuPont, Chairman; R. A. Laudise, Bell Laboratories; R. J. Maurer, University of Illinois; W. W. Meinke, National Bureau of Standards; and A. Wold, Brown University.

In 1933 G. E. F. Lundell, then the Chief Chemist of the National Bureau of Standards, wrote a very candid article entitled, "The Chemical Analysis of Things as They Are."[(4)] He said he used that title because so many articles on analysis appearing in the literature dealt with the "chemical analysis of things as they are not."

Much the same problem exists in the materials analysis field today. The scientific literature contains numerous references to materials of 5-9's and 6-9's purity. Such statements are misleading and of questionable validity because the estimates are in many cases derived from resistivity measurements supplemented by emission spectrographic analyses (with a sensitivity of only 1–10 ppm for most elements). To establish that a sample of material contained less than 1 ppm total impurities would require analyses for all elements present by techniques with sensitivities and accuracies in the range 1–20 ppb.

The magnitude of effort required is illustrated by the classic work on zone-refined aluminum by Albert of the CNRS Laboratory in Paris.[(5)] Samples were analyzed for over 60 elements plus the rare earths by high-sensitivity neutron activation using detailed radiochemical procedures. Elements such as carbon, oxygen, and nitrogen were determined by photonuclear or charged-particle activation. This procedure required the efforts of a four-man team for 12 hr, an additional person for nine days, and another person for two weeks to analyze the rare earths. After the amounts of individual contaminants were totaled (for many there were only experimental upper limits of 1–10 ng), it was possible to establish that a particular sample of aluminum contained less than 2 ppm total impurities (i.e., it was not quite 6-9's pure).

Unfortunately, little confidence can be placed in most general statements of purity, such as 6-9's. It is imperative that analytical information be developed from more discriminating techniques.

2.2. General Overview

The following sections contain short summaries of a number of different analytical techniques that are useful in the determination of chemical composition. The first techniques include those that usually require dissolution of the sample, with measurements being made on the resulting solutions. Then techniques that can be applied directly to the sample with or without some modification of the sample itself are described. Finally, techniques that provide chemical structure information as well as information on the localization of elements and on surfaces are discussed. The major emphasis is on the analysis of inorganic materials, although a number of techniques that give information on the rapidly developing field of organic materials have also been included.

For each section several definitive references enable the reader desiring more information to go directly to authoritative literature that describes the technique and defines some of its important characteristics and limitations.

There are several general references that are applicable to many of the areas described. Kolthoff and Elving's *Treatise on Analytical Chemistry*[6] is a many-volume work that treats in detail most of the individual techniques described in the chapter as well as many applications of these techniques of interest to the materials scientist. The proceedings of the first NBS Institute for Materials Research Symposium on Trace Characterization, Chemical and Physical[7] contains a number of definitive papers. Furthermore, each year the journal *Analytical Chemistry* publishes a set of reviews in many areas of analytical chemistry that should be consulted for information on specific materials problems. In the even years these appear as fundamentals reviews,[8] while in the odd years they are applications reviews.[9]

Although a serious attempt has been made to be selective and candid in the references, the reader should be aware that measurements at or near the limits mentioned may prove very difficult unless the analyst has had considerable experience in applying the method to the matrix described.

Experience at NBS in the certification of high-purity Standard Reference Materials has illustrated the point that on most real samples, trace element analysis below the ppm level gives values of questionable accuracy.[10,11] All trace methods of analysis except activation analysis must deal realistically and directly with problems of "blank" contamination, i.e., contamination introduced into the sample through preliminary handling and/or subsequent processing in the analytical procedure.

Furthermore, interpretation of the analytical results must include critical consideration of the "sampling" procedure, e.g., how representative the analyzed sample is of the total material being used for materials studies.

2.3. Analytical Techniques: Present Status

2.3.1. Wet Chemistry—Classical Gravimetry and Titrimetry

Gravimetry,[12,13] or quantitative analysis based on the weight of a reaction product, is the oldest of the analytical techniques and one of the most useful for major and minor constituents. It is capable of high accuracy, especially when corrections for solubility are made. Much of the early atomic weight work, with precisions as high as 0.001 %, was based on this technique.

Titrimetry,[12,13] or quantitative analysis based on the amount of a standard solution consumed in a reaction, is another valuable technique for the analysis of both major and minor constituents. It is capable of high accuracy, especially when the amount of reagent consumed is determined by weight, and in such cases accuracies of 0.01 % have been obtained. Volumetric titrimetry can readily yield results of 0.1 % accuracy.

Classical gravimetric and titrimetric procedures based on reactions in aqueous or liquid media have been designated as "wet chemical methods" to distinguish them from instrumental analytical methods. In a recent discussion of the criteria for the development of gravimetric methods it was pointed out that, with some exceptions, it is difficult to find an instrumental method that does not use classical analyses to provide the compositions of standards required for calibrating the instruments.[14]

The relatively high accuracy of these gravimetric methods stems from the ability to assess systematic errors from theoretical equilibria calculations,[15] and from the relatively small corrections applied to the final weighed products. Present-day automated balances have greatly simplified the weighing process. Gravimetric methods, when applied to the newer radiochemical, stable-isotope mass spectrometric, and other instrumental methods, play a significant role in current analytical procedures for chemical characterization of a wide variety of materials. The continuous growth in use of these techniques arises from the extensive applications of complexometric reactions that assure greater specificity.[16,17]

Titrimetric (or volumetric) methods depend on the detection of an "end point" of the stoichiometry of a reaction. Strictly interpreted, the classical methods are limited to visual indicators of the end point. There are, however, other instrumental modes that include a variety of electrometric, spectrophotometric, spectrofluorimetric, and enthalpic techniques. These latter methods are also amenable to automation and computer interfacing. As indicated in the literature,[18,19] this field offers selective and sensitive methodology for major and trace components of many different materials.

2.3.2. Coulometry

Coulometric measurement involves the quantitative electrochemical conversion of a constituent in solution from one initial oxidation state to another, well-defined oxidation state. The quantity measured is the charge necessary to perform this conversion. Since electrons are essentially being used as the measured reagent, this method is capable of high precision and accuracy.

In an electrolysis system the conversion can be performed in two experimentally controlled modes: controlled-potential mode and constant-current mode. In the former case one selects the potential to be maintained at such a value that only the desired reaction can proceed. In the second mode, the constant-current mode of electrolysis, a "depolarizer" introduced into the electrolysis system reacts at the electrode to produce a titrant for the species in question. Thus the name "coulometric titration" has been attached to this mode of electrolysis.

Coulometry can be classified as a macro method, involving grams of sample.[20] The use of submilligram samples in coulometry, on the other hand, classifies it equally well as a micro method.[21] High-precision coulometry has been widely used for the determination of major constituents of various materials.[22] Determination of trace impurities, e.g., dopants such as chromium in ruby lasers, has been performed at the nanogram level.[21] It has been used successfully for the determination of stoichiometry of materials such as GaAs,[23] for measurement of physical constants such as the faraday[24] and atomic weights,[25] for determination of major constituents of Standard Reference Materials,[26] and for the analysis of important research materials such as separated isotope solutions.

2.3.3. Ion-Selective Electrodes

Ion-selective electrodes[27,28] can be used for analytical determinations over an extremely wide range of solution concentrations (from 10^{-5} to 10^{-7} M to saturation) and are therefore applicable to trace as well as to major-constituent analysis. There are presently about two dozen ion-selective electrodes commercially available for cations, including the heavy metals; for anions such as fluoride, sulfide, and the halides; and for gases such as oxygen, carbon dioxide, and ammonia.[29]

These sensors are usually used in one of two operational modes, direct potentiometry or potentiometric titrations. Direct potentiometry (similar to a *p*H determination with a *p*H meter), which is based on the correlation of the electrode emf to a standard or calibration curve, is normally limited to a precision of, at best, 0.5 %. It is, nonetheless, a technique that is very simple, rapid, and convenient to use.

Greater precision can be attained with a titration procedure but with a concomitant increase in experimental sophistication. Typically, precisions on the order of 0.1 % are achieved, while under optimum conditions this may be improved by one order of magnitude. A titration procedure has an additional advantage in that the number of species that can be determined is greatly increased by using indirect titration methods.

A good, working knowledge of wet chemical procedures, solution equilibria, and electrode limitations is necessary for the reliable application of these sensors to analytical problems. Although analogous to electrometric *p*H determinations in theory and in operation, ion-selective electrodes are considerably more subject to chemical interferences and prone to nonideal behavior.

2.3.4. Polarography

This versatile solution technique[30,31] covers the entire concentration spectrum from trace quantities through major constituent levels. In principle, any element capable of electrooxidation or electroreduction can be determined, and the polarographic behavior of some 80 elements has been described. Polarographic half-wave potentials are ordinarily sufficiently different to make possible the simultaneous determination of several elements in the same solution, thus minimizing the need for preseparations in many instances. Sensitivities in the range 0.1–0.001 ppm have been reported, depending upon whether conventional or modified techniques such as cathode-ray,[32] pulse, or anodic stripping[33] polarography are employed. Precision near the limit of detection is about 20 %, increasing to about 2 % in favorable concentration ranges. Differential techniques that have been developed make possible the determination of microamounts of major constituents with standard deviations of 0.02–0.05 %.[34] Polarographic techniques have made important contributions to the trace analysis of a number of reference materials, including trace glass samples,[35] lunar samples,[36] and botanical samples.[37]

2.3.5. Spectrophotometry

Absorption spectrophotometry is one of the most widely used analytical techniques for solutions. For example, of the 10^9 clinical analyses made in the U.S. in 1972 more than 90% used spectrophotometry as the final measurement step. It is used in many laboratories for trace analysis[38] because of its simplicity, modest apparatus requirements, sensitivity, and good precision (1–5%). Although it is primarily a trace method, differential techniques also permit the determination of major constituents with a relative standard deviation of 0.1%. By the use of photon counting,[39] a precision of 0.01% should be attainable.

While in most cases the color reactions used are not specific, numerous procedures can be made selective by the proper choice of *p*H and the addition of masking agents. Recent developments[40,41] in double-wavelength spectrophotometry have further increased the specificity of many direct methods. Separation and preconcentration techniques can be applied, but these impose a restriction on the detection limit based on the reproducibility and magnitude of the blank. Theoretical and practical considerations presently set the limit of detection on the most sensitive spectrophotometric methods at 5–20 ng.

In addition to the analysis of solutions, spectrophotometry has also found unique applications in the nondestructive determination of doped impurities in single crystals such as, for example, the determination of chromium in ruby.[42]

2.3.6. Spectrofluorimetry

Since the 1950's fluorescence has enjoyed increased use and acceptance as an analytical measuring tool because of increased instrumental refinements and the method's inherent sensitivity and selectivity. High detection sensitivity results from the fact that the signal-to-noise ratio is large and can be electronically amplified, while selectivity results from the fact that either the emission or excitation spectra can be used to characterize the fluorescing species. In addition, analytically useful information such as sample purity, concentration, metal oxidation state, and number of binding sites can be obtained from consideration of quantum efficiencies and lifetime and polarization measurements.

This is a rapidly developing field.[43] Analytical procedures can be established by several methods: specie fluorescence; fluorescence quenching; chemically induced fluorescence (e.g., chelation of non-fluorescent metal ions with fluorescent ligands); and enzymic reactions that produce fluorescent products.[44] Sample concentrations and identities can be determined in solution, on powders, or on glasses.

Fluorimetry is essentially a trace method, since the emitted light is proportional to specie concentrations only at low concentration levels (10^{-4}–10^{-9} M). Concentration quenching usually occurs at higher levels, and linearity is destroyed. At the limits of detection, fluorescence can be drastically affected by the presence of fluorescing impurities, by any process which quenches fluorescence, or by interferences by Raman or Rayleigh scattering.

Microspectrofluorimetry is also being used to determine luminescence in heterogenous systems such as powders[45] or cells,[46] with an ultimate sensitivity on the order of 10^{-14} g of material.

2.3.7. Organic Microanalysis

Microanalysis, the detection and identification of materials present in small size but relatively high concentration, is distinct from trace analysis, which is concerned with the characterization of small concentrations of material. Organic microanalysis is usually taken to mean elemental analysis (primarily C, H, O, N, P, S, Cl, Br, I, and Si), and functional group analysis (acetyl, carboxyl, benzoyl, amino, nitro, hydroxy, etc.) on samples usually 1–10 mg in size. The semiquantitative results, accurate to about 10%, serve as a measure of impurities, or inhomogeneity, or for structure determination in solid organic substances. Accurate results of 1% or better may be expected when large (1 g) samples are taken for analysis and the entire chemical apparatus is scaled upward in size. However, small samples take less time to analyze, so the micro methods are more popular than macro methods.

Recent texts[47,48] describe modern methods for organic elemental and functional group analysis that still depend heavily on classical wet chemical procedures. Automated instruments are now available for laboratories with heavy work loads of routine C, H, O, and N analyses. A rapid, semiquantitative analytical procedure for organic functional groups and for inorganic elements in organic and inorganic materials is afforded by the Weisz ring-oven.[49] This is an instrument for separating the components in a drop (0.05 ml) of solution, in much the same way as in thin-layer and paper chromatography. The separated components are not in the form of spots, however, but are concentrated into sharp lines by a heated platen (the oven) that evaporates the solvent and deposits all the solute at one specific location on the paper or other chromatographic substrate for further handling. Ring-oven separations of organic materials may be followed by color spot tests,[50] or the rings containing the solutes may be removed from the substrates, as in thin-layer chromatography, and characterized by sensitive instruments.

The identification of a completely unknown organic substance is a very difficult undertaking because of the very large number of different organic materials presently known. Most often there is at least a partial knowledge of the composition of the material, and additional information on purity or structure is needed. Then organic microanalysis can help complete the picture.

2.3.8. Thin-Layer Chromatography (TLC)

This semiquantitative method for the analysis of mixtures of nonvolatile organic or inorganic substances combines a simple means for separating the constituents of the dissolved mixture with their detection by the application of chemical and physical tests.[51] If the mixture contains unknown substances, identification can be made by molecular characterization of extracted constituents; somewhat less assurance is afforded through correlations with the TLC behavior of known compounds.

In practice, TLC can be used to analyze 300–1000 μg of a mixture applied to the plate as a single spot. The chemical and physical properties of each separated constituent, or those of the variety of products into which the constituent may be converted to enhance detection, provide its limit of detection. In favorable cases as little as 10–30 μg of a compound may be detected; commonly the limit is 100–1000 μg; occasionally more. Preconcentration of specific constituents of a mixture by liquid chromatography or thick-layer (preparative) chromatography can provide improved sensitivity.

Quantitative analysis may be performed directly on the thin-layer plate in a variety of ways: by visual comparison; by measurement of spot areas; by the transmittance of spots that are colored, are charred, or that absorb ultraviolet light; by reflectance; or by fluorescence. Alternatively, the zone of absorbent containing the constituent can be scraped from the plate and extracted. The extract is then analyzed by any of a variety of techniques appropriate for the substance. An extensive variety of organic and inorganic materials has been studied by TLC, and recent compendia[51,52] provide excellent sources of reference to specific applications.

2.3.9. Gas Chromatography*

This technique[53] assumes that the sample components of interest may be volatilized, have a vapor pressure of 0.1 mm or greater at a temperature less than 400°C, and are thermally stable at that temperature. In this manner, major and minor components for both organic and inorganic analyses can be separated and analyzed. The range of analysis nominally extends from major components to impurities present in concentrations as low as 10^{-12} g. Solids, liquids, and gases with molecular weights from 2 to 200,000 may be analyzed, although special sampling techniques may be required for nonvolatile compounds. These techniques include derivatization,[54] pyrolysis,[55] and indirect analysis of reaction products.[56] The time required for gas chromatographic separations is normally a few minutes; however, sample separation, quantative data reduction, and sample identification usually extend the total analysis time to several hours, depending upon the complexity of the sample.

Chemical characterization by gas chromatography is usually directed toward trace organic analysis. The accuracy of the analysis is limited by the sampling and injection procedures, chromatographic resolution, detector calibration, peak area measurements, and the availability of known standards. The relative precision is highly dependent upon the analyst's experience and understanding of the technique and on the concentration level being determined. The best possible precision that can be obtained under the most favorable conditions is 0.1% for major components in near-ideal samples, increasing to $\geq 10\%$ below the 10 ppm level.[57]

Identification of chromatographic effluents is most definitively accomplished by coupling to a mass spectrometer. Virtually every instrumental method of analysis, however, has been employed either on-line or off-line for peak identification.

2.3.10. Liquid Chromatography

High-performance liquid chromatography is emerging as a "new" separation and analysis technique for the characterization of volative

* A more correct designation is "gas–liquid chromatography."

and nonvolatile components.[58] The emphasis is directed primarily toward organic analysis because of the many potential applications in the analysis of complex materials. The sample must be soluble in a nonreactive, nonviscous solvent for separation on the basis of adsorption, partition, ion exchange, or molecular size.

The moving or mobile phase is a liquid of well-defined composition and is pressurized to 500–3000 psi in order to force the sample through 1-m columns. The chromatographic column is packed with particles ranging in diameter from 5 to 100 μm and analyses can be carried out in a few minutes for pesticides or in a few hours for urine separations. Stable and tailored column materials have only recently become available and are directly responsible for the resurgence of interest in liquid chromatography.[59] However, these materials are expensive ($5–$500 per column) and special low-volume (10 μl) detector cells are required. Analysis and detection is primarily by UV or differential refractive index measurement. These detectors are not particularly sensitive, and auxiliary methods are required for identification. Lower limits of detection for the UV detector depend on the molar absorptivity of the analyte.[60] Many compounds may be measured with this detector in the ppm to 0.001 ppm concentration range, while others will be best analyzed by the refractive index detector, with a lower limit of detection of approximately 1 ppm. This sensitivity may be extended by a judicious choice of mobile phase.

2.3.11. Activation Analysis

This technique[61–63] is particularly useful for ultra-trace analysis of solid samples in that it eliminates most of the problems of reagent "blanks" and/or sample contamination during analysis.[11] The large number of variables (such as type and energy of irradiating particle, time of irradiation and measurement, type of detector used, etc.) that can be judiciously altered for particular analyses make the technique relatively free from serious systematic errors or biases. A typical value for the random errors is $\pm 5\%$ at the 100-ng level.

The use of high-resolution Ge(Li) gamma-ray detectors allows limited nondestructive multielement analysis, but when the number of elements determined is large (>10) precision and accuracy for many of the elements may be expected to suffer ($> \pm 20\%$). This nondestructive approach,[64] i.e., "activation spectrometry,"[11] usually requires two or more separate irradiations with corresponding decay periods prior to measurement of up to 30 days or even longer for some matrices. This long analysis time is unacceptable for many applications.

Many laboratories employ radiochemical separation techniques to isolate groups of elements from the sample after irradiation. The combination of simple group separations and high-resolution gamma-ray spectroscopy reduces substantially the analysis time and analytical costs, and makes the technique nearly a true "multielement" method. Additional effort is being expended to automate these group separations to further decrease analytical costs.

Activation analysis with neutron generators (fast neutrons) provides a nondestructive method for the determination of oxygen at the 10–100-μg level. Carbon, nitrogen, and oxygen can be determined in submicrogram quantities with high-energy photons[65] (bremsstrahlung from electron linear accelerators, LINAC's) although no useful nitrogen determinations have been demonstrated in practical matrices. The short half-life (124 sec) of ^{15}O makes this measurement difficult, and methods development for a given matrix is necessary before the method can be applied routinely. Carbon analyses have been performed routinely with the LINAC, with excellent sensitivity and accuracy.

Activation analysis with charged particles, although essentially a surface technique, has been used to determine carbon, nitrogen, and oxygen, especially in high-purity silicon,[66] and offers excellent sensitivity and good reliability when done carefully.

2.3.12. Vacuum Fusion, Inert Gas Fusion, and Extraction

The effects of trace amounts of interstitial elements (H_2, O_2, and N_2) on the chemical and physical properties of materials have been widely recognized. From among the many techniques—activation analysis, mass spectrometry, spectroscopy, wet chemical, vacuum fusion, and inert gas fusion—the latter two have probably found widest application. Horton and Carson[67] describe advances in instrumentation for solid samples that have increased the daily output from a few samples to 60 samples. This in turn permits better evaluation of the experimental parameters and leads to the improvement of precision. However, the accuracy is still dependent upon the availability of appropriate standards.[68] Furthermore, when the concentration of gases is very low, below 1 ppm, one must distinguish carefully between that portion present on the surface and that in the bulk of the material.[69] Systems with computer-controlled interfacing have recently been described[70] that provide fully automated analysis.

2.3.13. Thermal Analysis

Thermal analysis gives information on the fundamental behavior and structure of materials based on their thermochemical and thermophysical properties. Differential thermal analysis (DTA), differential scanning calorimetry (DSC), thermogravimetry (TG), dilatometry, and other related dynamic thermal methods serve as analytical tools for characterizing a wide variety of solid materials.[71] Information obtainable by these methods includes phase relationships, identification and measurement of impurities in high-purity materials, "fingerprint" identifications, thermal histories of the material, and dissociation pressures.

Thermogravimetry offers accurate measurements of reactions involving weight changes and can be used in assessing the purity of materials. On the other hand, DTA measurements can be made to closely approximate the thermodynamic conditions of a process and elucidate the mechanism of reactions. Proceedings of several recent conferences,[72–74] as well as review articles,[75] describe a wide range of applications, from fundamental thermodynamics and kinetics to industrial process control.

2.3.14. X-Ray Fluorescence Spectrometry

This technique[76,77] is useful for the determination of major and minor constituents in solid samples with a typical sensitivity of 20–200 ppm, depending upon the atomic number of the element and the nature of the specimen. In favorable cases, or when preconcentration methods are used, the sensitivity may be extended to as low as 0.1 ppm.[78] Commercial instruments are available for analysis with X-ray wavelengths shorter than 10 Å; these permit the determination of any element heavier than sodium (atomic number 11).

The precision of X-ray spectrometric analysis (0.1 % in good cases; more typically 0.5 %) is frequently competitive with quantitative "wet" chemical analysis and often superior to other "instrumental" techniques. For X-ray analytical techniques to be accurate, standards of composition close to that of the samples, or suitable correction procedures for any existing interelement effects, must be available. X-ray analysis is widely used in industry because of its accuracy and because it lends itself to automated measurements and data evaluation procedures. Of additional advantage, it is usually rapid and often nondestructive. Samples that can be analyzed cover a broad range, including alloys, powders, solutions, slag, slurries, and ores.[79]

With the development of solid-state detectors of improved resolution, the application of energy-dispersive X-ray fluorescence analysis, using radioisotope or X-ray tube excitation, is extending rapidly. Sensitivities of a few ppm on simple matrices (e.g., organic materials) have been obtained.[80] The energy-dispersive method offers economical means of analysis for a wide range of materials, including metals, ores, and solid atmospheric pollutants.

2.3.15. Spark Source Mass Spectrometry (SSMS)

This instrument can detect ~85 elements at concentrations as low as 1–10 ng/g. Conducting and semiconducting solids are the ideal sample forms.[81] Powders, insulators, liquids, etc. can also be analyzed with various techniques, but with possible loss in sensitivity and/or introduction of contamination. Interferences between elements will exist for all samples, depending on the elements present and their concentrations. The precision of an SSMS analysis at the 1 μg/g concentration level is ±5–20% for photographic detection and ±2–5% for electronic detection.[82]

The accuracy of an analysis can approach these values when Standard Reference Materials of the same matrix are available or when a synthetic standard can be applied. In the absence of standards, concentrations can be estimated to within a factor of 3–10 of the true value. This method of analysis is ideal for a general survey covering all possible trace elements. The results of such survey analyses can then be used to assign the analysis of specific impurities to other trace techniques when more accurate values are required.

2.3.16. Isotope Dilution—Spark Source Mass Spectrometry

Isotope dilution analysis with the SSMS can be used with almost all of the polynuclidic elements. Accuracies of ±5–10% are obtained on 0.1–1 μg of a trace element using photographic detection;[83] better accuracies should be obtainable with electronic detection. Analysis of trace impurities at 0.01 ng/g concentrations has been possible[84] using SSMS isotope dilution on samples amenable to preconcentration techniques. Isotope dilution requires dissolution of samples before the separated enriched isotope (spike) of each element being determined is added and equilibrated. Chemical separation steps are then required to obtain the elements in a measurable form. However, since this technique is capable of determining many elements in a single sample simultaneously ($\sim$17 elements in acids), general group separations can be used.[84]

2.3.17. Isotope Dilution Mass Spectrometry

The availability of separated isotopes in the late 1940's made possible analyses by isotope dilution mass spectrometry.[85] Geologists were the first to capitalize on the new tool, and by the mid-1950's the field of geochronology was well established. Extremely precise and accurate concentration determinations down to the sub-ppm level are required for meaningful age calculations.

Geologists and lunar scientists now use this technique to measure not only elements like lead, uranium, thorium, rubidium, strontium, argon, and potassium, but they also now include barium and ten of the rare earths in geologic analysis. The nuclear industry measures lithium, boron, uranium, and plutonium routinely. The technique has a potential for broad application. Major and minor phases of such elements as molybdenum (55%) and rhenium (0.1%) can be determined because accuracy is possible even in the presence of major interferences. At impurity levels of 1–500 ppm almost all of the above-mentioned elements can be determined, and, in addition, such elements as magnesium, chromium, copper, silver, calcium, nickel, and cesium.[36,86,87]

It is possible to analyze approximately 40 elements by thermal ionization mass spectrometry and another ten elements by electron impact (including gaseous compounds of elements such as nitrogen, oxygen, carbon, and silicon). When isotope dilution is applicable, it usually is by far the most accurate method for the determination of impurities and traces. The leverage for precision and accuracy comes from the limited number of quantitative steps necessary for the analysis.

Certification of the elements important to geochronologists in the NBS Trace Glass Standard Reference Materials[35] has been accomplished in large part by this technique. Measurements of these elements at the 500-ppm level are certified at an accuracy of $\sim$0.2%; at the 50-ppm level at an accuracy of 0.2–0.5%; at the 1-ppm level at an accuracy of 0.2–0.8%; and at the 0.02-ppm level at an accuracy of

1.5–3.0%. Such measurements, especially at the lower two concentrations, require extreme care, special clean facilities, special clean reagents and containers, and a large amount of time and effort.[11]

2.3.18. Emission Spectroscopy

Qualitative analysis by emission spectroscopy[88,89] permits rapid, simultaneous detection of some 70 metallic and metalloid elements. Only a few milligrams of a solid sample are required. An estimate of the concentration of an element, often within one order of magnitude or somewhat better, can be made from intensities of the spectral lines. The experimental limit of detection of each element depends on the nature of the sample, the excitation procedure, and the properties of the spectrograph and detector.[90] Detection limits cited in the literature range down to nanogram levels for many elements. However, on complex practical samples few reliable quantitative measurements below the 1–10 ppm level are reported in the literature.

Recent developments with controlled spark discharges, induction-coupled plasmas, capillary arcs, and laser excitation have considerably expanded the capabilities for quantitative analysis by emission spectroscopy.[91] The detection of nonmetallic elements, including carbon, nitrogen, oxygen, and the halogens, has been reported, often at concentrations as low as 10 ppm.[88]

The unavailability of suitable standards can severely restrict the quantitative applications of this method. For some sample forms standards can be prepared from the sample by the method of additions. Preconcentration methods that utilize chemical separation are also useful for spectrographic analysis; these methods facilitate the preparation of standards because the matrix has been simplified. On the other hand, such preconcentration methods or the preparation of standards by the method of additions increase the likelihood of contamination of the sample. "Common matrix" methods, involving dilution of the samples with a material such as germanium or gallium oxide, are not applicable to the analysis of the purest materials, since the dilution process raises the limits of detection.

2.3.19. Flame Emission, Atomic Absorption, and Atomic Fluorescence Spectrometry

These three spectroscopic methods[92–95] employ rather similar equipment, and some commercial apparatus offers interconvertibility.

The methods are complementary in that elements with poor detection limits by one technique can often be determined with high sensitivity by the other. They also share many problems, particularly with respect to interferences.

The sensitivity of detection of some 70 elements by flame emission or absorption spectrometry can, in many instances, be very low.[96,97] The level of detection of sodium by flame photometry is often limited by the residual content of the water or other solvent used for the sample; e.g., 1 ng/g of sodium can be detected without difficulty. Detection limits reported for other elements range from 1 to 1000 ng/g by one or the other of these two techniques. The potential capabilities of these methods have been considerably improved by the introduction of nonflame techniques.[98]

All of these methods require that the sample be dissolved, which necessarily causes some dilution and a consequent loss of sensitivity. In addition, matrix effects can cause serious error unless standards similar in composition to the sample can be prepared. The use of organic solvents, rather than water, often improves the limits of detection, occasionally by an order of magnitude or more; flame emission and absorption measurements are often especially useful in the determination of impurities following solvent extraction.

Atomic fluorescence signals have been observed for 35 elements in flame and nonflame devices, and preliminary work[92] indicates that this method may provide very low limits of detection. Matrix effects may, however, cause difficulties in the analysis of practical samples.

2.3.20. Mössbauer Spectroscopy

This analytical method[99,100] has been used to great advantage in the determination of chemical structure.[101] It complements the other hyperfine interaction spectroscopies (such as nuclear magnetic resonance) in several ways. The technique observes the changes in two energy levels within the nucleus as a function of chemical environment, as contrasted to changes in the ground state only, e.g., in NMR. The Mössbauer technique, in contrast to NMR, is primarily used in the solid state. The broadening of NMR spectra caused by solid-state interactions is absent in Mössbauer spectroscopy because recoilless emission in Mössbauer spectroscopy demands no solid-state (phonon) interaction.

The technique has proven to be a powerful tool for measuring properties such as magnetic and electric field interactions in the solid state.[102] It has also proven useful in the fields of metallurgy, solid-state physics, coordination chemistry, and biochemistry, and in the theory of the hyperfine interaction.

At present the elements with which this effect can be observed with high practicability are limited to iron, tin, iodine, gold, and a few rare earths.

The inherent excitation linewidth guarantees specificity of elemental detection. Most Mössbauer measurements have been made on materials in which the Mössbauer resonating element is a macroconstituent in the matrix. Some work has been done in which the matrix investigated constitutes the source of excitation for Mössbauer spectroscopy and in this sense constitutes an analysis of a trace impurity in the matrix.[103]

2.3.21. Nuclear Magnetic Resonance (NMR)

NMR is a well-established method[104–106] for identifying and studying structure of molecules in the liquid state. Since solids are in general characterized by broad resonance lines with little fine structure, application of NMR to the characterization of solids with respect to composition is limited and highly specialized. Sensitivity varies with the kind of nucleus but generally does not exceed 10^{19} nuclei/cm^3 or 0.1 at.%.

In metals with cubic lattice symmetry, determinations of impurities at concentrations of 0.1 at.% have been made using quadrupole interactions and changes in the Knight shift. Another approach involves study of the coupling between the nuclear spin system and the lattice through measurement of the spin–lattice relaxation time.[107,108] In nonmagnetic solids paramagnetic impurities affect the relaxation time dramatically, and these effects can be detected for impurities in the 10–100-ppm range. However, these latter methods are not specific for a single impurity but are influenced by valence, size, or some other property that may be common to a number of impurities.

2.3.22. Electron Spin Resonance (ESR)

This technique[109] is useful for studying conduction electrons in metals and semiconductors as well as bonding of transition metal ions and other paramagnetic species. In general, it is more suitable for determination of environment than for identification of elements and measurement of concentration. Sensitivity varies with the paramagnetic species, and for transition elements under optimum conditions is in the range of 10^{10} spins/cm^3. Therefore it can be used to detect trace impurities in special cases; e.g., free radicals in an organic matrix[110] or transition metals in a diamagnetic solid.[111]

2.3.23. Residual Resistance Ratio (RRR)

This method[112] measures the sum of the "electrically active" elements and is highly sensitive. It does not identify impurities, and different impurities affect residual resistance in varying degrees; some contaminants have little or no appreciable effect. RRR finds wide application in the characterization of metals,[113,114] but from an analytical viewpoint it is of value primarily to check the consistency of other analytical data for high-purity materials.

2.3.24. Electron-Probe Microanalysis

In many types of solid samples the distribution of the elements present is as important as the average composition. This is particularly true for many alloys, minerals, ceramics, and semiconductor devices. The development of the electron probe, capable of investigating the distribution (on the μm scale) of all elements of atomic number above five at concentrations as low as 0.1 %, and frequently 0.01 %, permits studies of microstructure to an unprecedented degree.[115]

The localization of structural features in the electron probe is aided by scanning techniques[116] that produce electron microprobe images of small sectors of the sample surface. It is also possible to obtain scanning images of element distribution. The most attractive feature of electron probe microanalysis is the fact that a quantitative analysis is possible, with errors of less than 3 % relative in most cases.[117,118] Data evaluation requires the use of a computer, but the operation can be performed in a time-share mode, or even on line with a small computer, and pure elements or simple compounds can be used as reference materials.

The in-depth resolution of the electron probe is about the same as the lateral resolution (1–3 μm). Normally, the analysis is restricted to regions as large or larger than these dimensions. However, the instrument is employed increasingly in the characterization of films of submicrometer dimensions, (determining both film thickness and composition[119]) as well as of small particles. A search of the literature shows that the electron probe has been applied to a wide range of materials.[120]

2.3.25. Ion-Probe Microanalysis

The limitations in sensitivity and in depth resolution of the electron-probe microanalyzer prompted the development of ion-probe microanalysis. This technique[121] is based on mass spectrographic analysis of the secondary ions emitted from a sample under the impact of a focused and accelerated primary ion beam. This type of analysis also offers, in comparison with the electron probe, the possibilities of isotopic analysis and the investigation of elements of low atomic number, including hydrogen, at trace concentrations.

Although the lateral resolution of the ion beam is similar to that of the electron beam (1–3 μm), the investigation of even a few atomic layers is possible.[122]

By gradual erosion of the sample with a primary beam, a sensitive investigation of surface layers and of distribution in depth of impurities can be performed. It has been possible, for instance, to demonstrate the enrichment of traces of aluminum in the *p–n* junction of a silicon transistor device.

The sensitivity of the method, for many elements and samples, is in the ppm range. The analysis of nonconductors is possible when negative primary ions are employed. Therefore the method can be applied to a wide range of materials, including metals and alloys, oxidized alloy surfaces, semiconductors, ceramics, and minerals.

The possibility of quantitative analysis with the ion probe is presently a matter of controversy. Although theories have been developed that aim to predict the strong matrix effects inherent in this technique and to correct for them, further work is required before the efficacy of these procedures of data evaluation can be definitely assessed.[123] It is possible with calibration procedures, however, to obtain information on individual particles of micrometer size.[124]

2.3.26. Photoelectron Spectroscopy

A very useful technique for determining the electronic states at the surface is photoelectron spectroscopy.[125,126] Photon excitation of a surface from induced X-ray emission in materials such as magnesium or aluminum produces photoelectrons with energy sufficient to escape from depths of not more than a few atomic layers. Moreover, those electrons that escape without energy loss have an energy linewidth sufficiently narrow to allow observation of differences in chemical binding of the order of 0.1 eV.[127] This means that in many materials differences in chemical structure on the surface can be observed.

Alteration of the photoexcitation energy can produce other types of electron emission (such as the Auger effect) which result in a modified spectrum of the emitted electrons, thereby providing a "different look" at the energy levels in a chemical structure.[128] All of these techniques combine to provide a total picture of the surface structure.

2.4. Precision and Sensitivity of Analytical Techniques

The precision, the sensitivity, and the area of application of the techniques described in the previous sections are summarized in Table 1.

3. Application of Current Techniques to Characterization of Materials

3.1. Characterization of Major Phase

3.1.1. Stoichiometry

Classical wet chemical techniques based on gravimetry, titrimetry, and electrochemistry are most commonly used for major element analyses and stoichiometry, i.e., the relative ratio of the elements in a compound. Conventional gravimetric and volumetric methods[12,13] are used for determinations at $\sim \pm 0.1\,\%$, and with refinements such as homogeneous precipitation, solubility correction, and the use of weight burettes, precisions can be extended to $\pm 0.01\,\%$. The most precise determinations of stoichiometry have been based on constant-current coulometry, where reliability at the 0.001–0.01 % level can be obtained. For example, Marinenko[23] has determined the ratio of gallium to arsenic in single-crystal gallium arsenide to $\pm 0.007\,\%$, using constant-current coulometric determinations for both elements.

With X-ray fluorescence stoichiometry can be determined to ± 0.01–$0.1\,\%$ when suitable standards are available. Precise lattice parameters and pycnometric density determinations have been used to determine deviations from stoichiometry or nonstoichiometry in Bertholide-type compounds.[151] Other instrumental techniques such as nuclear magnetic resonance and comparison of ferroelectric Curie temperatures have also found applications.[2]

TABLE 1

Precision and Sensitivity of Analytical Techniques

Technique	Applications	Sensitivity[a]	Precision	References
Wet chemistry gravimetry	Major- and minor-phase concentration	100 mg to 1 g	0.003–0.01%	15, 129
		1–10 mg	0.1%	15, 129
Wet chemistry titrimetry	Major- and minor-phase concentration; also impurities	10^{-2} M in solution	0.01%	15, 16, 129, 130
		10^{-5} M in solution	0.1%	15, 16, 129, 130
		10^{-6}–10^{-7} M in solution	0.2–1.0%	15, 16, 129, 130
Coulometry	Major-phase concentration	–	0.001–0.005%	20, 22, 26, 131
Ion-selective electrodes (direct)	Major- and minor-phase concentration; also impurities	10^{-3} M–saturation in solution	0.5–2%	27, 132
		10^{-5}–10^{-3} M in solution	1–5%	27, 132
		$<10^{-5}$ M in solution	2–30%	27, 132
Polarography	Impurities, major and minor constituents	All in solution:		
		10–100 ppm	0.1–2%	34
		0.1–10 ppm	2–10%	37
		0.001 ppm (with anodic stripping preconcentration)	5–10%	33
Spectrophotometry	Impurities	10–100 ppm in solution	1–5%	133–135
		0.005–0.1 ppm in solution	5–10%	133–135
Spectrofluorimetry	Impurities; organic	0.002–10 ppm	0.5–20%	45, 46
	Impurities; inorganic:			
	Rare earths	1–10^4 ppm	1–50%	45, 46
	Non rare earths	0.001–10 ppm	1–10%	45, 46
Organic micro-analysis	Major-phase impurities	10 μg	±0.6%	47, 136

TABLE 1 (continued)

Technique	Applications	Sensitivity[a]	Precision	References
Thin-layer chromatography	Minor constituents, trace impurities	10–1,000 μg	5–50%	51
Gas chromatography	Major- and minor-phase concentration; also impurities; organic analysis	Major component	0.1%	53, 57
		1–10%	0.2–0.5%	53, 57
		0.1–1%	0.5–1%	53, 57
		0.01–0.1%	1–5%	53, 57
		10–100 ppm	5–10%	53, 57
		< 10 ppm	≥ 10%	53, 57
Liquid chromatography	Major- and minor-phase concentration; organic analysis	0.001–1 ppm	1–20%	58, 137
Activation analysis	Impurities:			
	Individual with radiochemistry	0.1–10 ppm	2–5%	62–64, 138
		0.001–0.01 ppm	2–10%	62–64, 138
	Nondestructive (survey)	0.1–10 ppm	10 to > 20%	62–64, 138
Vacuum fusion-mass spectrometry	Impurities: O_2	0.07 ppm	20%	139, 140
	N_2, H_2	100 ppm	5%	139, 140
Thermal analysis	Change of phase; Impurities	10–100 μg	1–15%	71, 141, 142
X-ray fluorescence spectrometry	Major and minor constituents	20–200 ppm generally;	0.1–0.5%	76, 77
		0.1 ppm with preconcentration	2–10%	78

Technique	Applications	Sensitivity[a]	Precision	References
Spark source mass spectrometry	Impurities (survey)	0.001–0.1 ppm	5–20%	81, 82
Spark source mass spectrometry; isotope dilution	Impurities	10^{-5}–0.1 ppm	5%	83, 84
Isotope dilution mass spectrometry	Major- and minor-phase concentration;	500 ppm–100%	0.1–0.2%	35
		1–500 ppm	0.2–0.5%	35, 36, 86
	Impurities	0.00001–1 ppm	0.5–50%	35, 36, 86
Emission spectroscopy	Impurities (survey)	0.1–100 ppm	5–10%	88, 90, 143
Flame emission	Impurities	0.1–10 ppm in solution	0.5–5%	92, 93, 95
		0.002–0.1 ppm in solution	5–10%	92, 93, 95
Atomic absorption	Impurities	0.1–10 ppm in solution	0.5–5%	92, 93, 95
		0.005–0.1 ppm in solution	5–10%	92, 93, 95
Atomic fluorescence	Impurities	0.1–10 ppm in solution	2–5%	92, 95, 144
		0.001–0.1 ppm in solution	5–15%	92, 95, 144
Mössbauer spectroscopy	Major-phase valence; also impurities valence	Down to 0.0001–0.1% depending upon density of matrix	Semiquantitative now	145, 146, 147
Nuclear magnetic resonance	Major- and minor-phase concentration; Impurities	10^{-4} M in solution	1%	148

TABLE 1 (concluded)

Technique	Applications	Sensitivity[a]	Precision	References
Electron spin resonance	Impurities	0.001–1 μg	Semiquantitative now	110, 149, 150
Residual resistance ratio	Impurities (survey integral)	10^{-9} g	1–5%	113, 114
Electron probe microanalysis	Microscopic; homogeneity of major phases and minor phases	0.01–0.1% over a 1–5 μm scan diameter	0.5%	115, 117, 118
Ion probe microanalysis	Microscopic; traces, surface layers	1 ppm	Semiquantitative now	123
Photoelectron spectroscopy	Surface analysis of major phase	0.1%	5%	—

[a] Provided purified reagents and clean room facilities are used. See note of caution in text.

3.1.2. Homogeneity (Including Impurity Distribution)

In addition to determining the *presence* of impurities in solid solution, it is also important to study the *distribution* of such impurities. The importance of knowing how the impurities are distributed, whether completely statistically at the atomic level or in a single "inclusion," is vital to every interpretation of properties measured on the phase. In this field, the last ten years have brought a step-function advance with the introduction of the electron microprobe.[115] Yet, even here spatial resolution is limited to about 1 μm, and sensitivity and precision are sometimes insufficient.

In some cases, cathodoluminescence in the probe provides a tool for detecting inhomogeneities at the ppm level with micrometer resolution. The scanning electron microscope, used judiciously, can provide (in favorable cases) data on impurity distribution at the 0.1-μm level. Here the ion probe holds promise both of sensitivity increased by a hundredfold and of sampling a few atom layers deep.

3.1.3. Oxidation State

Once exact stoichiometry has been determined it is conventional to represent the oxidation state (valence) of a multivalent element in a compound by rounding off to the nearest whole number and assigning formal charges so as to preserve electrical neutrality. This presents difficulties when more than one multivalent element is present in the same compound or when several oxidation states are possible for a given element in the compound, e.g., iron in magnetite, Fe_3O_4.

If it is possible to dissolve the compound without altering the oxidation states, the classical wet chemical techniques of gravimetry, titrimetry, and electrochemistry can define the oxidation state to a precision of about 0.01 %. For some materials, such as $K_2Cr_2O_7$, such dissolution is possible with no measurable change in oxidation state to at least one part in 100,000. For other materials there is at least a partial change in oxidation state on dissolution, and one must resort to instrumental methods. For iron, tin, iodine, gold, and some rare earths Mössbauer spectroscopy is applicable with a precision of about 0.1 %.[2] High-resolution X-ray spectra have also been used to determine oxidation state.

3.2. Characterization of Minor Phases and Impurities

Many impurities can be detected at the 1–10 ng/g level in simple systems. However, the limit of quantitative determination in practical samples is generally much higher. The accuracy of quantitative analysis by many instrumental methods is frequently limited by a lack of suitable standards. Detection and determination of organic impurities in an organic matrix is an especially difficult problem.

3.2.1. Survey Methods

Emission spectroscopy is the most generally applicable of the survey methods although there are limitations on its sensitivity. It is widely used for characterization of solids, powders, liquids, and gases and has the capability of detecting up to 70 elements by direct current arc excitation. Determination of nonmetallic elements is also possible with emission spectroscopy, but this requires special techniques that are infrequently used.

Spark-source mass spectrometry can detect all elements, with sensitivities often as good as 1 ng/g. Residual gases in the vacuum system restrict detection limits for carbon, oxygen, and nitrogen to 0.1–1 μg/g.

X-ray spectroscopy can also be employed for survey analysis of impurities and has the advantage of being nondestructive. Lower limits of detection are rarely better than 10–100 ppm, unless preconcentration is used, and the elements of the first period cannot be detected at low concentrations.

Electrical measurements are useful for determining the total content of electrically active impurities in conductors and semiconductors. The shape of the freezing curve gives considerable information on total impurities in a material with a suitable melting point. Special techniques must be used for most metals that melt at high temperatures[114] or for materials that decompose on melting.

Gas chromatography is a useful survey technique for organic analysis because a very broad range of samples may be characterized very quickly with columns that are nonpolar (for fixed gases and aliphatics), of intermediate polarity (for aromatics), and highly polar (for alcohols and esters). High-pressure liquid chromatography should be regarded as a complementary organic survey technique since it separates high-molecular-weight compounds. Ionic, polymeric, and thermally labile compounds may be separated by this method, although its sensitivity is presently limited to the μg range.

3.2.2. Quantitative Methods

Many of the techniques listed in Section 2.3 under analytical techniques and summarized in Table 1 can be applied to the quantitative determination of impurities. The most serious limitations of these methods are the relatively poor detection limits for some, the need for standards of similar composition for the more sophisticated instrumental methods, and the hazard of contamination in methods that require preliminary dissolution and/or considerable handling or preparation of the sample.

Activation analysis is especially suitable for trace characterization, since most of the sample handling occurs after the activation step, and contamination problems are minimized. Equipment and operational costs, although high, are comparable to those required of many modern analytical techniques.

Spectroscopic methods (emission, mass, X-ray) are useful for determination of many elements in small absolute amounts, but are limited in the amount of sample that can be analyzed; they are also useful methods for determining groups of elements after chemical separation or preconcentration. These techniques (except for isotope dilution procedures) are limited by the availability of suitable standards.

Most of the other methods listed in Section 2.3 are restricted to the determination of a single element or a few elements at a time; they are most useful when there is critical interest in specific contaminants in the samples.

Organic impurities in organic matrices can frequently be separated by thin-layer chromatography, paper chromatography, liquid chromatography, or gas chromatography and can be determined by integration of the detector response curves. Detection limits depend heavily on the nature of the organic materials and impurities and on the signal/noise ratio of the chromatographic or spectroscopic method used to resolve the impurities or their signals. With practical samples detection limits are usually no better than 10 ppm.

The various techniques that give information on oxidation state and location all suffer from relatively poor sensitivity and are therefore more suitable for the characterization of the major phase than of the impurities. Magnetic resonance techniques (NMR and ESR) have low limits of detection, but both are limited in applicability.

Homogeneity of the sample is often a limiting factor in the analysis for a trace constituent.

3.3. Characterization of Surfaces

Undoubtedly, among the least understood and most poorly characterized features of a solid is the surface. Electronic "mapping" of solid surfaces on an atomic scale is now in progress with such tools as field emission microscopy, surface infrared and nuclear magnetic resonance spectroscopy, low-energy electron diffraction, electron-probe techniques, and surface conductance measurements. New insights into the nature and properties of surfaces become possible with the application of techniques such as ESCA,[125] secondary ion mass spectrometry,[121,122,152] and ion scattering.[153] Further discussion of these methods will be found in Volume 6.

4. Utilization of Existing Techniques

4.1. Literature Examples

Published information on characterization of seven different materials has been gathered from the literature to serve as a guide to the use of existing analytical techniques. These seven materials are copper, silicon, gallium arsenide, potassium chloride, zinc sulfide, anthracene, and trace element glasses. Collectively they represent an extremely broad spectrum of analytical problems and illustrate both strong and weak points in the application of the present state of the art of measurement of composition.

The results, summarized in Tables 2–6, do not necessarily represent optimum performance, but they are fairly typical of published information on practical samples of highly purified materials. The notes of Tables 3–6 are quite representative of experience with different techniques on materials with impurities at levels of 500, 50, 1, or 0.02 ppm.

4.2. Factors Determining Use

These and other literature examples demonstrate the pronounced tendency of most investigators to use only a few, often only one, of the existing analytical techniques in characterization of materials. In the past the principal information on purity has been provided by emission spectroscopy, which was popular because it was broadly applicable and was a survey technique that required relatively small expense in terms of time, money, and sample size. Unfortunately, information at ppm levels derived by emission spectroscopy (see Table 2) is now often of marginal interest as impurity concentrations of 0.1–0.01 ppm and lower become important. The spark source mass spectrometer can now

TABLE 2

Published Data on Purified Materials

Materials	Method of preparation	Impurities (ppm)	Method of analysis
Copper[154]	Zone refined	Sb,* Cr(1), Co(60), Fe(10), Mn,* Ni(10), Si,* Ag,* Sn*	Emission spectroscopy
Copper[155]	Zone refined	Fe(0.5), As(<0.0001), Ag(0.01–0.02), Sb(<0.0001), Mn(<0.0001), Zn(<0.05), Na(<0.001), Cr(0.002–0.003), Ag(0.04), Se(<0.001), P(0.03–0.01)	Activation analysis
Silicon[156]	Thermal decomposition of silane	Fe(<0.017), Cu(0.011), Bi(<0.0016), As(0.001), P(<0.001), Tl(<0.00091), Zn(<0.00041), Ni(<0.00027), In(<0.00032), Ga($<6.2 \times 10^{-6}$), Sb($<3.7 \times 10^{-6}$), Mn($<6.2 \times 10^{-7}$)	Activation analysis
Silicon[157]	(Semiconductor material)	Fe, Tl, Bi, Pb, Zn, In, Cd, Ni, Cu (all at ~0.01 ppm)	Polarography
Gallium arsenide[134]	(Semiconductor material)	Si(0.03), Te(0.1)	Spectrophotometry
Gallium arsenide[23]	(Semiconductor material)	Stoichiometric within 0.007% (i.e., Ga/As = 1.000)	Coulometry
Potassium chloride[158]	Single-crystal growth from melt	Br(<1), C(<20), Fe(<1), I(<1), N(2), Na(<0.8) OH(0.01), P(<1), Pb(0.06), Rb(1), S(<1)	Spectroscopy, flame emission, atomic absorption, wet chemical, activation analysis

TABLE 2 (concluded)

Published Data on Purified Materials

Materials	Method of preparation	Impurities (ppm)	Method of analysis
Potassium chloride[159]	(Chemical reagent)	Bi, Pb, Ni, Cu, Fe, at 0.1 ppm level	Spectrophotometry
Zinc sulfide[160]	Single-crystal growth from vapor phase	Al(1.2), Ca (not detected), Cu(3), Fe(1.5), Mg(2), Mn(0.2), Ni(<1), Pb,* Sb(<10)	Emission spectroscopy
Anthracene[161]	Gas chromatography, zone refining, gradient sublimation, scavenging	Tetracene(<1), Anthraquinone(<1), Fluorene(<1), Phenanthrene(<1), Carbazole(1–10), Cr(<0.1), Fe(<0.5), Mg(<0.2), Pb(<0.1), Si(<0.5), Mn(<0.05), Cu(<0.1), Ag(<0.1), Al(<0.1)	Fluorescence, optical absorption, emission spectroscopy, gas chromatography
Trace element glasses[35] (a) 500 ppm (b) 50 ppm (c) 1 ppm (d) 0.02 ppm	Homogeneous batches of doped glass drawn into canes by modified Czochralski technique	(a) See Table 3 (b) See Table 4 (c) See Table 5 (d) See Table 6	Isotope dilution mass spectrometry, activation analysis, polarography, spectrophotometry, spark source mass spectrometry, flame emission, atomic absorption

* Not detected.

TABLE 3

Composition Values for NBS-SRM 610, 611[(35)]; Trace Elements in a Glass Matrix, 500 ppm from Certificate of Analysis (Revised August 8, 1972)[a]

Element	Value[b]	Notes[c]
Boron	(351)	1
Cobalt	(390)	2
Copper	(444 ± 4)	3
Gold	(25)	4
*Iron	458 ± 9	5
*Lead	426 ± 1	6
*Manganese	485 ± 10	7
*Nickel	458.7 ± 4	8
Potassium	(461)	1
*Rubidium	425.7 ± 0.8	9
Silver	(254 ± 10)	10
*Strontium	515.5 ± 0.5	11
Thallium	(61.8 ± 2.5)	12
*Thorium	457.2 ± 1.2	13
Titanium	(437)	14
*Uranium	461.5 ± 1.1	15
Zinc	(433)	16

[a]The present status of the analytical certification is given in the table. An asterisk before the element indicates a certified concentration for that element. The indicated limits on the concentration are equal to the entire range of observed results among sample points and/or the 95% confidence interval, whichever is larger. Values in parentheses are interim for the reasons given in the notes. Nominal composition of the support matrix is 72% SiO_2, 12% CaO, 14% Na_2O, and 2% Al_2O_3.

[b]All values given in table are in ppm by weight.

[c]Notes: 1. Isotope dilution: interim value because of high blank. 2. Two independent sets of analyses by neutron activation disagree. 3. Isotope dilution: limits dictated by an observed trend in element concentration, well outside the precision of the method, 4. Spectrophotometry and neutron activation give grossly different results; value included only to indicate that the gold was not all lost in the processing of the glass rods. 5. Pooled value from data by spectrophotometry and polarography. 6. Pooled value from data by isotope dilution at two independent laboratories: NBS and USGS. 7. Value by spectrophotometry, substantiated by neutron activation. 8. Isotope dilution data accepted for certification, substantiated by spectrophotometry and polarography. 9. NBS isotope dilution data accepted for certification, cooperating analysts' data have a much larger uncertainty statement (range). 10. Isotope dilution: interim results because of questionable result on Rod No. 78 (8 ppm above average, not included in average). Neutron activation data have much larger range. 11. Pooled data: NBS isotope dilution data accepted and substantiated by USGS and Australian National University. The normalized $^{87}Sr/^{86}Sr$ ratio = 0.7094 ± 0.0002. 12. Isotope dilution: one method only, large uncertainty statement (range) is the result of a high value for Rod No. 2, which gave results 1.5 ppm higher than the average and was not included in the reported average. 13. Pooled isotope dilution data: NBS data accepted for certification and substantiated by USGS. 14. Polarographic: one method only. 15. Isotope dilution: NBS substantiated by USGS. Uranium in glass depleted in ^{235}U. The atom per cent ^{235}U = 0.2376. 16. Atomic absorption only: systematic error unknown.

TABLE 4

Composition Values for NBS-SRM 612, 613[35]; Trace Elements in Glass Matrix, 50 ppm from Certificate of Analysis (Revised August 8, 1972)[a]

Element	Value[b]	Notes[c]
Barium	(41)	1
Boron	(32)	2
Cerium	(39)	1
Cobalt	(35.5 ± 1.2)	3
Copper	(37.7 ± 0.9)	4
Dysprosium	(35)	1
Erbium	(39)	1
Europium	(36)	1
Gadolinium	(39)	1
Gold	(5)	5
*Iron	51 ± 2	6
Lanthanum	(36)	1
*Lead	38.57 ± 0.2	7
Manganese	(39.6 ± 0.8)	8
Neodymium	(36)	1
*Nickel	38.8 ± 0.2	9
Potassium	(64)	10
*Rubidium	31.4 ± 0.4	11
Samarium	(39)	1
*Silver	22.0 ± 0.3	12
*Strontium	78.4 ± 0.2	13
Thallium	(15.7 ± 0.3)	4
*Thorium	37.79 ± 0.08	14
Titanium	(50.1 ± 0.8)	15
*Uranium	37.38 ± 0.08	16
Ytterbium	(42)	1

[a] The present status of the analytical certification is given in the table. An asterisk before the element indicates a certified concentration for that element. The indicated limits on the concentration are equal to the entire range of observed results among sample points and/or the 95% confidence interval, whichever is larger. Values in parentheses are interim for the reasons given in the notes. Nominal composition of the support matrix is 72% SiO_2, 12% CaO, 14% Na_2O, and 2% Al_2O_3.

[b] All values given in table are in ppm by weight.

[c] Notes: 1. Isotope dilution: interim data from only two sample points. 2. Nuclear track counting plus two sample points by isotope dilution, insufficient precision and accuracy for certification. 3. Neutron activation: one method only. 4. Isotope dilution: one method only, observed range caused by sample variability. 5. Spectrophotometry and neutron activation give grossly different results; value included only to indicate that the gold was not all lost in the processing of the glass rods. 6. Pooled value from data by spectrophotometry and polarography. 7. Pooled isotope dilution data: NBS and USGS data weighed equally. 8. Spectrophotometry: one method only. 9. Isotope dilution data accepted for certification substantiated by spectrophotometry. 10. Interim data: isotope dilution and atomic absorption (both troubled with high blanks). 11. NBS isotope dilution data accepted for certification; cooperating analysts' data have a much larger uncertainty statement (range). 12. NBS isotope dilution data accepted for certification, substantiated by neutron activation. 13. Pooled data: NBS isotope dilution data accepted and substantiated by USGS and Australialian National University. The normalized $^{87}Sr/^{86}Sr$ ratio = 0.7089 ± 0.0002. 14. Pooled isotope dilution data: NBS data accepted for certification and substantiated by USGS. 15. Polarographic: one method only. 16. Isotope dilution: NBS substantiated by USGS. Uranium in glass depleted in ^{235}U. The atom per cent ^{235}U = 0.2392.

TABLE 5

Composition Values for NBS-SRM 614, 615[35]; Trace Elements in a Glass Matrix, 1 ppm from Certificate of Analysis (Revised August 8, 1972)[a]

Element	Value[b]	Notes[c]
Antimony	(1.06)	1
Boron	(1.30 ± 0.2)	2
Cadmium	(0.55)	3
Cobalt	(0.73 ± 0.02)	4
*Copper	1.37 ± 0.07	5
Europium	(0.99 ± 0.04)	4
Gallium	(1.3)	1
Gold	(0.5)	1
Iron	(13.3 ± 1)	6
Lanthanum	(0.83 ± 0.02)	4
*Lead	2.32 ± 0.04	7
Nickel	(0.95)	6
*Potassium	30 ± 1	8
*Rubidium	0.855 ± 0.005	9
Scandium	(0.59 ± 0.04)	4
*Silver	0.42 ± 0.04	10
*Strontium	45.8 ± 0.1	11
Thallium	(0.269 ± 0.005)	12
*Thorium	0.748 ± 0.006	13
Titanium	(3.1 ± 0.3)	6
*Uranium	0.823 ± 0.002	14

[a] The present status of the analytical certification is given in the table. An asterisk before the element indicates a certified concentration for the element. The indicated limits on the concentration are equal to the entire range of observed results among sample points and/or the 95% confidence interval, whichever is larger. Values in parentheses are interim for the reasons given in the notes. Nominal composition of the support matrix is 72% SiO_2, 12% CaO, 14% Na_2O and 2% Al_2O_3.

[b] All values given in the table are in ppm by weight.

[c] Notes: 1. Neutron activation: one method only with an apparently large systematic error (> 10%). for this element because of poor correlation among measurements at various concentrations. 2. Nuclear track counting: one method only, but with very good correlation interpolating between concentrations. 3. Spark source isotope dilution: one method only. 4. Neutron activation: one method only. 5. Pooled data: spark source isotope dilution and the thermal ionization isotope dilution. 6. Polarographic: one method only. 7. Pooled isotope dilution data: value from USGS accepted because of smaller error limits and substantiated by NBS. 8. Pooled data: NBS flame emission data substantiated by NBS isotope dilution. 9. Isotope dilution data substantiated by flame emission which has a much larger uncertainty statement (range). 10. Pooled data: spark source isotope dilution plus neutron activation. 11. Pooled data: NBS isotope dilution data accepted and substantiated by USGS and Australian National University. The normalized $^{87}Sr/^{86}Sr$ ratio = 0.7083 ± 0.0002. 12. Isotope dilution: one method only with good correlation interpolating between concentrations. 13. Pooled isotope dilution data: value from NBS accepted because of smaller error limits and substantiated by USGS. 14. Isotope dilution: NBS isotope dilution data used, substantiated by USGS isotope dilution and NBS nuclear track counting data, which both had slightly higher uncertainties. Uranium in glass depleted in ^{235}U. The atom per cent ^{235}U = 0.2792.

TABLE 6

Composition Values for NBS-SRM 616, 617[35]; Trace Elements in a Glass Matrix, 0.02 ppm from Certificate of Analysis (Revised August 8, 1972)[a]

Element	Value[b]	Notes[c]
Antimony	(0.078 ± 0.007)	1
Boron	(0.20 ± 0.02)	2
Copper	(0.80 ± 0.09)	3
Gallium	(0.23 ± 0.02)	1
Gold	(0.18 ± 0.01)	1
Iron	(11 ± 2)	4
Lanthanum	(0.034 ± 0.007)	3
*Lead	1.85 ± 0.04	5
*Potassium	29 ± 1	6
Rubidium	(0.100 ± 0.007)	7
Scandium	(0.026 ± 0.012)	3
*Strontium	41.72 ± 0.05	8
Thallium	(0.0082 ± 0.0005)	9
*Thorium	0.0252 ± 0.0007	10
Titanium	(2.5 ± 0.7)	4
*Uranium	0.0721 ± 0.0013	11

[a] The present status of the analytical certification is given in the table. An asterisk before the element indicates a certified concentration for that element. The indicated limits on the concentration are equal to the entire range of observed results among sample points and/or the 95% confidence interval, whichever is larger. Values in parentheses are interim for the reasons given in the notes. Nominal composition of the support matrix is 72% SiO_2, 12% CaO, 14% Na_2O and 2% Al_2O_3.

[b] All values given in table are in ppm by weight.

[c] Notes: 1. Neutron activation: one method only with an apparently large systematic error (> 10%) for this element because of poor correlation among measurements at various concentrations. 2. Nuclear track counting: one method only, but with very good correlation extrapolating down from 1 ppm glass. 3. Neutron activation: one method only with good correlation extrapolating down from 1 ppm glass. 4. Polarographic: one method only with good correlation extrapolating down from 1 ppm glass. The large uncertainty is due to a large chemical blank. 5. Pooled isotope dilution data: value from USGS accepted because of smaller error limits and substantiated by NBS. 6. Pooled data: NBS flame emission data substantiated by NBS isotope dilution. 7. Isotope dilution: one method only with good correlation extrapolating down from 1 ppm glass. The uncertainty statement (range) is the result of one sample point which gave a result 0.005 ppm higher than the average. 8. Pooled data: NBS isotope dilution data accepted and substantiated by NBS flame emission. The normalized $^{87}Sr/^{86}Sr$ ratio = 0.7080 ± 0.0002. 9. Isotope dilution: one method only with good correlation extrapolating down from 1 ppm glass. 10. Pooled isotope dilution data: value from NBS accepted because of smaller error limits and substantiated by USGS. 11. Pooled data: NBS and USGS isotope dilution data and NBS nuclear track counting all weighted equally. Uranium in glass depleted in ^{235}U. The atom per cent ^{235}U = 0.616.

TABLE 7

Current Use of Analytical Techniques

	Use			Required for greater use			
Technique	Low	Int.	High	Better training and education	More trained personnel	Greater convenience	Other
Wet chemistry—gravimetry	—	—	×	—	—	—	Automation (with standards)
Wet chemistry—titrimetry	—	—	×	—	—	—	Automation (with standards)
Coulometry	×	—	—	×	×	—	Greater publicity
Ion-selective electrodes	×	—	—	×	×	—	Development of interference-free procedures
Polarography	—	×	—	×	×	—	—
Spectrophotometry	—	—	×	—	—	—	—
Spectrofluorimetry	×	—	—	—	×	—	Less sophisticated instrumentation
Organic microanalysis	×	—	—	×	—	×	—
Thin-layer chromatography	—	×	—	—	—	—	Improved means of quantitation
Gas chromatography	—	—	×	—	—	—	Characterization of column materials for specific application
Liquid chromatography	×	—	—	—	×	—	Choice of column conditions
Neutron activation	—	×	—	—	×	×	Development of multielement approach

TABLE 7 (continued)

Technique	Use			Required for greater use			
	Low	Int.	High	Better training and education	More trained personnel	Greater convenience	Other
Vacuum fusion—Mass Spectrometry	—	×	—	—	×	×	—
Thermal analysis	—	×	—	×	—	—	Standards
X-ray fluorescence	—	—	×	—	—	—	Better standards
Spark source Mass spectrometry	—	×	—	—	—	×	More standards
Spark source mass spectrometry, Isotope dilution	×	—	—	×	—	—	—
Isotope dilution Mass spectrometry	×	—	—	×	—	—	General availability of better instrumentation
Emission spectroscopy	—	—	×	×	—	—	Better standards
Flame emission	—	×	—	×	—	—	Improvement in atomization; multielement approach
Atomic absorption	—	—	×	×	—	—	Improvement in atomization; multielement approach

TABLE 7 (concluded)

Technique	Use			Required for greater use			
	Low	Int.	High	Better training and education	More trained personnel	Greater convenience	Other
Mössbauer spectroscopy	×	—	—	—	×	×	Application to more elements
Nuclear magnetic resonance	—	×	—	—	×	—	—
Electron spin resonance	×	—	—	—	×	—	—
Residual resistance ratio	×	—	—	—	×	—	General availability of instrumentation
Electron probe microanalysis	—	×	—	—	×	×	Automation
Ion probe microanalysis	×	—	—	×	×	—	Development of lower cost instruments
Photoelectron spectroscopy	×	—	—	×	×	—	General availability of better instrumentation

be expected to assume many of these survey functions to provide concentration information at the sub-ppm level.

There are only a modest number of research organizations with large analytical chemistry groups staffed with skilled professionals with the capability for comprehensive characterization of materials. Similarly, there are only a few analytical service laboratories capable of providing reliable composition values for materials research at a prescribed level of accuracy. The meaningful measurement of chemical composition is expensive. The materials investigator must plan to invest a significant portion of his resources in the determination of this "benchmark" analytical information that will enable him to reproduce his materials for future studies.

The availability of techniques, the availability of trained personnel, and the cost and time required for meaningful analytical measurements are all major factors in determining current utilization of existing analytical methods. Table 7 summarizes the best estimates on current use of these techniques.

Acknowledgments

The MAB Report that supplied the format and scientific approach for this chapter was the product of much thoughtful discussion among the members of the MAB Panel on Composition. Each panel member contributed to the original MAB Report in areas of his own expertise. Chairman Monroe Sadler edited these components into a final report that was reviewed by the panel for consistency and appropriateness.

For this chapter the original MAB Report contributions of the many NBS experts in specific analytical techniques have been revised and updated to represent the situation in January 1973. I would like to thank these many NBS contributors for their perceptive comments. The assistance of Mrs. Rosemary Maddock in the overall coordination of this manuscript, in particular in the preparation of the bibliography, is especially appreciated.

References

1. N. B. Hannay, Trace characterization and the properties of materials, in *Trace Characterization, Chemical and Physical* (W. W. Meinke and B. F. Scribner, eds.), pp. 5–38, NBS Monograph 100, U.S. Government Printing Office, Washington, D.C. (1967).

2. R. A. Laudise, Opportunities for analytical chemistry in solid state research and electronics, in *Analytical Chemistry: Key to Progress on National Problems* (W. W. Meinke and J. K. Taylor, eds.), pp. 19–64, NBS Special Publication 351, U.S. Government Printing Office, Washington, D.C. (1972).
3. The Committee on Characterization of Materials, Materials Advisory Board, Division of Engineering, National Research Council, National Academy of Sciences, National Academy of Engineering, *Characterization of Materials*, MAB-229-M, U.S. Clearinghouse for Federal Scientific and Technical Information, Springfield, Virginia (1967).
4. G. E. F. Lundell, The chemical analysis of things as they are, *Ind. and Eng. Chem. (Anal. Ed.)* **5**, 221–225 (1933).
5. P. Albert, A combination of chemical and physiochemical methods for a systematic separation of large numbers of radioisotopes on one experimental analysis of aluminum, iron, and zirconium by radioactivation, in *Modern Trends in Activation Analysis* (Proc. 1961 Int. Conf. on Modern Trends in Activation Analysis, College Station, Texas, December 1961), pp. 86–94, Texas A & M University, College Station, Texas (1962).
6. I. M. Kolthoff and P. J. Elving (eds.), *Treatise on Analytical Chemistry, A Comprehensive Account in three parts, Part I: Theory and Practice* (10 vols.), *Part II: Analytical Chemistry of the Elements* (14 vols.), *Part III: Analysis of Industrial Products* (2 vols.), Interscience, New York (1959–1971).
7. W. W. Meinke and B. F. Scribner (eds.), *Trace Characterization, Chemical and Physical* (Proc. 1st Materials Research Symp., October 1966), NBS Monograph 100, U. S. Government Printing Office, Washington, D.C. 20402 (1967).
8. Analytical reviews 1972, Fundamentals, *Anal. Chem.* **44**(5), 1R-572R (1972).
9. Analytical reviews 1971, Applications, *Anal. Chem.* **43**(5), IR-388R (1971).
10. W. W. Meinke, Is radiochemistry the ultimate in trace analysis?, Proc. Int. Conf. on Analytical Chemistry, Kyoto, Japan, April 1972, *Pure and Appl. Chem.* **34**, 93–104 (1973).
11. W. W. Meinke, The ultimate contribution of nuclear activation analysis, Proc. of 4th Int. Conf. on Modern Trends in Activation Analysis, Saclay, France, October 1972, *J. Radioanalytical Chem.* (in press).
12. W. F. Hillebrand, G. E. F. Lundell, H. A. Bright, and J. I. Hoffman, *Applied Inorganic Analysis*, 2nd ed., Wiley, New York (1953).
13. I. M. Kolthoff and P. J. Elving (eds.), *Treatise on Analytical Chemistry*, Part 1, Vol. 1; Part 2, Vols. 1–14, Wiley—Interscience, New York (1959–1971).
14. L. Erdey, L. Pólos, and R. A. Chalmers, Development and publication of new gravimetric methods of analysis, *Talanta* **17**, 1143–1155 (1970).
15. E. B. Sandell, Errors in chemical analysis, in *Treatise on Analytical Chemistry* (I. M. Kolthoff and P. J. Elving, eds.), Part 1, Vol. 1, pp. 19–46, The Interscience Encyclopedia, Inc., New York (1959).
16. A. Ringbom, *Complexation in Analytical Chemistry*, Interscience, New York (1963).
17. J. R. Bacon and R. B. Ferguson, Gravimetric and coulometric analysis of beryllium samples using 2-methyl-8-quinolinol, *Anal. Chem.* **44**, 2149–2152 (1972).
18. R. S. Danchik, Analytical reviews 1971/Applications: Nonferrous metallurgy. 1. Light metals: aluminum, beryllium, titanium, and magnesium, *Anal. Chem.* **43**(5), 109R–145R (1971).

19. R. P. Buck, Analytical reviews 1972/Fundamentals: Ion-selective electrodes, potentiometry, and potentiometric titrations, *Anal. Chem.* **44**(5), 270R–295R (1972).
20. G. Marinenko and C. E. Champion, High-precision coulometric titrations of boric acid, *J. Res. NBS* (*U.S.*), **75A** (*Phys. and Chem.*), 421–428 (1971).
21. C. E. Champion, G. Marinenko, J. K. Taylor, and W. E. Schmidt, Determination of submicrogram amounts of chromium by coulometric titrimetry, *Anal. Chem.* **42**, 1210–1213 (1970).
22. G. Marinenko and J. K. Taylor, High-precision coulometric iodimetry, *Anal. Chem.* **39**, 1568–1571 (1967).
23. G. Marinenko, Gallium arsenide stoichiometry, in *Electrochemical Analysis Section, Summary of Activities, July 1970 to June 1971* (R. A. Durst, ed.), pp. 24–29, NBS Technical Note 583, U. S. Government Printing Office, Washington, D.C. (1973).
24. G. Marinenko and J. K. Taylor, Electrochemical equivalents of benzoic and oxalic acid, *Anal. Chem.* **40**, 1645–1651 (1968).
25. G. Marinenko and R. T. Foley, A new determination of the atomic weight of zinc, *J. Res. NBS* (*U.S.*), **75A** (*Phys. and Chem.*), 561–564 (1971).
26. K. M. Sappenfield, G. Marinenko, and J. L. Hague, *Standard Reference Materials: Comparison of Redox Standards*, NBS Special Publication 260-24, U.S. Government Printing Office, Washington, D. C. (1972).
27. R. A. Durst (ed.), *Ion-Selective Electrodes* (Proc. of a Symp. on Ion-Selective Electrodes, January 1969), NBS Special Publications 314, U.S. Government Printing Office, Washington, D.C. (1969).
28. J. Koryta, Theory and applications of ion-selective electrodes, *Anal. Chim. Acta* **61**, 329–411 (1972).
29. R. A. Durst, Ion-selective electrodes in science, medicine, and technology, *Am. Scientist* **59**, 353–361 (1971).
30. L. Meites, *Polarographic Techniques*, 2nd ed., Interscience, New York (1965).
31. D. D. Gilbert, Electroanalytical methods, in *Guide to Modern Methods of Instrumental Analysis* (T. H. Gouw, ed.), pp. 393–431, Wiley—Interscience, New York (1972).
32. E. J. Maienthal, Polarographic analysis at NBS, *Am. Laboratory* **4**(6), 12–21 (1972).
33. T. M. Florence, Anodic stripping voltammetry with a glassy carbon electrode mercury-plated *in situ*, *J. Electroanal. Chem.* **27**, 273–281 (1970).
34. E. J. Maienthal and J. K. Taylor, Improvement of polarographic precision by a comparative technique, *Mikrochim. Acta* **1967**, 939–945.
35. Certificates of analysis (provisional): Trace elements in a glass matrix, standard reference materials 610 and 611 (trace element concentration 500 ppm); 612 and 613 (trace element concentration 50 ppm); 614 and 615 (trace element concentration 1 ppm); and 616 and 617 (trace element concentration 0.02 ppm); August 5, 1970, revised August 8, 1972; available from Office of Standard Reference Materials, National Bureau of Standards, Washington, D.C.
36. I. L. Barnes, B. S. Carpenter, E. L. Garner, J. W. Gramlich, E. C. Kuehner, L. A. Machlan, E. J. Maienthal, J. R. Moody, L. J. Moore, T. J. Murphy, P. J. Paulsen, K. M. Sappenfield, and W. R. Shields, Isotopic abundance ratios and concentra-

tions of selected elements in Apollo 14 samples, *Proc. Third Lunar Science Conf., Geochim. Cosmochim. Acta*, Supplement 3, Vol. 2, pp. 1465–1472, MIT Press, Cambridge, Mass. (1972).

37. E. J. Maienthal, Analysis of botanical standard reference materials by cathode ray polarography, *J. Assoc. Official Analytical Chemists* **55**, 1109–1113 (1972).
38. T. S. West, Chemical spectrophotometry in trace characterization, in *Trace Characterization, Chemical and Physical* (W. W. Meinke and B. F. Scribner, eds.), pp. 215–301, NBS Monograph 100, U.S. Government Printing Office, Washington, D.C. (1967).
39. J. D. Ingle, Jr. and S. R. Crouch, Pulse overlap effects on linearity and signal-to-noise ratio in photon counting systems, *Anal. Chem.* **44**, 777–783 (1972).
40. S. Shibata, M. Furukawa, and K. Goto, Dual-wavelength spectrophotometry. Part II. The determination of mixtures, *Anal. Chim. Acta* **53**, 369–377 (1971).
41. T. J. Porro, Double-wavelength spectroscopy, *Anal. Chem.* **44**(4), 93A–103A (1972).
42. D. M. Dodd, D. L. Wood, and R. L. Barns, Spectrophotometric determination of chromium concentration in ruby, *J. Appl. Phys.* **35**, 1183–1186 (1964).
43. C. E. White and R. J. Argauer, *Fluorescence Analysis*, Marcel Dekker, New York (1970).
44. R. Mavrodineanu, J. I. Shultz, and O. Menis (eds.), Accuracy in spectrophotometry and luminescence measurements, Part 2. Luminescence, Proc. Conf. on Accuracy in Spectrophotometry and Luminescence Measurements, March 1972, *J. Res. NBS (U.S.)*, **76A** (*Phys. and Chem.*), 547–654 (1973); also NBS Special Publication 378, U. S. Government Printing Office, Washington, D.C. (1973).
45. C. A. Parker, Spectrophosphorimeter microscopy: an extension of fluorescence microscopy, *The Analyst* **94**, 161–176 (1969).
46. S. Udenfriend, *Fluorescence Assay in Biology and Medicine*, Vols. I, II, Academic, New York (1962, 1969).
47. J. P. Dixon, *Modern Methods of Organic Microanalysis*, Van Nostrand, Princeton, New Jersey (1968).
48. G. Tölg, *Ultramicro Elemental Analysis*, Wiley–Interscience New York (1970).
49. H. Weisz, *Microanalysis by the Ring-Oven Technique*, 2nd ed., Pergamon, New York (1970).
50. F. Feigl, *Spot Tests in Organic Analysis*, 7th ed., Elsevier, Amsterdam (1966).
51. E. Stahl, *Thin-Layer Chromatography*, Springer-Verlag, New York (1969).
52. A. Niederwieser and G. Pataki (eds.), *Progress in Thin-Layer Chromatography and Related Methods*, Vol. I (1970), Vol. II (1970), Vol. III (1972), Ann Arbor–Humphrey Science Publishers, Ann Arbor.
53. S. Dal Nogare and R. S. Juvet, Jr., *Gas—Liquid Chromatography, Theory and Practice*, Interscience, New York (1962).
54. A. E. Pierce, *Silylation of Organic Compounds*, Pierce Chemical Co., Rockford, Illinois (1968).
55. R. W. McKinney, Pyrolysis gas chromatography, in *Ancillary Techniques of Gas Chromatography* (L. S. Ettre and W. H. McFadden, eds.), pp. 55–87, Wiley–Interscience, New York (1969).
56. M. Beroza and M. N. Inscoe, Precolumn reactions for structure determination, in *Ancillary Techniques of Gas Chromatography* (L. S. Ettre and W. H. McFadden, eds.), pp. 89–144, Wiley–Interscience, New York (1969).

57. A. J. Raymond, D. M. G. Lawrey, and T. J. Mayer, Acquisition and processing of gas chromatographic data using a time-shared computer, *J. Chromatog. Sci.* **8**, 1–12 (1970).
58. J. J. Kirkland, *Modern Practice of Liquid Chromatography*, Wiley–Interscience, New York (1971).
59. J. J. Kirkland, Columns for modern analytical liquid chromatography, *Anal. Chem.* **43**(12), 36A–48A (1971).
60. H. Veening, Liquid chromatography detectors, *J. Chem. Ed.* **47**, A549–A568, A675–A686, A749–A762 (1970).
61. P. Kruger, *Principles of Activation Analysis*, Wiley–Interscience, New York (1971).
62. J. R. DeVoe and P. D. LaFleur (eds.), *Modern Trends in Activation Analysis* (Proc. 1968 Int. Conf. on Modern Trends in Activation Analysis, October 1968), NBS Special Publication 312, Vols. I and II, U.S. Government Printing Office, Washington, D.C. (1969).
63. G. J. Lutz, R. J. Boreni, R. S. Maddock, and J. Wing (eds.), *Activation Analysis: A Bibliography Through 1971*, NBS Technical Note 467, U.S. Government Printing Office, Washington, D.C. (1972).
64. D. DeSoete, R. Gijbels, and J. Hoste, *Neutron Activation Analysis*, Wiley, New York (1972).
65. G. J. Lutz, Photon activation analysis—a review, *Anal. Chem.* **43**, 93–103 (1971).
66. E. A. Schweikert and H. L. Rook, Determination of oxygen in silicon in the sub-part-per-million range by charged-particle activation analysis, *Anal. Chem.* **42**, 1525–1527 (1970).
67. W. S. Horton and C. C. Carson, Gas analysis: Determination of gases in metals, in *Treatise on Analytical Chemistry* (I. M. Kolthoff and P. J. Elving, eds.), Part I, Vol. 10, Section E, Chapter 103, pp. 6017–6144, Wiley, New York (1972).
68. O. Menis and J. T. Sterling, *Standard Reference Materials: Determination of Oxygen in Ferrous Metals—SRM 1090, 1091 and 1092*, National Bureau of Standards Misc. Publ. 260-14, U.S. Government Printing Office, Washington, D.C. (1966).
69. K. W. Guardipee, Two methods for separation of surface and bulk gases in vacuum-fusion analysis of metals, *Anal. Chem.* **42**, 469–473 (1970).
70. J. W. Frazer, Digital control computers in analytical chemistry, *Anal. Chem.* **40**(8), 26A–40A (1968).
71. P. D. Garn, *Thermoanalytical Methods of Investigation*, Academic, New York (1965).
72. R. F. Schwenker, Jr. and P. D. Garn (eds.), *Thermal Analysis*, Vol. 1, *Instrumentation, Organic Materials, and Polymers*, Vol. 2, *Inorganic Materials and Physical Chemistry*, Academic, New York (1969).
73. O. Menis (ed.), *Status of Thermal Analysis* (Proc. Symp. on the Current Status of Thermal Analysis, April 1970), NBS Special Publication 338, U.S. Government Printing Office, Washington, D.C. (1970).
74. H. G. Wiedemann (ed.), *Thermal Analysis*, Vol. 1, *Advances in Instrumentation*, Vol. 2, *Inorganic Chemistry*, Vol. 3, *Organic and Macromolecular Chemistry, Ceramics, Earth Science*, Birkhäuser, Basel, Switzerland (1972).
75. C. B. Murphy, Analytical reviews 1972/Fundamentals: Thermal analysis, *Anal. Chem.* **44**(5), 513R–524R (1972).

76. H. A. Liebhafsky, H. G. Pfeiffer, E. H. Winslow, and P. D. Zemany, *X-Ray Absorption and Emission in Analytical Chemistry*, Wiley, New York (1960).
77. R. O. Müller, *Spectrochemical Analysis by X-Ray Fluorescence* (K. Keil, transl.), Plenum, New York (1972).
78. C. L. Luke, Determination of trace elements in inorganic and organic materials by x-ray fluorescence spectroscopy, *Anal. Chim. Acta* **41**, 237–250 (1968).
79. L. S. Birks, Analytical reviews 1972/Fundamentals: X-ray absorption and emission, *Anal. Chem.* **44**(5), 557R–562R (1972).
80. R. D. Giauque and J. M. Jaklevic, Rapid quantitative analysis by x-ray spectrometry, in *Advances in X-ray Analysis* (K. F. J. Heinrich, C. S. Barrett, J. B. Newkirk, and C. O. Ruud, eds.), Vol. 15, pp. 164–175, Plenum, New York (1972).
81. R. Brown, M. L. Jacobs, and H. E. Taylor, A survey of the most recent applications of spark source mass spectrometry, *Am. Laboratory* **4**(11), 29–40 (1972).
82. R. A. Bingham and R. M. Elliott, Accuracy of analysis by electrical detection in spark source mass spectrometry, *Anal. Chem.* **43**, 43–54 (1971).
83. P. J. Paulsen, R. Alvarez, and C. W. Mueller, Spark source mass spectrographic analysis of ingot iron for Ag, Cu, Mo, and Ni by isotope dilution and for Co by an internal standard technique, *Anal. Chem.* **42**, 673–675 (1970).
84. E. C. Kuehner, R. Alvarez, P. J. Paulsen, and T. J. Murphy, Production and analysis of special high-purity acids purified by sub-boiling distillation, *Anal. Chem.* **44**, 2050–2056 (1972).
85. C. C. McMullen and H. G. Thode, Isotope abundance measurements and their application to chemistry, in *Mass Spectrometry* (A. McDowell, ed.), pp. 375–441, McGraw-Hill, New York (1963).
86. I. L. Barnes, E. L. Garner, J. W. Gramlich, L. J. Moore, T. J. Murphy, L. A. Machlan, W. R. Shields, M. Tatsumoto, and R. J. Knight, The determination of lead, uranium, thorium and thallium in silicate glass standard materials, *Anal. Chem.* **45**, 880–885 (1973).
87. Certificates of analysis (provisional): Orchard leaves, standard reference material 1571 (October 1, 1971); Bovine liver, Standard Reference material 1577 (April 15, 1972); available from the Office of Standard Reference Materials, National Bureau of Standards, Washington, D.C.
88. B. F. Scribner and M. Margoshes, Emission spectroscopy, in *Treatise on Analytical Chemistry* (I. M. Kolthoff and P. J. Elving, eds.), Part I, Vol. 6, Chapter 64, pp. 3347–3461, Interscience, New York (1965).
89. E. L. Grove (ed.), *Analytical Emission Spectroscopy*, Vol. I, Part I (Analytical Spectroscopy Series, Vol. II, 1972; Vol. III, to be published), Marcel Dekker, New York (1971).
90. V. G. Mossotti, Emission spectroscopy including dc arc, spark, and other methods, in *Techniques of Metals Research* (R. F. Bunshah, ed.), Vol. III, Part 2, pp. 533–572, Interscience, New York (1970).
91. R. M. Barnes, Analytical reviews 1972/Fundamentals: Emission spectrometry, *Anal. Chem.* **44**(5), 122R–150R (1972).
92. J. A. Dean and T. C. Rains, *Flame Emission and Atomic Absorption Spectrometry*, Vol. 1, *Theory*, Vol. 2, *Components and Techniques*, Marcel Dekker, New York (1969, 1971).

93. R. Mavrodineanu (ed.), *Analytical Flame Spectroscopy, Selected Topics*, Macmillan, London (1970).
94. E. E. Pickett and S. R. Koirtyohann, Emission flame photometry—A new look at an old method, *Anal. Chem.* **41**(14), 28A–42A (1969).
95. D. P. Hubbard, *Annual Reports on Analytical Atomic Spectroscopy 1971*, Vol. 1, The Society for Analytical Chemistry, London (1972).
96. G. D. Christian and F. J. Feldman, A comparison study of detection limits using flame-emission spectroscopy with the nitrous oxide—acetylene flame and atomic-absorption spectroscopy, *Appl. Spectr.* **25**, 660–663 (1971).
97. J. W. Robinson and P. J. Slevin, Recent advances in instrumentation in atomic absorption, *Am. Laboratory* **4**(8), 10–18 (1972).
98. G. F. Kirkbright, The application of non-flame atom cells in atom-absorption and atomic-fluorescence spectroscopy, a review, *The Analyst* **96**, 609–623 (1971).
99. V. I. Goldanskii and R. H. Herber, *Chemical Applications of Mössbauer Spectroscopy*, Academic, New York (1968).
100. L. May, *An Introduction to Mössbauer Spectroscopy*, Plenum, New York (1971).
101. G. Stevens, J. C. Travis, and J. R. DeVoe, Analytical reviews 1972/Fundamentals: Mössbauer spectrometry, *Anal. Chem.* **44**(5), 384R-406R (1972).
102. R. L. Mössbauer, Gamma resonance spectroscopy and chemical bonding, *Angew. Chem. Internat. Ed.* **10**, 462–472 (1971).
103. J. J. Spijkerman and P. A. Pella, A review of selected highlights of Mössbauer spectrometry, *Crit. Rev. Anal. Chem.* **1**, 7–45 (1970).
104. R. M. Lynden-Bell and R. K. Harris, *Nuclear Magnetic Resonance Spectroscopy*, Nelson, London (1969).
105. J. W. Emsley, J. Feeney, and L. H. Sutcliffe, *High Resolution Nuclear Magnetic Resonance Spectroscopy*, Vols. 1, 2, Pergamon, New York (1965, 1966).
106. F. A. Bovey, *Nuclear Magnetic Resonance Spectroscopy*, Academic, New York (1969).
107. C. A. Poole, Jr. and H. A. Farach, *Relaxation in Magnetic Resonance; Dielectric and Mössbauer Applications*, Academic, New York (1971).
108. J. I. Kaplan, Numerical solution of the equation governing nuclear magnetic spin–lattice relaxation in a paramagnetic-spin-doped insulator, *Phys. Rev.* **B3**, 604–607 (1971).
109. J. E. Wertz and J. R. Bolton, *Electron Spin Resonance: Elementary Theory and Practical Applications*, McGraw-Hill, New York (1972).
110. F. Gerson, *High Resolution Electron Spin Resonance Spectroscopy*, Wiley, New York (1970).
111. A. Abragam and B. Bleaney, *Electron Paramagnetic Resonance of Transition Ions*, Oxford, Clarendon Press, London (1970).
112. L. R. Weisberg, Electrical measurement for trace characterization, in *Trace Characterization, Chemical and Physical* (W. W. Meinke and B. F. Scribner, eds.), NBS Monograph 100, U.S. Government Printing Office, Washington, D.C. (1967).
113. V. A. Deason, A. F. Clark, and R. L. Powell, Characterization of high purity metals by the residual resistivity ratio, *Mat. Res. and Std.* **1971**(8), 25–28.
114. A. F. Clark, V. A. Deason, J. G. Hust, and R. L. Powell, *Standard Reference Materials: The Eddy Current Decay Method for Resistivity Characterization of High Purity Metals*, NBS Special Publication 260-39, U.S. Government Printing Office, Washington, D.C. (1972).

115. L. S. Birks, *Electron Probe Microanalysis*, 2nd ed., Wiley—Interscience, New York (1971).
116. K. F. J. Heinrich, *Scanning electron probe microanalysis*, NBS Technical Note 278, U.S. National Technical Information Service, Springfield, Virginia (1967).
117. K. F. J. Heinrich, *Quantitative Electron Probe Microanalysis*, NBS Special Publication 298, U.S. Government Printing Office, Washington, D.C. (1968).
118. K. F. J. Heinrich, Errors in theoretical correction systems in quantitative electron probe microanalysis—A synopsis, *Anal. Chem.* **44**, 350–354 (1972).
119. R. Tixier and J. Philibert, Analyse quantitative d'echantillons minces, in *Proc. 5th Int. Congress on X-Ray Optics and Microanalysis* (G. Möllenstedt and K. H. Gaukler, eds.), pp. 180–186, Springer-Verlag, Berlin (1968).
120. W. J. Campbell and J. V. Gilfrich, Analytical reviews 1970/Fundamentals: X-ray absorption and emission, *Anal. Chem.* **42**(5), 248R–268R (1970).
121. A. J. Socha, Analysis of surfaces utilizing sputter ion source instruments, *Surface Sci.* **25**, 147–170 (1971).
122. A. Benninghoven, Beobachtung von Oberflächenreaktionen mit der statischen Methode der Sekundärionen-massenspektroskopie; I. Die Methode, *Surface Sci.* **28**, 541–562 (1971); A. Benninghoven, Surface investigation of solids by the statistical method of secondary ion mass spectroscopy (SIMS) *Surface Sci.* **35**, 427–457 (1973).
123. C. A. Anderson, Progress in analytical methods for the ion microprobe mass analyzer, *Int. J. Mass Spectry. Ion Phys.* **2**, 61–74 (1969).
124. J. A. McHugh and J. F. Stevens, Elemental analysis of single micrometer-size airborne particulates by ion microprobe mass spectrometry, *Anal. Chem.* **44**, 2187–2192 (1972).
125. K. Siegbahn, C. Nordling, A. Fahlman, R. Nordberg, K. Hamrin, J. Hedman, G. Johansson, T. Bergmark, S. Karlsson, I. Lindgren, and B. Lindberg, *ESCA: Atomic, Molecular and Solid State Structure Studied by Means of Electron Spectroscopy*, Nova Acta Regiae Societatis Scientiarum Upsaliensis, Series IV, Vol. 20, Almqvist and Wiksells Boktryckeri AB Uppsala (1967).
126. D. A. Shirley (ed.), *Electron Spectroscopy*, North-Holland, Amsterdam (1972).
127. D. M. Hercules, Analytical reviews 1972/Fundamentals: Electron spectroscopy. II, X-ray photoexcitation, *Anal. Chem.* **44**(5), 106R–112R (1972).
128. K. Siegbahn, D. Hammond, H. Fellner-Feldegg, and E. F. Barnett, Electron spectroscopy with monochromatized x-rays, *Science* **176**, 245–252 (1972).
129. W. H. McCurdy, Jr. and D. H. Wilkins, Analytical reviews 1966/Fundamentals: Volumetric and gravimetric analytical methods for inorganic compounds, *Anal. Chem.* **38**, 469R–478R (1966).
130. A. L. Underwood, Photometric titration, in *Advances in Analytical Chemistry and Instrumentation* (C. N. Reilley, ed.), Vol. 3, pp. 31–104, Interscience, New York (1964).
131. G. Marinenko and J. K. Taylor, Precise coulometric titration of dichromate, *J. Res. NBS (U.S.)*, **76A** (*Phys. and Chem.*), 453–459 (1963).
132. T. M. Florence, Ion-selective electrodes, *Proc. Roy. Austral. Chem. Inst.* **37**, 261–270 (1970).
133. T. S. West, Some sensitive and selective reactions in inorganic spectroscopic analysis, *The Analyst* **91**, 69–77 (1966).

134. J. A. Roberts, J. Winwood, and E. J. Millett, The spectrophotometric determination of sub-microgram amounts of impurities in semiconductor materials, in *Proc. Soc. Analytical Chemistry Conf., Nottingham, 1965*, pp. 528–538, Heffer & Sons, Cambridge (1965).
135. I. P. Alimarin, Progress and problems of trace determination in pure substances, *Zh. Analit. Khim.* **18**, 1412–1425 (1963).
136. M. Vecera and J. Horska, A study of the accuracy and precision of methods for the determination of carbon and hydrogen in organic compounds, *Pure Appl. Chem.* **21**(1), 47–84 (1970).
137. N. Hadden, F. Baumann, F. MacDonald, M. Munc, R. Stevenson, D. Gere, F. Zamaroni, and R. Majors, *Basic Liquid Chromatography*, Varian Aerograph, Walnut Creek, California (1971).
138. J. M. A. Lenihan and S. J. Thomson (eds.), *Advances in Activation Analysis*, Vol. 2, Academic, New York (1972).
139. J. P. Bruch, Determination of gases in steel and application of the results, Iron and Steel Institute Special Report No. 131, *Determination of Chemical Composition—Its Application and Process Control*, Iron and Steel Institute, London (1971).
140. W. Schwarz and H. Zitter, Determination for oxygen content in steel by hot extraction, *Berg. Hutten. Monatsh.* **113**, 1–10 (1968).
141. C. Mazieres, Differential thermal microanalysis, physical chemical applications, *Bull. Soc. Chim. France* **1961**, 1695–1701.
142. G. V. Davis and R. S. Porter, Application of the differential scanning calorimeter to purity measurements, *J. Thermal Anal.* **1**, 449–458 (1969).
143. N. W. H. Addink, *DC Arc Analysis*, Macmillan, London (1971).
144. M. D. Amos, P. A. Bennett, K. G. Brodie, P. W. Y. Lung, and J. P. Matousek, Carbon rod atomizer in atomic absorption and fluorescence spectrometry and its clinical application, *Anal. Chem.* **43**, 211–215 (1971).
145. P. A. Pella and J. R. DeVoe, Determination of tin in copper-base alloys by Mössbauer spectroscopy, *Anal. Chem.* **42**, 1833–1835 (1970).
146. L. H. Schwartz, Quantitative analysis using Mössbauer effect spectroscopy, *Int. J. Nondestruct. Test.* **1**, 353–381 (1970).
147. P. A. Pella and J. R. DeVoe, International standardization in Mössbauer spectrometry, *Appl. Spectry.* **25**, 472–474 (1971).
148. T. C. Farrar and E. D. Becker, *Pulse and Fourier Transform NMR*, Academic, New York (1971).
149. R. S. Alger, *Electron Paramagnetic Resonance: Techniques and Applications*, Section 3.3, Sensitivity, pp. 69–91, Interscience, New York (1968).
150. C. P. Poole, Jr., *Electron Spin Resonance: A Comprehensive Treatise on Experimental Techniques*, pp. 523–600, Interscience, New York (1967).
151. R. F. Gould, *Nonstoichiometric Compounds*, Advances in Chemistry Series 39, American Chemical Society, Washington, D.C. (1963).
152. A. Benninghoven, Mass spectrometric analysis of monomolecular layers of solids by secondary ion emission, in *Advances in Mass Spectrometry* (A. Quale, ed.), Vol. 5, pp. 444–447, Elsevier, New York (1971).
153. D. P. Smith, Analysis of surface composition with low-energy backscattered ions, *Surface Sci.* **25**, 171–191 (1971).
154. E. D. Tolmie and D. A. Robins, The zone-refining of impure copper, *J. Inst. Metals* **85**, 171–176 (1957).

155. M. Cuypers, Systematic analysis of high purity copper, following its irradiation by thermal neutrons, *Ann. Chim. (Paris)* **9**, 509–540 (1964).
156. C. H. Lewis, M. B. Giusto, H. C. Kelly, and S. Johnson, The preparation of high-purity silicon from silane, in *Ultrapurification of Semiconductor Materials* (M. S. Brooks and J. K. Kennedy, eds.), pp. 55–56, Macmillan, New York (1962).
157. F. A. Pohl and W. Bonsels, Zur spurenanalyse sehr reinen siliciums, *Mikrochim. Acta* **1960**, 641–649.
158. C. T. Butler, J. R. Russell, R. B. Quincy, Jr., and D. E. LaValle, A method for purification and growth of KCl single crystal, Oak Ridge National Laboratory Technical Report ORNL-3906, U.S. Atomic Energy Commission Technical Information Center, Oak Ridge, Tennessee.
159. A. Glasner and P. Avinur, Spectrophotometric methods for the determination of impurities in pure and analytical reagents—III. The determination of six ions in KCl, *Talanta* **11**, 775–780 (1964).
160. A. Kremheller, Growth and heat treatment of zinc sulfide single crystals, *J. Electrochem. Soc.* **107**, 422–427 (1960).
161. G. J. Sloan, Studies on the purification of anthracene; determination and use of segregation coefficients, *Molecular Crystals* **1**, 161–194 (1966).

8

Structural Characterization of Solids

R. E. Newnham and Rustum Roy
Materials Research Laboratory
The Pennsylvania State University
University Park, Pa.

1. Introduction

In principle, a solid is characterized absolutely if one can list all the atoms or ions present, their spatial distribution, and the bonds holding them together. The dynamics of the atoms, and even possibly the electrons, might be added to this list. The goal of structural characterization is, in fact, no more than the description of the three-dimensional arrangements of the atoms (and electrons) in a solid. In so doing, it spans the range from the grossest features (such as the external morphology and cracks visible to the naked eye) down to the finest detail of atomic arrangement (such as the omission of single ions from their expected location in a periodic array).

The simplest conceptualization of structural characterization is as follows: Visualize a real material, perhaps a cube 1 cm on edge, on which some very sophisticated physical measurements are to be performed. If such measurements are to be of any value, they must be reproducible elsewhere on the "same" material. This brings us up against the problem of adequately describing the material—chemically and structurally—for reproducibility. Clearly, the description is in two parts: a determination of the chemical composition and its variation throughout the crystal with spatial resolution approaching atomic dimensions, and a determination of the structure or atomic arrangement and its variation in space. In this review we are concerned exclusively with the latter, the structural characterization, elucidating the various parameters describing the location of atoms and electrons within the solid.

TABLE 1

Structural Characterization and Measurement Techniques*

Phase identification	Optical microscopy, X-ray diffraction; electron microscopy and diffraction for small amounts
Distinguish amorphous and crystalline materials	Microscopy, X-ray diffraction
Changes of structure (transitions)	Differential thermal analysis, microscopy, or *in situ* X-ray diffraction
Macroscopic morphology, crystal system information	Goniometry, optical microscopy
Unit cell and space group	X-ray (neutron, or electron) diffraction
Site symmetry in crystalline and amorphous materials	Visible and infrared absorption, X-ray emission or diffraction, resonance methods
Position of atoms, thermal vibration amplitudes	X-ray (neutron, or electron) diffraction
Imperfections	Microscopy, X-ray topography, indirect and resonance methods, absorption spectroscopy
Spin configuration	Neutron diffraction and magnetic properties, magnetic resonance methods, Mössbauer techniques
Bonding (electron dynamics)	Electron density by X-ray diffraction; spin density by neutron diffraction; dynamics by indirect methods

* Listed in order of universality of application.

Table 1 presents the kind of information one needs to obtain about real specimens, together with the methods which are available to obtain each particular type of information. The sequence of listing is intended to convey a series of steps in obtaining increasingly detailed structural information. In other words, the first few steps are quite essential for *every* piece of solid-state work; the next group is necessary for *most* scientific studies, and the last two or three may only be required when certain elaborate physical measurements are to be made on the material.

The first aim in characterizing a solid would certainly be to determine whether the material is single phase or not. This can be accomplished both by chemical and structural means of characterization. Here, the polarizing microscope and powder X-ray diffraction can in a matter of minutes provide answers in the case of many real solids. From the same instruments, we would know whether or not the phase(s) were crystalline or noncrystalline; the rapid increase in the number of devices based on glasses or other noncrystalline solids makes this an important problem. Next, we have listed a simple examination designed to determine the existence of phase changes within the expected variations of temperatures and pressure. Since property measurements are often made at elevated temperatures, many errors and difficulties can be avoided by obtaining this information *early*. The next group of measurements is concerned with the description of symmetry (or its absence), from the macroscopic scale of crystal morphology to the microscopic symmetry of the first coordination sphere of anions surrounding a cation of interest. The final group concerns the determination of the exact position of the atoms, the orientations of the magnetic moments of the electron spins, the nature of the bonding between the atoms, and the principal types of defects.

Structural characterization is performed by instruments and operators trained in a particular instrumental approach, and a survey in depth of their present capability and research needs will be most effective if conducted in these same categories. Below, in turn, each of the classes of methods by which the "structure" of a material can be determined is treated in detail.

2. Structural Characterization by Optical Techniques

Optical examination is almost inevitable in materials characterization; clearly, the first and most obvious thing to do with a new material is to look at it! However, with the wide gamut of sophisticated techniques available for materials studies, what was in earlier times a long, hard look, has now frequently degenerated to a mere cursory glance; and much useful, sometimes vital, information that could be easily obtained from a more perceptive optical study is completely missed. Often, the only equipment necessary for such studies would be a metallurgical microscope, or a petrographic instrument equipped with objectives for incident illumination.

In the following, we summarize the types of information that can be obtained from a thorough optical study. Methods such as absorption spectroscopy which depend primarily on the wavelength of light are discussed in a later section. The remaining studies can be grouped under four headings, morphology, surface studies, bulk optical properties, and scattering phenomena.

2.1. Morphology

The external morphology of well-formed crystals frequently provides information on the crystal symmetry, a useful precursor to a full X-ray, electron, or neutron diffraction study of the structure. Both naked eye and goniometric information are useful, although the latter technique is now only rarely employed. In regard to physical measurements, morphology provides a rapid method for crystal orientation.

In principle all 32 crystal classes (point groups) can be identified from crystal morphology, though in practice the required forms seldom develop. Under these circumstances the apparent symmetry is higher than the true symmetry, as illustrated for quartz in Figure 1. There are also circumstances when the apparent symmetry is lower than the true symmetry because of crystal growth conditions inhibiting the equal development of equivalent faces. As noted by Abbe Haüy centuries ago, facial areas may vary from crystal to crystal, but the angles between faces are fixed.

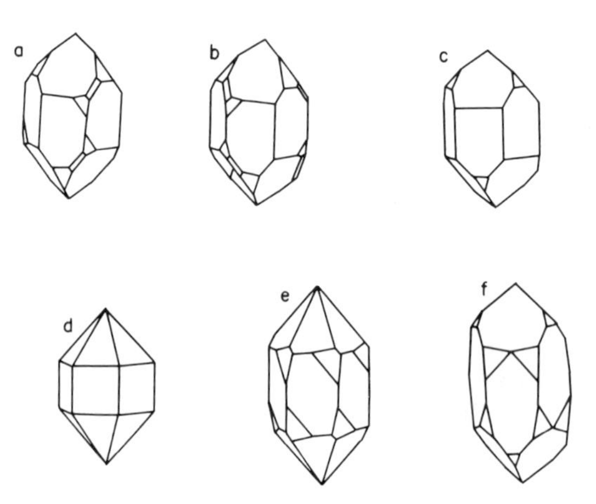

Fig. 1. General form development in (a) right-handed quartz, and (b) left-handed. Enantiomorphic pairs can be distinguished by crystal morphology. Both (a) and (b) display the true symmetry of quartz, class 32. The quartz crystals in (c) and (d) show higher symmetries because the general form failed to develop. In (c) the two rhombohedral special forms develop differently, giving $\bar{3}m$ as the apparent symmetry. When developed equally, (d), the apparent symmetry is $6/mmm$. Twinned quartz crystals are shown in (e) and (f). Right- and left-handed regions are intergrown in Brazil twins (f) with apparent symmetry $\bar{3}m$, while Dauphine twins are composed of intergrown crystals of the same handedness. Interpenetrating right-handed crystals with apparent symmetry 622 are shown in (e).

Twinned crystals can sometimes be recognized by the peculiar morphological development resulting from the twinning operation. Two types of twinning observed in quartz are shown in Figure 1.

The angles between faces give axial ratios and interaxial angles. In quartz the angle between the normals to the (10.0) prism face and (10.1) rhombohedral face is 38° 13′ which is equivalent to $\tan^{-1}(\sqrt{3}a/2c)$. Classical crystallography consisted of the determination of crystallographic data and symmetry from crystal morphology. The techniques are summarized in Phillips's book.[1] Morphological studies were extremely fashionable in the Victorian era, culminating in Groth's monumental series, *Chemische Krystallographie*, published around 1910, which remains an important reference work.[2]

2.2. Bulk Optical Properties

In this category are methods which can be applied to transparent solids, such as refractive index measurement, refraction anisotropy, and rotary dispersion. Details regarding the use of the polarizing microscope may be found in many excellent textbooks, such as those by Hartshorne and Stuart[3] and Wahlstrom.[4]

The procedure generally followed in the optical characterization begins by separating the material into its constituent phases, whenever practicable. This is often achieved by sieving, flotation, chemical attack, or even with tweezers. The crystals are then examined in ordinary light, noting the color, the morphology, interfacial angles, growth steps, inclusions, and cleavage traces, using several observation directions. Stereographic projections are useful in interpreting the angular data. Immersing the crystal in a liquid of matching refractive index eliminates surface scattering, making interior flaws more visible.

Polarized light gives additional information. Color often depends on polarization direction, producing dichroism in uniaxial crystals and trichroism in biaxial crystals. Orthoscopic examination between crossed Nicols with a polarizing microscope gives the extinction directions and angles. Birefringence can be measured with a wedge, and the fast and slow directions identified. Measurements of refractive indices and dispersion are usually accomplished with index liquids. Phase transitions and twinning are usually apparent in polarized light. Hot stages are useful in measuring transition temperatures and melting points and in studying crystallization phenomena.

The optical character (uniaxial or biaxial) and sign (positive or negative) can be ascertained by conoscopic examination between crossed Nicols. The angle between optical axes ($2V$) and dispersion of optical axes are also readily determined under these conditions.

Precise measurement of the refractive index, or, more frequently, of the three principal indices of the indicatrix, is frequently used to characterize transparent materials. High precision is possible using monochromatic light with temperature-controlled immersion methods, giving the index to $n \pm 0.0005$. The method is tedious and not completely unequivocal as an identification tool. However, routine immersion techniques are capable of determining the refractive indices in the higher-symmetry systems to ± 0.002 in a matter of minutes and are an invaluable simple characterization tool. The determinative tables collected by Winchell[(5)] and Berman[(6)] are helpful in identifying unknown phases.

Detailed evaluation of the optical indicatrix can give useful information on the crystal symmetry. Cubic crystals and amorphous solids are optically isotropic; trigonal, tetragonal, and hexagonal crystals are uniaxial; orthorhombic, monoclinic, and triclinic crystals are biaxial. Extinction directions vary with wavelength in monoclinic and triclinic crystals.

These data together with the refractive indices provide one of the most effective structural characterizations of a solid. Optical methods are particularly sensitive in detecting pseudosymmetric forms of a higher symmetry phase and, therefore, for detecting phase changes with temperature in ferroelectric, antiferroelectric, and transparent ferrimagnetic crystals.

Because of the optical anisotropy in lower-symmetry crystals, twinning can frequently be detected immediately in the polarizing microscope. The effects of superposed twinned regions may give rise to very complex and often beautiful optical patterns. Ferroelectric domains are visible in polarized light when accompanied by reorientation of the optical indicatrix (Figure 2). The motion of domain walls under applied electric fields can be studied with the polarizing microscope.

Magnetic domains in transparent crystals show optical rotation in polarized light, the sign of rotation changing with the direction of the magnetization vector. The Faraday effect has been useful in studying bubble domains and their motions. Infrared converters are sometimes employed for iron oxides which are opaque throughout most of the visible range.

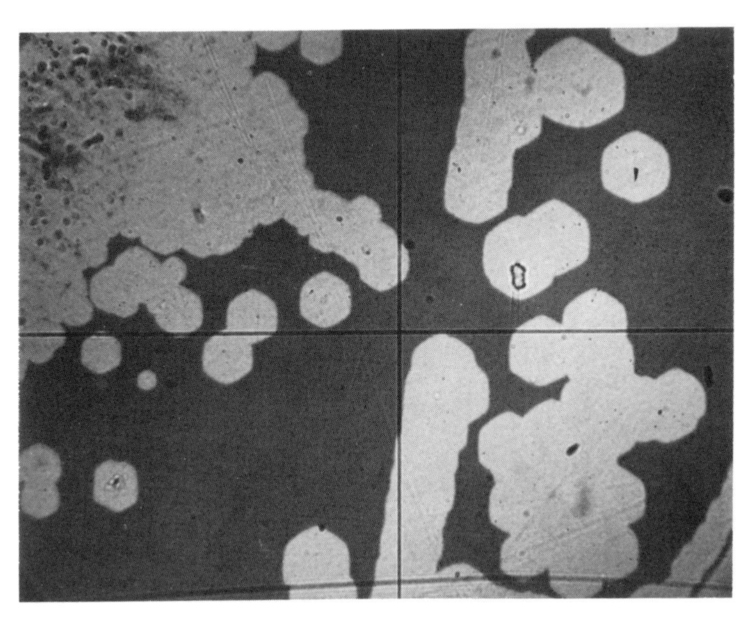

Fig. 2. Domain structure in $Pb_5Ge_3O_{11}$, an optically active ferroelectric. 180° domains in a c-cut crystal are clearly visible in plane-polarized white light with analyzer positioned at 87.5° to the polarizer. The crystal thickness is 0.5 mm and the small hexagonal domains are about 0.1 mm in diameter.

Phase separation may be detected optically before there is any evidence in an X-ray powder study. Optical study is especially valuable in critical applications, where only a small amount of a second phase is present. Concentrations of second phase far below 0.01 % can be detected, compared to the approximate 1 % limit for X-ray methods.

In certain cases, specific structural questions may be answered from the bulk property data. Thus, the orientation of the planar CO_3 groups in carbonates can be inferred from the refractive indices, while the presence of optical activity is usually an indication of enantiomorphism in the structure.

Special optical tests have been developed with applicability over a limited range of problems. Photoelasticity, the change in refractive indices with stress, has been used to measure strain in engineering structures. Precise measurement of birefringence also provides a measure of internal stress in imperfect crystals, and an index of the efficiency of annealing techniques. In special cases, dislocations may be observed directly from the stress field around the core.

In some transparent crystals of lower symmetry, useful data can be obtained from the detection of harmonic generation. The matrix of the higher-order optical coefficients is similar to that of the piezoelectric constants, and relates in the same manner to the crystal symmetry. The presence of second harmonic light indicates the absence of a center of symmetry. Kurtz and Perry[7] have recently developed an SHG test for polycrystalline materials which is more convenient than piezoelectric tests for noncentosymmetric crystals.

2.3. Scattering Studies

Scattering studies are essentially adapted to the investigation of heterogeneous systems where there is a distribution of fine particles (the scatterers) in a transparent matrix. Two types of experiment are possible: scattering from (1) distributed centers and (2) individual centers, both of which provide useful information on imperfections in transparent crystals.

The total scattered intensity is measured in the first experiment. No effort is made to separate contributions from the individual scattering centers. When the total scattered intensity is measured as a function of angle and of the wavelength of incident light, the mean particle size and number of scattering centers may be deduced. If there is a preferred orientation of anisotropic scattering centers, some information may be obtained on the particle anisotropy. Commercial equipment is available for measuring molecular weight in polymers from total scattered intensities.

Two techniques are used for investigating scattering from individual centers. One is the ultramicroscope, which measures scattering at 90° to the incident beam; the other is the dark-field microscope, which measures small-angle scattering close to the illumination direction. Very recently, ultramicroscopy has been greatly improved with the use of coherent light sources, and it is possible now to detect microheterogeneities on a very small scale. For example, using a 1-mW gas laser, it is possible to detect 900-Å Latex spheres in water.[8] With currently available high-power laser systems and with photographic recording, it would appear that "resolution" limits of the order of 100 Å are possible in ruby laser rods and other solid-state device crystals.

2.4. Surface Characterization

In this category, we consider those methods of characterization which rely on examination of the surface topology of a natural, cleaved, or etched surface. Such studies can be carried out on all solids with well-defined surfaces and provide information regarding growth mechanisms, crystal symmetry, crystal perfection, surface layers, domain structure, twinning, and grain boundaries. Techniques available for the evaluation of solid surfaces include surface roughness measurements, electron and optical microscope studies, and several diffraction techniques. The low-energy electron diffraction method used in the direct determination of surface structures is discussed elsewhere in this series (Volume 6).

In both metals and nonmetals, reflection from a polished and etched section is a standard method for delineating grain boundaries. Many methods are available for assessing mean grain size from measurements on such sections. The technique is also of particular value in revealing evidence of second phases which frequently segregate at the grain boundary and are therefore very difficult to detect by X-ray methods.

Dye tests and radioactive penetrants are useful in characterizing surface integrity. Polishing sometimes produces surface damage allowing a colored or fluorescent dye to penetrate cracks or checks. Liquid dyes are applied to the solid and the surface layer is then removed by rubbing or washing. Large cracks are made very apparent, and numerous grain boundaries cause a broad discoloration. Radioactive penetrants work on the same principle, using radiation counters as detectors.

Grinding and polishing can also change the surface stress patterns, distortions which could later lead to crack formation and fracture. X-ray diffraction line profile analysis appears to be the best method for determining residual stresses, a method which can be applied to polycrystalline materials as well as single crystals.

Twin boundaries and low-angle grain boundaries can often be observed directly on intersection with the surface. Frequently, the small tilts associated with internal twinning (as in antiferroelectrics) may be observed by reflection microscopy or multiple-beam interferometry. The etch technique used to identify optical and electrical twins in quartz is of considerable practical importance in fabricating piezoelectric oscillator plates.[9]

Optical surface studies are useful in judging crystal perfection. Etched surfaces can provide quantitative information on internal surfaces and on line defects. Emerging dislocations show etch pits at the surfaces, and etch pit counting is a standard method for determining dislocation densities in metals. In copper, the method is extremely reliable when etching is done on $\{111\}$, though results on $\{100\}$ and $\{110\}$ are less certain. Dislocation densities in the range $1\text{–}10^8/\text{cm}^2$ can be determined microscopically if the etching is done on surfaces close to (111). One limitation is the size of the etch pits. Visible pits are at least 1 μm in size, large enough to contain several dislocations.

In certain circumstances, etch pits are sensitive to the orientation of the dislocation line and its Burgers vector, making it possible to distinguish edge and screw dislocations. They are also sensitive to impurities, and not all etch pits are associated with dislocations. Etch pit densities observed on heating and cooling aluminum are hundreds of times larger than the actual dislocation concentrations. It is imperative that etching results be calibrated with more direct methods such as transmission electron microscopy or X-ray topography.

The symmetries of etch pits can give useful additional information on the crystal symmetry and crystal orientation since they can be classified into the ten two-dimensional point groups. A useful table of etchants and polishes for solid-state materials has recently been compiled.[10]

Ferroelectric domains etch at different rates according to the orientation of the polarization vector with respect to the surface. Etching is a standard method for observing 180° domains in perovskite ferroelectrics. Ferromagnetic domains can be viewed by decorating the flux emergent at the wall region by a colloidal magnetic powder deposited from a suitable suspending liquid. In III–V semiconductor compounds and other materials with the zincblende structure, the etch method provides a rapid method for identifying polar faces. This is important in making electrical contacts to the crystals.

Obst[11] has developed a decoration technique for investigating normal and superconducting regions in superconductors. In the intermediate state in lead, the phase boundaries are oriented along cube edges. Other type I superconductors also show correlations between orientation of the flux-line lattice and the crystal lattice.

Though somewhat less versatile than normal microscopic examination, multiple- and two-beam interferometry provide a very precise measurement of the surface topography, revealing features down to 5 Å in height.[12] Such extreme accuracy in elevation is accomplished at the expense of horizontal resolution. For extremely sharp fringes, the resolution in the plane of the surface is only about five wavelengths because of sideways displacement during multiple reflection. Optical interferometry has been used in studying crystal

TABLE 2

Typical Resolutions of Surface-Microtopographic Instruments (after Young[15])

Instrument	Approximate vertical resolution, Å	Approximate horizontal resolution, Å
Transmission electron microscope	1500	50
Scanning electron microscope	1000	100
Optical interference	5	25,000
Stylus instrument	25	10,000
Topografiner	30	4,000

growth, electropolishing, cleavage, and various imperfections such as slip bands.[13]

In past years, a contacting-stylus instrument similar to a phonograph needle has been the standard method for surface-profile measurements. A fine diamond stylus coupled to a transducer scans the surface, providing closely spaced profiles that can be displayed on an x–y recorder. The elevation sensitivity is excellent but unfortunately the stylus sometimes obliterates the original surface features. The diameter of the stylus point determines the width and depth of the crevices it will measure. Table 2 compares the vertical and horizontal resolving powers of several microtopography techniques. The other electron techniques mentioned in Table 2 will be discussed in later sections.

Ellipsometry has been used in studies on ultraclean surfaces; the formation of surface films can be followed by measurement of the azimuth and ellipticity of reflected polarized light. For thicker films (above 20 Å), the method allows a separation of the thickness and refractive index, and thickness may be measured for films to a fraction of an angström in average thickness. Recent ellipsometric measurements[14] on single-crystal silicon surfaces subjected to polishing, etching, sputtering, cleavage, and annealing illustrate the power of the method. Information concerning the optical properties and average thickness of the damaged surface layer is obtained as well as those of the oxide film and silicon substrate.

It recently has been shown that modulation of the reflectance of a solid surface by periodic electric and elastic stress fields can give details on surface structure. These special techniques are, in general, relatively expensive, and for effective exploitation may require expensive auxiliary equipment, such as ultrahigh vacua, sophisticated electronics, or other features. One of the important advantages of all optical methods is their nondestructive nature.

2.5. Particle Size and Shape

The vast majority of solid materials occurring in nature or produced by technology are polycrystalline, and usually polyphasic. It is therefore obvious that one of the principal tasks of characterization of a solid would be to describe the size, shape, and relative orientation of the separate units in any powder, aggregate, or formed piece. Examples run from the description of a loose sand, to a boron-epoxy composite, to a polished section of a maraging steel. As the crucial role of the grain boundary in metals and ceramics has come to be recognized, as has simultaneously the importance of the surface in all insulator and semiconductor applications, the characterization of particle size and shape has received a great impetus in the last two decades. Particle *shape* has also assumed a newly important role in application ranging from fibrous composites, through fine particle magnets, to sintering studies. In the past, the qualification of shape was an extensively tedious process and hence rarely carried out for a large number of particles. Only with automated methods and computerized data retrieval has shape analysis become possible. Indeed, the appearance of the scanning electron microscope and automated quantitative microscopy[16] heralds probably the biggest single advance in characterization in the seventies.

In Table 3 are listed the spectrum of methods used and the size range of effectiveness and the properties measured in each case. Clearly there is neither a single measure of "size" nor a single tool with which to measure it.

2.5.1. Computerized Automated Quantitative Microscopy

Over the last year or two, true second-generation automated instruments have been developed where the analysis of the data obtained by electronic means is fed directly into a small computer with much of its logic for processing several standard suites of information already built into its circuits.

The following independent measurements can be made on any field of particulate matter (Bausch and Lomb QMS): (A) Entire field measurements: (1) total particle count, (2) lower positive tangent

TABLE 3

Techniques Used in Particle Size Determination

Property measured	Method	Size application, μm
Length (and area, etc.)	Optical microscopy	0.5–2000
	Scanning electron microscopy	0.2–500
	Transmission electron microscopy	0.002–10
Minimum length	Sieving	10–6000
Projected surface area	Light scattering	0.3–30
	Automatic image analyzing microscopy or SEM	0.5–2000
		0.002–1 (e.s.d.)
Active surface area	Gas adsorption	(0.001 m^2/g and up)
Volume	Electrical resistivity	0.5–400
	Sedimentation	0.5–200
	Elutriation	5–100
	Centrifugation	0.01–30
	Impaction	0.02–50
Crystallite volume	X-ray diffraction line broadening	0.015–0.050

count, (3) lower negative tangent count, (4) intercept count, (5) oversize feature count, (6) oversize positive tangent count, (7) oversize negative tangent count, (8) oversize intercept count, (9) total projected length, (10) total area, (11) average area, and (12) average projected length. (B) Selective particle measurement: (13) Feret's diameter, (14) longest dimension, (15) area including holes, (16) area excluding holes, (17) projected length. (C) Entire field multiple measurement/count ratios: (18) percent of area, (19) area per positive tangent count, (20) projected length per positive tangent count, (21) area per negative tangent count, (22) projected length per negative tangent count, and (23) average intercept length.

2.5.2. Computer-Evaluated Scanning Electron Microscopy

While the total capability of the scanning electron microscope (SEM) will be treated later in this chapter, since one of the SEM's major applications is in quantitative powder characterization, we will refer here to this application alone. Clearly, the automated quantitative micrography instruments referred to in the last section can process any optical image, including transmission and scanning electron microphotographs. Of course, it is inefficient to convert a rastered electronic signal into an optical image in photography and then process this

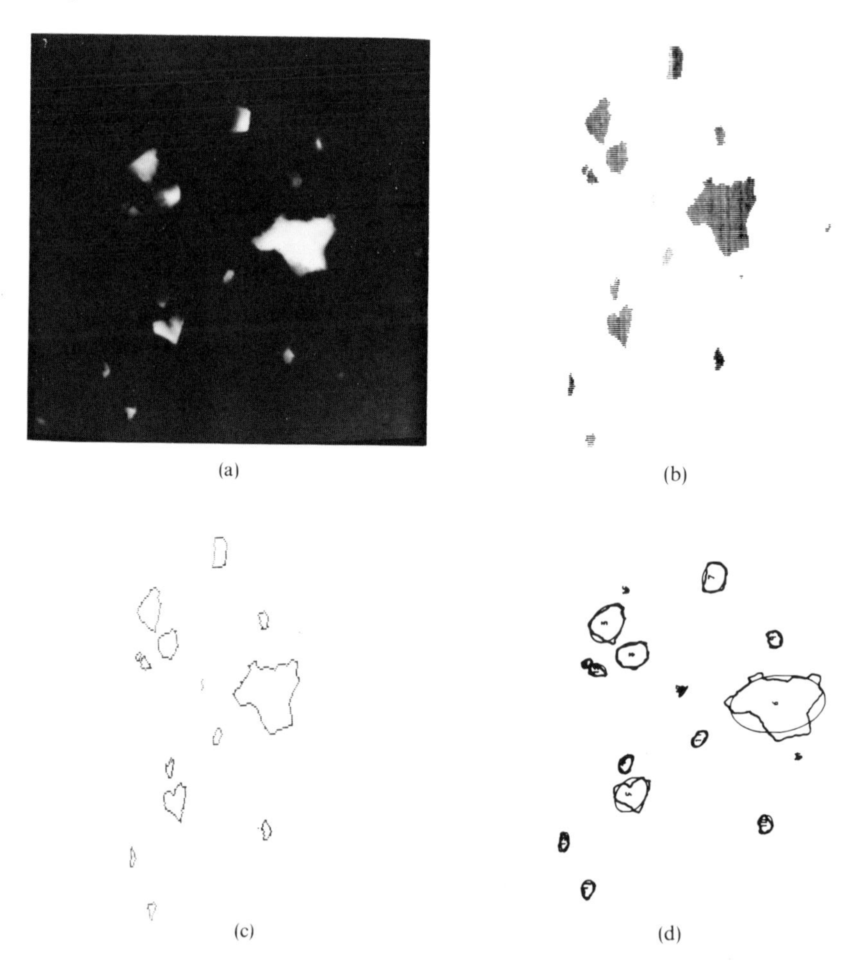

Fig. 3. Computer-evaluated scanning electron microscope images.[17] A secondary electron image of a fine-grained alumina powder is shown in (a), and reproduced in a binary-coded map in (b) to give particle areas. A perimeter printout is shown in (c), and with fitted ellipses in (d).

photograph. White and collaborators[17] were able to develop direct image processing and computer evaluation of the same even before the optical instruments were available (Figure 3).

The SEM has, of course, one additional feature which renders it capable not only of particle size and shape analysis but also of simultaneous chemical analysis. The electron beam generates not only the secondary electrons which form the SEM image, but also X rays characteristic of the elements present in each particle. It is therefore possible, in principle, to "tag" each particle by semiquantitative analysis during the electron scan. The experimental difficulty is associated with the scan rate required to generate a sufficiently strong X-ray signal, which is considerably slower than that required for the SEM image.

In summary, therefore, it is now routinely possible to obtain quantitative automated analyses of the shape, size, and, simultaneously, the principal compositional characteristics of any powder or work surface with particles from 0.1 μm up. A sufficient number of particles can be handled in about 1 hr to give a great deal of statistical validity to the data obtained.

3. Structural Characterization by X-Ray Diffraction

X-ray diffraction is without question the most important initial method for characterizing materials in the solid state because the results are derived rather directly from the atomic or molecular arrangement of the substance, and the methods are relatively routine and simple. X-ray methods are used in research, development, and quality control in metallurgy, ceramics, and the mineral, chemical, and pharmaceutical industries, and practically anywhere chemical and physical properties must be related to crystal structure. A modern materials program thus requires extensive use of X-ray diffraction analysis. This section outlines the use of the method for the study of materials in the form of powders and single crystals.

Polycrystalline specimens are used for identification of the phase and for lattice parameter determination. Well-crystallized single crystals are used in structure determination and X-ray topographic study of perfection. The capability at room temperature and pressure will be treated first, followed by sections dealing with high- and low-temperature studies and high-pressure work.

3.1. X-Ray Powder Methods

Counter tube detectors were introduced in commercial X-ray apparatus at the end of World War II. This development launched a vast increase in the use of X-ray methods. Although film techniques are still widely used, the bulk of X-ray powder work is now done with counter tube diffractometers.

By present standards, it is possible to define a reasonable standard of a "good" X-ray diffractometer pattern against which performance can be judged. Such a pattern is one of the most important single pieces of physical data for the characterization of a material. "Good" means that (1) the resolution is adequate to resolve most of the important lines down to about $d = 1.5$ Å ($60°\ 2\theta$ with Cu $K\alpha$), (2) the precision is sufficient to determine the reflection angle to about $0.01°\ 2\theta$, (3) the line profile shape is sufficiently free of instrumental broadening factors to recognize gross lattice imperfections (e.g., stacking disorder and small particle size), (4) the intensities and peak-to-background ratios are adequate for good counting statistics ($\pm 10\%$ for the principal lines), and (5) the methods of recording the data do not introduce significant errors. A further step in characterization is achieved if the pattern has been indexed and all lines related to the unit cell parameters. The success of the crystallographic indexing is dependent largely on the precision with which the interplanar spacings are measured. Computer programs have been developed for indexing both directly from the powder pattern as well as from unit cell dimensions derived from single crystals using X-ray or electron diffraction methods. Single-crystal patterns are generally required for low-symmetry crystals.

3.1.1. Phase Identification

Qualitative analysis of polycrystalline materials is the most widely used application of X-ray diffraction. The specimen may be in the form of a loose, fine powder, a solid polycrystalline aggregate, a coating, a foil, or even a frozen liquid. The complexity (i.e., the number of lines) of a powder pattern increases with increasing unit cell dimensions and decreasing crystallographic symmetry, and is essentially independent of the chemical composition. The use of X-ray diffractometer patterns as a means of "fingerprint" identification is practiced widely, by comparing the spacings and intensities of the lines of the unknown with those of knowns in the JCPDS (formerly ASTM) Card Files.[18] There is no doubt that this method is the best routine, solid-state characterization tool now available. It is limited only by the number and precision of the standard patterns in the card indexes and by the precision of the data collected for the unknown. The present JCPDS index has data for about 20,000 substances, and is growing at the rate of 2000 patterns per year. Each pattern is cross-indexed according to its eight strongest diffraction lines. Computer retrieval programs are available for routine phase identification, as are graphical representations of the X-ray patterns.

The relative amounts of two (or more) phases present in a mixture may also be obtained from powder X-ray data. The accuracy is dependent on the specimen preparation, the quality of standard patterns, and the care taken to reduce systematic and random errors. The lower limits of detectability depend on the complexity of the pattern, the relative X-ray absorption of the phases, and related factors. The range of sensitivity of detection of the second phase is very great. For instance, while casual examination of routine patterns will probably have a lower detection limit between 1–5 % of a second phase, 0.1 wt % of silicon in a tungsten matrix and 0.01 % in a lithium fluoride matrix may be detected in reasonable counting times, providing the proper care and techniques are used. However, even if the impurity is known, the detection limit is increased when the matrix has a very complex pattern, for example, magnetite in a granite rock. Amorphous and segregated phases may be overlooked by this method, making optical microscopic studies advisable.

3.1.2. Lattice Parameter Measurement

The lattice parameters are the averages of millions of unit cell dimensions in the specimen and are important in characterizing materials. A precision of 1–0.1 % can be obtained rapidly using routine methods. Higher precision measurements of the order of 0.01 % or better are required to determine the coefficients of thermal expansion and in a variety of other studies. The indexing of complex powder patterns requires precise measurements of many reflections in the low-angle region.

There have been many claims in the literature of very high accuracy, often exceeding 0.001 %, based upon the reproducibility of powder data, but these measurements do not usually indicate the real accuracy. The difficulties of handling the systematic errors and the various correction factors limit the accuracy. The observed X-ray diffraction profiles recorded with a diffractometer are a convolution of the profiles arising from wavelength and intensity distributions of the incident X-ray beam, the various instrument factors, diffraction geometry, and specimen properties. All these factors combine to cause small distortions and shifts of the observed profiles, thereby introducing systematic errors and problems in interpretation when the highest accuracy is sought. Figure 4 summarizes the results of an international round-robin collaborative study[19] by 16 laboratories using the same sample of elemental silicon powder. The composite mean value of 25 average values submitted was 5.43054 Å at 25°C, standard deviation ±0.00017 Å, and the agreement calculated from the highest and lowest means was 0.012 %.

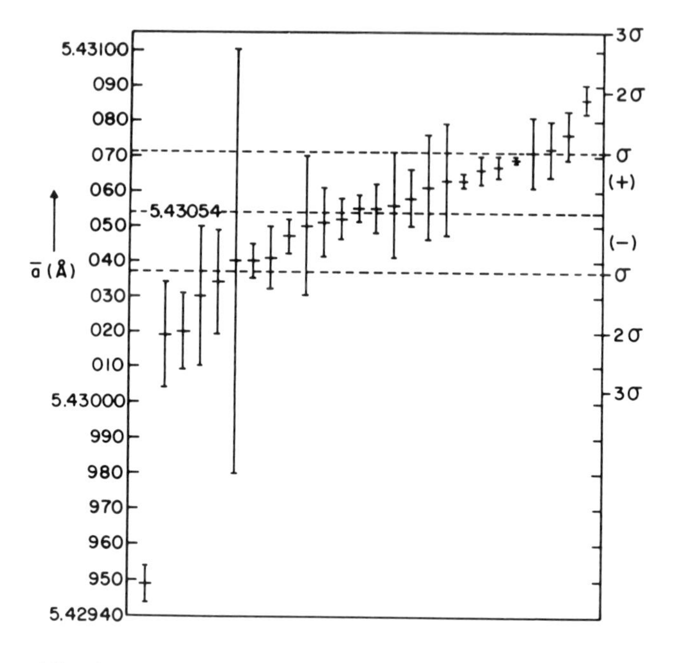

Fig. 4. Precision determination of the lattice parameter of silicon by 16 different laboratories. Individual mean values (short horizontal lines) are shown together with reported error limits (vertical lines). Dashed lines show the composite mean and its standard deviation.

Relative lattice parameter values may be determined with much higher precision. Bond[20] has described a relatively simple method for single-crystal plates in which a precision of a few parts per million was attained. The Kossel technique has been reported to have an accuracy better than one part in 50,000 provided certain precautions are taken. The elegant "centroid" method is limited at present to high-symmetry substances of small unit cell dimensions and by the need for better X-ray spectral data.

Since there is usually a fairly simple relationship (approaching linearity) between lattice parameter and composition of solid or crystalline solutions, precision measurements of the separation and relative intensities of the lines provide an important and widely used method of materials study and characterization. If proper precautions are observed, one can determine chemical compositions from these structural data; sensitivities of 0.1% are obtained in favorable cases.

3.1.3. Line Broadening

X-ray patterns provide a view not only of the average structure but of deviations from the average as well. The influence of various imperfections on the diffraction pattern is summarized in Table 4. Line broadening has been used extensively to study the substructure of polycrystalline materials, measuring two types of deviation from a perfectly periodic atomic arrangement: microstrain and coherent domain size. Microstrain is a measure of the variation of the interplanar d spacing about its average value. Because of this variation, the Bragg condition is satisfied for a range of 2θ angles and the peak is broadened. Any component of atomic displacement lying in the diffracting plane does not contribute to the measured microstrain.

The second type of broadening arises because diffraction takes place from small domains within the sample. For angles different from the Bragg angle, destructive interference takes place between the X-ray beams scattered from planes near the surface and those scattered from planes deeper within the crystal. The effect of the finite size of the coherently diffracting domains is to produce incomplete destructive interference for angles only slightly different from the Bragg angle. This is similar to the effect observed in spectroscopic diffraction gratings,

TABLE 4

Types of Disorder and Their Diffraction Transforms (after Bernal[21])

Degree of crystallinity	Diffraction pattern	Examples
Large perfect crystal	Sharp spots at reciprocal lattice points; primary extinction	Perfect crystals, diamond
Very small, equidimensional crystals	Equidimensional diffuse spots	Colloids, gold sol
Very thin, needlelike crystals	Spots extended along layer planes	Fibers, whiskers
Thin, platy crystals	Spots extended along row lines	Clays, carbons
Thermal motion	Weakened Bragg peaks; interspot diffuse scattering	All crystals at high temperatures
Point imperfections	Perfect crystal pattern plus low-angle scattering	Irradiated crystals
Line imperfections, dislocations	Mosaic-type diffraction; secondary extinction	Metals; plastically deformed crystals
Planar imperfections, stacking faults	Interspot diffraction streaks and spikes	Twinned and rapidly grown crystals

where the width of the diffraction maxima increases as fewer lines are used in the grating.

Each of the single crystals in a polycrystalline sample is made up of a large number of mosaic blocks. These blocks are regions of the crystal which are so perfect that the dynamical theory of X-ray diffraction holds within them. The coherent domain size measured in the Warren–Averbach method[22] is the average length of the mosaic blocks measured in a direction perpendicular to the diffracting planes.

In order for the dynamical theory of X-ray diffraction to hold, successive reflecting planes would have to be parallel to each other to within a few seconds of arc. This is the case within the mosaic blocks which comprise the crystal. Imperfections produce a warping of the lattice which causes different mosaic blocks to have a misorientation of the order of minutes of arc. The boundaries of these mosaic blocks can be interpreted as consisting of arrays of dislocations. The various ways in which arrays of dislocations can form boundaries are illustrated in the article by Hirsch.[23] The average thickness of the mosaic blocks measured in a particular direction is referred to as the coherently diffracting domain size.

3.1.4. Disordered and Noncrystalline Solids

Since disordered solids retain varying degrees of periodicity in their atomic distribution, they can be studied by diffraction techniques. The principal tool in such studies has been X-ray diffraction, supplemented by electron and neutron diffraction. The degree of faulting or strain in otherwise well-ordered materials can be estimated from diffraction experiments, along with the ratio of crystalline to amorphous material, and radial electron density distributions.

Instrumentation for studies of this nature are usually variations of the normal powder X-ray diffractometer. Except for faulting and strain in single crystals, which are better treated as defects, the very nature of the material limits studies to powders or aggregates. X-ray powder patterns of simple metals can be analyzed to yield information on particle size, deformation fault probability, mean-square strain, and twinning. The theory and techniques used to study diffraction line broadening, peak shifts, and line profile asymmetry have been derived and applied by Warren,[24] and Warren and Averbach.[25] To assess faulting probability, certain drastic assumptions are necessary, reducing the detectability limit to approximately one faulted layer in 200.

In the case of a crystalline phase in an amorphous matrix, a rough measure of their ratio can be obtained if the compositions are known. This is accomplished by summing the powder diffraction line intensities, subtracting the amorphous (background) intensity, and correcting for Compton-modified scattering. The fact that there is a continuous spectrum of order from well-crystallized to noncrystalline phases is an inherent limitation to the method.

Just as crystals are described by a unit cell and a set of atomic positions, amorphous materials have a structure defined by a radial distribution function. The radial distribution function (RDF) gives the number of atoms per unit volume at a distance r from the reference atom. The source of such information is an X-ray scattering curve from which the Compton-modified component has been subtracted. This is accomplished by measuring scattered intensity as a function of angle using monochromatic radiation. Only simple amorphous materials such as carbon black and silica can be analyzed in detail. More complex glasses are beyond the scope of present theory and techniques, although a combination of neutron and X-ray diffraction intensity data helps in identifying interatomic vectors. The RDF method can only be used to eliminate or corroborate a selected model; equally good agreement may be determined for several models.

Studies on poorly crystalline and noncrystalline materials have not been particularly widespread because of theoretical and experimental difficulties. Today, however, with the advent of thin-film technology and materials such as neodymium laser glass and amorphous semiconductors, there is great need for such information. The nature of these studies precludes precise results, but the numerical data can be correlated with properties and provide an insight into materials preparation phenomena.

For three or more decades, low-angle X-ray scattering has been applied to the detection of inhomogeneities on a scale of 10–1000 Å in liquids, glasses, and crystals.[26] The most favorable cases are those in which particulate phase differs markedly in atomic number from the matrix. Naturally, the calculations require a model of the composition of the two phases, which is almost always a zeroth-order approximation. Combined X-ray and optical low-angle scattering experiments may provide an approach to the simultaneous determination of composition and size fluctuations. As a characterization tool in the general cases, X-ray scattering can detect incipient phase separation, but has proved rather disappointing in its quantitative aspects. Transmission microscopy is proving to be a more reliable and direct tool for most such studies of inhomogeneitics or second-phase distribution.

3.2. Single-Crystal X-Ray Methods

The determination of crystal structures is one of the outstanding triumphs of 20th century science and dates back to the discovery of X-ray diffraction in crystals by von Laue in 1912. Several Nobel prizes have been awarded for the X-ray determination of protein and vitamin structures, testifying to the power of the method in determining complex atomic arrangements. Less well publicized are the numerous crystallographic studies of smaller molecules and solid-state materials. Thousands of crystal structures have been worked out, and the results are of inestimable value in understanding the properties of materials.

3.2.1. Strategy of Structure Analysis

Two quantities are measured for each reflection on the X-ray diffraction pattern, the angular position θ and the integrated intensity I. The size and shape of the unit cell are determined from the angles, and the intensities yield the locations of atoms within the unit cell. In addition to the atomic coordinates, less precise information concerning thermal vibration amplitudes and electron density are also derived from the intensity data.

The procedure followed in structure analysis is illustrated schematically in Figure 5. Beginning with the diffraction pattern, the reflections are indexed to give the unit cell dimensions and the interaxial angles. Film techniques are generally employed during preliminary stages since the entire diffraction pattern can be quickly surveyed using rotation, Weissenberg, and precession cameras. Indexing is especially easy with precession photographs because the reflections are arrayed in a pattern resembling the reciprocal lattice. Lattice constants are refined using high-angle data and either least squares or graphical extrapolation procedures, until the agreement between observed and calculated θ values is satisfactory.

The powder method is difficult to apply in the determination of crystal structures except in simple cases because the overlapping reflections prevent unequivocal indexing and measurement of intensities. Although this is a fundamental limitation that cannot be overcome by better instrumentation, the powder method is nevertheless useful for a variety of structure studies.

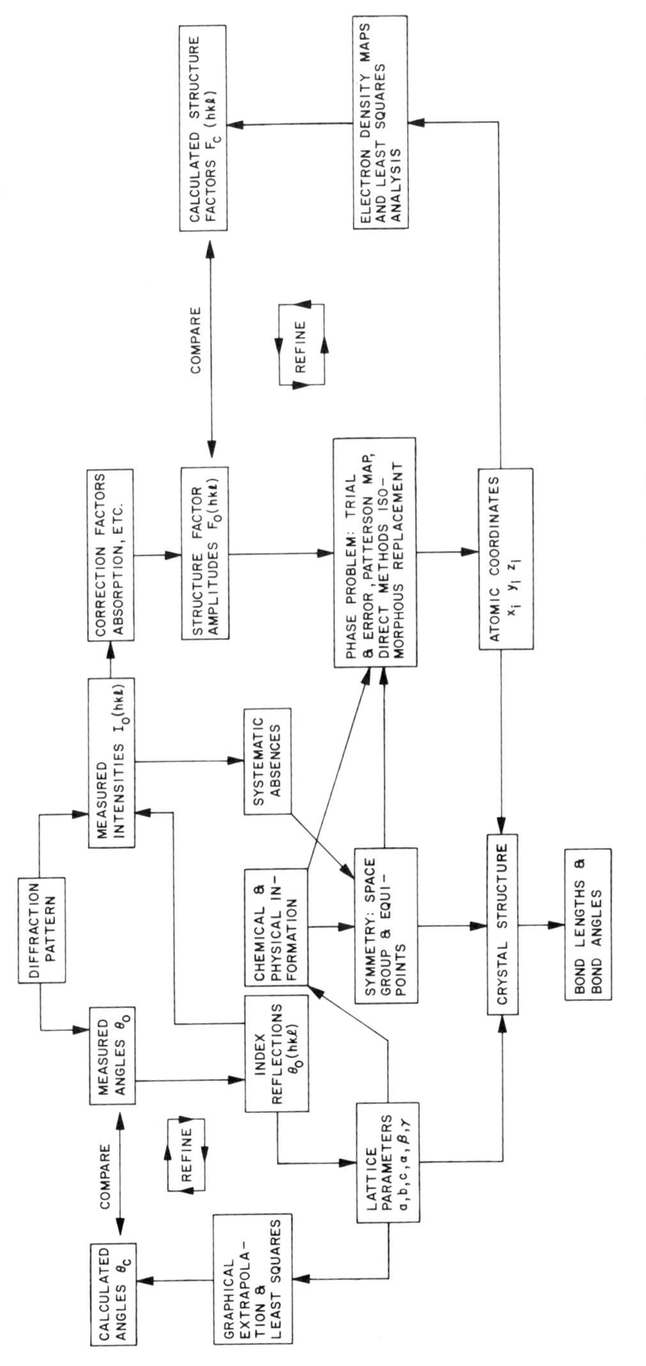

Fig. 5. Procedure followed in determining crystal structures from X-ray diffraction patterns.

The number of molecules per unit cell Z is obtained from the cell volume V, the molecular weight M, the measured density ρ, and Avogadro's number $N_0 : Z = \rho V N_0/M$. From Z, we know how many atomic coordinates (x_i, y_i, z_i) must be determined from the intensities $I(hkl)$.

Since symmetry usually reduces the number of independent coordinates, it is helpful to determine the space group. In olivine (Mg_2SiO_4), for example, Z is 4 so that there are 28 atoms per unit cell. But of the $28 \times 3 = 84$ coordinates, only 11 are crystallographically independent, simplifying the problem considerably. Space groups are determined chiefly from X-ray data by noting systematic absences and equivalences among reflections.[27] Equivalences determine the Laue group, and absences the diffraction symbol. To continue the example: The olivine X-ray pattern can be indexed on an orthorhombic cell and the reflections hkl, $hk\bar{l}$, $h\bar{k}\bar{l}$, $\bar{h}kl$, $h\overline{kl}$, $\bar{h}k\bar{l}$, $\overline{hk}l$, and $\overline{hkl}$ are all of equal intensity, so the Laue group is *mmm*. The $h0l$ reflections are absent when $k + l$ is odd and $hk0$ reflections are missing when h is odd, giving diffraction symbol *Pn.a*. Only two of the 230 space groups are therefore possible: *Pnma* and $Pn2_1a$. Statistical analysis of the intensity data or physical properties such as the piezoelectric effect can be used to select the correct space group. After the space group is determined and the atoms assigned to equipoints, the remaining independent coordinates are determined from the measured intensities.

The intensities are first converted to structure factor amplitudes by correcting for absorption, Lorentz-polarization factors, and multiplicity. In general the structure factors are complex, but only the amplitude can be measured experimentally, and not the phase angle. The so-called "phase problem" has prevented X-ray crystallography from becoming a routine procedure, although the development of powerful analytic techniques makes this less true now than in the past.

There are four important techniques[28,29] used in surmounting the phase problem: trial-and-error, Patterson maps, direct methods, and isomorphous replacement. Most of the simple structures have been determined by trial and error. Structure factors are calculated from an assumed set of coordinates $F_c = \sum_n f_n \exp[2\pi i(hx_n + ky_n + lz_n)]$ and compared with experiment using the R factor as a measure of agreement:

$$R = \sum_{hkl} \| F_0(hkl)| - |F_c(hkl)\| / \sum_{hkl} |F_0(hkl)|$$

A low R factor means good agreement between calculated and observed structure factors, and therefore a correct structure.† As a rough guide, R factors less than 0.2 indicate that a structure is basically correct and R factors less than 0.1 mean the coordinates have been well-refined. With good counter data and accurate corrections, many crystallographers are achieving R factors less than 0.05 at present. In the trial-and-error procedure, structure factors are calculated for numerous models until R is minimized. Crystal symmetry and crystal chemistry are helpful in selecting trial structures.

The Patterson function

$$P(X, Y, Z) = (1/V) \sum_{hkl} |F(hkl)|^2 \exp[-2\pi i(hX + kY + lZ)]$$

is also useful in solving the phase problem, since it can be evaluated directly from the intensity data and gives a map of vector space when plotted as a function of position throughout the cell. For every pair of atoms located at r_n and r_m, there are Patterson peaks at $r_n - r_m$ and $r_m - r_n$ centered at the origin of the cell. Patterson maps are difficult to interpret when the number of atoms N in the unit cell is large because the number of Patterson peaks is proportional to N^2. Superposition methods, implication diagrams, and other techniques have been developed to aid the interpretation.[31]

The presence of a heavy atom greatly simplifies the structure determination since the prominent Patterson peaks are those corresponding to vectors between heavy atoms. Once the heavy atoms are located, the other atoms can be positioned from Fourier electron density maps using the phase angles determined by the heavy-atom coordinates. The ease of solving such structures is partly responsible for the rapid growth of metal organic chemistry.

Direct methods[32] constitute the third important approach to the phase problem. Certain physical properties of crystals (such as the fact that the electron density is always positive) place restrictions on the magnitudes and phases of the structure factors. For centrosymmetric crystals, the structure factors are real, and the phase problem is therefore one of sign determination. Sayre's equation for intense reflections is an example of sign determination by direct methods:

$$s(hkl) = s(h', k', l')s(h + h', k + k', l + l')$$

The sign s is ± 1. Thus if the 110, 211, and 321 reflections are all strong,

† The question of homometric structures—two different structures giving the same diffraction pattern—has never been adequately investigated. In practice, most crystallographers are overjoyed at obtaining one correct structure, and do not pursue others. Homometric structures are undoubtedly rare but at least one example exists.[30]

there is a relation between their phases. Sign relations such as these can be quickly surveyed by computer and the phases determined. More than half the structure analyses reported recently employed direct methods.

Patterson maps and direct methods are generally successful for structures containing fewer than 50 atoms per asymmetric unit. They sometimes work for larger structures like vitamin B 12 as well. For proteins and other really large structures, however, the most successful method is that of chemical substitution. The phase problem is solved by comparing the X-ray patterns for two heavy-atom derivatives of the biological molecule. Fourier analysis then gives the shape of the molecule.

Crystal structures are generally refined by least squares analysis, although Fourier maps are required if one or more atoms are badly misplaced. In the least squares method, atomic coordinates are varied until the differences between observed and calculated structure factors is minimized. Anisotropic temperature factors and site occupancies are also determined. The latter is especially important in disordered materials. The least squares technique also provides a rapid error analysis so that standard deviations of the atomic coordinates and bond lengths can be quickly assessed.

The results of crystal structure determinations are published in *Acta Crystallographica*, *Zeitschrift fur Kristallographie* and, occasionally, in other journals such as *American Mineralogist* and the *Journal of Metals*. Collected reviews of crystal structure determination are published in a series of books by the International Union of Crystallography[33] and by Wyckoff[34]; efforts are being made to bring these abreast with the current literature.

Data collection and analysis are no longer time-consuming steps, so that under favorable circumstances, structure determinations can be completed in weeks rather than years. The increased capabilities have resulted in substantial increases in the number of structures solved and in the size of the structures attempted. In 1959, the structure analyses reported in *Acta Crystallographica* averaged ten atoms in the asymmetric unit; in 1969, the average was 19.

The replacement of film by automatic counter methods provides a potential increase in speed and accuracy. The American Crystallographic Association recently sponsored a project[35] to obtain a quantitative comparison of the absolute accuracy and assess the various systematic errors involved in X-ray intensity measurement. Seven well-known crystallographers measured intensities from the same crystal, a small CaF_2 sphere, using all major types of counter tube apparatus and methods. Each experimenter used Mo $K\alpha$ radiation, measured the integrated intensities of every reflection in the (hhl) zone to $\sin\theta/\lambda = 1.00$ Å^{-1}, made absorption corrections, and calculated structure factors. The agreement between experiments was stated to be remarkably good, although there was no evidence that any lab obtained F^2 values to better than 2%. The largest systematic error was θ dependent and probably resulted from extinction. Among the other factors affecting intensity measurements are absorption, spectral distribution of the X-ray beam, and variations in background scattering caused by thermal diffuse scattering and other effects. Time also becomes an important factor in the counting statistical accuracy since thousands of reflections must be measured. The stability of the crystal, the X-ray source, and counting circuits, and the reproducibility of crystal settings may become important problems.

Typical of the excellent results which can be obtained with an automatic single-crystal diffractometer is Ladell's work on topaz.[36] The unit cell parameters are $a = 4.6499$ Å, $b = 8.7968$ Å, and $c = 8.3909$ Å, and the space group is *Pnma*. Fifteen levels containing 6117 reflections were measured with Mo $K\alpha$ radiation over a period of 17 days. Fifteen positional parameters, 36 anisotropic temperature factors, a scale factor, and an extinction parameter were refined using 2230 independent reflections with the calculations giving an R of 0.04. If only those reflections with a counting statistical error less than 1% are used, $R = 0.017$ based on 23 reflections per parameter. Positional coordinates were accurate to ± 0.00005 Å, and the bond lengths to ± 0.0008 Å. An earlier analysis by Alton and West[37] in 1929, based on photographic data, gave positions accurate to ± 0.05 Å.

The determination of the distribution of unlike atoms over structurally equivalent lattice sites is of considerable importance in solid-state studies as, for example, in the ferrites.[38] The distribution may be determined from accurate intensity measurements of those reflections most sensitive to the ordering effects. By using X-ray wavelengths close to the absorption edge of one of the atoms, it is possible to enhance the difference of the atomic scattering factors for the two atoms. Anomalous scattering is also useful in single-crystal studies. The (111) and $(\overline{111})$ faces of III–V semiconductor crystals can be identified in this way.[39]

3.2.2. Electron Density Maps

Increased accuracy in structure determination requires good instrumentation and methods, as well as the willingness to explore the various factors which may affect the intensity data. Accurate intensity measurements (2% or better) are especially important in deducing bonding information from electron density maps. Present accuracy of electron density maps, under the most favorable conditions, using the best techniques, and making all the known corrections, is of the order of 1–5% of the atomic number of the atom or, say, 0.1 electron/Å^3 in an organic structure. Generally speaking, the accuracy of the map is proportional to the accuracy of the intensity measurements. The number of electrons per unit volume at a position (x, y, z) is given by

$$\rho(x, y, z) = (1/V) \sum_{h} \sum_{k} \sum_{l} F(hkl) \exp[-2\pi i(hx + ky + lz)]$$

(all three sums run from $-\infty$ to $+\infty$)

In the equation, $F(hkl)$ is the structure factor for reflection hkl and V is the unit cell volume. The summation is over all reflections, but only a finite number are measured experimentally, introducing series termination errors. Choosing the X-ray wavelength as short as possible (Ag $K\alpha$) extends the series to high orders, and cooling the specimen eliminates thermal vibrations tending to weaken high-order reflections. As discussed previously, measurement of the Bragg intensities gives $|F(hkl)|^2$. To obtain the series coefficients $F(hkl)$ from the intensities, it is necessary to solve the phase problem first. The phase problem becomes trivial in simple centric structure types where the atoms are in special positions. It is therefore feasible to obtain accurate electron density maps when the intensities are carefully measured.

The determination of electron distributions in crystals has recently been reviewed by Brill.[40] Ionic, covalent, metallic, and molecular bonding have been investigated by the technique. Regarding bonding, there are three points of interest in the electron density maps: the sphericity of the atoms, the total charge at each atom, and the position of the minimum between neighboring atoms. The degree of ionization, the number of d electrons, and the asymmetry of the electron cloud are some of the features which can be estimated from the maps. Further improvements in experimental accuracy and elimination of extinction and other systematic errors promises to elucidate problems of chemical bonding and to check theoretical predictions.

The map for NaCl shown in Figure 6(a) gives electron density contours projected on (100). Such maps require very accurate X-ray intensity data, and although the atom centers can be located with considerable precision, the contours of the bonding electrons are rather more uncertain. The difference in sphericity between NaCl and

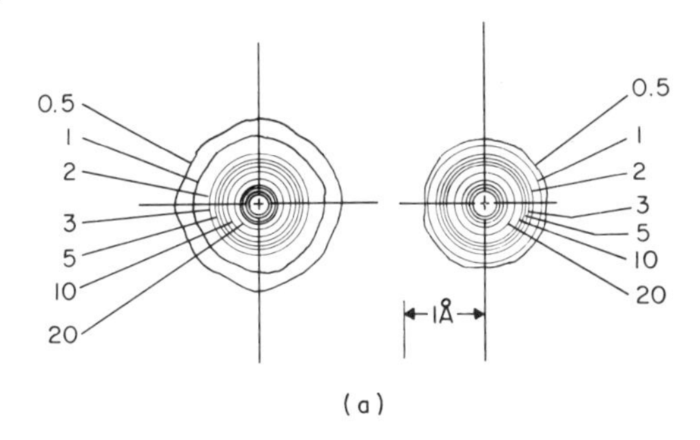

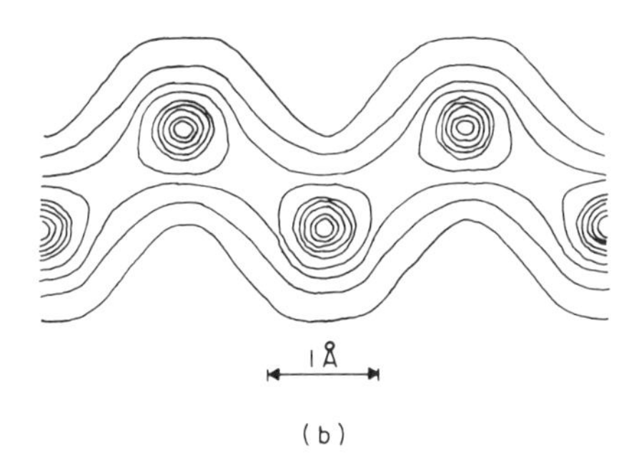

Fig. 6. Fourier syntheses of (a) rocksalt[41] and (b) diamond.[42] Contours are in one electron/Å^3 intervals, except for the lowest level of 0.5 electrons/Å^3. The three-dimensional Fouriers are projected on (100) for NaCl and (110) for diamond.

diamond can be seen by comparing Figures 6(a) and 6(b). The electron density midway between neighboring atoms is very much less in rocksalt, providing one quantitative and absolute (if imprecise) measure of ionicity. Moreover, the atoms are more nearly spherical in rocksalt despite series termination errors, while the electron density in diamond is definitely concentrated along the bonds linking nearest neighbors and corresponding to the electrons involved in the covalent bond.

The total charge associated with each atom is estimated by computing the charge around each atom, though it is difficult to know where to draw the boundary between atoms. Likewise, the size of each ion can be estimated from the position of the minimum and compared with ionic and atomic radii. Sanderson[43] has done this for several maps with the results listed in Table 5.

The results indicate that the atoms are not fully ionized—not surprising perhaps—and that ionic radii appear to be consistently in error. Anions are not as large as the usual ionic radius scale, and cations are bigger than expected. For example, Cl^- is 0.1–0.3 Å smaller, and cations are 0.2 Å bigger on the average. It should be noted that not all authors agree on the interpretation of the electron density maps. Brill[40] argues that NaCl is fully ionized instead of 0.67 as in Table 5.

TABLE 5

Integrated Charge and Radii Obtained from Electron Density Maps (after Sanderson[43])

			Radius, Å		
Compound	Calculated charge atom in compound		X-ray "ionic"	Ion	Ion
LiF	F	−0.74	1.09	1.33	F^-
CaF_2	F	−0.47	1.10	1.33	F^-
CuCl	Cl	−0.29	1.25	1.81	Cl^-
NaCl	Cl	−0.67	1.64	1.81	Cl^-
KCl	Cl	−0.76	1.70	1.81	Cl^-
CuBr	Br	−0.25	1.36	1.96	Br^-
MgO	O	−0.42	1.09	1.45	O^{2-}
LiF	Li	0.74	0.92	0.68	Li^+
NaCl	Na	0.67	1.18	0.98	Na^+
KCl	K	0.76	1.45	1.33	K^+
CuCl	Cu	0.29	1.10	0.96	Cu^+
CuBr	Cu	0.25	1.10	0.96	Cu^+
MgO	Mg	0.42	1.02	0.65	Mg^{2+}
CaF_2	Ca	0.94	1.26	0.94	Ca^{2+}

Weiss[44] has discussed the applications to metals. The number and disposition of 3*d* and 4*s* electrons in transition metals have been determined by Fourier methods.

Freund[45] has proposed a new, nonlinear X-ray diffraction method to determine the valence electron charge density in a wide variety of covalently bonded materials. The technique is based on the spontaneous parametric down conversion of a source of X rays interacting with a suitably oriented crystal and involve measurements comparable in difficulty with X-ray Raman scattering.

3.2.3. Orientation of Crystals

It is often necessary to orient crystals before making physical measurements. Before cutting a specimen from the crystal, it is oriented relative to a fixed support, usually by X-ray methods. Special specimen holders are required to maintain the orientation during the cutting and polishing processes. In high-symmetry crystals, there are a number of equivalent planes and directions, and the shape of the specimen often makes one more advantageous than another.

The crystal may be less than 0.1 mm in diameter, or as large as several inches. It may be transparent or opaque, and may or may not exhibit recognizable crystal faces. Crystal morphology, etch pits, or cleavage traces are helpful in locating crystal axes, and polarized light can be used to establish extinction directions in transparent crystals.

For crystals larger than a 1 mm in size, back-reflection Laue cameras and X-ray goniometers are the most commonly used instruments. Smaller crystals can be oriented from Weissenberg or precession camera photographs. Interplanar angles can be measured directly from Laue photographs, and compared with angles calculated from the lattice parameters to give the crystal orientation. High-symmetry directions can often be identified by the symmetry of the Laue pattern, without recourse to numerical calculation. With care, the Laue method is accurate to about 1° in orientation.

Single-crystal goniometers are similar in design to X-ray diffractometers. Knowing the d spacing of the desired plane, the counter is set at twice the Bragg angle, and the crystal orientation is adjusted until a signal is detected. Careful maximization of the diffracted intensity can give crystal orientations accurate to within a few minutes of arc. Preliminary work can be done with a Laue camera if the approximate orientation is not known.

Details of the methods are given by Wood.[46]

3.3. Temperature and Pressure Experiments

Many materials are used at high temperatures or high pressures, and their characterization under these conditions can be even more important than the usual room-temperature study. Crystal structures are sensitive to temperature because of phase transitions, thermal vibrations, and thermal expansion. Crystalline materials have been studied by X rays over the temperature range from 1.5°K to 3000°C, and bibliographies on high-[47] and low-temperature[48] X-ray diffraction have been compiled.

3.3.1. High Temperature

X-ray diffraction studies at elevated temperatures comprise a major source of data on thermal expansion, high-temperature phase transformations, and reactions such as the oxidation of metals. To obtain such information, it is necessary to maintain the specimen at high temperatures for extended times without any changes other than those under study. This may require vacuum conditions or an inert or an oxidizing atmosphere to prevent dissociation together with fragile X-ray-transparent windows. It is of course necessary to maintain a homogeneous temperature over the sample and to measure it accurately and to observe changes in structure and lattice dimensions over a wide range of diffraction angles.

The two principal instrumental techniques for obtaining high-sample temperatures involve (1) a wire-wound, oven-type resistance heater surrounding the specimen, and (2) a metal-foil resistance heater in direct contact with the sample. Ribbon-type contact heaters have been used to 2600°C and are fast and convenient, but the temperature gradients are large (sometimes in hundreds of degrees) and it is difficult to measure temperatures to better than 10°C. Oven-type heaters have smaller temperature gradients ($\sim 0.2°$) across the sample and temperatures can be controlled to about 1° at 1000°C. Other high-temperature X-ray methods include: induction heating; arc image, solar, or electron-beam heating; and self-heating by passing electric current directly through the sample.

High-temperature measurements are normally made on powders or polycrystalline aggregates, as are most phase transition studies, although a few studies of atomic parameters and phase transitions have been made on single crystals. High-temperature measurements generally have a lower accuracy than those made at room temperature because the apparatus is more complicated and specimen preparation is more difficult. The position of the specimen is often difficult to locate and to maintain because of shielding. There may also be grain growth causing poorer crystallite-size statistics, as well as thermal gradients at the specimen. Unit cell dimensions can be measured to about 0.01 % at 1000°C by the powder diffraction method.

X-ray diffraction lines (or spots) from a specimen at high temperature have a lower intensity and higher background than those from specimens at room or low temperature. Thermal vibration of the atoms limits the accuracy of determining the atomic positions to about 0.01 Å at 1000°C.

3.3.2. Low Temperature

X-ray diffraction studies are conducted at low temperatures to observe phase transitions; to investigate materials which would be liquids or gases under standard conditions; to obtain coefficients of thermal expansion; and to obtain better atomic positional parameters through increased quantity of diffraction data and improved intensity measurements afforded by lessening of thermal vibrations.

Low-temperature techniques are applied to powders and to single crystals using both film and counter methods. There are two commonly used procedures for obtaining low temperatures. In the first method, the sample is immersed in a cold stream of nitrogen near its boiling point of 77°K. The sample is usually a small single crystal mounted on a standard goniometer head in a standard camera diffractometer. Temperature control is rather uncertain ($\sim 5°$) in this experiment, but it is suitable for structure analysis work. In the second method, powder specimens are placed on a metal block which is in contact with a cryogenic medium such as liquid nitrogen or liquid helium. Temperatures are controlled to about 0.1°K from 4 to 20°K and to better than 1°K at higher temperatures. Control is obtained by admixing warm gas in the cooled stream or by a small heater in or near the metal block.

Although low-temperature X-ray diffraction, because of smaller vibrations, permits greater accuracy in the determination of lattice parameters and atomic positions, this has yet to be achieved in practice because of the difficulties in measuring intensities for a specimen surrounded by a cryostat. Lattice parameters can be measured with a precision of better than 0.01 % at liquid nitrogen and liquid helium temperatures. Atomic coordinates have been measured to 0.0002 Å at 78°K; very few accurate X-ray structure analyses have been attempted at lower temperatures.

3.3.3. Thermal Expansion and Thermal Vibration Amplitudes

Temperature produces several changes in crystal structure which can be investigated by X-ray methods: the lattice parameters, atomic coordinates, thermal vibration amplitudes, frequency spectrum for atomic vibrations, bonding configuration, and the defect concentration all vary with temperature.

The changes in lattice parameters obtained from accurate Bragg angle measurements can be analyzed to give thermal expansion coefficients. The accuracy of the X-ray method is comparable to classical methods for measuring thermal expansion, and considerably faster. An important advantage of the X-ray technique compared to bulk thermal expansion measurements is that anisotropic thermal expansion coefficients can be measured from polycrystalline samples.

For simple structures such as rocksalt, the changes in interatomic distances with temperature depend only on unit cell dimensions, but in general the fractional atomic coordinates x, y, z also depend on T. Except for very anisotropic materials, the changes are small, however, making them difficult to measure. Atomic coordinates are obtained from the intensities of the Bragg peaks, as explained previously. Both lattice parameters and fractional atomic coordinates are sensitive to high-angle data. As temperature increases, high-angle diffraction peaks become broader and weaker, disappearing into the background. Thus, the accuracy of crystallographic measurements decreases as temperature increases, becoming especially poor near the melting point.

Thermal vibration amplitudes are determined from intensity measurements, replacing the atomic scattering factor f_n with $f_n e^{-M_n}$. The temperature factor M is sometimes a nuisance in structure refinement but it offers a way of measuring the mean square vibration amplitude of each atom:

$$M_n = 2\pi^2 \overline{u_n^2}/3d^2$$

$\overline{u_n^2}$ is the mean square displacement of the nth atom normal to the diffracting plane with interplanar spacing d. In general, M is not a scalar but a second-rank tensor, giving anisotropic temperature factors. Typical values of the displacement are compiled in Volume III of the *International Tables for X-ray Crystallography.*[49] In NaCl, the root-mean-square vibration amplitude is about 0.2 Å at room temperature. Temperature factors are difficult to measure accurately because of various systematic errors in the intensity data which are also smooth functions of the Bragg angle.

Measurement of the temperature factors M_n and the $\overline{u_n^2}$ values gives a time and space average which does not distinguish high- and low-frequency vibration modes. Under favorable circumstances, however, amplitudes can be analyzed in terms of molecular translations and librations.[50]

3.3.4. Thermal Diffuse Scattering

Temperature diffuse scattering patterns are of three main types.[51] Hard materials such as diamond and tungsten show only small, isolated diffuse spots which vary little with temperature. Graphite, hexamethylbenzene, and other layer structures give patterns with large, isolated diffuse spots whose intensities vary strongly with temperature. The fact that the spots are isolated indicates that long waves are of greater amplitude than high-frequency vibrations with wavelengths comparable to interplanar spacings. The third type of pattern is also temperature dependent, and consists of streaks and extended regions of diffuse scattering, implying the existence of large-amplitude, high-frequency thermal waves. Such waves are often in well-defined directions, as in urea, $OC(NH_2)_2$. The diffuse X-ray pattern of urea shows strong peaks along the [110] directions corresponding to molecular librations and vibrations. Metallic sodium also shows strong diffuse streaks in particular reciprocal lattice directions.

Quantitative measurements of the diffuse intensities yield the frequency spectrum of elastic waves, though the analysis is rather lengthy. The elastic spectrum of aluminum[52] shown in Figure 7 is one of the few complete determinations by X-ray methods. Temperature-diffuse scattering experiments are performed on single crystals using monochromatic radiation. Intensities are measured throughout reciprocal space and corrected for Compton-modified scattering and second-order temperature-diffuse scattering. The procedures are described by Warren,[53] who pioneered the field.

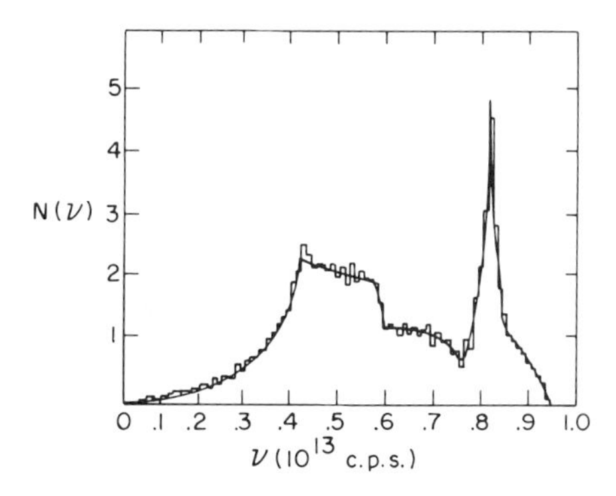

Fig. 7. Frequency distribution of elastic waves in aluminum at room temperature. $N(\nu)\,d\nu$ is the number of waves with frequencies in the range ν to $\nu + d\nu$. The histogram is obtained from 8373 calculated eigenfrequencies (after Walker[52]).

3.3.5. Clustering and Ordering

In a substitutional solid solution, atom A substitutes for B in substantially the same crystallographic site. Elements A and B need not be distributed randomly. If the two elements tend to segregate separately, clustering is said to occur. Clustering is found in certain materials which form solid solutions at high temperatures, followed by precipitation or spinodal decomposition on cooling. As unmixing progresses, A–A bonds and B–B bonds predominate over A–B bonding.

The opposite effect takes place in solid solutions showing a tendency toward order. A and B atoms prefer adjacent sites so that A–B bonding dominates. Short-range ordering occurs in brasses and other alloys with order–disorder phase transformations. As the temperature decreases, the distribution changes gradually from a random arrangement to states showing short-range order, and then abruptly to a state with substantial long-range order. Not all materials showing short-range order convert to a long-range ordering.

Ordering and clustering are described by an ordering coefficient $a = 1 - (P_{AB}/M_A)$, where P_{AB} is the probability of finding an A atom next to a B atom, and M_A is the overall atomic fraction of A atoms. For complete ordering, $a = -1$. Complete segregation is plus one and random occupation is zero. Clustered states lie between zero and one, and for short-range order, a ranges from zero to one.

Ordering parameters can be determined from X-ray diffraction experiments in which diffuse scattering is measured as a function of the Bragg angle θ. The intensity measurements require monochromatic radiation and corrections for Compton-modified scattering and temperature-diffuse scattering. Fourier transform methods are used to analyze the intensity data in evaluating ordering clustering.

Long-range ordering affects the Bragg peaks, changing the intensities and sometimes the space group. In CuAu, the appearance of new superstructure lines signifies the onset of long-range order, but in spinels, the ordering of cations in the tetrahedral and octahedral sites give rise to intensity changes only. Structure factor calculations are used to determine the type and degree of long-range ordering.

3.3.6. High-Pressure Studies

X-ray diffraction studies of substances maintained at high pressures are undertaken to identify new high-pressure phases and to determine crystal structures. Compressibilities are measured by determining the dependence of lattice dimensions on pressure. Instrumentally, the problem is to contain a sample in a vessel which is mechanically strong, yet transparent to X rays. Up to approximately 10 kbar (1 kbar $\simeq$ 1000 atm), the problem is relatively simple. Above such pressures, instrumentation becomes increasingly difficult, at 100 kbar and above, high-pressure X-ray diffraction involves an exceedingly difficult experimental problem. This is a relatively new area of experimental physics, one which requires considerable developmental instrumentation. Crude diffraction patterns, however, have been obtained at pressures up to 500 kbar; in a few studies, temperatures up to 1000°C have been applied at 100 kbar.

With very few exceptions, all high-pressure X-ray diffraction studies have involved powder rather than single-crystal specimens. This is due in part to the instrumental difficulties in maintaining quasi-hydrostatic conditions, and in part to the fact that relatively few materials remain as single crystals after passing through a phase transition.

In the range of 10 kbar, a gas-filled bomb with beryllium windows suffices. Since 1949, a number of miniature pressure vessels have been designed in which both the bomb and sample are bathed in the X-ray beam. Polycrystalline beryllium, single-crystal beryllium, and single-crystal diamond "containers" have been used for X-ray transparency. More recently, diamond and WC anvils have permitted pressures of 200–500 kbar, respectively. Details of such instrumentation have been reviewed by Klement and Jayaraman.[54] These techniques employ the Debye–Scherrer geometry and use mostly film recording, though counter tubes have been used in some of the anvil procedures. Cell dimension accuracy is limited to about 0.5% under these conditions and structure refinements are hampered by the difficulties associated with intensity measurement.

One of the more promising new techniques is the high-pressure time-of-flight neutron diffraction apparatus which has been applied to heavy elements.[55] Nondispersive X-ray and neutron experiments are an attractive alternative to conventional powder pattern methods. The constant-angle geometry simplifies the design of high-pressure dies. Freud and LaMori[56] obtained cell constants using tungsten white radiation together with lithium-drifted germanium detectors and a multichannel analyzer, but the method has not developed rapidly.

3.4. X-Ray Topography and Interferometry

Topographic methods[57] reveal the details of stacking faults, grain boundaries, long-range strain fields, and precipitates. Analysis of the direct image of individual dislocations gives the direction but not the sign of the Burgers vector. Dislocations densities up to $10^6/\text{cm}^2$ can be mapped but the most useful range is 0–$10^4/\text{cm}^2$.

There are three experimental methods (Figure 8). The back-reflection Berg–Barrett camera is useful in examining imperfections near the crystal surface at depths of 10–100 μm. A well-collimated monochromatic X-ray beam incident on the crystal is diffracted by planes (hkl) and recorded on film. Because of extinction effects, the diffracted intensity from imperfect regions exceeds that of the perfect regions. Dislocations appear as bands of contrasting darkness, 5–50 μm wide. Film and crystal are moved in synchronization to map out

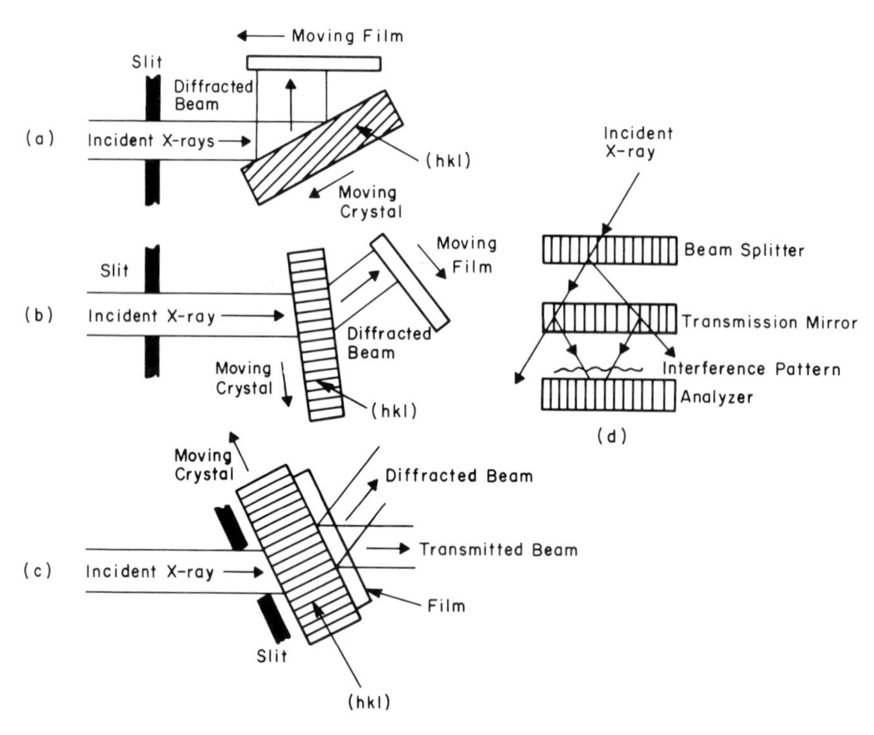

Fig. 8. Three X-ray topographic methods: (a) back-reflection Berg–Barrett, (b) transmission Berg–Barrett, (c) Borrmann anomalous transmission. In the Laue case X-ray interferometer (d) two coherent beams are produced by the beam splitter; the transmission mirror recombines the beams to form one interference pattern at the analyzer.

crystalline imperfections. Magnifications of 500 × are achieved using high-resolution X-ray photographic emulsions and light microscopy.

The transmission Berg–Barrett or Lang camera operates on the same principle, using transmission geometry rather than reflection. A stereoscopic view of dislocations is obtained by recording images from two different diffracting planes. The Borrmann method requires exceedingly perfect crystals, using fairly thick specimens of about 1 mm. Interior dislocations and their Burgers vectors can be viewed stereoscopically.

X-ray topography lacks the simplicity of etch patterns and the high resolution of transmission electron microscopy. However, the latter investigations are confined to thin films, and etch pits are not always associated with dislocations. Topographic techniques provide an excellent method for characterizing dislocation content of high-quality crystals up to a few millimeters thick.

Crystals oriented to satisfy Bragg's law are used as optical components in X-ray interferometry.[58] The interference patterns are sufficiently sensitive to small displacements to allow the observation of strains and defects in crystals. Lattice rotations of 10^{-8} rad or dilations $\delta d/d$ of 10^{-8} can be detected. Inhomogeneous strain produces moiré fringe patterns showing dislocations and the strain-associated impurities in concentrations below one in 10^{7}.

Although the physics of these devices is simple, construction and operation of the X-ray interferometer had to await the arrival of large, perfect single crystals such as semiconductor silicon. The interferometer makes use of the fact that the refractive index for X rays is slightly less than unity, about $1 - 10^{-5}$. This makes conventional optical techniques difficult because the focal length of a lens would be several kilometers, but for interferometry, where optical paths are balanced, the physical dimensions need only be accurate to $\lambda(1 - n)^{-1} \sim 50\ \mu\text{m}$.

Most applications to date make use of Laue case interferometers (Figure 8d) which consists of a beam splitter, transmission mirror, and analyzer. The analyzer crystal transforms the atomic scale interference pattern into a macroscopic pattern which can be observed with counters or film.

Bragg reflection occurs over a very narrow angle ($\sim 10^{-5}$ rad) for perfect crystals. The degree of perfection determines the relative phases of beams generated by the X-ray interferometer. This sensitivity permits measurement of crystal strains as well as shifts caused by objects placed in the path.

In addition to characterization work, interferometers can be used to measure X-ray scattering factors, to do phase contrast microscopy, and for absolute measurements of lattice spacings.

4. Electron Methods for Materials Characterization

Electron microscopy and diffraction are indicated whenever one dimension of the particle is 2000 Å or less, or when fine surface or internal structure is in question. It is therefore preeminent for thin films or crystals and for very fine-particle characterization where other methods fail or are of limited utility. Bulk materials can also be examined by thinning down to foils. The chief limitation is that the electron beam can, in general, give information on materials only less than about 2000 Å in thickness by transmission. Where samples are not thinned, information on surface structure can be gained by replica techniques and by reflection diffraction. Several books have appeared on the subject and these provide details on the various applications and capabilities of the technique.[59–62]

4.1. Electron Microscopy

4.1.1. Magnification, Precision, and Resolution

Most micrographs are taken at from 4000 × to 16,000 ×, but it is possible to work from 300 × to about 250,000 ×. Reported magnifications in electron microscopy have an accuracy of about 10 %, which is satisfactory for most problems. At higher magnifications, accuracy may be considerably reduced.

Modern microscopes can give a resolution of about 5 Å. With a great deal of time and effort, it is possible to show resolution of about 2 Å using 200-kV electrons. Several Japanese microscopists claim 1.8 Å on metals and Heidenreich[59] has obtained 2.1 Å on graphite.

However, there are three different tests for resolution, and there is no agreement as to which method is best. Further, very few materials are suitable for resolution tests. In most problems, there is no need for resolutions of better than 5 Å, and with most samples it would be difficult to prove resolution of below 10 Å. Resolutions of 2–3 Å were recently achieved for intercalation complexes of superconducting TaS_2, allowing direct observation of the crystalline lattice and its imperfections.[63] Computer enhancement of high-resolution electron micrographs of stained and unstained catalase crystals show amazing

Fig. 9. Contoured intensities of the catalase crystal structure using computer-enhanced electron microscope images.[64]

detail[64] (Figure 9). Photographic images obtained with the electron microscope can be sharpened using laser techniques to remove the fuzziness. In its most spectacular application, the holographic filter[65] improved the resolution of electron micrographs to 2.5 Å and revealed, for the first time, the helical structure of a virus.

For high resolution of the order of 2–3 Å, contrast is often the limiting factor. The image contrast depends on a phase relationship which is not known with exactitude. At this level of resolution, a compromise has to be made between the various image defects, of which spherical aberration is probably the most serious. However, astigmatism and alignment have to be adjusted exactly to obtain such values. Practice has fortunately led theory in high-resolution microscopy and, repeatedly, microscopists have achieved higher resolution than predicted by theory. However, if the theory were better, it might be possible to image atom positions in suitable thin specimens. If the objective can be improved so that the spherical abberation is reduced by a small factor, then it may well be possible to obtain direct atomic resolution. A superconducting lens offers distinct possibilities for the future.

Resolution in reflection microscopy is about 100 Å at best, but 200 Å is more common. In replicas, resolution is about 20 Å, at best, with direct replicas. Two-stage replicas give about 50 Å and three-stage replicas about 100 Å resolution. Decoration replicas show, indirectly, faults that may be of unit cell size.

4.1.2. Experimental Difficulties and Future Prospects

Specimens should be less than 1000 Å in thickness for transmission electron microscopy, and this remains the main experimental difficulty in electron microscopy. Pure materials can be thinned from the bulk by special methods, but in specimens with precipitates or with more than one phase, ordinary thinning methods often do not work. There is as yet no general method which can be used for a wide variety of materials. It is possible that ion bombardment at low angles will provide a general method, but the latest results suggest that the influence of the ion beam on the structure may itself be significant.

A serious equipment limitation is the poor vacuum in electron microscopes; less than 10^{-4} mm Hg is common at the specimen. This means that many problems cannot be attempted or give ambiguous answers.

In regard to sources, point filaments with a double condenser give beams of less than 0.5 μm spot size, with the possibility that cold cathode sources will further improve resolution. One difficulty is that point sources have a very small emitting surface. It would be best to work at high magnifications of about 250,000× for high resolution, and at this figure, beam intensity is low. A cold cathode may give the high intensity required and a smaller source size than is presently available.

Several high-voltage microscopes have been built which operate in the megavolt range. Energetic electrons penetrate thicker samples, cause less background, and are capable of higher resolutions. Megavolt machines have been in operation for a decade and have been rather difficult to operate, expensive, and have not produced outstanding results to date. Smaller machines in the 500,000-V range have given resolution of about 2 Å. These appear to be the best compromise for more routine characterization, since they are much more easily operated than the 10^6-V instruments. However, the cutting edge of the field is definitely moving to higher voltages, with several machines in the 1–10-MV range in the testing or construction stage. These are of special importance in biological systems where *in vivo* studies become feasible.

Cryogenic microscopes with superconducting lens systems are under construction. Atomic resolution may be feasible in these instruments with their improved objectives and better vacuums.

4.1.3. Imperfections, Small Particles, and Amorphous Materials

Particle size can be observed directly and more exactly by electron microscopy than by X-ray methods, especially if the particles are single crystals. This is very evident in the size range 500–5000 Å where X-ray methods are difficult. Contrast limits particle size determination in the region below 100 Å, although electron microscopy can resolve particles of gold or platinum as small as 10 or 20 Å. Materials which are poorly crystalline or amorphous give low contrast. These may have sizes of 30 Å and be unresolved by direct microscopy, although electron diffraction can give an estimate of crystallite sizes below 100 Å. The relation between crystallite size and line broadening is sometimes difficult to utilize in practice. Below 40 Å, the effect of particle size and poor crystallinity merge, making interpretation difficult.

As the theory of image contrast has improved, more and more information has been gained by direct observation of metal foils by electron microscopy.[62] X-ray topography used in conjunction with electron microscopy has been very valuable, especially with dislocation problems. Stacking faults can be resolved directly, determining their size, density, shear vector, and fault plane.

Dark-field electron microscopy is of tremendous importance in the determination of defects in thin foils, and in distinguishing crystalline from noncrystalline regions. In bright field, defects give a subtractive contrast image, and in dark field, an additive contrast image. Often the dark-field image has better contrast, because in very thin foils, only a small fraction of the intensity is diffracted out in the normal image.

Recent technological applications of poorly crystalline and glassy materials as laser hosts and amorphous semiconductors has pointed to the need for characterization methods to supplement diffraction techniques. After differential etching, electron microscopy can be used to detect substructure in glass at the 15–30-Å level when both components are noncrystalline. This overlaps the region in which low-angle X-ray scattering can be utilized for the same information. Light scattering has been of value in the 100-Å range using laser sources.

Under special conditions, percentage crystallinity can be evaluated with accuracy by electron microscopy. For example, in a phase transformation from amorphous to crystalline, contrast between the two phases is sometimes sufficient for an accurate analysis. Where, however, contrast difference is slight, as with organic materials, the problem is rather difficult and is better done by X-ray methods.

4.1.4. Ferromagnetic and Ferroelectric Materials

Since the electron beam is deflected at ferromagnetic domain boundaries, depending on whether the beam converges or diverges, a light or dark line appears on the micrograph, and the convergent boundaries show interference fringes. Bloch, Néel, and hybrid "cross-tie" walls can be distinguished, and the width of the walls measured from micrographs. Fine magnetic ripple is also evident and is orthogonal to the magnetization vector. In this way, the local magnetization vector in various parts of a single domain can be measured. The magnetization configuration within the walls can be determined and wall motions observed under applied magnetic fields. Films can be made to switch and pinning and reversal can be studied directly, showing that holes or second phases rather than dislocations or stacking faults are the usual pinning sites. During these studies, the film may be imaged at high magnification, or selected-area diffraction patterns taken. Under applied mechanical stress, the effect on strain-sensitive magnetic materials can be studied, and on heating, a rough indication is given of the Curie temperature.

Ferroelectric domains have also been observed by electron microscopy, in barium titanate; as the domain forms, the lattice constant changes, giving a contrast difference between domains. Moreover, the electron beam can charge the surface and switch domains, making switching processes apparent.

4.2. Electron Diffraction

Electron diffraction[66,67] is generally done with commercial electron microscopes using three techniques: (i) transmission with a diffraction camera accessory, (ii) selected-area diffraction, and (iii) reflection diffraction. In determining unit cell dimensions, the precision decreases in the order given above. Claims have been made of lattice parameter determinations accurate to 0.1 % by electron diffraction, but this is unrealistic, and a figure of about 1 % is more typical.

With good samples, the transmission camera is capable of 0.3 % using an internal standard. Selected-area diffraction gives about 0.5 %, and reflection diffraction about 1 %. Electron diffraction patterns are sensitive to surface defects, particle size, sample position and orientation, and refraction effects, so that most materials are unsuited to precision measurements. Normally, electron diffraction is a poor second to X-ray diffraction as far as accuracy is concerned.

4.2.1. Phase Identification

Electron diffraction should be used for routine identification only when the results cannot be obtained with X rays. Electron microscopy is superior to X-ray methods for thin films, or when very small amounts of a second phase are present, or if the particle size is less than 1000 Å. Electron microscopy and electron diffraction can identify impurities and show that amorphous materials are present at levels well below the sensitivity of X rays and should be used for this purpose. Differences in the shape and habit of crystals in a mixture are readily perceived and each phase can then be identified by selected-area diffraction. In favorable cases where a low-level impurity has a distinctive morphology (e.g., needles mixed with cubes), one can detect and positively identify less than 0.1 % impurities, although it is extremely difficult to give an overall percentage for the impurity.

The principle of the selected-area diffraction technique is illustrated in Figure 10. The diffraction pattern is obtained on the viewing screen of a conventional electron microscope by adjusting the diffraction lens. The pattern is focused from the back-focal plane of the objective lens onto the object plane for the projector lens, which then

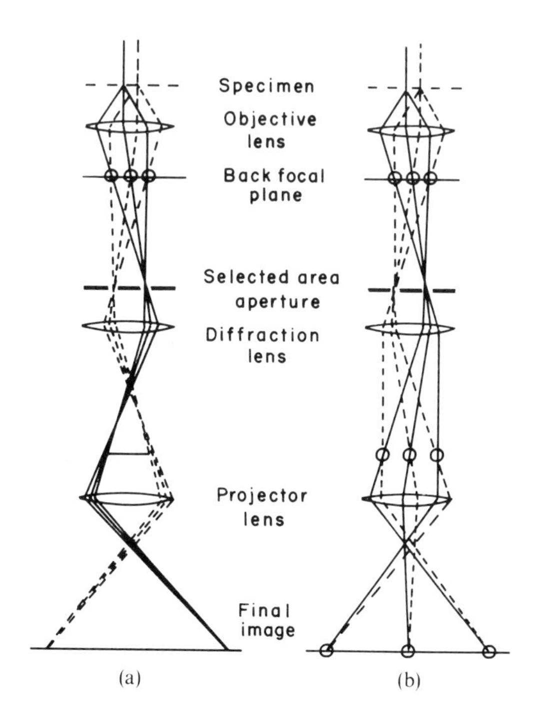

Fig. 10. The ray paths in an electron microscope (a) to produce a high-magnification image, and (b) to produce a diffraction pattern of a selected area of the specimen.[67]

magnifies it. An aperture positioned in the plane of the first image ensures that the electrons contributing to the diffraction are those forming the selected-area image. Spherical abberation of the objective lens limits the areas to 1 μm or greater in diameter.

For polycrystalline materials, electron methods can be used to supplement and clarify X-ray results. For example, with X-ray powder patterns, it is often difficult or even impossible to index lines in a diffraction pattern if mixtures are present, whereas, using electron diffraction, the problem can be solved by obtaining unit cell data for each phase from selected-area diffraction patterns. Likewise, with complex powders of new materials, the indexing of X-ray patterns is especially difficult if the unit cell is large and of low symmetry. Electron diffraction patterns from small crystals reveal the reciprocal lattice, giving information on the principal axes and crystallographic symmetry. Many new materials are prepared as small samples of tiny crystals suitable only for electrons microscopy, so that electron diffraction is often the best way to obtain crystallographic data on the material.

Electron diffraction and microscopy have been used to examine glass, organic waxes, smokes, and clays. Materials which are "amorphous" by normal X-ray diffraction sometimes display electron diffraction patterns characteristic of very small crystals. Information on the incipient particle size and crystallinity is given by the diffraction pattern, although identification may be uncertain. Characterization of new crystalline material is very difficult if only two or three broad rings are present, and the interplanar spacings are inaccurate. Radial distribution analyses are seldom performed in electron diffraction, although there has been some recent work on biological materials and organic polymers.

Electron microscopic hot and cold stages are commercially available as standard items, capable of maintaining temperatures from −180°C to about 1000°C; and some microscopes have been made with liquid helium stages and with hot stages to 2000°C. Several very interesting problems on phase transformations and chemical reactions have been investigated. However, as a tool for studying phase transi-that electron methods should be used only when there is a good reason.

4.2.2. Structure Analysis

It is generally faster and more accurate to do structure analysis by single-crystal X-ray methods than by electron diffraction. While

Russian workers[66] claim solutions of crystal structure by electron methods alone, very little of this work is done elsewhere. Most crystallographers consider the Russian claims exaggerated and point out that in many cases the structure had been solved previously by X-ray methods.

In electron diffraction, the problem of intensities is much more complex than for X-ray diffraction. Compared to the structure factor calculations used for X-ray intensities, the theoretical treatments are complicated and only approximate. In many cases, it is questionable whether crystal structures could have been deduced from electron diffraction intensities if X-ray models were not already available. Although Russian workers take a contrary view, it is the general opinion that crystal structure work is more easily and better done by X-ray than by electron diffraction.

Fortunately, the simple kinematic theory is sufficient and semi-quantitative for amorphous and small crystals (below about 100 Å). In the kinematic theory, it is assumed that the electron has only one interaction and that the scattered and incident beams do not interact. For thicker crystals, the two-beam dynamical theory gives reasonable results in image interpretation even though it is approximate. Here the theory considers a single diffracted wave and incident wave, but allows the two beams to interact. However, present theory, although giving a reasonable quantitative picture, is not nearly as well developed as for X rays and needs further refinement.

For structure analyses based on the arc or ring patterns from polycrystalline materials, the available evidence suggests the following conclusions.[67] The two-beam approximation gives satisfactory results for structures containing light atoms. Except for strong inner reflections, the intensity errors are less than 20%. The two-beam calculation is less reliable for heavy-atom structures and for single-crystal diffraction patterns.

One solution to the problem may be the use of n-beam calculations during the refinement stages of structure analysis. Given approximate phases and amplitudes for the strong reflections, n-beam effects can be calculated for the weak reflections if the crystal dimensions and orientation are accurately known. An alternative approach is the use of Bethe's dynamical potentials, which appear accurate for powder patterns from simple metals.

Higher accelerating voltages decrease the atomic scattering factors and make kinematic theory more accurate. However, the situation is complicated by relativistic effects which become important above 200 kV.

4.2.3. Low-Energy Electron Diffraction

Low-energy electron diffraction (LEED) can provide structural information on the first two or three atomic layers.[68] Although it has given important information on the structure of material at surfaces, its use for the characterization of real materials is limited because the method of specimen preparation is so specialized and necessarily alters the nature of the original surface. Further, a strict interpretation is often difficult and as yet not generally accepted, so that the intensity data cannot be inverted directly to give details of the atomic structure of the surface. Instead, one relies on trial models, and comparison between measured and calculated intensities. The surface structure is refined by adjusting atomic coordinates to give the best agreement with observed intensity data. Calculations are based on multiple scattering theory[69] or on the matching of wave functions.[70]

Recent advances in electron diffraction techniques and in high-vacuum technology have led to a resurgence of interest in the field. The Ehrenberg detection system incorporates a post-diffraction acceleration grid which eliminates most of the inelastically scattered electrons.

This is a field of enormous potential. The structure determination of surfaces is in its infancy, comparable to the X-ray crystallography of half a century ago. Thousands of real surfaces await structure determination, surfaces of importance in catalysis, corrosion, oxidation, epitaxy, and lubrication, and although most of these are not likely to be highly ordered, LEED will no doubt play a role in determining the atomic arrangement of the outermost surface layer in "standard" materials.

4.3. Scanning Electron Microscopy[71]

If X-ray diffraction is the single most important tool for characterization, scanning electron microscopy is certainly the second, and together they are so much more general and universally applicable that they must be put in a category by themselves. The scanning microscope combines some of the best features of the electron microprobe, the electron microscope, and the optical microscope into an instrument of outstanding performance, high reliability, and ease of operation. It is uniquely suited to a large variety of research problems, and in five years caused a revolution in the study of topographical, morphological, and even electrical features of surfaces and the underlying bulk material. Its uses include the examination of metal insulator and semiconductor powders or surfaces of any compact, phosphor layers, magnetic tape coatings, epitaxial layers, and evaporated films, where depth of focus, light scattering, or stereo views are critical. It is being used in catalysis, surface wear, and a host of materials processing problems. It is the only practical tool for the nondestructive examination of MOS transistors, integrated electronics, and microdevices as they are working, and for investigating their failures and defects. Its application to problems in biology is far-reaching since it expands the resolution for useful taxonomy manyfold.

The scanning electron microscope is basically a closed-circuit television system with four principal components: a scanning electron beam (or probe), a sample which generates the secondary electrons, a detector, and amplifiers to feed an electrical signal to the display system. A beam of electrons 50–100 Å in diameter scans the surface of the specimen in a raster pattern and the secondary electrons from the specimen are then selected out, detected, and amplified. The resulting signal is used to modulate the light intensity of a cathode-ray tube screen which is scanned in registration with the electron probe scan. Besides the secondary electron, many other types of radiation are emitted simultaneously and each of these can be used to provide more data on the sample. The backscattered electrons, X rays, and light are often detected and imaged. Also the target current, the electron-beam-induced current, and the electron-beam-induced-target current provide different types of useful information.

Magnifications range between $15\times$ and $10^5\times$, bridging the gap between light and electron microscopes. The ease of changing magnification makes it easy to zoom from gross image to fine detail. Scanning electron microscope manufacturers guarantee resolutions of 100–200 Å, about midway between the 2 Å resolution obtained with the transmission electron microscope and the 2000 Å for the light microscope.

Compared to conventional electron or light microscopes, scanning electron microscope pictures exhibit remarkable three-dimensional effects due to the extraordinary depth of field made possible by the wavelength and apertures involved (Figure 11). Depth of field is 1 cm at $100\times$ and 1 μm at $10{,}000\times$. The scanning microscope is superior to the transmission microscope because it records secondary electrons from the surface rather than primary transmitted electrons. Thin samples are therefore unnecessary and the depth of focus gives the topography directly.

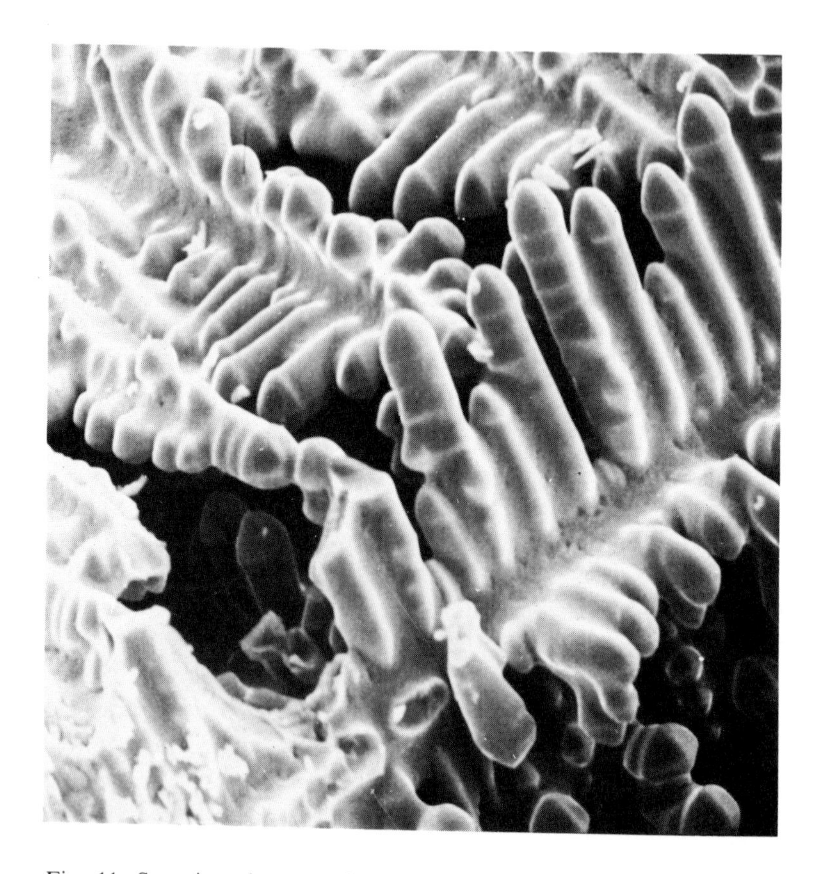

Fig. 11. Scanning electron microscope image of water-quenched Ge–Te crystallites. Magnification 1000×.

Sample preparation is easy compared to the transmission electron microscope. Specimens of nearly any size can be examined since it is unnecessary to thin specimens to 100 Å thickness. For insulators, a thin layer of gold or carbon about 50 Å thick is evaporated on the surface to prevent charge accumulation, although even this often proves to be unnecessary. Finally, the ease of operation and the correspondence of the image to normal vision put this instrument into a unique class.

Connecting a computer directly to the SEM makes it possible to do quantitative petrography on a scale and with a precision not conceived a decade ago. The particle size, shape, and chemical composition can be automatically determined on either a loose powder or aggregate or a surface of a polycrystalline polyphase body at a speed of roughly 100 sec per frame.[72]

The *electron mirror microscope* is still a research instrument. It has a rather low resolution (about 800 Å), and the electrostatic lens used gives problems. It is particularly suited for examination of electric and magnetic patterns, but is not yet an instrument of general utility.

The *field emission microscope* and the *field ion microscope* are used primarily in the study of surface phenomena.[73] With the field emission microscope, one can determine how the electron work function varies with crystallographic orientation, and how it is affected by the absorption of foreign materials. Magnifications of better than a million make it possible to observe effects on an atomic scale. Individual atoms and adsorbed molecules can be located with the field ion microscope. Very recently Müller[74] has adapted a mass spectrometer to the field ion microscope so that, in principle, single atoms may be imaged, evaporated, and analyzed. Unfortunately, the field methods are severely limited with respect to the materials that can be studied. Only highly conducting metals can be used, and then they must be thinned to a very sharp point. Some work has been done on molecules adsorbed on the points but this is also very limited.

5. Neutron Scattering from Solids

The use of neutron diffraction in establishing details of solid structures has developed as a complement to X-ray diffraction for specialized and restricted applications. Stated very simply, one can say that neutron methods should never be considered because of expense and intensity considerations if X rays can possibly do the job. It is only for those cases where the use of neutron radiation offers additional or unique information that the power of the neutron technique becomes significant. Some of the advantages and the limitations of the method are outlined in the following discussion.[75]

5.1. Neutron Sources

At room temperature, thermal neutrons have energies near 0.03 eV, velocities of 2×10^5 cm/sec, and wavelengths of 2 Å. All known sources are characterized by being continuous spectrum sources with a spectrum resembling a Maxwellian distribution. Most diffraction techniques utilize a monochromatic beam, and for the neutron case this means a selection of a monochromatic band of radiation by diffraction from a crystal, or by some alternative mechanical system. This has two implications: (a) a large fraction of the neutron intensity is automatically discarded, and (b) the monochromatic band is of finite width, in no case as sharply defined as the X-ray $K\alpha$ line. On a quantum intensity comparison, the specific neutron quantum intensity available in such a monochromatic band is many orders of magnitude less than that available from a copper X-ray tube. Since the absolute scattering amplitudes of an atom for the two types of radiation are comparable, the diffracted neutron intensities are smaller than equivalent X-ray intensities. Thus, the accuracy of a neutron intensity measurement is practically always limited by statistical considerations. Cost is a second disadvantage. Nuclear reactors serve as the most intense neutron source, and these are expensive to build and to operate.

Because of the slow rate of neutron data collection, automatic operation of spectrometers is a necessity; and it is fair to say that the neutron technology is as advanced here as in X-ray methods. Very elaborate, computer-controlled neutron spectrometers have been designed and are in operation at many reactor installations.

Among the new techniques which are being developed in the neutron diffraction field, perhaps the most interesting and potentially useful is the new full-spectrum, pulsed-source method of collecting diffraction data. In the usual diffraction method, a monochromatic portion of the neutron spectrum is selected by a monochromator crystal, and this is scattered by a powder or crystal with intensity measurement at various scattering angle positions. With a pulsed source of neutrons, however, one can use the full spectrum incident on the sample with intensity measurement at a fixed angle with flight-time analysis. Thus, all details in a powder pattern are displayed as a function of neutron wavelength at a fixed scattering angle. In essence, through use of multichannel analyzers, time resolution replaces angular resolution, and the advantage of this system comes through use of the complete neutron intensity in the beam from the source. Thus, comparing a fixed-power neutron source and a pulsed neutron source of the same average power level, it is expected that the latter system can supply diffraction data at an improved rate by two orders of magnitude.

Pulsed systems offer additional technical advantages for some types of investigations. A fixed scattering angle system is convenient with high-pressure sample cells and in high magnetic field studies, where pulsed magnetic fields can be applied to specimens synchronized with the neutron bursts. Short-period relaxation phenomena become amenable to study through this technique.

Because of the great transparency of most materials for neutron radiation, it is rather easy to obtain diffraction data under conditions where the specimen must be contained in temperature or pressure enclosures. Thus, temperature-controlling radiation shields present no particular problem in neutron technology, and both low- and high-temperature apparatus are commonly used. Low-temperature cryostats, some in conjunction with simultaneous magnetic or electric field application, have been particularly popular in the field; and it can be said that many more neutron investigations have been carried out in the liquid helium temperature region than have been performed in the X-ray field. Similar technical advantages occur in the high-temperature areas but have not been particularly exploited by neutron investigators. Likewise, the relatively easy introduction of neutron radiation into high-pressure sample containers has not been exploited. Only a few neutron investigations of structure modifications at high pressure have been reported.

5.2. Interactions with Matter

Before discussing applications of neutron methods, it is worthwhile to outline some neutron properties and the types of interactions that a neutron can experience in its passage through matter.

Nuclear force interactions: Neutrons, being fundamental nuclear particles, interact with nuclei of atoms through the little-understood nuclear force interaction. The strength of this interaction is not amenable to calculation, and recourse must be had to experiment in evaluating its magnitude. Unlike the X-ray case, there is no regular variation with atomic number, with neighboring atoms or nuclei sometimes exhibiting drastic differences in both magnitude and sign of the scattering amplitude. Different isotopes of the same atom sometimes differ significantly in their scattering characteristics, and there is also observed a dependence upon the spin alignment of a particular isotope. Because of the short range of the nuclear force interaction, nuclear scattering is always isotropic with no form factor dependence, unlike the X-ray case. There is, however, a temperature factor caused by thermal vibration of the nucleus.

Magnetic interaction with electrons: Neutrons possess magnetic dipole moments which interact with electronic magnetic moments or electronic magnetic fields. The interactions can be calculated from first principles, and turn out to be comparable in magnitude to the nuclear force interaction. The magnetic scattering amplitude obviously depends upon the electronic structure of the atom and upon the absolute direction of the atomic magnetization. It is characterized with a form factor because of the spatial distribution of the unpaired electrons within the atom.

Neutron momentum–energy properties: Slow neutrons, by definition, are those obtained after thermal moderation in a reactor, and hence they possess kinetic energy comparable to the thermal energy levels in a solid. At the same time, the neutron momentum is also comparable to that transported in the thermal oscillations or phonons. This favorable relationship can be exploited in obtaining phonon dispersion curves and frequency distributions. In such experiments, it is necessary to determine the energy of the scattered radiation and special spectrometer techniques have been developed to measure this inelastic scattering.

5.3. Structure Analysis with Neutrons

Both powder and single-crystal procedures have been developed, but because of expense, applications are normally considered only where the use of neutron radiation is clearly advantageous. The principal areas of investigation are those in which the neutron amplitudes are favorable, such as hydrogen-containing crystals. X-ray and electron scattering factors are proportional to the number of electrons in the atom, making it difficult to locate light atoms, but the neutron scattering power of hydrogen is comparable with that of other atoms so that it becomes feasible to perform complete structure analyses of hydrides or of organic crystals. Scattering factors for Pb and H are about 100:1 for X rays and 2:1 for neutrons. Powder methods are sometimes used in assessing hydrogen atom positions in simple structures, but, as with X-ray diffraction, the more complicated structures are handled by single-crystal methods. Representative of the type of hydrogen structures now being studied is the work by Brown and Levy[(76)] at Oak Ridge on sucrose, which contains 45 nonequivalent atoms in the unit cell. Using a neutron spectrometer with automatic data collection, the intensities of 5000 reflections were measured over a period of about three months. The resultant structure model showed an agreement factor of 3% considering all reflections, as good as the *R* factors obtained in X-ray studies.

The location of light atoms is only one example where the neutron scattering amplitudes are useful in establishing structural detail. Ordering with the alloy FeCo can be easily established with neutrons because the scattering amplitudes are favorable for enhancing the superlattice reflections. The Mn and Fe distributions in manganese ferrite can be studied with neutrons for the same reason. A convenient tabulation of these neutron scattering amplitudes can be found in the review book by Bacon.[75] Many applications of neutron diffraction are discussed in this reference book.

5.4. Magnetic Structure Analysis

This area has developed into one of great activity because of the unique interaction occurring between neutrons and electrons and its usefulness in supplying information on electron spin distributions in materials. Since magnetic scattering depends upon the unpaired spin distribution among the atoms in the unit cell, a new type of magnetic crystallography becomes available, and many ferromagnetic, ferrimagnetic, and antiferromagnetic substances have been explored by this technique. In addition to the interest in such materials for device applications, magnetic structure studies are of fundamental interest due to their bearing on basic electron interactions among the atoms in the unit cell. The magnetic properties of materials are to a large extent dictated by these exchange interactions. X-ray methods provide information on the spatial distribution of electron charge in and between atoms, and the neutron studies supplement this with information on the electron spin distribution.

The magnetic structures that have been determined range from simple ferromagnetic structures to very much more complicated helical spin structures with periodicities as large as 30 Å. Both polycrystalline and single-crystal methods have been utilized, sometimes with temperature and magnetic field control on the scattering sample. In such work it is, of course, necessary to separate the magnetic scattering from the nuclear scattering, which is always present; auxiliary temperature and field control are helpful in accomplishing this (Figure 12).

One of the great triumphs of neutron diffraction was the determination of antiferromagnetic structures; antiparallel spin arrays had been hypothesized from susceptibility measurements but diffraction patterns provided the first direct demonstration. Not all magnetic structures consist of simple collinear spin arrays. Some possess canted

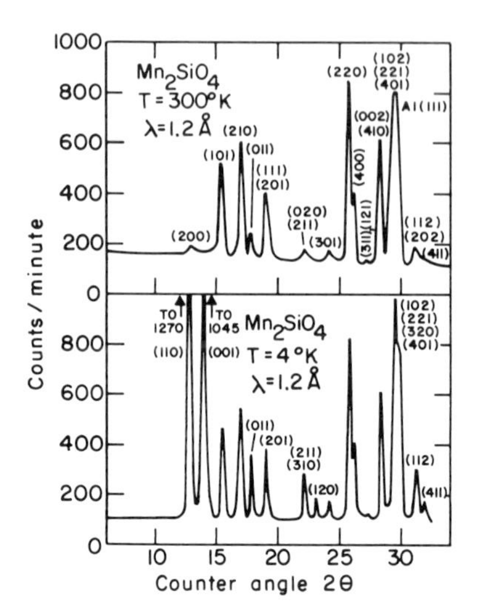

Fig. 12. Neutron diffraction powder patterns of tephroite, Mn_2SiO_4. The room-temperature pattern shows Bragg peaks from nuclear scattering and a rather high background, caused partly by magnetic scattering. At 65°K, Mn_2SiO_4 undergoes a paramagnetic–antiferromagnetic phase transition. Both magnetic and nuclear Bragg peaks are present in the low-temperature pattern, which has an appreciably lower background. Analysis of magnetic intensities yields a canted spin structure.[78]

superstructures, and others, like the rare earth metals, have modulated helical spin structures whose periodicity varies with temperature and is not a multiple of the chemical unit cell. The symmetry description of such structures has been a difficult question for crystallographers.

Polarized neutron radiation in which the neutron spins are parallel is helpful in measuring weak magnetic scattering. Recent illustrations have occurred in studies of electron spin pairing in superconductors and of unpaired spin density in simple ferromagnetic elements such as iron and nickel. Spin density maps are obtained from neutron diffraction data in the same manner as ordinary electron density maps are obtained from X-ray diffraction intensity data. Spin density contours in ferromagnetic iron are not spherically symmetric about the atoms and are extremely important in elucidating magnetic interactions and chemical bonding.[77]

Information about magnetic structures can be obtained from studies of small-angle scattering and magnetic diffuse scattering. The disruption of long-range magnetic order at the transition temperatures gives rise to critical small-angle scattering around reciprocal lattice points. At temperatures above the Curie point, magnetic order disappears and the small-angle scattering expands into a distributed diffuse scattering. Study of both the intensity and energy transfer in the diffuse scattering yields not only the spatial correlation of the spins but elucidates time correlations as well. Along the same line, analysis of the magnetic diffuse scattering resulting from the introduction of either magnetic or nonmagnetic impurity atoms into magnetic lattices has been very useful in assessing the spin correlation range.

5.5. Lattice and Spin Dynamics

Because of the favorable energy–momentum relationship with slow neutrons, many details of thermal energy excitations in solids may be studied by neutron scattering. In such experiments, neutrons of known energy and momentum are inelastically scattered by crystals in all directions. Energy analysis of the scattered neutrons gives the energy and momentum of the phonon responsible for the scattering. In this way, phonon dispersion curves for the lattice may be obtained, and these may be used in a very fundamental way in assessing interatomic force constants. Because of the necessity of energy analysis of the scattered radiation, the measured intensity is low, and work in this field has naturally concentrated at research centers with high-power reactor sources. Nevertheless, there is an increasing research effort being devoted to those problems—perhaps 50 compounds have been studied to date by such methods.

Similar effects of magnetic origin are to be found with magnetic substances. Electronic spins are aligned in magnetic materials with interaction energies comparable to atomic bonding, and magnetic inelastic scattering of neutrons yields information on the strength, directional character, and range of the exchange interaction. It is, of course, necessary to separate the magnetic effects from the phonon effects, and this has been accomplished by applying a magnetic field to the specimen, and using polarized neutron beams. Excellent reviews of the inelastic scattering studies performed with slow neutrons have been given by Brockhouse[79] and Egelstaff.[80]

5.5.1. Displacive Transitions

A number of neutron investigations of displacive phase transitions have been reported.[81] Phase transformations often result from an instability in a normal vibration of the lattice. In ferroelectrics, the displacive transition at the Curie point involves the condensation of a phonon which is both polar and of long wavelength, giving rise to spontaneous polarization. Phonon instabilities involving shorter-wavelength phonons are accompanied by more subtle property changes. The α–β transition in quartz at 573°C is an example. The displacements linking the two structures (Figure 13a) are neither Raman- or infrared-active, explaining why there has been little evidence of the existence of soft modes or critical behavior at the transition. An advantage of neutron spectroscopy over optical spectroscopy is that the selection rules are less stringent. Temperature-dependent inelastic neutron scattering (Figure 13b) has been observed in the β-quartz phase above 573°C. By measuring the integrated intensity associated with the inelastic scattering about many reciprocal lattice points, the nature of the atomic displacements can be deduced. In this case, the scattering has been interpreted as arising from an overdamped phonon.[81]

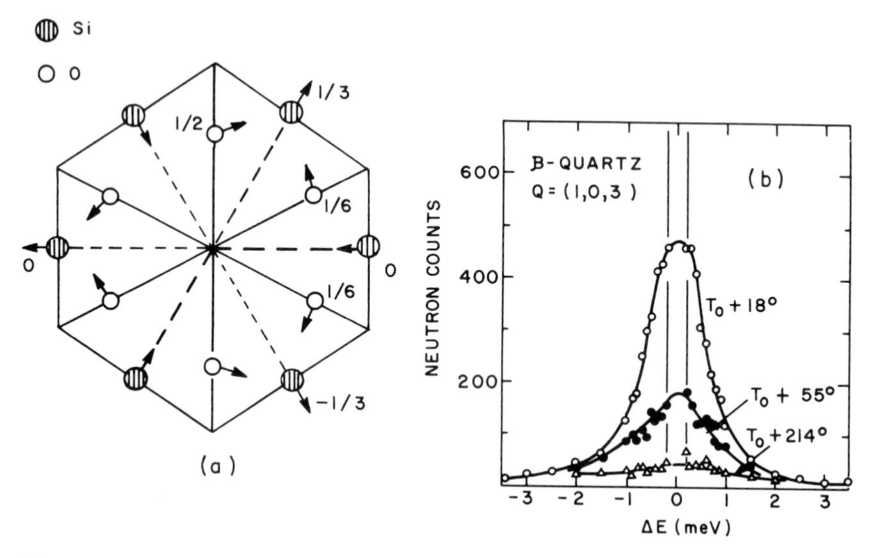

Fig. 13. (a) A projection on the basal plane of the unit cell of β-quartz. The arrows indicate the shifts in atomic positions between β-SiO_2 and α-SiO_2. (b) Energy analysis of scattering about the (103) reflection in β-SiO_2 at several temperatures above T_0. The strong central Bragg peak is superimposed upon weaker inelastic critical scattering.[81]

5.5.2. Inelastic Scattering of Low-Energy Neutrons

A useful characteristic of slow neutron sources is the availability of neutron radiation of sufficiently low energy, or sufficiently long wavelength, that Bragg scattering processes from crystals are no longer possible. This permits a direct study of incoherent, randomly distributed scattering events arising from the presence of impurity centers, vacancies, or other defects. In principle, a study of the differential scattering cross section, or the energy dependence of the total cross section, can yield the "Fourier transform" of such defect centers. A few studies in this area have been reported; for instance, those dealing with the defects arising from radiation dosage on glass. By way of increasing the intensity of such long-wavelength neutron radiation, experiments in some cases have inserted cold moderating regions into the reactor.

Intermolecular motions of H_2O molecules in crystals can be determined by the inelastic scattering of low-energy neutrons. Numerous spectra from powder samples have been recorded with a cold-neutron time-of-flight spectrometer.[82,83] In this experiment, the neutrons are filtered through refrigerated beryllium so that only neutrons of energy less than 40 cm^{-1} (0.005 eV) are incident on the sample. Neutrons are sometimes scattered elastically (no change in energy) and sometimes inelastically, gaining energy from thermally populated energy states in the crystal. The energy spectrum of the scattered neutrons is measured by chopping the scattered beam and measuring the time of flight at a scattering angle of 90°. The energy range covered is from 0.003 eV (24 cm^{-1}) to 0.13 eV (1040 cm^{-1}). Any transition in this range manifests itself in the neutron spectrum as a peak whose width depends on instrumental resolution and phonon lifetime, and whose height depends mostly on the Boltzmann factor.

Hydrogen dominates the inelastic scattering because of its large, incoherent scattering cross section. Several types of motions can be observed for H_2O molecules. All three libration modes (wagging, twisting, and rocking) are accessible to the neutron scattering experiment. The observed transitions are not limited by selection rules as in other types of spectroscopy. Study of H_2O neutron spectra in various crystals provides information on the degree of rotational and vibrational freedom, the distribution of H_2O molecules among various lattice sites, and the clustering of molecules. A typical spectrum is shown in Figure 14.

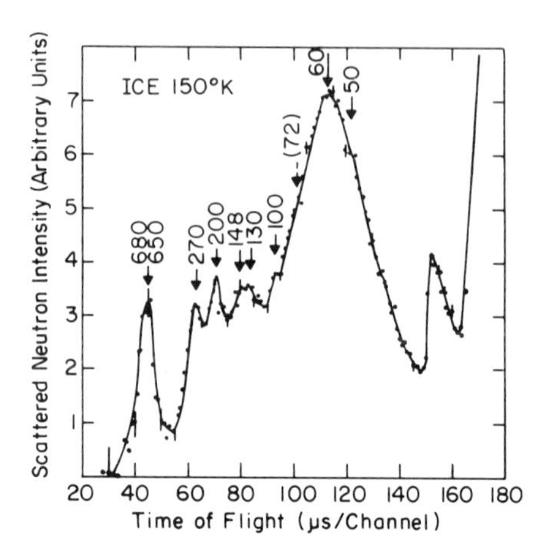

Fig. 14. Time-of-flight spectrum of neutrons scattered inelastically by a polycrystalline sample of ice at 150°K and at 65° scattering angle. The energy transfer in cm^{-1} is indicated above each peak. The incident neutrons scattered elastically correspond to the line at channel 166. The ordinates are proportional to the differential scattering cross section.

6. Spectroscopy and Local Symmetry

Total structural characterization describes the location of the nearest neighbors, next-nearest neighbors, and so on, in principle, to infinity. Diffraction methods are the only ones which provide information on the long-range order (say >10 Å) which exists in a solid. The nearest-neighbor configuration or local symmetry, however, appears to be more easily correlated with, and hence studied by, the secondary methods of physical property measurements (see Volume 2 for more detailed discussion of these). Measurements of local symmetry are all based on the strategy of obtaining energy differences between different states of atoms or complexes, usually electronic states or vibrational states. In contrast to molecules, rotational states are seldom excited in a solid. Since there are, in general, many excited states, these measurements provide a spectrum of the many resonant energies. A successful experiment requires not only the resolution of these energies, each of which has a finite linewidth, but also the assignment of each absorption band to a particular transition. The identification is accomplished by starting with a theoretical model for the possible local symmetries and a knowledge of the selection rules. Altering the experimental conditions with applied magnetic, elastic, or stress fields or varying the plane of polarization of the incident radiation removes degeneracies and changes the selection rules. Since the vibrational modes of the molecular units of a glass or the energy levels of localized electrons in a crystal are determined by the local symmetry, successful matching of a theoretical model with an observed spectrum of resonant energies provides positive identification of the local symmetry. Localized electrons in crystals are most commonly associated with transition metal or rare earth atoms, but they may also occur in defect traps or at impurity atoms.

There are two important restrictions for this type of measurement. The sample must be transparent to the radiant energy used to excite the system, and second, it must be possible to both resolve and identify all the resonant energies. If there are too many types of symmetry centers, the energy spectrum becomes hopelessly complex.

6.1. Absorption Spectra in the Visible Range

Commercially available spectrometers are generally used in the visible and near-infrared ranges. Radiation is dispersed by prisms or diffraction gratings and detected photoelectrically. If a sample is placed in the beam, absorption maxima occur at resonant energies $\hbar\omega$ corresponding to allowed transitions. It is possible to resolve energy separations as small as $1\ \mathrm{cm}^{-1}$. If the incident radiation is polarized and the plane of polarization is varied relative to the crystal orientation, it is sometimes possible to alter the selection rules, aiding in the identification of the observed resonant energies. Emission spectra are obtained by pumping with a high-energy radiation and analyzing the fluorescence. This is a useful supplement to absorption spectroscopy.

Among the types of structural information that can be obtained are the determination of the local site symmetry and ionization state for transition metal or rare earth ions present as dilute impurities in optically transparent host crystals.[84] This type of study is important not only as a measure of the strength and precise symmetry of the internal crystalline field operated typically by a polyhedron of anions or a contained cation, but it also provides data on the influence on multiplet splittings, and a measure of the broadening of certain transitions as a result of excitations to orbitals of different mean radial extension. Such information has an immediate practical application in the evaluation of host crystals for optically pumped lasers. It has also been used to identify transitions from static to dynamic Jahn–Teller distortions of the local site symmetry.

Crystal-field spectra are sensitive to interatomic distances. The Al_2O_3–Cr_2O_3 solid solution series is an interesting example. Ruby (Al_2O_3:Cr^{3+}) is red, but isomorphous Cr_2O_3 is green. The Cr–O distances in chromia are about 4% larger than in ruby. Crystal field calculations[85] show that the compression of the CrO_6 octahedron shifts the optical absorption bands by an amount sufficient to explain the color difference. Electronic transitions involving the lone-pair $6s^2$ electrons color many salts of divalent lead.

In molecular solids, the π electrons of hydrocarbons and the nonbonding electrons in carbonyl groups absorb radiation in the visible or near-ultraviolet region. Another type of electronic rearrangement gives rise to charge transfer spectra. Absorption of a photon results in the transfer of an electron to a neighboring atom or molecule. Charge transfer spectra are responsible for the deep brown color of I_2 complexes and the pleochroic colors of tourmaline. Magnetite (Fe_3O_4) and the other mixed-valence oxides are generally dark for this reason.

Band gaps can, of course, be determined optically, as can the energy levels of dopants.[86] The distribution of local symmetries of impurities in semiconductors can also be determined optically. In these experiments, stress is applied to remove degeneracies, and the plane of polarization of radiant energy is rotated with respect to the stress directions in order to vary the selection rules. Surface-state energies can be measured relative to the edge of the conduction band in semiconductors by placing the sample in an electrolyte between the plates of a condenser. If electrons are injected into a *p*-type region (or vice versa), direct electron–hole recombination gives rise to luminescence. However, the presence of surface states within the gap of forbidden energies offers an alternative, nonradiative path for hole–electron recombination. Therefore, the amplitude of the luminescence varies with the position of the surface states relative to the band edge, which can be altered by the application of an external field.

Pure alkali halide crystals are uncolored but it is possible to color them by heating in alkali vapor or by radiation damage.[87] *F* centers (electrons trapped at anion vacancies) and other light-absorbing defects are introduced in this manner. Orientations of the aspherical electron distribution in *F* centers can be detected by observing the change in resonant absorption with the plane of polarization of light.

6.2. Infrared Absorption Spectroscopy

Infrared spectroscopy is a useful tool for the characterization of solids because of the ease of sample handling, the ability to work with finely divided polycrystalline materials, and the wide availability of first-rate spectrometers. A spectral range from 2 to 300 μm is available commercially.

At the lowest level of sophistication, this method offers a method for recognizing molecular entities in solid materials.[88] For example, hydroxyl ions or residual carbonate and sulfate groups in glasses are easily recognized because of their characteristic spectra. The hydroxyl ion, in particular, can best be characterized by infrared means and, because of a systematic frequency shift, the lengths of the hydrogen bonds and thus the location of the OH group can be inferred.

The primary coordination number of a cation is strongly reflected in the IR spectrum. In simple structure types with one cation, coordination numbers can be determined quite accurately for oxides and fluorides. When cations are present on several different sites, the interpretation becomes more equivocal, and only the smallest coordination numbers can be determined with certainty. Characteristic frequency ranges for most of the common cations in four and six coordination have been determined. There is a general relationship between the square of the vibrational frequency and the inverse cube of the interatomic distance that gives first approximations to these distances.

At a higher level of sophistication, the behavior of the spectrum during polymorphic and order–disorder transitions can be predicted by group-theoretic arguments. Infrared spectroscopy is a powerful tool for order–disorder study because dramatic changes sometimes occur in the IR spectrum in systems where the ordering reflections in the X-ray diffraction pattern may be weak. In spinel,[89] for instance, a fully ordered normal spinel has four IR-active and five Raman-active bands. A disordered inverse spinel shows the same number of bands, with only a frequency change accompanying the order–disorder transformation, but other types of ordering, such as the 1 : 3 octahedral ordering found in $LiAl_5O_8$, result in a great increase in the number of allowed modes and a far more complex spectrum.

The symmetry of normal modes and the selection rules governing transitions can be predicted by factor group analysis. Only a knowledge of the space group and the population of the equipoints is required. For a structure of any complexity, the first step is to determine the invariance conditions for each sublattice. The reducible representation

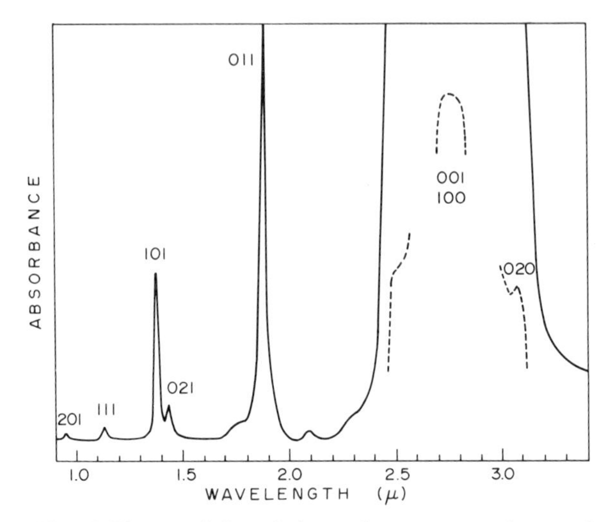

Fig. 15. The near-infrared absorption spectrum of trapped water molecules in pollucite. Peak assignments refer to the three principal H_2O vibration modes.[91]

is then calculated by summing over the sublattices. After distributing the degrees of freedom among the irreducible representations of the factor group, the selection rules are determined by noting the transformation properties of a dipole moment operator (for IR activity) and the polarizability tensor (for Raman activity). The selection rules include the polarization dependence for noncubic crystals. Details of the factor group method together with a number of illustrative examples were described in a recent review paper.[90] After identifying the peaks in the infrared spectrum, atomic force constants can be estimated from the measured frequencies, providing information about the bonding.

Foreign molecules are sometimes trapped in crystal structures. This is rather common in minerals since many silicates have open structures with cages and channels large enough to accommodate water molecules, carbon dioxide, or inert gases. Chemical analysis of gem-quality beryl ($Be_3Al_2Si_6O_{18}$) crystals often shows more than 1 wt. % H_2O. Trapped molecules can be identified from the infrared spectrum, as illustrated for pollucite in Figure 15. Pollucite, ideally $CsAlSi_2O_6$, contains water molecules loosely bonded to the aluminosilicate framework. The near-infrared spectrum is dominated by intense absorption near 2.7 μm caused by the O–H stretching fundamental. Most of the smaller peaks occurring at shorter wavelengths can be identified as combination bands involving the symmetric (v_1) and antisymmetric (v_3) stretching modes and the bending mode (v_2). Vibrational quantum number assignments (v_1, v_2, v_3) for the various excited states are listed in Figure 15. In noncubic crystals, the polarization dependence of the infrared absorption bands sometimes makes it possible to determine the preferred orientation of the trapped molecules.

6.3. Raman Spectra

In Raman spectroscopy, the sample is illuminated by a monochromatic light source of any convenient wavelength; laser sources are used in most modern spectrometers. Scattered light is observed at right angles to the incident beam. A spectrum of discrete lines is observed when the scattered light is dispersed with a high-resolution grating. Incident light of frequency ν_0 induces transitions in the solid, and the photon gains or loses energy. A vibrational transition of frequency ν_i gives rise to Raman lines of frequency $\nu_0 \pm \nu_i$ in the scattered beam. The principal selection rule for Raman vibrational spectra states that the nuclear motions must produce a change in polarizability. The procedure for obtaining the Raman-active modes was outlined in the preceding section.[90]

Raman scattering from polycrystalline ceramics contains most of the essential features of single-crystal Raman spectra. Linewidth measurements provide a means of following compositional changes through a solid-solution region, detecting stoichiometry variations. The method makes use of the broadening of Raman-active modes which occurs when the translational symmetry is reduced by deviations from simple atomic ratios. Stoichiometry variations of less than 0.5% in $LiNbO_3$ have been measured by Raman powder spectroscopy.[92]

Laser sources make it possible to investigate inelastic light scattering phenomena associated with *F* centers, phonons, magnons (spin waves), polaritons (mixed electromagnetic–mechanical excitations), and plasmas and Landau levels in semiconductors. Porto[93] has discussed the physical origins of the scattering mechanisms together with some interesting examples.

6.4. Soft X-Ray Spectra

Soft X rays (wavelength 5–500 Å) are probably the most generally applicable of all spectroscopic tools, yet they have not been widely used for the characterization of solids. These methods are likely, however, to dominate the field in the near future, especially for the light elements, due to the trends in laboratory instrumentation. Most laboratories working in materials science have access to an electron microprobe or scanning electron microscope. A peripheral capability of both these instruments is the determination of the X-ray emission spectra from samples as small as 10–20 μm^3. Furthermore, recent improvements in electron microprobe and X-ray spectrographic instrumentation have made it possible to record the X-ray spectrum for all elements from boron to the end of the periodic table. Soft X-ray spectroscopy techniques are used to measure the wavelength and shapes of emission bands, which are related to the distribution of electrons in the valence bands of solids. Since soft X-ray emission bands contain electrons involved in chemical bonding, it follows that the X-ray spectra will depend on how an element is chemically combined in a material. Unfortunately, the present theory cannot utilize the observable fine structure of soft X-ray spectra.

Over the last decade, several investigators have correlated changes in soft X-ray spectra with crystal chemical parameters such as valence, coordination number, bond type, interatomic distance, and changes in coordinating anion species.(94) Although most of the results are preliminary, it appears that one can predict a great deal about the crystal chemistry of an unknown substance from these relatively simple measurements. Thus, for example, one can distinguish the valence states of Cl or S (ranging from $+7$ to -1 and $+6$ to -2, respectively). The first sphere coordination number of, say, Al or Si can be determined in a matter of minutes. What is even more encouraging is that this can be done on extremely small samples (see above). Where only a very small fraction of one element substitutes for another (e.g., Al in a silicate), one can determine the nearest-neighbor coordination of that element. Furthermore, the cation–anion distance in silicates and aluminates can be predicted with a precision comparable to that of X-ray structure determinations from the measured shift in position of the K peak.

An important advantage of the soft X-ray method is that spectra appear unaffected by the degree of long-range order; thus, the technique can be applied to the study of crystalline and amorphous materials with equal facility.

6.4.1. X-Ray Absorption-Edge Spectroscopy

X-ray absorption methods are sometimes used for routine chemical analyses, although in most cases, X-ray fluorescence is preferred for accuracy and convenience. Measurements of the absorption coefficient in the region of the absorption edge have, however, also shown that certain features can be used in structural characterization.[95] For example, the oxygen content of chromium can be predicted with a 5–10% precision from the height of a small peak at the leading edge, and the valence of iron can be estimated from the shift in energy of the half-height of the main edge. The precision is about 10% of the amount present.

Absorption-edge methods also provide information about the bond type, as judged from spectral profiles for covalent, metallic, and ionic compounds. It is too early to predict accurately the extent to which the technique can be adopted for routine characterization of the crystal chemistry of materials.

6.5. Electron Spin Resonance

If unpaired electrons are localized on a paramagnetic ion or a defect, then the local symmetry can be obtained by measuring the microwave absorption while subjecting the sample to a variable magnetic field. Resonant energies resolved by this technique cover the range $0.001 < \hbar\omega < 3\ \text{cm}^{-1}$. The fact that microwave measurements are generally performed in cavities fixes the resonant frequency. Therefore, in order to obtain an energy spectrum, it is necessary to sweep the applied field, altering the energy level separations. Resonant absorption is observed when the splittings of allowed transitions match the cavity frequency. The observed splittings are influenced not only by the external field, but also by the crystal-field and spin–orbit splittings.

With good resolution of the resonance lines, it is also possible to observe hyperfine splittings caused by interactions between the electron spin and the nuclear spin, or even superhyperfine splittings due to interactions of the electron spin with nuclear spins on neighboring atoms. There are also quadrupole effects that influence the hyperfine splittings.

Although the fundamental theory is well known, and only computational difficulties remain in matching the observed spectrum to microscopic model, ESR is generally restricted to single crystals ($1\ \text{mm}^3$) containing a dilute concentration of paramagnetic ions. This restriction is due to the line broadening caused by internal fields which differ from site to site in amorphous or powdered materials. Thermal line broadening caused by spin–lattice interaction can be avoided by the use of low temperatures (liquid nitrogen or lower). Since the technique is limited to crystals that are transparent to microwaves, ESR is applicable to all insulators with low dielectric loss.

A few examples of the type of information that can be obtained from spin resonance experiments are listed below.[96] Microwave measurements have been used to establish the orientation of the planar rings in diphenyl single crystals. A molecule of phenanthrene is similar to a diphenyl molecule in both size and shape, differing only in the addition of two coplanar carbon atoms. If a trace of phenanthrene is added to a diphenyl crystal, it substitutes for diphenyl molecules and is readily excited to the triplet state by incident light. Using plane-polarized microwaves, absorption measurement as a function of crystal orientation specifies the orientations of the planar molecules to within about 1°. Determination of the valence state and location and position of chromium ions in CdS crystals provides a second example. The problem was to distinguish the chromium ionization, and whether the chromium impurities substitute at tetrahedral sites or enter interstitially. It was determined that the chromium is present as Cr^{2+} in interstitial sites with slightly distorted octahedral symmetry. Measurement of resonant fields at a single, fixed frequency gives an ambiguous result, in this case, because there are several alternative models with unknown splittings resulting from localized Jahn–Teller distortions. However, the problem was resolved by making measurements at a number of microwave frequencies.

The observation of transition temperatures with an S-state ion probe illustrates the usefulness of ESR in characterization problems. Pressure transforms CdS to the rocksalt structure, and the high-pressure phase can be maintained at atmospheric pressure when cooled at liquid nitrogen temperatures. Probes using Mn^{2+} can be substituted for the Cd^{2+} ions. This ion has a ^6S ground state, which is not split by crystalline fields or spin–orbit couplings. However, the hyperfine splitting (~ 75 G) is sensitive to the amount of covalence, and, therefore, to whether the Mn^{2+} ions are in octahedral or tetrahedral sites. In the case of CdS with the rocksalt structure, formation of the new phase under pressure produces a powder specimen, so that the details of the hyperfine structure are not resolved, but it is possible to observe a transition from the rocksalt to the wurtzite structure at about -80°C. The Gd^{3+} (^8S) ion has been substituted into the perovskites $SrTiO_3$ and $BaTiO_3$ to provide a similar probe for the ferroelectric transitions in these compounds.

Ordinarily, the resolution of superhyperfine splitting requires highly symmetric uniform sites for the localized electrons in order to reduce the number of lines. The electron–nuclear double resonance (ENDOR) technique, discussed in a later section, may be used to improve resolution. Sometimes, however, the superhyperfine splittings are large. For example, $SnO_2:V^{4+}$ and $SnO_2:Nb^{4+}$ exhibit splittings of 170 and 400 G, respectively, whereas in $K_2PtCl_6:Ir$ and $CdS:Mn^{2+}$, the splittings are only about 3 G. The single d electron at V^{4+} and Nb^{4+} interacts symmetrically with the spins of the near-neighbor tin nuclei, indicating that these ions enter substitutionally for tin in SnO_2. The anomalously large splittings indicate that the singly occupied orbitals have an extremely large radial extension, compatible with the fact that transition metal oxides with the rutile structure tend to be metallic.

Although ESR experiments are generally carried out on transition metal ions, all that is really required is a resultant angular momentum. Hence molecules such as oxygen and nitrous oxide give signals, as do CH_3 radicals, color centers, and conduction electrons. Point defects in concentrations as low as 10^{11} spins/cm^3 can be identified under favorable circumstances.

6.5.1. Resonance in Exchange-Coupled Systems

Ferromagnetic, ferrimagnetic, and antiferromagnetic resonance[97] are similar to paramagnetic resonance (ESR), except for the presence of strong exchange forces. As a result, the atomic moments tend to precess coherently about their equilibrium orientation in the presence of an applied field H. If H is large enough to remove the domain structure, the process can be viewed as the precession of the total moment of a ferromagnet, or as the total sublattice moments of a ferrimagnet or an antiferromagnet. Since the smallest change in angular momentum corresponds to the reversal of single electronic spin, the difference in energy between adjacent states is $g\mu_B H_e$, where H_e is an effective magnetic field. In the case of a ferromagnet, H_e depends upon the applied field H, the demagnetizing fields due to sample shape, and the crystalline anisotropy. The g factor is the component of a tensor that depends upon spin–orbit coupling and, hence, upon the symmetry and strength of the crystalline fields via their influence on the multiplet splittings.

In a collinear antiferromagnet, the effective field is approximated by $(H_K H_E)^{1/2}$, where H_E is the intersublattice exchange field and H_K is the anisotropy field. In conjunction with dc magnetic measurements, this measurement is useful in determining the strength of the inter-sublattice exchange, magnetic anisotropy, and the g tensor. Since AFMR is usually done without an applied field, it is possible to use powder specimens.

6.6. Nuclear Magnetic Resonance

In a nuclear magnetic resonance (NMR) experiment,[98] electromagnetic radiation couples directly to the nuclear spin I. In this case, the resonant energies are given by $\hbar\omega = \gamma_n H_n$, where H_n is the local field at the nucleus. The nuclear gyromagnetic ratio $\gamma_n = ge/2Mc$ leads to a resonant frequency that is smaller than that found in ESR experiments by roughly the ratio of the electron mass to the nuclear mass M. In an applied field of about 5 G, the resonant frequencies for protons are in the radiofrequency range, where it is convenient to vary the frequency, keeping a fixed magnetic field. The local field H_n consists of several terms, including a chemical screening factor arising from induced spin density at the nucleus (core polarization, covalence, or conduction-electron polarization). In ionic compounds, the shift in the resonant frequency due to the screening factor is called the chemical shift. In metals, it is generally larger and is called the Knight shift. A second term in H_n arises from the coupling between the nuclear spin and unpaired-electron spin. This term vanishes in the paramagnetic temperature range. Additional terms in H_n appear because of coupling with the spins of neighboring nuclei, and because of quadrupole effects, a topic discussed in the next section.

The NMR technique has three general restrictions. The sample must be transparent to radio frequencies, which restricts it to nonmetals, powdered metals (or metal sponge), and metallic films. Second, there must be a large number of identical sites to provide sufficient signal but without broadening the lines. Sensitivity depends upon a homogeneous field and the absence of paramagnetic impurities, as well as on the nuclear species being measured. And, of course, the atomic nucleus must have a spin. In regard to the source, not only must the frequency range span the resonant frequencies, but the powder level must also be correct since too much power saturates the system.

Four examples of the type of structural information that can be obtained from NMR experiments are as follows.

1. Determination of valence state and local environment: The chemical shift differs for the same atom in different valence states and local environments. In addition, the coupling of the nuclear spin with neighboring nuclei gives a different fine structure for different environments. This method may be particularly useful for the identification of the many molecular-like clusters in amorphous materials.

2. Spin-configuration can be determined because of the interaction between the electron spins and nuclear spins. The NMR frequency depends upon the direction of the electron spin when magnetic ordering is present, giving the angle between noncollinear spins. For example, in the spinel $Mn^{2+}Cr_2^{3+}O_4$, the spin configuration at low temperatures is a complex magnetic spiral. Neutron diffraction does not provide a unique solution, and it would be helpful to specify the angle between the chromium spins. This can be obtained from the fact that spin direction varies with the angle between the applied magnetic field and the spin.

3. In a rigid structure, dipole–dipole interactions introduce variations in local fields that broaden the resonance lines. Line narrowing with increasing temperature may signal the onset of additional vibrational or rotational degrees of freedom. In crystalline NH_4Cl, the proton linewidth widens at temperatures below 135°K where the proton positions become ordered.

4. Interatomic distances can sometimes be determined from NMR linewidths. Proton–proton distances in gypsum ($CaSO_4 \cdot 2H_2O$) and other hydrates have been measured to an accuracy of 0.02 Å.

6.6.1. Nuclear Quadrupole Resonance

The presence of quadrupole charge distribution Q in the nucleus leads to a quadrupole interaction energy $Q \cdot \nabla E$, with resonant frequencies in the radiofrequency range. Absorption lines resulting from transitions between quadrupole states are interpreted in terms of the electron distributions causing the electric field gradients ∇E. Quadrupole spectra provide information concerning the positions of atoms and molecules in the unit cell and their motional behavior.[99] Experimental techniques are very similar to those for NMR. Applied magnetic fields produce Zeeman splittings of the quadrupole spectra, and in some cases,[100] the quadrupole interaction energy is only a small perturbation on the interaction between the applied magnetic field and the nuclear magnetic dipole moment.

The quadrupole interaction energy is zero if the nuclear spin $I \leq \frac{1}{2}$, or if the nucleus is in a high-symmetry field such that $\nabla E = 0$.

Measurement of resonant frequencies of a single crystal with plane-polarized radiation gives the orientation of unique directions in the local field. Not all nuclei possess quadrupole moments. Most NQR studies have been performed on compounds containing certain isotopes of nitrogen and the halogens Cl, Br, and I, or with the following metals: Li, Na, K, Rb, Cs, Cu, Be, Hg, B, Al, Ga, In, As, Sb, and Bi.

Boron NQR resonances have been used successfully to determine the fraction of borons that are four-coordinated in borate glasses.[101] The remaining three-coordinated borons can be further subdivided into those that have all three oxygens bridging and those that have at least one nonbridging oxygen. If the boron is tetrahedrally coordinated to four oxygens, then the field gradient at the boron nucleus is small. Substantially larger terms in the electric field gradient tensor are obtained for triangularly coordinated boron. The boron NMR spectrum in alkali borate glasses (Figure 16) shows peaks originating from four-coordinated boron, and both bridging and nonbridging three-coordinated sites. The fractional numbers in each site can be determined within 5% under favorable circumstances. Computer simulation techniques[102] greatly simplify the analysis of the spectra.

Impurities in the sample containing the resonant nuclei sometimes lead to broadening and may smear the quadrupole resonance beyond detection. Both charge and size effects play a role in impurity-

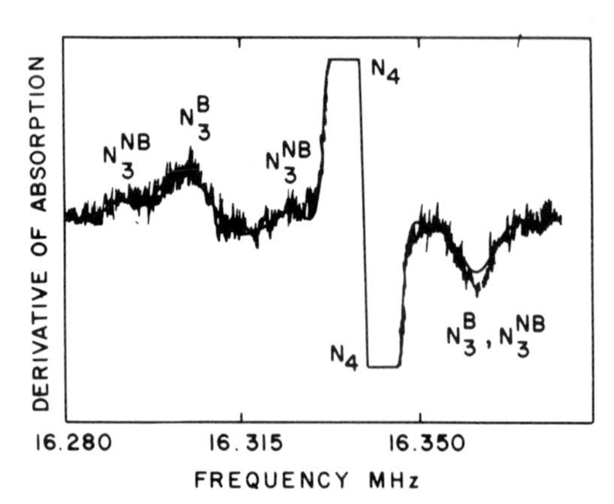

Fig. 16. NMR derivative spectrum of ^{11}B in 40 mole % Cs_2O–60 mole % B_2O_3 glass.[102] Transition is $m = 1/2 \rightarrow m = -1/2$. Major contributions of tetrahedrally coordinated boron (N_4) are compared with three-coordinated sites, both bridging ($N_3{}^B$) and nonbridging (N_3^{NB}).

broadening. Line broadening also occurs in solid solutions and in irradiated specimens.

6.6.2. Double Resonance (ENDOR)

Double resonance is a combination of ESR and NMR that permits excellent resolution of the hyperfine or superhyperfine structure while retaining good sensitivity. The technique is to pump at a microwave frequency with sufficient power to saturate the population of the upper state. If this is done, there is no resonant absorption, since nearly as much energy is being emitted as absorbed. The populations within the two main energy levels responsible for the ESR transition can be altered by the additional application of an rf excitation. The rf field produces NMR between the fine-structure levels, removing the saturation condition, and causing the ESR absorption to reappear. Moderate NMR power levels are used to saturate the NMR signal, thus maximizing the effect. This technique has been employed successfully to resolve the superhyperfine structure of electrons trapped at *F* centers and at donor atoms in semiconductors. The superhyperfine structure provides information about the point symmetry of the near neighbors to a trapping site.

ENDOR has a number of applications but is especially useful in probing the surroundings of a paramagnetic center in an insulating solid.[103] Centers of this type are generated by irradiation with photons, by heat treatment, or by doping a diamagnetic crystal with paramagnetic impurities. Many such materials (phosphors, solid-state masers, extrinsic semiconductors) are technologically important, increasing the interest in the precise nature of the centers responsible for their properties.

In identifying dopants and determining their positions, ENDOR is far more suitable than the usual techniques for structure analysis.[103] X-ray diffraction is not convenient, because the centers have low density and irregular atomic arrangement. Isoelectronic ions such as O^{2-} and F^{-} are easily differentiated by resonance techniques—an almost impossible task by X rays.

In the ENDOR technique, the nuclear magnetic resonances of nuclei adjacent to a paramagnetic center are scanned, giving a spectrum which depends on the local fields created by the paramagnetic ion. In this way, it is possible to identify and position each of the surrounding nuclei with nonzero nuclear magnetic moments, providing a detailed map of the environment of the paramagnetic center.

To observe ENDOR, the concentration of paramagnetic ions should be less than $10^{19}/cm^3$; otherwise the resonance lines broaden because of interactions between neighboring ions. The NMR signals are small for typical sample volumes ($\sim 10^{-2}$ cm^3) but increased sensitivity is attained by observing changes in the EPR signal. Experiments are performed with a modified EPR spectrometer in which a klystron supplies microwave power to a resonant cavity containing the sample. For ENDOR experiments, the cavity is altered to include a large rf magnetic field at the NMR resonant frequency. Double resonance is detected by measuring changes in the reflected microwave power while sweeping the applied rf frequency through various resonances.

Many of the experiments performed thus far have been on CaF_2 doped with paramagnetic rare earth ions such as Ce^{3+} or Yb^{3+}. The Yb^{3+} ion is slightly smaller than Ca^{2+}, causing nearby fluorine ions to collapse around the impurity ion. Displacements of about 0.03 Å were measured for next-nearest neighbors. Substitution of Yb^{3+} for Ca^{3+} requires charge compensation. One common compensation mechanism is interstitial F^- ions. ENDOR spectra clearly identify charge-compensating ions and the resulting distortions of the crystal structure.

6.7. Mössbauer Effect

Mössbauer measurements are relatively easy to perform for certain isotopes and give a surprising amount of information.[104] The experiments involve determination of the counting rate of γ rays emitted by a source and passed through an absorber. The counting rate is measured as a function of the Doppler motion of the source and absorber. Doppler velocities are low, generally less than 5 cm/sec.

The principal limitation of the method is the lack of suitable isotopes. It is unfortunate that elements such as carbon, oxygen, and silicon do not show a Mössbauer effect. In addition to iron, Mössbauer effects are readily observed in isotopes of tin, xenon, iodine, europium, gold, and a few others.[105] Isomer shifts make it possible to distinguish the valence states of iron and tin, but most of the other nuclei have greater resonance linewidths, making it difficult to observe chemical effects.

The Mössbauer effect is a γ-ray absorption phenomenon with a frequency precision sufficient to resolve hyperfine splittings of 10^{-4}–10^{-5} eV. In order to have resonant absorption, the energy of the incident γ rays from an emitter atom must equal the energy of the absorbed γ ray. A free-atom emitter or absorber suffers a recoil energy $R = \Delta E^2/2Mc$, where M is the mass of the emitting (or absorbing) nucleus and ΔE is the energy difference between two nuclear states. This means that the emitter γ ray has an energy $\Delta E - R$, whereas the absorbed γ ray has the energy $\Delta E + R$. In order to have resonance, it is necessary to have $R \approx 0$, a condition sometimes satisfied when the absorber and emitter are bound in a solid, either crystalline or amorphous.

The experimental problem is to obtain a monochromatic source and to sweep the frequency. The frequency is changed by moving the emitter relative to the absorber. Relative motions of ~1 cm/sec provide sufficient Doppler shift to sweep through the desired frequency range. The isotope ^{57}Fe has a γ-ray energy of 14.4 keV with a lifetime of about 10^{-7} sec and is therefore convenient. Cobalt-57 in stainless steel, where there is no magnetic order and no quadrupole splitting, makes an excellent monochromatic source. Cobalt-57 decays to an excited state of ^{57}Fe, which in turn decays to the ground state of ^{57}Fe accompanied by the emission of a γ ray of the proper frequency to excite a ^{57}Fe nucleus in the absorber. Figure 17 illustrates how the

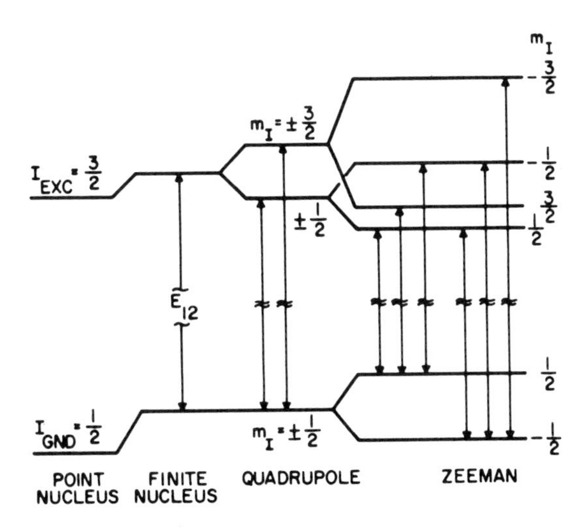

Fig. 17. Nuclear hyperfine interaction for ^{57}Fe. Effect of finite nucleus in absorber is different from that in source so that E(absorber) $\neq$ E(emitter). The energy difference is called the isomer shift. Selection rule: $\Delta m_I = 0, \pm 1$.

nuclear energy levels are split by the field at the nucleus. From the selection rule $m_I = 0, \pm 1$, a characteristic six-line spectrum is obtained if there is an internal field due to magnetic ordering.

The isomer shift provides information on the electronic charge density at the nucleus, while the quadrupole splitting is related to the symmetry of the surrounding electron charge distribution. The changes which occur in these parameters when the Mössbauer nuclei are in different chemical environments can be interpreted in terms of changes in hybridization and the partial ionic character of the chemical bonds.

Examples of the type of structural information obtained from Mössbauer spectra include determination of valence states, local symmetry, field gradients, and magnetization directions. The isomer shift gives the valence state and local symmetry. Since the nuclear size of the excited state differs from that of ground state, the contact terms contributing to H_n are different for E_1 and E_2. Therefore, these terms enter the energy difference ΔE to produce an isomer shift that is analogous to the chemical shift in NMR. This shift is sensitive to the valence state and local environment of the atom. For example, the garnet structure of $Y_3Fe_5O_{12}$ contains tetrahedral and octahedral Fe^{3+} ions, giving two different isomer shifts; the spinel $Fe^{3+}[Fe^{2+}Fe^{3+}]O_4$ gives three different isomer shifts, corresponding to tetrahedral and octahedral Fe^{3+} ions and octahedral Fe^{2+} ions. Positive identification of $Fe^{3+}V^{3+}O_3$ was possible with this technique.

Quadrupole coupling with the ^{57}Fe nucleus splits the Mössbauer spectrum into a doublet, with the magnitude of the splitting proportional to the product of the nuclear quadrupole moment and the electric field gradient at the nucleus. The quadrupole splitting in Fe^{2+} is considerably larger than in Fe^{3+}, making it easy to distinguish the two more common oxidation states of iron.

In principle, the fraction of γ rays emitted without recoil is also related to interatomic forces, but a useful relationship has yet to be established. Some work has been initiated on the determination of lattice dynamics from this effect.

Zeeman splitting of the nuclear energy levels occurs in alloys and compounds with long-range magnetic ordering. For these materials, the effective magnetic fields acting at the nuclei have been measured from the magnitude of the splittings. In magnetized samples, the intensities of the Mössbauer lines depend upon the angle between the net field at the nucleus and the incident γ ray. Monitoring the intensity as a function of angle of incidence on a single crystal provides the direction of spontaneous magnetization.

6.8. Electron Spectroscopy

Structure determinations by electron spectroscopy are comparable to those by infrared and NMR techniques. In electron spectroscopy, ionizing radiation impinges on a sample causing electron ejection. The energies and intensities of the ejected electrons are measured with an electron monochromator and detector. Electron spectrometers require vacuums better than 10^{-5} Torr and residual magnetic fields of $< 10^4$ G. The He I and He II resonance lines at 21.22 and 40.8 eV are excellent photon sources for ejecting outer electrons, while the $K\alpha$ lines of Mg and Al are suitable for higher-energy sources.

Electron spectroscopy involves four closely related phenomena between an atom A, photons $h\nu$, and electrons e^-:

Photoionization	$A + h\nu \rightarrow A^{+*} + e^-$
Electron bombardment	$A + e^- \rightarrow A^{+*} + 2e^-$
X-ray emission	$A^{+*} \rightarrow A^+ + h\nu$
Auger emission	$A^{+*} \rightarrow A^{2+} + e^-$

Photoionization and electron bombardment are one-step processes by which a photon (or electron) ejects an electron from an atom. When photoionization occurs, the energy of the photon goes into liberating the photoelectron and providing it with kinetic energy, giving the relation

$$h\nu = E_B + E_K$$

among the binding energy E_B and the measured energies of the photon and the photoelectron. If the photoelectron originates from an outer shell, E_B corresponds to valence binding energies of the various molecular orbitals. For inner electrons, the variation of binding energy with chemical environment gives a characteristic chemical shift. For example, the photoelectron spectrum of ethyl trifluoroacetate,[106] $(CF_3)(CO)O(CH_2)(CH_3)$ shows fine structure for the carbon 1*s* electrons. The four types of carbon give rise to four peaks of nearly equal intensity with binding energies between 285 and 295 eV.

Correlation charts can be developed for particular types of coordination. Nitrogen 1*s* electrons from nitrates, nitrites, and azides are readily identified in this way since the chemical shifts differ by several electron-volts.[107] Electron spectroscopy is also useful in determining ionization potentials and in surface chemistry. The average penetration of a photoelectron is in the range 20–50 Å.

X-ray fluorescence and Auger spectroscopy are two-step processes in which an inner electron (not necessarily the innermost) is ejected by a photon or an electron. The structural capabilities of X-ray fluorescence have been described in another section. Auger spectroscopy can be used in chemical problems[108] but the spectra are more complex to analyze than photoionization patterns.

The applications of both Auger spectroscopy and ESCA have so far been largely found in the surface chemistry of organic molecules, where they have proved to be enormously powerful. In the characterization of inorganic materials, one limitation compared to, say, X-ray emission is the large sample required, while the advantage is the sensitivity with respect to thickness (10–50 Å). As the characterization of surfaces becomes increasingly important, the next few years should demonstrate the relative effectiveness of the various competing techniques.

6.9. Acoustic Spectroscopy

Since mechanical radiation is sensitive to both electronic and structural inhomogeneities, it is a particularly useful tool for examining defects and dynamic processes in condensed matter. Using the sonar principle, ultrasonic waves can be used to locate flaws inside metal castings and other large objects. In some cases, the technique is more sensitive than conventional X-ray radiography. Another acoustic method measures attenuation rather than reflection, relating sound absorption to dislocation damping and other loss mechanisms.[109] The use of ultrasonics for the determination of elastic constants is described in the next section.

Piezoelectric and magnetostrictive transducers have been the conventional method of generating and detecting ultrasonic waves. However, it is also possible to generate sound by nonlinear electrostrictive coupling via the intense electromagnetic excitation of a laser beam, an effect known as stimulated Brillouin scattering.

Elastic waves have an advantage over optical waves, in that they can traverse opaque media and may be strongly coupled to atomic spins. Because of the strong interaction with the atomic array, even quite minor imperfections can produce severe scattering of the elastic waves, scattering which is easily detected. If elastic waves of sufficiently high frequency can be generated, it is possible to excite a paramagnetic ion from one electronic state to another. Spin–phonon interactions are especially strong if spin–orbit coupling is present, as in octahedrally coordinated Fe^{2+} ions. As the elastic wave frequency approaches the frequency of an allowed spin transition, the wave suffers strong attenuation and dispersion. In the case of Fe^{2+} ions, the coupling of sound waves (phonons) to spins is stronger and of a different type than the coupling of electromagnetic waves to spins. This has two consequences: (1) it may be easier to observe such resonances by sound waves than by electromagnetic waves, and (2) transitions that are forbidden to electromagnetic waves may be allowed to ultrasonic waves.

7. Physical Properties as Characterization Tools

7.1. Introduction

In previous sections, we have dealt with those methods—optical microscopy and X-ray, electron, and neutron diffraction—that serve as tools of the scientist interested in determining directly the position of the atoms in a solid, i.e., structurally characterizing the solid. In practice, during the last two decades in which the physics of solids experienced an explosive growth, it very often happened that the only "characterization" which was actually performed was the monitoring of the physical property of greatest interest for the understanding of the device. Resistivity ratios, for example, were used extensively as indices of purity. Obviously such physical property measurements provide only an indirect or secondary means of characterization. Indeed, they can only be used for characterization when some correlation has been experimentally established between the given property and a structural or compositional variation. However, the fact remains that such secondary characterization methods are exceedingly common and will continue to be used. In this section, we will show the structural basis for secondary characterization by several commonly used physical property measurements (see also Volume 2).

In characterizing solids by the use of physical properties, the principal drawback is the ambiguity in interpreting the results. A change in electrical resistivity, for instance, might be caused by deviations in stoichiometry, or by impurities, or by imperfections which act as scattering centers. The ambiguity can sometimes be removed by supplementary experiments, but other characterization methods such as electron microscopy are often more straightforward. On the other hand, the advantage of certain property measurements is their great sensitivity. Resistivity changes of many orders of magnitude are observed in semiconductors with hardly sensible changes in composition; and in ferroelectrics, the changes in dielectric and optical properties at the Curie point are usually more sensitive than X-ray lattice parameter measurements. From a structural viewpoint, another limitation of the use of secondary or physical property characterization is related to the fact that it is difficult for such methods to handle two-phase (or multiphase) materials. Further, as will be seen below, space-group-related characterization cannot be secured from a secondary property, although point group information can be obtained from many of the property tensors, including charge-carrier mobility, dielectric constant, magnetostriction, piezoelectricity, and piezooptical and elastic constants. It is also obvious that since crystallography was not originally concerned with spin orientation at all, and since only neutrons are able to determine spin states, it is in the magnetic area especially that the secondary characterization becomes important.

In the following, very brief summary of the use of a vast array of measurements used to characterize a material, we have attempted to present the structure-related framework into which each measurement fits.

7.2. Some Crystal Physical Generalizations

Property tensors(110) give information about the point group (crystal class) rather than the space group because macroscopic property measurements are insensitive to the microscopic translations which differentiate the various space groups belonging to the same point group. A net plane or a glide plane restricts the electrical conductivity tensor in the same way as a mirror plane. Some properties are more useful than others in symmetry determination. Odd-rank polar tensors (piezoelectricity) and all axial tensors (magnetoelectricity) are null effects which disappear for certain symmetries. In such cases, a qualitative measurement is sufficient to establish the absence of certain symmetry elements. Crystallographers use the Geibe–Schiebe test for piezoelectricity to identify acentric crystals. A similar test has been developed by exposing the material to laser light and measuring the second harmonic generated by the sample.(7) The symmetry restrictions for SHG are almost identical to piezoelectricity.

In general, the higher-rank tensors give more precise point group information than do the simpler properties. The number of nonzero coefficients and the number of independent coefficients for various tensors are listed in Table 6. Representative physical properties are given in Table 7. A drawback with most property measurements is that they require crystals with dimensions of millimeter size or larger, whereas X-ray diffraction analyses use single crystals about 0.1 mm in size, or, where powders can be used, they apply to crystals in the 1–10 μm range.

TABLE 6

Tensor Properties for the 32 Crystal Classes*

Crystal class	1	2	3	4	Example
$1 = C_1$	3 (3)	9 (6)	18 (18)	36 (21)	Kaolinite $Al_2Si_2O_5(OH)_4$
$\bar{1} = C_i$	0	9 (6)	0	36 (21)	Copper sulfate $CuSO_4 \cdot 5H_2O$
$2 = C_2$	1 (1)	5 (4)	8 (8)	20 (13)	Sucrose $C_{12}H_{22}O_{11}$
$m = C_s$	2 (2)	5 (4)	10 (10)	20 (13)	Potassium nitrite KNO_2
$2/m = C_{2h}$	0	5 (4)	0	20 (13)	Orthoclase $KAlSi_3O_8$
$222 = D_2$	0	3 (3)	3 (3)	12 (9)	Iodic acid HIO_3
$mm2 = C_{2v}$	1 (1)	3 (3)	5 (5)	12 (9)	Sodium nitrite $NaNO_2$
$mmm = D_{2h}$	0	3 (3)	0	12 (9)	Forsterite Mg_2SiO_4
$3 = C_3$	1 (1)	3 (2)	13 (6)	24 (7)	Nickel tellurate Ni_3TeO_6
$3 = C_{3i}$	0	3 (2)	0	24 (7)	Ilmenite $FeTiO_3$
$32 = D_3$	0	3 (2)	5 (2)	18 (6)	Low-quartz SiO_2
$3m = C_{3v}$	1 (1)	3 (2)	8 (4)	18 (6)	Lithium niobate $LiNbO_3$
$\bar{3}m = D_{3d}$	0	3 (2)	0	18 (6)	Corundum Al_2O_3
$4 = C_4$	1 (1)	3 (2)	7 (4)	16 (7)	Iodosuccinimide $C_4H_4INO_2$
$\bar{4} = S_4$	0	3 (2)	7 (4)	16 (7)	Boron phosphate BPO_4
$4/m = C_{4h}$	0	3 (2)	0	16 (7)	Scheelite $CaWO_4$
$422 = D_4$	0	3 (2)	2 (1)	12 (6)	Nickel sulfate $NiSO_4 \cdot 6H_2O$
$4mm = C_{4v}$	1 (1)	3 (2)	5 (3)	12 (6)	Barium titanate $BaTiO_3$
$\bar{4}2m = D_{2d}$	0	3 (2)	3 (2)	12 (6)	Potassium dihydrogen phosphate KH_2PO_4
$4/mmm = D_{4h}$	0	3 (2)	0	12 (6)	Rutile TiO_2
$6 = C_6$	1 (1)	3 (2)	7 (4)	12 (5)	Nepheline $NaAlSiO_4$
$\bar{6} = C_{3h}$	0	3 (2)	6 (2)	12 (5)	Lead germanate $Pb_5Ge_3O_{11}$
$6/m = C_{6h}$	0	3 (2)	0	12 (5)	Apatite $Ca_5(PO_4)_3F$
$622 = D_6$	0	3 (2)	2 (1)	12 (5)	High-quartz SiO_2
$6mm = C_{6v}$	1 (1)	3 (2)	5 (3)	12 (5)	Zincite ZnO
$\bar{6}m2 = D_{3h}$	0	3 (2)	3 (1)	12 (5)	Benitoite $BaTiSi_3O_9$
$6/mmm = D_{6h}$	0	3 (2)	0	12 (5)	Beryl $Be_3Al_2Si_6O_{18}$
$23 = T$	0	3 (1)	3 (1)	12 (3)	Sodium chlorate $NaClO_3$
$m3 = T_h$	0	3 (1)	0	12 (3)	Pyrite FeS_2
$432 = O$	0	3 (1)	0	12 (3)	Manganese β-Mn
$\bar{4}3m = T_d$	0	3 (1)	3 (1)	12 (3)	Zincblende ZnS
$m3m = O_h$	0	3 (1)	0	12 (3)	Rocksalt NaCl

* Zero means the effect is absent. Other numbers indicate the number of nonzero coefficients, and the number of independent nonzero coefficients is given in parentheses. Third- and fourth-rank tensors are referred to the shortened two-subscript matrix notation.

TABLE 7

Some Crystal Properties Represented by Polar Tensors

First-rank tensors
- Electric polarization caused by temperature change (pyroelectricity)
- Electric polarization caused by hydrostatic pressure
- Heat evolved by applying an electric field (electrocaloric effect)

Second-rank tensors
- Electric polarization produced by an applied electric field (electric susceptibility)
- Magnetization produced by an applied magnetic field (magnetic susceptibility)
- Electric current density produced by an applied electric field (electrical conductivity)
- Heat flux produced by a temperature gradient (thermal conductivity)
- Elastic strain caused by temperature change (thermal expansion)
- Elastic strain caused by hydrostatic pressure
- Heat evolved by applying mechanical stress (piezocaloric effect)
- *Thermoelectric effect relating the gradient in electrochemical potential to the temperature gradient

Third-rank tensors
- Electric polarization produced by an applied mechanical stress (direct piezoelectric effect)
- Mechanical strain caused by an applied electric field (converse piezoelectric effect)
- *Change in refractive indices with applied electric field (electrooptic effect)

Fourth-rank tensors
- Mechanical strain caused by mechanical stress (elastic compliance)
- *Variation of refractive indices with mechanical strain (photoelastic effect)
- *Second-order effect relating strain to applied electric field (electrostriction)

* Asymmetric tensors, with additional coefficients.

7.3. Dielectric Measurements

In general, the dielectric constant is a complex quantity, $\kappa = \kappa' - i\kappa''$. The dielectric constant κ' is controlled by the conductivity at very low frequencies, and by various polarization mechanisms at higher frequencies.[111] Electronic polarization and molecular dipole moments can be estimated from the refractive indices and dielectric constants. The dissipation factor $D = \kappa''/\kappa'$ of a dielectric is sensitive to charge carriers and reorientable dipoles. Accurate loss measurements are capable of detecting free-carrier concentrations of $10^{15}/\text{cm}^3$. Various contributions to the loss can be separated by determining the dependence on temperature and frequency.

Ionic crystals are sometimes ferroelectric, antiferroelectric, or paraelectric. Ferroelectric crystals belong to one of the ten polar classes (1, 2, *m*, *mm*2, 3, 3*m*, 4, 4*mm*, 6, 6*mm*) and are therefore pyroelectric and piezoelectric. The spontaneous polarization of a ferroelectric can be switched from one orientation to another with an electric field, giving rise to hysteresis loops analogous to those in ferromagnetics. Antiferroelectrics and paraelectrics are defined in analogy to antiferromagnetic and paramagnetic substances. Paraelectric materials exhibit Curie–Weiss behavior in the electric susceptibility, and in antiferroelectrics, the electric dipoles (ion displacements from a prototype structure) cancel, as do the magnetic dipoles in an antiferromagnetic. Ferroelectrics and antiferroelectrics generally show a phase transition to a paraelectric region at high temperatures.

A convenient method for determining the phase transition temperatures is to measure the relative permittivity κ' as a function of temperature, using a capacitance bridge. The ratio of the capacitance of an electrical condenser with and without the dielectric between the condenser plates gives the permittivity. The value of κ' approaches a maximum at the transition temperature, where it changes discontinuously. Hysteresis loops, spontaneous polarization, and coercive fields are obtained with a Sawyer–Tower circuit and displayed with an oscilloscope.

7.4. Electrical Characterization of Solids

All physical properties depend on defect concentration but some are far more sensitive than others. Electrical measurements are most often used in studying point defects. Transport properties tend to be more sensitive than equilibrium properties since defects often control diffusion or charge transport. Thus *electrical conductivity* is capable of detecting electrons, holes, and atomic point defects down to $10^8/cm^3$, far better than can be done by density measurements. In the latter, atomic defect concentrations of $10^{18}/cm^3$ can be detected by comparing hydrostatic density with the value computed from the measured lattice parameters.

Hall effect, cyclotron resonance, thermoelectric power, ionic conductivity, and self-diffusion measurements are also capable of detecting defect concentrations of 10^8–$10^{11}/cm^3$, but all these methods suffer from a lack of specificity: While the measured quantities are sensitive to defect *density*, the observations do not *identify* the defects. Combining the results of several different measurements sometimes removes the ambiguity. For example, combining conductivity measurements with the Hall coefficient and its temperature dependence yields values for the charge carrier concentration and the sign of the carriers as well as mobility values. The Hall effect is just one of many galvanomagnetic and thermomagnetic phenomena.[112] When a transverse magnetic field is applied to a current-carrying conductor, an electric field appears in the direction perpendicular to both the current flow and the magnetic field, giving the Hall coefficient $R = E_z/B_y J_x$.

In semiconductors, electrons and holes often recombine via impurities, point defects, and dislocations. The *lifetime* characterizing the recombination rate is extremely structure sensitive, making it a useful test of crystal perfection. Numerous lifetime studies have established recombination parameters such as the energy levels associated with various imperfections and capture cross sections for holes and electrons. In germanium samples, lifetimes range from 10^{-2} to 10^{-8} sec, so that no single experimental method is sufficient for all conditions. One class of experiments involves the integrated detection of the total number of excess carriers in the sample. A second type utilizes localized detection and is capable of measuring the spatial distribution of carriers. Both steady-state and transient measurements are possible.

Thermoelectric power is also influenced by changes in carrier density and in scattering mechanisms. When an electric current I flows between two points in a conductor with a temperature difference ΔT, a quantity of heat $\mu I \, \Delta T$ is emitted or absorbed in addition to Joule heating. The multiplier μ is called the Thomson coefficient. The absolute thermoelectric power is given by $\int_0^T (\mu/T) \, dT$, and is usually measured in reference to lead, which has a very small Thomson coefficient at low temperatures. Thermoelectric power is extremely sensitive to small impurity additions. Concentrations of 0.0026 at. % tin in copper change the thermoelectric power from $+0.3$ to -60 μV/°C. Defects introduced by cold working, strain, and irradiation also produce measurable changes.

Magnetoresistance is defined as the change in electrical resistivity under applied magnetic field divided by the zero-field resistivity. It can be measured with the magnetic field parallel to current flow (longitudinal magnetoresistance) or perpendicular to it (transverse magnetoresistance). In metals, magnetoresistance is a second-order effect caused by deviations from free-electron behavior. It is a first-order effect in semiconductors, where comparison of the longitudinal and transverse magnetoresistance measured in various crystallographic directions determines effective masses and portions of the energy surface in **k** space.

In *cyclotron resonance*, a metallic or semiconductor sample is mounted in a microwave cavity and illuminated by a light beam shining through a hole in the cavity wall. Modulation of the light beam produces modulation of the free-carrier concentration. The carriers spiral around an applied dc magnetic field, satisfying resonance conditions when the microwave frequency $\omega = eH/m^*c$ and yielding the effective mass of the carriers m^*.

7.5. Magnetic Measurements

The magnetic susceptibility $\chi = dM/dH$ and the spontaneous magnetization M_s below a Curie temperature T_C are obtained by measuring[113] the magnetization M of a sample in an applied field H. The advent of superconducting magnets makes dc fields of about 60 kG available to most investigators, while pulsed fields up to 250 kG are available to a few laboratories. Vibrating-sample magnetometers can be purchased commercially, and when aligned carefully are capable of a sensitivity of 10^{-7} emu/g.

7.5.1. Paramagnetism

In the case of localized-electron atomic moments, the induced magnetization is $M = N\mu$, where μ is the component of the average atomic moment parallel to the applied magnetic field for a volume containing N atoms. In general, there are two contributions to μ, one arising from the localized permanent moments aligned by the applied field H, and the other from an induced atomic moment opposed to H. These represent the paramagnetic and diamagnetic contributions, respectively. Although the diamagnetic contribution is generally small (unless the sample is superconducting), it is necessary to estimate it and subtract it from the observed susceptibility before interpreting the paramagnetic contribution. The theory has been worked out for localized electrons, so that from the temperature dependence of χ, it is possible to determine the magnitude of the localized moment. For energy-level separations Δ comparable to kT, the magnitude of Δ can be estimated also. Susceptibility measurements can be performed on polycrystalline samples since the anisotropy in the paramagnetic state is usually small.

In the case of broadband collective electrons (bandwidth $\gg kT$), there is an induced, paramagnetic spin contribution to μ if the bands are partially filled. This effect is temperature independent, as with the diamagnetic contribution, and somewhat reduced because of an induced, diamagnetic orbital contribution. There is also a diamagnetic contribution from the atomic core electrons that may be of comparable magnitude in heavy metals.

Narrow band electrons (bandwidth $\leq kT$) have induced, paramagnetic spin contributions to μ that are temperature dependent, and in the narrowband limit, this temperature dependence approaches that for localized electrons. It is not, however, interpretable on the basis of localized-electron theory.

A variety of structural information can be inferred from magnetic susceptibility data.[114] The site symmetry of a paramagnetic ion can be determined from the localized atomic moment. The orbital contribution to atomic moment is strongly influenced by the crystalline fields, and therefore magnetic susceptibility measurements provide information about the local symmetry of the ions. For example, Co^{2+} ions have a nearly spin-only atomic moment of $3\mu_B$ in a tetrahedral interstice, whereas octahedrally coordinated Co^{2+} has a moment of about $3.7\mu_B$.

Many transition metal ions are found in high-spin and low-spin ion states. If intraatomic exchange splitting is greater than the crystal-field splitting, then the ion is in a high-spin state in accordance with Hund's highest-multiplicity rule for a free atom. However, if the reverse is true, then the ion may be in a low-spin state. Octahedrally coordinated Co^{3+}, for example, has a spin moment of $4\mu_B$ in the high-spin state, but is diamagnetic with zero spin in the low-spin state. In $LaCoO_3$, the high-spin and low-spin energy levels are separated by less than kT, giving rise to a plateau in the susceptibility between 125 and 370°K. This behavior signals the onset of ordering of the two types of ions at higher temperatures, resulting in a subtle change in symmetry from $R\bar{3}c$ to $R\bar{3}$. A similar effect is observed in MnAs, a metallic compound that exhibits an anomaly in χ about 100° above a ferromagnetic Curie temperature. The anomaly is caused by a change in the spin state of the manganese atoms, creating a small distortion in structure from hexagonal to orthorhombic.

Susceptibility measurements are especially useful in determining magnetic ordering temperatures and in identifying the type of magnetic order. At the Néel temperature where a disordered spin array transforms to long-range antiferromagnetic order, χ shows a sharp maximum with a discontinuous change in slope. Plotting χ_m^{-1} against temperature gives Curie–Weiss behavior for the high-temperature data. The intercept with the temperature axis θ is usually negative.

In the case of a ferromagnet, θ is generally positive but less than the Curie point T_C. Below T_C, the susceptibility becomes a many-valued function because of hysteresis. Ferrimagnets also show hysteresis below T_C, but θ is negative, as in an antiferromagnet. A first-order phase change at T_N or T_C signals a structural change accompanying long-range magnetic ordering.

Localized and collective *d*-electron configurations can be identified from magnetic data since collective electrons usually exhibit no spontaneous atomic moment. As an example, isolated Ni^{3+} ions in oxides are paramagnetic with a spin moment of 1 μ_B, often showing Jahn–Teller distortions of the local site symmetry at low temperatures. The rhombohedral perovskite $LaNiO_3$, on the other hand, is metallic and exhibits temperature-independent Pauli paramagnetism.

Changes in interatomic exchange interactions accompany changes in structure. The orthorhombic perovskite $LaMnO_3$ exhibits a cooperative Jahn–Teller distortion about the Mn^{3+} ions below 850°K. The resulting electronic order introduces ferromagnetic coupling between Mn^{3+} ions in the pseudotetragonal basal plane with antiferromagnetic coupling between planes. At high temperatures, all magnetic interactions are ferromagnetic. The intercept θ of the Curie–Weiss law reflects the magnitude and sign of the interatomic-exchange interactions, so that a distinct anomaly in the reciprocal susceptibility curve occurs at the phase change.

Disproportionation has been studied in PdF_3, reported to have the ReO_3 structure. Were this so, the compound might contain Pd^{3+} ions with a spin contribution of either 3 μ_B (high-spin state), or 1 μ_B (low-spin state), or 0 μ_B (metallic state). These could be distinguished by susceptibility data from the actual situation: Pd^{4+} (0 μ_B) + Pd^{2+} (2 μ_B).

Paramagnetic impurities or chemical inhomogeneities can be detected in a diamagnetic or Pauli paramagnetic host from low-temperature magnetic measurements. The ordered compound FeAl, for example, is a Pauli paramagnetic metal. However, every disordered iron atom has a localized atomic moment, as in α–Fe, and the character of the low-temperature susceptibility curve indicates how much disorder is present. In some compounds, such as sintered $LaCoO_3$, the chemical inhomogeneities segregate into microphases that are magnetically ordered at low temperatures. The moment of a microphase cluster may be much larger than an atomic moment, and in a diamagnetic host, this gives rise to *superparamagnetism.*

Phase boundaries between a ferromagnetic phase and another phase can be accurately determined from magnetic data. Within solid-solution regions, the Curie temperature T_C varies smoothly with the composition.

7.5.2. Magnetic Order

Below the critical temperature, the ordered magnetic structure often consists of collinear atomic spins as in simple antiferromagnets, although occasionally the spins adopt helical or canted arrangements. Measurements of spontaneous magnetization, magnetic anisotropy, magnetostriction, and magnetic susceptibility provide information about the magnetic order and sometimes about the magnitude of the spontaneous atomic moments. Single-crystal measurements are usually required, in contrast with measurements of paramagnetic susceptibility, where polycrystalline samples are generally adequate.

If the atomic moment is due to localized electrons, it is possible to obtain the magnitude of the spontaneous atomic moment from susceptibility data above the magnetic ordering temperature. Spontaneous moments caused by collective electrons in narrow bands cannot be interpreted from a conventional Curie–Weiss law, but if the sample is a collinear ferromagnet at low temperatures, the magnitude of the average atomic moment can be measured directly.

Measurement of the direction of easy magnetization requires a single crystal. A ferromagnet or collinear spin ferrimagnet usually has several equivalent directions of easy magnetization in which the magnetic moments are stabilized in the absence of an applied field. In spherical samples, where shape is unimportant, the magnetization directions are determined by crystalline anisotropy. For a multicomponent system, the anisotropy may be caused by local ordering, which can often be modified by annealing just below the Curie temperature in the presence of a magnetic field. This technique has important practical applications since heat treatment may cause a polycrystalline material to act magnetically like a single crystal.

Except for weak parasitic effects, an antiferromagnetic compound carries no net moment. Nevertheless, information about the direction of the atomic moments can be inferred from low-temperature susceptibility data. For collinear antiferromagnets, the susceptibility is nearly temperature independent when the applied field is perpendicular to the direction of the atomic moments. The component of χ parallel to the easy axis falls to zero at $T = 0°K$.

Frequently, the two sublattice moments of an antiferromagnet are canted from a collinear axis to give a small ferromagnet component. Cant angles of about 1° may be induced by the local symmetry if the atomic moments are due to localized electrons, as in the orthorhombic perovskite $LaCrO_3$. They may also occur as a result of anisotropic exchange. In either case, *parasitic ferromagnetism* is only compatible with a restricted set of symmetry conditions. For example, α-Fe_2O_3 has the corrundum structure and is antiferromagnetic. Just below the Néel temperature, the spins lie in the basal plane, and the symmetry conditions permit spin canting, producing parasitic ferromagnetism. Below a spin-flip temperature, the spins are collinear with an orientation which is not compatible with spin canting. Therefore, the spin-flip transition is accompanied by the disappearance of parasitic ferromagnetism.

Magnetization measurements have been used to determine the ionic distribution in magnetic materials. The atomic moments of various ions with unpaired d or f electrons are known experimentally and theoretically. The ionic distribution in a collinear ferrimagnetic can therefore be inferred from the saturation magnetization at low temperatures. For example, the ferrospinels have tetrahedral-site moments aligned antiparallel to octahedral-site moments. Simple ferrospinels have the chemical formula $M_x^{2+}Fe_{1-x}^{3+}[M_{1-x}^{2+}Fe_{1+x}^{3+}]O_4$, where the octahedral ions are enclosed by brackets and are twice as numerous as the tetrahedral ions. The net magnetic moment is a function of the ordering parameter x. In chromium spinels, the spins are not collinear, so that the measurement is not applicable. Measurement of the susceptibility in large pulsed fields has been used to test whether the spin system is collinear or canted.

Critical temperatures separating two ordered magnetic phases are readily detected. Antiferromagnetic–ferromagnetic or antiferromagnetic–ferrimagnetic phase transitions, as in MnP and $Mn_{2-x}Cr_xSb$, are usually accompanied by significant structural changes as a result of magnetostrictive effects.

Garnets and certain other ferrimagnets exhibit a compensation temperature where the net magnetization vanishes. The spontaneous magnetization of a collinear ferrimagnet is $M_s = M_{As} - M_{Bs}$, where M_{As} and M_{Bs} are the spontaneous magnetizations of the two antiparallel sublattices. Since the temperature dependences of M_{As} and M_{Bs} may be different, it is possible to have a compensation temperature T_{cm} at which $M_s = 0$. The M_s changes sign on passing through T_{cm}, making it possible to demonstrate the effect directly, by suspending a sample on a fiber in a magnetic field and varying the temperature. As the temperature passes through T_{cm}, the sample rotates 180°.

Hysteresis loops sometimes reveal the simultaneous presence of antiferromagnetic and ferromagnetic phases, especially if the structures are similar and epitaxial overgrowth occurs. An applied magnetic field will reverse the magnetization in the ferromagnetic phase, which has a net moment, without reversing the moments in the antiferromagnetic phase. Exchange coupling between the two phases makes it easier to return the ferromagnetic phase to its original position, offsetting the B–H hysteresis loop along the H axis. Unidirectional anisotropy was first observed in Co particles covered by a surface layer of cobalt oxide. Since then, the phenomenon has proven useful in interpreting the magnetization of many partially disordered alloys, such as disordered Ni_3Mn.

An indirect way to characterize materials is to investigate the B–H hysteresis loop. Coercivity and remanence are both extremely structure sensitive. In polycrystalline materials, the grain boundary is a surface of discontinuity in the magnetization vector unless it is possible to align the directions of easy magnetization in neighboring grains. Alignment can be accomplished in three ways: (1) rolling metallic sheets to align a crystallographic easy axis along the rolling direction, (2) magnetic annealing, and (3) subjecting magnetostrictive materials to tensile stress.

A second phase also generates a surface discontinuity in M_s, making magnetic processes sensitive to the existence of a foreign phase. Chemical inhomogeneities with a similar crystal structure but a slightly different magnitude for M_s also offer a surface of discontinuity in M_s. Undetectable by X rays, they may be observed as intragranular nucleation centers for domains of reverse magnetization. It is thought that these may be a necessary imperfection for creation of square hysteresis loops in the polycrystalline ceramics used as memory cores in digital computers.

7.5.3. Magnetic Domains

There are four methods of directly observing magnetic domains: Bitter patterns, the Kerr effect, Faraday rotation, and electron microscopy.

Bitter patterns are observed by placing a magnetic colloid containing tiny Fe_3O_4 particles on the surface of the sample. The surface should be polished and etched to remove surface strains. If the surface contains an easy direction of magnetization, magnetic flux emerges only at domain boundaries and the magnetic colloid is attracted to the emergent flux. In bulk material, the magnetization within a boundary rotates within a plane forming equal angles with the magnetization direction in the domains on either side. This is called a Bloch wall. In thin films, the magnetization direction in a boundary rotates within the plane of the film and is called a Néel wall. Films with thickness comparable to a boundary thickness have complex rotations within the boundary, and the Bitter patterns show short cross-walls spaced periodically along the principal boundary. Cross-tie walls fade away as they penetrate the domains.

The reflection of light from the surface of a magnetic material depends on the permeability, a magnetooptic phenomenon called the Kerr effect. Since the rotation of reflected plane-polarized light depends upon the angle between the magnetization and the **H** vector of the incident light, it is possible to distinguish different domains by viewing the crystal through an analyzer. This technique can be used to observe dynamic processes, but intense light sources are required. Laser sources could profitably be developed for this type of study.

Another way of making domains visible is to analyze the transmitted light, since the plane of polarization is rotated differently in oppositely directed domains. Faraday rotation provides a three-dimensional view of domains in the bulk but can only be used for transparent magnetic materials.

7.6. Calorimetric Measurements[115]

One of the most important characterizations of a solid is the determination of the existence of any phase transitions which may occur in the temperature or pressure range of experimentation. Calorimetric measurements provide a rapid and accurate determination of the existence of a phase transition and its order, as well as the temperature, or temperature range, over which it occurs. These measurements may be dynamic as in differential thermal analysis (DTA), or adiabatic, as in specific heat measurements.

Heating and cooling curves are used in determining phase diagrams, since all (first-order) phase changes give rise to thermal arrests (the temperature remains constant for a period of time). Differential thermal analysis (DTA) is a refinement of the time–temperature method. In a DTA experiment, the sample is heated simultaneously with a standard sample, usually an inert material such as α-Al_2O_3. Thermocouples are inserted in both specimens to measure the sample temperature and the temperature difference between the two specimens. Differential temperature is graphed as a function of temperature at heating rates of about 10°C/min. The curve remains fairly flat as long as no phase change occurs, but the temperature arrest associated with a phase transition in the sample produces a peak in the DTA thermogram. Polymorphic transitions, magnetic transitions, decomposition reactions, and melting points can be located by DTA. It is used routinely in the determination of phase diagrams and in the identification and purity measurement of organic compounds and pharmaceuticals. Several types of instruments are commercially available, and some have a sensitivity of better than 10^{-3} cal/mole-deg. Controlled-atmosphere and cryogenic instrumentation is also available.

Quantitative DTA systems can make accurate, reproducible calorimetric measurements that involve energy changes as small as 1 mcal and temperature changes of less than 0.001°C. The area under the peak is proportional to ΔH_{tr}, the energy absorbed or released during the transition. These data can be used to *estimate* specific heat, heats of fusion or vaporization, and transition energies, and to calculate the kinetic factors controlling reaction rates.

Thermogravimetric analysis (TGA) is another dynamic method in which weight changes are recorded during heating or cooling. TGA is useful in studying oxidation and reduction reactions, volatility effects, and hydration phenomena, where the chemistry of a material may be changed within the excursion of the experiment designed.

The heat capacity of a body is the amount of heat required to produce a temperature rise of 1°C. Specific heat is the heat capacity per unit mass, and generally decreases with lowering temperature, falling to zero at 0°K. Anomalies in the specific heat curve can be related to the molecular origins of phase transitions. Beta brass, for instance, exhibits a lambda point at 470°C due to an order-disorder transition. The specific heat anomaly is referred to as a lambda point because of the resemblance to the Greek letter λ.

Most methods for measuring specific heat are based directly on its definition. The heat content is changed by a known amount and the corresponding temperature rise is measured. Heat is usually added electrically by resistance heating, allowing the sample to reach thermal equilibrium at each step. Allowance must be made for the exchange of heat with the surroundings. In general, the accurate measurement of specific heat is a research problem requiring specialized techniques and considerable experience. Calorimeter designs depend on the thermal characteristics of the sample and the temperature range.

Acknowledgment

This paper owes a great deal to a Materials Advisory Board Report on the Characterization of Materials (#229-M, March 1967), and specifically to the report of the Panel on Structure, chaired by one of us (R.R.). Panel members Dr. J. B. Goodenough, M. D. Coutts, W. Parrish, and C. G. Shull have therefore contributed materially to this effort.

References

1. F. C. Phillips, *An Introduction to Crystallography*, Longmans, Green and Co., London (1946).
2. P. Groth, *Chemische Krystallographie*, Wilhelm Engelmann, Leipzig, Vol. I (1906), Vol. II (1908), Vol. III (1910), Vol. IV (1917), Vol. V (1919).
3. N. H. Hartshorne and A. Stuart, *Crystals and the Polarizing Microscope*, Edward Arnold, London (1960).
4. E. E. Wahlstrom, *Optical Crystallography*, Wiley, New York (1951).
5. A. N. Winchell, *Elements of Optical Mineralogy, Part II. Descriptions of Minerals*, Wiley, New York (1947).
6. E. S. Larsen and H. Berman, Microscopic Determination of the Nonopaque Minerals, U.S.G.S. Bulletin 848 (1934).
7. S. K. Kurtz and T. T. Perry, A powder technique for the evaluation of nonlinear optical materials, *J. Appl. Phys.* **39**, 3798–3813 (1968).
8. V. Vand, K. Vedam, and R. Stein, The laser as a light source for ultramicroscopy and light scattering by imperfections in crystals, *J. Appl. Phys.* **37**, 2551–2557 (1966).
9. R. A. Heising, *Quartz Crystals for Electrical Circuits*, D. Van Nostrand, New York (1946).
10. M. F. Ehman, Surface Reactivity and Basic Etching Mechanisms, Ph.D. Thesis, The Pennsylvania State University (1970).
11. B. Obst, Anisotropie in kubischen Supraleitern, *Phys. Stat. Sol.* **45**, 453–482 (1971).
12. S. Tolansky, *Multiple-Beam Interference Microscopy*, Academic, New York (1970).
13. O. S. Heavens, *Optical Properties of Thin Solid Films*, Academic, New York (1955).
14. K. Vedam and S. S. So, Characterization of real surfaces by ellipsometry, *Surface Sci.* **29**, 379–395 (1972).
15. R. D. Young, *Surface Microphotography*, *Physics Today* **24**, 42–49 (1971).
16. G. A. Ratz, The mike for automated quantitative microscopy, *Metals Progr.* **1968** (August), 153–156.

17. E. W. White, K. Mayberry, and G. G. Johnson, Jr., Computer analysis of multichannel SEM and X-ray images from fine particles, *Pattern Recognition* **4**, 173–193 (1972).
18. Joint Committee on Powder Diffraction Standards, *Index to the Powder Diffraction File*, Swarthmore, Pa. (1972).
19. W. Parrish, *X-Ray Analysis Papers*, Centrex, Eindhoven (1965).
20. W. L. Bond, Precision lattice constant determination, *Acta Cryst.* **13**, 814–818 (1960).
21. J. D. Bernal, Order and disorder and their expression in diffraction, *Z. Krist.* **112**, 4–21 (1959).
22. B. E. Warren and B. L. Averbach, The effect of cold-work distortion on X-ray patterns, *J. Appl. Phys.* **21**, 595–599 (1950).
23. P. B. Hirsch, Mosaic structure, *Progr. Metal Phys.* **6**, 236–339 (1956).
24. B. E. Warren, X-ray studies of deformed metals, *Prog. Metal Phys.* **8**, 147–202 (1959).
25. B. E. Warren and B. L. Averbach, in *Imperfections in Nearly Perfect Crystals* (W. Shockley, J. N. Holloman, R. Maurer, F. Seitz, eds.), pp. 152–172, Wiley, New York (1952).
26. A. Guinier and G. Fournet, *Small Angle Scattering of X-rays*, Wiley, New York (1955).
27. *International Tables for X-ray Crystallography*, Vol. I, *Symmetry Groups*, Kynoch Press, Birmingham, England (1952).
28. M. J. Buerger, *Crystal-Structure Analysis*, Wiley, New York (1960).
29. H. Lipson and W. Cochran, *The Determination of Crystal Structures*, Bell, London (1953).
30. L. Pauling and M. D. Shappell, The crystal structure of bixbyite and the modification of the sesquioxides, *Z. Krist.* **75**, 128–142 (1930).
31. M. J. Buerger, *Vector Space and Its Application in Crystal-Structure Investigation*, Wiley, New York (1959).
32. M. M. Woolfson, *Direct Methods in Crystallography*, Oxford University Press, London (1961).
33. *Structure Reports*, Oosthoek, Utrecht (annual volumes).
34. R. G. Wyckoff, *Crystal Structures*, 2nd ed., Interscience, New York (1965).
35. S. C. Abrahams, L. E. Alexander, T. C. Furnas, W. C. Hamilton, J. Ladell, Y. Okaya, R. A. Young, and A. Zalkin, American Crystallographic Association Single-Crystal Intensity Project Report, *Acta Cryst.* **22**, 1–6 (1967).
36. J. Ladell, Refinement of the topaz structure, *Norelco Reporter* **12**, 34–39 (1965).
37. N. A. Alston and J. West, The Structure of Topaz, $[Al(F, OH)]_2SiO_4$, *Z. Krist.* **69**, 149–167 (1928).
38. J. Smit and H. P. J. Wijn, *Ferrites*, pp. 140–146, Wiley, New York (1959).
39. R. Srinivasan, Application of X-ray anomalous scattering in structural studies, *Advances in Structure Research by Diffraction Methods* (W. Hoppe and R. Mason, eds.), Vol. 4, pp. 105–197, Pergamon, Oxford (1972).
40. R. Brill, Determination of electron distribution in crystals by means of X-rays, *Solid State Phys.* **20**, 1–35 (1967).
41. H. Witte and E. Wölfel, Röntgenographische Bestimmung der Elektronenverteilung in Kristallen II, *Z. Physik Chem.* [*N.F.*] **3**, 296–329 (1955).

42. S. Gottlicher and E. Wölfel, Röntgenographische Bestimmung der Elektronemverteilung in Kristallen, *Z. Elektrochem.* **63**, 891–901 (1959).
43. R. T. Sanderson, The nature of ionic solids, *J. Chem. Ed.* **44**, 516–523 (1967).
44. Richard Weiss, *X-ray Determination of Electron Distribution*, Wiley, New York (1966).
45. I. Freund, Nonlinear x-ray diffraction. Determination of valence electron charge distributions, *Chem. Phys. Letters* **12**, 583–588 (1972).
46. E. A. Wood, *Crystal Orientation Manual*, Columbia University Press, New York (1963).
47. H. J. Goldschmidt, *High-Temperature X-ray Diffraction Techniques*, International Union of Crystallography (1964).
48. B. Post, *Low-Temperature X-ray Diffraction*, International Union of Crystallography (1964).
49. *International Tables for X-ray Crystallography*, Vol. III, Kynoch Press, Birmingham (1959).
50. D. W. J. Cruickshank, The determination of the anisotropic thermal motion of atoms in crystals, *Acta Cryst.* **9**, 747–756 (1956).
51. K. Lonsdale, Experimental studies of atomic vibrations in crystals and of their relationship to thermal expansion, *Z. Krist.* **112**, 188–212 (1959).
52. C. B. Walker, X-ray study of lattice vibrations in aluminum, *Phys. Rev.* **103**, 547–557 (1956).
53. B. E. Warren, *X-Ray Diffraction*, Addison-Wesley, Reading, Mass. (1968).
54. W. Klement and A. Jayaraman, Phase relations and structures of solids at high pressures, in *Progress in Solid State Chemistry* (A. Reiss, ed.), Vol. 3, Chapter 7, pp. 289–376, Pergamon Press, New York (1967).
55. R. M. Brugger, R. B. Bennion, T. G. Worlton, and W. R. Myers, Neutron diffraction at high pressures, *Trans. Amer. Cryst. Assoc.* **5**, 141–154 (1969).
56. P. J. Freud and P. N. LaMori, Non-dispersive high pressure—high temperature X-ray diffraction analysis, *Trans. Am. Cryst. Assoc.* **5**, 155–162 (1969).
57. W. W. Webb, X-ray diffraction topography, in *Direct Observations of Dislocations* (J. B. Newkirk and J. H. Wernick, eds.), Interscience, New York (1962).
58. M. Hart and V. Bonse, Interferometry with X-rays, *Phys. Today* **23**(8), 26–31 (1970).
59. R. D. Heidenreich, *Fundamentals of Electron Transmission Microscopy*, Interscience, New York (1964).
60. G. Thomas, *Transmission Electron Microscopy of Metals*, Wiley, New York (1962).
61. S. Amelinckx, *Direct Observations of Dislocations*, Academic, New York (1964).
62. R. B. Nicolson, P. B. Hirsch, A. Howie, D. W. Pashley, and M. J. Whelan, *Electron Microscopy of Thin Crystals*, Butterworth, London (1965).
63. H. Fernández-Morán, M. Ohstuki, A. Hibino, and C. Hough, Electron microscopy and diffraction of layered superconducting intercalation complexes, *Science* **174**, 498–500 (1971).
64. R. Nathan, Image processing for electron microscopy: I. Enhancement procedures, in *Optical and Electron Microscopy*, Vol. 4, pp. 85–125 (R. Barer and V. E. Cosslett, eds.) (1971).
65. G. W. Stroke, Spectroscopic implications of new holographic imaging methods, *Physica* **33**(1), 253–267 (1967).

66. Z. Pinsker, *Electron Diffraction*, Butterworth, London (1953).
67. J. M. Cowley, Crystal structure determination by electron diffraction, *Prog. Mat. Sci.* **13**, 267–321 (1968).
68. J. J. Lander, Low energy electron diffraction and surface structural chemistry, in *Progress in Solid State Chemistry* (H. Reiss, ed.), Vol. 2, pp. 26–116, Pergamon Press, Oxford (1965).
69. J. L. Beeby, The density of electrons in a perfect or imperfect lattice, *Proc. Roy. Soc. A* **302**(1468), 113–136 (1967).
70. E. G. McRae, Multiple-scattering treatment of low energy electron diffraction intensities, *J. Chem. Phys.* **45**, 3258–3276 (1966).
71. C. W. Oatley, W. C. Nixon, and R. F. W. Pease, Scanning electron microscopy, in *Advances in Electronics and Electron Physics*, Vol. XXI (L. Marton, ed.), Academic, New York (1956).
72. R. E. McMillan, G. G. Johnson, Jr., and E. W. White, Computer processing of binary maps of SEM images, in *Proc. Scanning Electron Microscopy Symp.* (O. Johari, ed.), pp. 439–444, IITRI, Chicago, Ill. (1969).
73. J. A. Becker, Study of surfaces by using new tools, *Solid State Phys.* **7**, 379–424 (1958).
74. E. W. Müller, Atom-probe field-ion microscopy, *J. Vac. Sci. Tech.* **8**, 1–89 (1971).
75. G. E. Bacon, *Neutron Diffraction*, Oxford Press, London (1962).
76. G. M. Brown and H. A. Levy, Sucrose: Precise determination of crystal and molecular structure by neutron diffraction, *Science* **141**, 921–923 (1963).
77. C. G. Shull and Y. Yamada, Magnetic electron configuration in iron, *J. Phys. Soc. Japan* **17**(Suppl. B-III), 1–6 (1962).
78. R. P. Santoro, R. E. Newnham, and S. Nomura, Magnetic properties of Mn_2SiO_4 and Fe_2SiO_4, *J. Phys. Chem. Solids* **27**, 655–666 (1966).
79. B. N. Brockhouse, *Phonons and Phonon Interactions*, Benjamin, New York (1964).
80. P. A. Egelstaff, *Thermal Inelastic Scattering*, Academic, New York (1965).
81. J. D. Axe, Neutron studies of displacive structure phase transformations, *Trans. Amer. Cryst. Assoc.* **7**, 89–106 (1971).
82. H. Boutin, G. J. Safford, and H. R. Danner, Low-frequency motions of H_2O molecules in crystals, I, *J. Chem. Phys.* **40**, 2670–2679 (1964).
83. H. J. Prask and H. Boutin, Low-frequency motions of H_2O molecules in crystals, II, *J. Chem. Phys.* **45**, 699–705 (1966).
84. K. Jorgenson, Absorption spectra and chemical bonding in complexes, Pergamon, New York (1962).
85. L. E. Orgel, Ion compression and the colour of ruby, *Nature* **179**, 1348 (1957).
86. D. L. Greenaway and G. Harbeke, *Optical Properties and Band Structure of Semiconductors*, Pergamon Press, London (1968).
87. J. J. Markham, *F*-Centers in alkali halides, Academic, New York, (1966).
88. N. L. Alpert, W. E. Keiser, and H. A. Szymanski, *IR Theory and Practice of Infrared Spectroscopy*, Plenum, New York (1970).
89. W. B. White and B. A. DeAngelis, Interpretation of the vibrational spectra of spinels, *Spectrochim. Acta* **23A**, 985–995 (1967).
90. B. A. DeAngelis, R. E. Newnham and W. B. White, Factor group analysis of the vibrational spectra of metals: A review and consolidation, *Am. Min.* **57**, 255–268 (1972).

91. R. E. Newnham, Crystal structure and optical properties of pollucite, *Am. Min.* **52**, 1515–1518 (1967).
92. B. A. Scott and G. Burns, Determination of stoichiometric variations in $LiNbO_3$ and $LiTaO_3$ by Raman powder spectroscopy, *J. Am. Ceram. Soc.* **55**, 225–230 (1972).
93. S. P. S. Porto, Light scattering with laser sources, in *Applied Solid State Physics* (W. Low and M. Schieber, eds.), Chapter 3, Plenum Press, New York (1970).
94. E. W. White and G. V. Gibbs, Structural and chemical effects on the Al $K\beta$ X-ray emission band among aluminum containing silicates and aluminum oxides, *Am. Mineral.* **54**, 931–936 (1969).
95. H. Friedman, X-ray spectroscopy, in *Advances in Spectroscopy* (A. W. Thompson, ed.), Vol. 2, pp. 57–100, Interscience, New York (1961).
96. C. P. Slichter, *Principles of Magnetic Resonance*, Harper & Row, New York (1963).
97. B. Lax and K. Button, *Microwave Ferrites and Ferrimagnetics*, McGraw-Hill, New York (1962).
98. G. E. Pake, Nuclear magnetic resonance, *Solid State Phys.* **2**, 1–92 (1956).
99. T. P. Das and E. L. Hahn, *Nuclear Quadrupole Resonance Spectroscopy*, Academic, New York (1968).
100. M. H. Cohen and F. Reif, Nuclear quadrupole effects in nuclear magnetic resonance, *Solid State Phys.* **5**, 321–438 (1957).
101. P. J. Bray and A. H. Silver, in *Modern Aspects of the Vitreous State* (J. D. MacKenzie, ed.), pp. 92–118, Butterworth, London (1960).
102. P. C. Taylor and P. J. Bray, Structural properties of glasses inferred from computer simulations of magnetic resonance spectra, *Bull. Am. Ceram. Soc.* **51**, 234–239 (1972).
103. J. M. Baker, E. R. Davies, and T. R. Reddy, Detailed mapping of atomic positions using electron nuclear double resonance (ENDOR), *Contemp. Phys.* **13**, 45–59 (1972).
104. G. K. Wertheim, *Mössbauer Effect: Principles and Applications*, Academic Press, New York (1964).
105. V. I. Goldanskii and R. H. Herber, *Chemical Applications of Mössbauer Spectroscopy*, Academic Press, New York (1968).
106. K. Siegbahn, *Atomic Molecular and Solid State Structure Studies by Means of Electron Spectroscopy*, Almquist and Wilksells, Uppsala (1967).
107. D. M. Hercules, Electron spectroscopy, *Anal. Chem.* **42**(1), 20A–40A (1970).
108. L. A. Harris, Auger electron emission analysis, *Anal. Chem.* **40**(12), 24A–34A (1968).
109. R. Truell, C. Elbaum, and B. B. Chick, *Ultrasonic Methods in Solid State Physics*, Academic, New York (1969).
110. J. F. Nye, *Physical Properties of Crystals*, Clarendon Press, London (1957).
111. A. R. von Hippel, *Dielectrics and Waves*, Wiley, New York (1954).
112. T. C. Harmon and J. M. Honig, *Thermoelectric and Thermomagnetic Effects and Applications*, McGraw-Hill, New York (1967).
113. A. H. Morrish, *Physical Principles of Magnetism*, Wiley, New York (1965).
114. J. B. Goodenough, *Magnetism and the Chemical Bond*, Interscience, New York (1963).
115. P. D. Garn, *Thermoanalytical Methods of Investigation*, Academic, New York (1965).

Index

Acoustic spectroscopy, 514–515
Activation analysis, 398–399, 415
Activity coefficients, 367–369
Alkali halides, 2–5, 19–21, 32, 35
  band structure, 85–86
  vacancy pairs, 370
Alkali metals, cohesive energies, 12–13
Alloys, 136–137, 150–157
  bonding in, 17, 28–30
  defects in, 285
  density of states, 96–100
  magnetic, 187–199, 218
  mechanical properties, 262
  phase stability, 115, 119–122, 128–129, 134–137, 149–161
  superconducting, 219–240
Amorphous Si, Ge, 101–105
Amorphous solids, magnetism in, 217–218
Analytical techniques, 387–426
  activation analysis, 398–399, 415
  chromatography, 396–398, 414–415
  coulometry, 392
  electron probe microanalysis, 406
  electron spin resonance, 405–406, 415
  gas analysis, 399–400
  gravimetry, 391–392
  ion-probe microanalysis, 407
  ion-selective electrodes, 392–393
  mass spectrometry, 401–403, 414–415
  microanalysis, 395–396
  Mössbauer spectroscopy, 404–405
  nuclear magnetic resonance, 405, 415
Analytical techniques *(cont'd)*
  optical spectroscopy, 403–404
  photoelectron spectroscopy, 407, 408
  polarography, 393
  resistance ratio, 406
  spectrofluorimetry, 394–395
  spectrophotometry, 394
  thermal analysis, 402
  titrimetry, 391–392
  x-ray fluorescence, 400–401, 414–415
Anderson problem, 96, 100–101
Antiferroelectric, 241–242
Antiferromagnetic interactions, 165
Antiferromagnetic resonance, 505–506
Antiferromagnetism, 176–218
Antistructure defects, 342, 344
Associated defects (*see* Defects, Defect equilibria)
Atomic orbitals, 9–10
Augmented plane wave (APW), 64–66

Band structures, 81–94
Bands (*see* Energy bands)
Bardeen–Cooper–Schrieffer (BCS) theory, 219–225
Block theorem, 50
Bonding, 1–40
  alloys, 17, 28–30
  atomic orbitals, 9–10
  compounds, 17–28
  intermetallic solutions, 28–30

Bonding *(cont'd)*
  ionic crystals, 2–5, 7, 17–22, 34–37, 124–127
  metals, 6, 11–17, 119–122, 128–129
  molecular crystals, 7
  semiconductors, 7, 22–28
  types of solids, 6–9
  valence compounds, 123–124
Boundaries
  antiphase, 319–320
  grain, 310–317
  stacking fault, 301, 317–318
  twin, 318
Bragg–Williams approximation, 367–368
Bravais lattice, 49, 50, 53
Brillouin zone, 49–50, 58–59, 79, 149–161
Brouwer's method, 356–360
Burgers vector, 293

Calorimetric measurements, 527–529
Cellular method, 63
Characterization, 387–426, 437–529
  absorption spectroscopy, 497–501
  acoustic spectroscopy, 514–515
  calorimetric methods, 527–529
  chemical composition, 387–426
  crystal structure analysis, 458–463, 482–483, 490–491
  defects, 350–351, 354–355
  electrical, 518–521
  electron diffraction, 480–484
  electron methods, 476–487
  electron microscopy, 476–480
  electron spectroscopy, 513–514
  electron spin resonance, 503–505, 509–510
  ferroelectric materials, 479–480
  ferromagnetic materials, 479–480, 491–493, 524–527
  homogeneity, 413
  impurities, 413–415
  infrared absorption, 499–501
  lattice dynamics, 493
  magnetic measurements, 521–527
  magnetic resonance, 503–510
  minor phases, 413–415
  morphology, 439–441
  Mössbauer effect, 510–512
  neutron scattering, 487–496
  nuclear magnetic resonance, 506–510
  optical methods, 439–451
  optical scattering, 444
  oxidation state, 413
Characterization *(cont'd)*
  particle size and shape, 448–451, 478–479
  phase identification, 452–453, 480–482
  Raman spectra, 501
  scanning electron microscopy, 484–487
  soft x-ray spectra, 501–502
  spectroscopic methods, 496–515
  stoichiometry, 408
  structural, 437–529
  surfaces, 415–416, 444–448, 484–487
  x-ray absorption edge spectroscopy, 502–503
  x-ray diffraction, 451–476
Chemical analysis, 387–426 (*see also* Analytical techniques)
Chemical bonds (*see* Bonding)
Chemical potential, 362
Chromatography, 396–398, 414–415
Clausius–Mosotti model, 21, 31–33
Clustering, measurement of, 472–473
Clusters, 323–329, 370
Coherent potential approximation (CPA), 98–100
Cohesive energy, 11–17, 37, 266
Complex defects, 323–329, 370
Compound semiconductors
  energy bands, 88–90
  impurities in, 374–376
  nonlinear optical properties, 253–257
Compounds
  bonding in, 17–28, 34–37
  defects in, 283–329, 341–383
  intermetallic, 140–146, 149–156, 187–199, 219–239, 285, 311–312, 319–320
Compressibility, 260–261, 265–267
Configurational entropy, 335–337
Coordination, 2–5, 17–30, 34–37, 129–133
Coulometry, 392
Covalency, 7–8, 13, 22–28, 34–38, 127–129 (*see also* Bonding)
Covalent energies, 34–36
Crystal field, 178–179
Crystal structure
  bonding as a basis for, 11–28, 119–172
  determination of, 458–463, 482–483, 490–491
  diagrams, 171–172
  stability of, 11–28, 119–172
  types, 129–133
Cu_3Au, 150–154, 189, 285, 319–320
Curie temperature, 246–247

Darken–Gurry theory, 29–39
Debye temperature, 221
Defect equilibria, 335–383
  anti-Frenkel, 348–349
  association, 376–378
  binary compounds, 341–383
  Brouwer's method, 356–360
  compounds, 374–376
  elemental crystals, 339–341, 372–374
  interstitials, 337
  intrinsic defects, 342–345
  ionization of defects, 351–353
  nonstoichiometry, 345–349, 358–360
  pressure effects, 345–351, 353–355
  vacancies, 337
  with an external phase, 345–348
Defects, 283–329, 335–383
  anti-Frenkel, 342, 344, 348–349
  antiphase domains, 319–320
  antistructure, 342, 344
  association of, 369–371
  characterization of, 350–351, 354–355, 442–443, 445–446, 455–458, 472, 474–476
  clusters, 323–329, 370
  complexes, 323–329, 370
  concentration of, 337–383
  dislocations, 286, 291–310
  divacancy, 323
  electrical conductivity studies, 354–355
  electronic, 338–339
  elemental crystals, 372–374
  energetics of formation, 339, 363–365
  equilibria (*see* Defect equilibria)
  extended, 380–383
  Frenkel, 284, 288–290, 337–338
  impurities, 286, 371–378
  interactions of, 367–371
  interstitials, 287–291, 337
  intrinsic, 342–345
  ionization of, 351–353, 365–367
  line, 286, 291–310
  native, 335–341
  nonstoichiometry, 345–349, 358–360
  planar, 310–322
  point, 284–291, 337–338
  precipitates, 311–312
  Schottky, 287–288, 342, 344–345, 356–360
  Schottky–Wagner, 289–290
  stacking faults, 301, 317–318, 322
  vacancies, 287–291, 323–328, 337
  volumetric, 323–329
Deformation, 267–271, 291–310
  magnetic effects, 194–199
deHaas–van Alphen effect, 81–83
Dehybridization, 9, 37, 127
Density of states, 54–56, 59–60, 219–240
  alloys, 96–100
Dielectric constant, 34–35, 77
Dielectric properties, 518–519
Dielectrics, 240–259
Dislocations, 286, 291–310
  annihilation, 308
  Burgers vector, 293
  climb, 303
  edge, 292
  energy of, 297
  Frank–Read source, 305
  interaction with impurities, 309
  interaction with vacancies, 303, 308
  motion of, 296
  partial, 300
  screw, 293
  slip, 291
  stress field, 297
Disordered solids, 94–106
Dispersion, 32–34
Divacancies, 323, 370
Double resonance (ENDOR), 509–510

Effective mass, 50–51, 69
Elasticity, 259–263
Electron density, determination of, 463–466
Electron diffraction, 480–484
Electron microscopy, 476–480
Electron probe microanalysis, 406
Electron spectroscopy, 513–514
Electron spin resonance, 405–406, 415, 503–505, 509–510
Electronegativity, 26–28, 35, 129, 256–257
Electrons, 338–339
Electrooptic effect, 257–258
Electrostatic valence rule, 126
ENDOR, 509–510
Energies
  cohesive, 11–17
  covalent, 34–36
  ionic, 34–36
  ionic crystals, 124–127
Energy bands, 10–11, 43–106, 146–161
  calculation of, 52–73
  effect on structure, 146–149
  independent-electron approximation, 45–48

Energy gap, 35
Exchange, 176–218
Extended defects, 380–383

Fermi energy, 51
Fermi level, 51
Fermi surface, 51, 73, 81–83, 146–161
Ferrimagnetism, 199–205
Ferroelectricity, 256, 257–258
Ferroelectrics, 240–249, 252, 257–258
Ferromagnetic resonance, 505–506
Ferromagnetism, 176–218
Force constants, 38
Frank–Read source, 305
Free energy, 335–337, 361
Frenkel defects, 284, 288–290, 337–338, 340

Garnets, 199–200, 204–205, 512
Gas analysis, 399–400
Goldschmidt radii, 18
Grain boundaries, 310–317
  tilt, 315
  twist, 315
Gravimetry, 391–392
Group IV semiconductors
  amorphous state, 101–105
  band structure, 86–88
  cohesive energies, 12–14
  impurities in, 372–374, 377–378
Guinier–Preston zones, 329

Hardness, 263, 267
Hartree approximation, 45–46
Hartree–Fock method, 46
Heat of atomization, 11
Holes, 338–339
Homogeneity, 413
Hume-Rothery phases, 156–157
Hume-Rothery rules, 28, 136, 137
Hund's rule, 16, 183
Hybridization, 8, 34

Imperfections, 283–329 (*see also* Defects)
Impurities, 371–378
  characterization, 413–415
Index of refraction, 34, 77
Interatomic force constants, 38
Intermetallic compounds, 140–146, 149–156, 187–199, 212–239, 285, 311–312, 319–320
Intermetallic solutions (*see* Alloys)
Interstitial phases, 263, 287–291, 337
Intrinsic defects, 342–345
Ionic crystals
  bonding in, 2–5, 7, 17–22, 34–37, 124–127
  defects in, 284–291, 322–329
  magnetic interactions, 176–179, 199–205, 206–209
  mechanical properties, 265–267, 270
Ionic energies, 34–36
Ionic radii, 18–22
Ionicity, 26, 36–38, 131, 255–257
Ionization of defects, 351–353, 365–367
Ion-probe microanalysis, 407
Ion-selective electrodes, 392–393

Jahn–Teller effect, 164, 165, 168–169, 202–203

KKR methods, 64–66, 72–73
k·**p** method, 70–72
Kramers–Kronig relation, 77

Lanthanides (*see* Rare earths)
Lattice dynamics, measurement of, 493
Lattice instability, 241
Lattice parameter measurement, 453–454
Lattice vibration, 220–240
Laves structures, 118, 142–144, 154
Law of mass action, 337–383
Line defects (*see* Distortions)
Local moments (*see* Magnetic moments)

Madelung constant, 18, 124, 126, 265
Magnetic alloys, 187–199, 218
Magnetic anisotropy, 194–199
Magnetic domains, observation of, 526–527
Magnetic interactions, 170, 176–218
Magnetic moments, 16–17, 176–202
  and superconductivity, 237–238
Magnetic oxides, 199–205, 206–209, 322–329, 353–355, 374–380, 512, 523, 525, 526
Magnetic phases, 320, 327–329
Magnetic resonance
  antiferromagnetic, 505–506
  double resonance, 509–510
  electron spin, 405–406, 415, 503–505, 509–510
  ENDOR, 509–510
  ferromagnetic, 505–506
  nuclear, 405, 415, 506–510
  nuclear quadrupole, 507–509

Magnetic semiconductors, 205–218
  chalcogenides, 209–217
  transition-metal oxides, 206–209
Magnetic structure characterization, 479–480, 491–493, 521–527
Magnetism, 176–218
  amorphous materials, 217–218
  effects due to deformation, 194–199
  linear, 214–217
  magnetic semiconductors, 205–218
  two-dimensional, 214–217
Magnetocrystalline anisotropy, 181, 193
Martensitic transformations, 170, 227
Mass spectrometry, 401–403, 414–415
Mechanical properties, 259–271
  dislocations, 286, 291–310
  elastic, 260–263
  ionic crystals, 265–267, 270
  metals, 260–263, 267–271
  nonmetals, 264–267
  plastic behavior, 267–271
  semiconductors, 264
  soft modes, 228–230
Mechanical strength, 286, 291
Melting point, 11, 38
Melting points, 262–263
Metal–insulator transitions, 206–209
Metallization, 9, 37
Metals
  band structure, 84–85
  bonding in, 6, 11–17
  mechanical properties, 260–263, 267–271
  phase stability, 119–122, 128–129
  rare earth, 182–187
  transition (3*d*), 178–182
Metamagnetism, 186
Metastable phases, 163
Minerals, 17–22
Molecular crystals, bonding in, 7
Moment methods, 67–69
Moments (*see* Magnetic moments)
Morphology, 439–441
Mössbauer effect, 510–512
Mössbauer spectroscopy, 404–405

Native defects (*see* Defects)
Nearly-free electron approximation, 52–53, 57–60, 69–70
Neutron scattering, 487–496
Nonlinear optical materials, 249–259
Nonstoichiometry, 234–236, 284–285, 289–290, 325–327, 345–349, 358–360, 378
  characterization of, 408
Nuclear magnetic resonance, 405, 415
Nuclear quadrupole resonance, 507–509

Optical properties, 34, 50, 73–81
  absorption, 77–80, 497–501
  characterization of solids, 439–451, 496–501
  nonlinear, 249–259
Optical susceptibility, 250, 253, 255–257
Ordering, measurement of, 472–473
Orthogonalized plane wave (OPW), 63–66
Oxidation state, 413

Paramagnetic impurities, in superconductors, 237–238
Partial ionic character (*see* Ionicity)
Particle size and shape, 448–451, 478–479
Pauling electronegativities, 26
Pauling radii, 18
Pauling rules, 21, 135
Peierls stress, 296
Periodic systems, 52–73
Perovskites, 242–247, 256, 523
Phase identification, 452–453, 480–482
Phase transition, 227–230, 236
  characterization of, 494
Phases, stability of, 115–172
Phonon (*see* Lattice vibration)
Photoelectron spectroscopy, 407–408
Photoemission, 80
Piezoelectrics, 245, 247–249
Planar defects, 310–322
Plasma energy, 5, 33, 34
Plasma frequency, 5, 33
Point defects, 284–291, 337–338, 361
Polarizability, 31–32
Polarization, 241–250
Polarography, 393
Precipitation, 311–312, 329
Pseudopotential theory, 44, 46, 48, 58, 60–62, 66, 69–70, 86, 157–161
Pyroelectric effect, 257–258

Radius ratio, 125, 142–143
Raman spectra, 501
Rare earth metals, 182–187
Reduced zone, 50 (*see also* Brillouin zone)
Resistance ratio, 406

Scanning electron microscopy, 484–487
Scattering, as characterization technique, 444

Schottky defects, 287–288, 342, 344–345, 356–360, 362
Schottky–Wagner defects, 289–290
Second harmonic generation, 250–252
Semiconductors (*see also* Group IV semiconductors, Compound semiconductors)
  band structures, 86–90
  bonding in, 7, 22–28, 34–37
  degenerate, 239–240
  magnetic, 205–218
  mechanical properties, 264
Semimetals, 8
Shell effects, 12–14
SiO_2, band structure, 90–92
Slip, 267–270, 291
Soft mode, 228–230
Soft x-ray absorption, 75–76, 501–502
Solid types, 6–9
Solubility, 29–30
Solutions, 28–30
s-p hybridization, 8, 22
Spectrofluorimetry, 394–395
Spectrophotometry, 394
Spinels, 199–202
Spin-orbit coupling, 169
Splat-cooling, 163, 236
Stacking faults, 301, 317–318, 322
Statistical thermodynamics, 361–367
Stoichiometry (*see* Nonstoichiometry)
Strength, mechanical, 259–260
Structure (*see* Crystal structure)
Superconductivity, 31, 219–240
  effect of electron concentration, 219–240
  effect of lattice instability, 225–236
  in degenerate semiconductors, 239–240
  paramagnetic impurities, 237–238
  stoichiometry, 234–236
  theory, 219–225
Superexchange, 202
Superstructure, 162
Surfaces, characterization, 415–416, 444–448, 484–487
Susceptibility, optical, 250, 253, 255–257

Thermal analysis, 402
Thermal expansion, measurement of, 469–470
Thomas–Fermi factor, 35
Tight-binding approximation, 52–56, 66
Titrimetry, 391–392
Transition metal compounds
  band structure, 92–94
  defects in, 353–355, 374–380
  ferroelectricity, 242–247, 257–258
  magnetic properties, 199–217, 522–527
  mechanical properties, 262–263
  metal–insulator transitions, 206–209
  phase stability, 225–240
  piezoelectricity, 247–248
  structure, 165–170
  superconductivity, 225–239
  volumetric defects in, 322–329
Transition metals
  band structure, 92
  cohesive energy, 14–17
  magnetic behavior, 176–182
  phase stability, 121–122
Transitions, metal–insulator, 206–209
Tungsten bronzes, 242–243, 256
Types of solids, 6–9

Vacancies, 287–291, 323–328, 337
Valence bonds (*see* Bonding)
Valence compounds, stability of, 123–124
Van der Waals forces, 7
Van der Waals radii, 133–136
Volume defects, 323–329

β-W phases, 220, 225–234
Wadsley phases, 320, 327–329
Wannier functions, 56–57
Wigner–Seitz cell, 49–50, 63–64
Wüstite, 284–285, 325–326

X-ray diffraction, 451–476
  disordered solids, 456–458
  electron density maps, 463–466
  high-pressure, 472–473
  high-temperature, 468
  interferometry, 474–476
  lattice parameter measurement, 453–454
  line broadening, 455–456
  low-temperature, 469
  noncrystalline solids, 456–458
  phase identification, 452–453
  powder methods, 451–458
  single crystal methods, 458–467
  topography, 474–476
X-ray fluorescence, 400–401, 414–415

Young's modulus, 259

Zintl border, 125